Digital Escape Room Designs in Education

Robertas Damaševičius
Vytautas Magnus University, Lithuania

Tatjana Sidekerskienė
Kaunas University of Technology, Lithuania

IGI Global
Scientific Publishing
Publishing Tomorrow's Research Today

Vice President of Editorial	Melissa Wagner
Managing Editor of Acquisitions	Mikaela Felty
Managing Editor of Book Development	Jocelynn Hessler
Production Manager	Mike Brehm
Cover Design	Phillip Shickler

Published in the United States of America by
IGI Global Scientific Publishing
701 East Chocolate Avenue
Hershey, PA, 17033, USA
Tel: 717-533-8845
Fax: 717-533-8661
Website: https://www.igi-global.com E-mail: cust@igi-global.com

Library of Congress Cataloging-in-Publication Data

Names: Damaševičius, Robertas, author. | Sidekerskiene, Tatjana, 1980- author.
Title: Digital escape room designs in education / Edited by Robertas Damaševičius, Tatjana Sidekerskiene.
Description: Hershey, PA : IGI Global Scientific Publishing, [2025] | Includes bibliographical references and index. | Summary: "This book is a comprehensive publication that delves into the world of digital escape room games. It explores the transition of escape rooms from physical to virtual environments, highlighting the influence of technological advancements in this transformation"-- Provided by publisher.
Identifiers: LCCN 2024058238 (print) | LCCN 2024058239 (ebook) | ISBN 9798369342190 (hardcover) | ISBN 9798369351147 (paperback) | ISBN 9798369342206 (ebook)
Subjects: LCSH: Educational games. | Escape room games--Design. | Education--Effect of technological innovations on. | Video games--Design. | Teaching--Aids and devices. | Internet in education.
Classification: LCC LB1029.G3 D26 2025 (print) | LCC LB1029.G3 (ebook) | DDC 371.33/7--dc23/eng/20241214
LC record available at https://lccn.loc.gov/2024058238
LC ebook record available at https://lccn.loc.gov/2024058239
British Cataloguing in Publication Data
A Cataloguing in Publication record for this book is available from the British Library.

Table of Contents

Preface

In an era where digital engagement is reshaping educational landscapes, Digital Educational Escape Rooms emerge as a pedagogical tool, merging gamification, storytelling, and interactive problem-solving with engagement and fun. This book, "Digital Escape Room Designs in Education," explores the theoretical foundations, design methodologies, technological innovations, and future directions of Digital Educational Escape Rooms, providing a structured approach for educators, researchers, and instructional designers to leverage the potential of these immersive learning experiences.

Education is increasingly influenced by digital transformation, where traditional classroom settings must evolve to engage a generation accustomed to interactivity and digital media. Digital Educational Escape Rooms offer a dynamic and motivational learning environment, combining cognitive challenges, cooperative problem-solving, and immersive storytelling to enhance knowledge retention, critical thinking, and student engagement.

As digital and remote learning environments gain prominence, Digital Educational Escape Rooms are gaining traction in both formal education (such as higher education and K-12 curricula) and corporate training (fostering team-building and professional development). This book bridges the gap between game-based learning and structured pedagogy, demonstrating how escape rooms can serve as effective educational tools in a variety of disciplines.

TARGET AUDIENCE

This book is intended for:

- Educators and instructional designers looking to incorporate Digital Educational Escape Rooms into their teaching strategies.

- Researchers and academics exploring gamification, immersive learning, and technology-enhanced education.
- Game developers and e-learning professionals interested in designing educational puzzles and interactive storytelling.
- Students and educational technology enthusiasts curious about how game-based learning can revolutionize education.

CHAPTER SUMMARIES

Each chapter of this book provides a comprehensive analysis of a key aspect of digital escape room design and implementation:

Chapter 1: Foundations of Digital Escape Rooms

This chapter traces the philosophical and historical roots of escape rooms, linking their evolution from ancient labyrinths, medieval riddles, and Enlightenment puzzles to modern physical and digital escape rooms. It explores how escape rooms are shaped by game design philosophy, cultural narratives, literature, cinema, and video games. The transition from physical to digital escape rooms is examined, alongside the impact of Artificial Intelligence (AI), virtual reality (VR), and gamification trends shaping the future of escape rooms.

Chapter 2: Educational Theories and Digital Games

This chapter contextualizes Digital Educational Escape Rooms within established educational theories, including constructivism, behaviorism, and cognitivism. It explores how gamification principles enhance motivation, engagement, and knowledge retention. The cognitive benefits of puzzle-solving—such as memory enhancement, problem-solving, and attention management—are discussed. The chapter also covers social learning theories by analyzing collaborative and multiplayer environments in digital escape rooms.

Chapter 3: Technology in Digital Escape Rooms

Focusing on the technological infrastructure of Digital Educational Escape Rooms, this chapter discusses how AI-powered narrative generation, interactive content creation tools, social communication platforms, and testing environments support the seamless design and implementation of digital escape rooms. The chapter provides a practical guide for educators and developers, detailing best practices for enhancing accessibility, usability, and integration of pedagogical goals into digital escape rooms.

Chapter 4: Designing a Digital Educational Escape Room

This chapter outlines the key principles and mechanics of designing effective Digital Educational Escape Rooms. It emphasizes the role of immersive storytelling, aligning game narratives with learning objectives, and balancing challenge with accessibility. The impact of gamification techniques on maintaining student motivation and cognitive skill development is explored. The chapter includes real-world examples of successful Digital Educational Escape Room implementations, illustrating effective narrative development and storyboarding techniques.

Chapter 5: Design Patterns and Integration Strategies for Implementing Digital Escape Rooms

This chapter delves into the structural design principles of Digital Educational Escape Rooms, covering narrative integration, puzzle-based learning, adaptive difficulty, and collaborative gameplay. It examines Mayer's Cognitive Theory of Multimedia Learning to demonstrate the importance of multimodal content presentation. Strategies such as scaffolding, feedback systems, and adaptive learning techniques are discussed to enhance personalized learning. The chapter also explores team-based challenges and how real-world relevance can increase engagement.

Chapter 6: Advanced Digital Escape Room Design Technologies in Educational Practice

This chapter investigates the integration of immersive technologies like Virtual Reality (VR) and Augmented Reality (AR) into Digital Educational Escape Rooms. It presents a conceptual framework based on experiential learning and constructivist principles, emphasizing the role of hands-on, interactive learning experiences. Design principles for narrative-driven, interactive, and feedback-rich Digital Educational Escape Rooms are explored. The chapter also highlights the cognitive and collaborative advantages of VR/AR-enhanced learning environments.

Chapter 7: Digital Educational Escape Room Design Using Genially Platform

This chapter provides a practical guide to creating Digital Educational Escape Rooms using the Genially platform. Through case studies in mathematics (differential equations and integral calculus), it demonstrates how interactive puzzles, adaptive learning mechanisms, and storytelling elements can be integrated into escape rooms. Educators will learn customization techniques, including embedding external code, to tailor game-based learning environments to their specific teaching needs.

Chapter 8: Future Directions and Innovations in Digital Educational Escape Rooms

The concluding chapter explores the future trajectory of Digital Educational Escape Rooms, focusing on emerging technologies such as Large Language Models, Generative AI, and adaptive learning environments. It discusses how AI-powered tools enable personalized learning paths and real-time feedback. The role of VR and AR in creating fully immersive educational simulations is highlighted, along with soft skill development through collaborative problem-solving. The chapter envisions the future of gamified education, addressing its potential to revolutionize STEM (Science, Technology, Engineering, Mathematics) learning and workforce training.

IMPACT AND CONTRIBUTION

This book contributes to the growing body of research on game-based learning by providing a structured framework for designing, implementing, and evaluating Digital Educational Escape Rooms. By integrating educational theory, instructional design, and technological innovation, it offers practical insights for educators, researchers, and game developers interested in enhancing digital learning experiences. As education continues to embrace immersive and interactive methodologies, this book serves as a valuable resource for those looking to redefine how students engage with learning through gamified experiences.

Chapter 1
Foundations of Digital Escape Rooms

ABSTRACT

This introductory chapter starts from the foundations of escape rooms by tracing their philosophical and historical roots. It explores the evolution of puzzles and labyrinths from ancient civilizations, where they symbolized spiritual journeys, to the logical puzzles of the Enlightenment. The chapter examines how these traditions influenced modern escape room design, drawing on game design philosophy, cultural, and artistic movements, including literature, cinema, and video games. Furthermore, it addresses the transition from physical escape rooms to digital ones, highlighting the rise of online and mobile platforms, and how game mechanics and narrative storytelling have evolved alongside technological advancements like AI and virtual reality (VR). The chapter also explores the psychological appeal of puzzle-solving and escapism, and their growing cultural impact and global reach. It concludes by looking at future trends, such as collaborative gameplay and the integration of immersive technologies, that are shaping the next generation of digital escape rooms.

HISTORICAL, PHILOSOPHICAL AND CULTURAL ROOTS OF ESCAPE ROOMS

Historical Evolution of Puzzles and Labyrinths

The labyrinth is one of the oldest symbols known to humankind, with its origins stretching back to ancient civilizations (Morrison, 2021). Archaeological evidence suggests that the labyrinth motif was widespread in various ancient cultures, from the complex pavement designs in Roman architecture to the stone carvings in Northern Europe. These early labyrinths were not designed as puzzles to be solved but as ritualistic paths, often used in religious ceremonies and meditative practic-

DOI: 10.4018/979-8-3693-4219-0.ch001

es. The most famous ancient labyrinth, often depicted in Greek mythology, is the Cretan Labyrinth (depicted artistically in Figure 1) that housed a bull-like monster, the Minotaur (Kotsonas, 2018). This labyrinth was not just a maze but a symbolic representation of a complex journey with spiritual and philosophical implications. The labyrinth in ancient times was a symbol of life's journey, a pathway of spiritual significance that represented the complicated and unpredictable path of life, death, and rebirth.

The Cretan Labyrinth tands as a powerful prototype of the modern escape room. This ancient structure was not merely a physical maze but a complex symbol of a journey filled with challenges, decisions, and the pursuit of a goal—in this case, the defeat of the Minotaur. Much like in an escape room, participants in the Labyrinth faced the unknown, requiring both wit and strategy to navigate its twists and turns. The Labyrinth's design embodied the essence of what modern escape rooms strive to replicate: an environment where problem-solving, critical thinking, and a structured path led to a final resolution.

Theseus' journey through the Labyrinth is particularly significant when viewed through the lens of education. The Labyrinth can be seen as a metaphor for the educational process, where the path is not always straightforward, and the goal is not simply to escape but to emerge transformed by experience. Theseus' journey represents the learning process, where each turn in the Labyrinth mirrors a step in acquiring knowledge or overcoming obstacles. His encounter with the Minotaur symbolizes the ultimate challenge that tests all the skills and knowledge he has gained along the way. In this sense, Theseus' journey is about more than just survival; it is an educational odyssey that demands courage, ingenuity, and persistence.

Figure 1. Theseus and the minotaur in the labyrinth by Edward Burne Jones

In modern educational theory, escape rooms are often used as tools for experiential learning as a process of human cognition (Fenwick, 2000), where participants engage with material in a hands-on, immersive environment. Theseus' story embodies this approach, as he learns through direct engagement with his environment, navigating the Labyrinth, deciphering its secrets, and ultimately succeeding through a combination of physical and intellectual effort. The journey through the Labyrinth, therefore, is not just a physical trial but an educational experience that equips Theseus with the skills and confidence needed to face future challenges. Theseus' journey reflects the essence of education: a transformative process that involves navigating complexities, solving problems, and emerging with newfound knowledge and abilities. This mythological narrative captures the spirit of modern escape rooms, where the true victory lies not only in the escape itself but in the learning and growth that occurs throughout the journey.

Medieval Labyrinths and Mazes

During the medieval period, labyrinths and mazes underwent significant evolution in both their design and purpose, reflecting the cultural and intellectual currents of the time. Unlike their ancient predecessors, which were often associated with mythological narratives or ritualistic practices, medieval labyrinths and mazes became deeply intertwined with the religious and philosophical ideals of the Christian faith. The complex floor designs found in medieval cathedrals, such as the labyrinth in Chartres Cathedral, exemplify this shift. These labyrinths were not merely architectural curiosities; they were profoundly symbolic, serving both educational and spiritual purposes that resonated with the medieval mindset. The labyrinth at Chartres Cathedral (Figure 2), one of the most famous examples, was constructed around the early 13th century. It is a circular design embedded in the floor of the cathedral's nave. Unlike a maze, which is designed to confuse and mislead, the Chartres labyrinth offers a single, non-branching path. This design choice is deeply symbolic, representing the journey of the soul towards God. The path is winding and complex, reflecting the challenges and trials that one might face in life, but ultimately, it leads to the center, symbolizing spiritual enlightenment and union with the divine.

Figure 2. The labyrinth at Chartres cathedral

Walking the labyrinth became a symbolic act of devotion, representing the pilgrim's journey to the Holy Land, or more broadly, the Christian's spiritual journey through life. The act of following the path was itself a form of prayer, meditation, and reflection, allowing the faithful to contemplate their spiritual journey and seek inner peace. This aligns closely with the purpose of modern escape rooms, which also engage participants in a journey—albeit one that is typically intellectual or problem-solving in nature, rather than spiritual.

The labyrinths in medieval cathedrals had an educational function. At a time when literacy was not widespread, and religious teachings were often conveyed through visual and symbolic means, the labyrinth served as a physical representation of Christian doctrine. The journey to the center of the labyrinth could be seen as a metaphor for the journey to understanding divine truth, with the winding path representing the moral and spiritual challenges that one must overcome to achieve enlightenment. This form of symbolic education is echoed in modern escape rooms, where participants often learn through the process of solving puzzles and overcoming challenges. The educational journey in an escape room, much like in a medieval labyrinth, is often about more than just the immediate task at hand; it involves developing critical thinking skills, learning to collaborate with others, and discovering new ways of approaching problems.

The role of labyrinths as tools for contemplation and education is particularly relevant when considering the design of escape rooms. Both labyrinths and escape rooms create a controlled environment where participants are guided through a sequence of challenges. In both cases, the journey is as important as the destination. In medieval labyrinths, the act of walking the path was a form of active meditation, where the physical movement through the space mirrored the spiritual movement towards enlightenment. In escape rooms, the process of solving puzzles and progressing through the room mirrors the intellectual journey of discovery and learning. The challenges faced in an escape room, much like the turns and twists of a labyrinth, are designed to provoke thought, encourage problem-solving, and ultimately lead to a deeper understanding or a satisfying conclusion.

The medieval labyrinth, particularly those found in cathedrals like Chartres, serves as a profound prototype for modern escape rooms. Both are carefully designed environments that guide participants through a journey—whether it be spiritual or intellectual—towards a final goal. The labyrinths' role in medieval society as tools for contemplation, prayer, and education mirrors the way escape rooms today are used not only for entertainment but also for experiential learning and personal development. The evolution of labyrinths during the medieval period highlights the enduring human fascination with journeys, both physical and metaphorical, and underscores the deep connections between the labyrinths of the past and the escape rooms of the present.

Renaissance Mazes

The Renaissance period marked a profound shift in European culture, characterized by a resurgence of interest in classical antiquity, humanism, and the arts. This cultural revival was reflected in various aspects of life, including architecture, garden design, and leisure activities. Among these, the development and flourishing of hedge mazes (such as Hedge Maze at Palace of Versailles shown in Figure 3) became a notable feature in the grand gardens of European royalty and nobility (Kaczmarczyk and Salvoni, 2016). These mazes, which emerged as symbols of intellectual challenge and entertainment, represented a departure from the spiritual and contemplative labyrinths of the medieval period and mirrored the Renaissance's emphasis on human ingenuity, exploration, and the complexities of life.

Unlike the medieval labyrinths, which were often associated with religious symbolism and spiritual journeys, Renaissance mazes were primarily designed for recreation and intellectual stimulation. These hedge mazes were complex and elaborate, crafted with precision and artistry to both delight and challenge those who entered them. Typically constructed from tall, manicured hedges, these mazes featured winding paths, dead ends, and sometimes complex, symmetrical designs that required careful navigation. The experience of moving through these mazes was both physical and mental, as participants had to engage their spatial reasoning and memory to find their way to the center or exit.

The purpose of these mazes extended beyond mere entertainment; they also served as a reflection of the intellectual currents of the time. The Renaissance was a period of exploration, not just in the geographical sense, with the discovery of new lands, but also in the intellectual and artistic realms. The mazes symbolized this spirit of exploration and the human desire to solve complex problems. They were a physical manifestation of the period's fascination with geometry, symmetry, and the application of mathematical principles to art and design. The maze, with its complex network of paths, represented the twists and turns of life, the challenges one might face, and the ultimate satisfaction of overcoming obstacles. In this way, the Renaissance maze can be seen as an early precursor to the escape room, where the journey itself—navigating through a series of challenges and reaching the goal—provides the primary pleasure.

Renaissance mazes were rich in symbolism, often representing the complexities of human life and the intellectual pursuit of knowledge. The maze could be interpreted as a metaphor for the mind itself, with its complex pathways symbolizing the complexity of thought and the intellectual journey towards enlightenment. This symbolic use of mazes aligns closely with the function of modern escape rooms, where the experience is designed to engage participants in a series of puzzles or challenges that require creative thinking, problem-solving, and sometimes even a deep dive into thematic narratives. The intellectual challenge posed by these mazes was a form of entertainment that appealed to the Renaissance ideals of wit and cleverness. Participants were expected to use their minds to outsmart the maze, much like how modern escape room participants must use their intelligence to solve puzzles and overcome obstacles. The sense of accomplishment upon finding the correct path or solving the final puzzle in a maze is akin to the satisfaction escape room players feel when they successfully complete the room.

The social aspect of Renaissance mazes is another important dimension to consider. These mazes were not just solitary experiences; they were often enjoyed in groups, where participants could engage in friendly competition or collaborate to find the solution. This social interaction is a key element of both Renaissance mazes and modern escape rooms. Escape rooms are often designed to encourage teamwork and communication, making them popular for group activities such as team-building exercises and social outings.

Figure 3. Plan of former hedge maze at Palace of Versailles (France)

The inclusion of mazes in the grand gardens of the Renaissance era reflects the period's broader cultural obsession with artifice, illusion, and play. The gardens themselves were designed as controlled environments that juxtaposed nature with human order, often incorporating elements like fountains, statues, and topiary in addition to mazes. These spaces were meant to delight and surprise, offering visitors an escape from the everyday into a world of beauty and intellectual challenge. This concept of creating an immersive, alternate reality is central to the design of escape rooms today. Just as visitors to a Renaissance garden might lose themselves in the twists and turns of a maze, escape room participants are transported into the world of the game, where they must navigate through a series of challenges to reach their goal.

The link between Renaissance mazes and modern escape rooms is evident in several aspects: the emphasis on intellectual challenge, the use of symbolism, the social dynamics, and the creation of immersive environments. Both experiences are designed to engage the participant on multiple levels—intellectually, socially, and emotionally. The maze, with its paths that seem to lead in circles, challenges participants to think critically and creatively, much like the puzzles in an escape room that require out-of-the-box thinking. In addition, both mazes and escape rooms serve as metaphors for life's challenges. The satisfaction derived from successfully navigating a maze or escaping from a room parallels the human experience of overcoming obstacles and solving problems. The journey through a Renaissance maze, like the journey through an escape room, is more than just reaching the end; it is about the process, the mental engagement, and the experience itself.

The design principles that guided the creation of Renaissance mazes—symmetry, geometry, and artful complexity—are mirrored in the design of escape rooms. Escape rooms often incorporate puzzles that rely on geometric patterns, spatial reasoning, and logical sequencing, all of which have their roots in the mathematical and artistic principles of the Renaissance. The evolution from these historical mazes to today's escape rooms highlights the enduring human fascination with challenges, puzzles, and the journey toward a goal, whether it be in the physical world or a constructed environment.

The Enlightenment and the Rise of Logical Puzzles

The Enlightenment period marked a significant shift in the nature of puzzles, with a growing emphasis on logic, reason, and intellectual challenge. This era witnessed the rise of logical puzzles that were designed to stimulate the mind and encourage analytical thinking. Puzzles such as riddles, brainteasers, and mathematical conundrums became popular forms of intellectual entertainment, reflecting the Enlightenment values of reason and empirical knowledge. One notable example from this period is the work of Lewis Carroll, known for his love of logical puzzles and paradoxes (Gardner, 1960). Carroll's puzzles were not just forms of amusement but also tools for exploring and understanding the complexities of logic and language.

The historical evolution of puzzles and labyrinths (Table 1) is a testament to the enduring human fascination with complexity, challenge, and the joy of discovery. From ancient spiritual symbols to tools of intellectual stimulation, these complex designs have mirrored the cultural and intellectual currents of their times, offering insights into the human psyche across ages.

Table 1. Historical evolution of escape games

Era	Key Developments
Ancient Times	Labyrinths as spiritual and symbolic journeys
Medieval Period	Labyrinths for meditation and religious practices
Renaissance	Hedge mazes as leisure and intellectual challenges
Enlightenment	Logical puzzles emphasizing reason and intellect
Modern Era	Emergence of physical escape rooms as entertainment
Digital Age	Development of digital and virtual escape games

Philosophical Theories Behind Game Design

The Concept of Play in Philosophy

The concept of philosophy, especially as it pertains to game design, is a profound and multi-faceted subject. Philosophical discussions about play often revolve around its intrinsic value, its role in human culture and individual development, and its relationship to creativity and freedom.

One philosophical perspective on play is derived from the work of Johan Huizinga, who in his seminal work "Homo Ludens," (Huizinga, 2014) proposes that play is a fundamental aspect of human culture. He argues that play is not just a form of entertainment but a crucial component of our social and cultural lives, offering a space for creativity, exploration, and the expression of freedom. This idea has profound implications for game design, suggesting that games, as structured forms of play, are not merely pastimes but essential tools for cultural expression and personal development.

Another important philosophical concept related to play is the notion of "the magic circle" introduced by Huizinga and later expanded by Salen and Zimmerman in "Rules of Play" (Salen and Zimmerman, 2003). The magic circle refers to the idea of a special 'space' created by the act of playing a game, where the usual rules and realities of the everyday world are suspended and replaced by the rules and fictional reality of the game. This concept is crucial in game design as it highlights the importance of creating an immersive environment that can transport players into an alternate reality.

Escapism

The concept of escapism in philosophy and game design is often viewed through a dual lens: on one hand, as a potentially negative desire to retreat from reality, and on the other, as a valuable means of exploring alternative realities and perspectives (Baeva, 2020). From a philosophical standpoint, escapism can be seen in the context of existentialism, particularly in the works of Jean-Paul Sartre and Albert Camus Lama et al. (2021). They suggest that escapism is a human response to the absurdity and alienation inherent in life. In game design, this existential perspective can be used to create games that offer not just a diversion from reality but also an opportunity to explore deeper questions about existence, choice, and identity.

Escapism in games can be viewed through the lens of positive psychology. Games offer a means of experiencing flow, a state of deep immersion and engagement, which can have therapeutic and stress-relieving benefits. From this perspective, escapism in game design is not about denying reality but rather about creating experiences that can provide relief, joy, and a sense of accomplishment. The philosophical theories behind game design, including the concept of play and escapism, are integral to understanding the deeper value and impact of games. These theories suggest that games are more than entertainment; they are important cultural artifacts that reflect and shape our understanding of ourselves and the world around us. Game design, therefore, becomes a significant medium for philosophical expression and exploration.

Simulacra and Simulation

The theory of Simulacra and Simulation by Jean Baudrillard (Baudrillard, 1981) provides a compelling lens to analyze the philosophical underpinnings of educational escape rooms. According to Baudrillard, our current society has replaced all reality and meaning with symbols and signs, and that human experience is of a simulation of reality.

Escape rooms are particularly fascinating examples of simulacra. They are designed environments that mimic real-world scenarios but are entirely artificial. For example, a room might be designed to look like an ancient tomb, a space station, or a detective office. Each setting, while artificial, is laden with symbols (e.g., hieroglyphs in the tomb, gadgets in the space station, clues in the detective's office) that suggest a certain reality. Participants interact with these symbols as though they were real, navigating the room's challenges using the logic that the room's design implies.

Baudrillard describes several stages of simulacrum ranging from faithful copies to distorted reality, to pretending to be a real thing, and finally to a simulacrum that bears no relation to any reality whatsoever. Escape rooms often operate at the level of pretending to be something real—they create a facade of reality, demanding players suspend disbelief and engage with the simulation as if it were genuine. This engagement reflects the blurring lines between the simulated and the real, central to Baudrillard's philosophy. In the context of educational escape rooms, the theory raises intriguing questions about learning and cognition:

- What is Learned? Are participants learning about the world itself, or are they learning to navigate symbols and representations of the world? In Baudrillard's terms, this could further the detachment from true reality, reinforcing knowledge of the simulacrum rather than the original.
- Nature of Knowledge: It also invites reflection on the nature of knowledge. Is knowledge itself a form of simulacrum, a series of signs and symbols detached from the "real" world? This can lead to a deeper philosophical inquiry into what it means to know something and how we come to understand our environment.

Escape rooms—simulations of reality—as educational tools-are intended to teach or reinforce real-world skills and knowledge through completely fabricated scenarios. This aligns with Baudrillard's notion of hyperreality, where the distinction between reality and simulation starts to blur, and the simulation becomes preferable to reality.

Cultural and Artistic Influences on Escape Rooms

The development of escape rooms as a form of entertainment and education has been profoundly influenced by various cultural and artistic elements. From literary inspirations to cinematic experiences and art movements, these influences (Table 2) have shaped the aesthetics, narratives, and overall experience of escape rooms.

Table 2. Thematic influences in escape games

Influence	Examples in Escape Games
Literature	Sherlock Holmes-inspired mysteries, Alice in Wonderland
Cinema	Suspense and thriller themes, dramatic narratives
Art Movements	Surrealism-inspired puzzles, minimalist aesthetics
Historical Contexts	Ancient tombs, World War II escape scenarios
Mythology	Labyrinths inspired by Greek myths
Video Games	Adaptations of "The Room" and similar puzzle genres

Literary Inspirations: From "Alice in Wonderland" to "Sherlock Holmes"

Literature has always been a rich source of inspiration for escape rooms. Classic works like Lewis Carroll's "Alice in Wonderland" and Arthur Conan Doyle's "Sherlock Holmes" series have particularly left an indelible mark on the design and thematic elements of many escape rooms. The fantastical, surreal world of "Alice in Wonderland" with its riddles and peculiar characters provides an ideal setting for escape rooms that focus on whimsy, puzzles, and the unexpected. On the other hand, "Sherlock Holmes" inspires escape rooms that emphasize deduction, logical reasoning, and mystery-solving, drawing players into a world of intrigue and intellectual challenge.

The influence of literature on escape rooms extends beyond thematic elements to the structuring of narratives and puzzles. Literary works often feature complex plotlines and complex characters, which can be adapted into compelling story arcs and challenges in escape rooms. This not only enhances immersive experience but also provides educational value, encouraging participants to engage with the narrative in a meaningful and interactive way.

Escape Rooms in Cinema and Theatre

The cinematic and theatrical arts have also significantly influenced the development of escape rooms. The concept of an escape room parallels many elements found in suspense and mystery films, where characters are often placed in confined spaces or challenging situations that require ingenuity and problem-solving skills to overcome. This cinematic influence is evident in the dramatic narratives, tension-building scenarios, and visually rich environments found in many escape rooms.

In theatre, immersive and interactive plays have contributed to the evolution of escape room experiences. Works that involve audience participation, non-linear storytelling, and the use of elaborate sets and props provide a template for the creation of escape room environments where players become an integral part of the narrative. Theatrical techniques such as lighting, sound effects, and set design are employed to enhance the immersive quality of escape rooms, making them more engaging and realistic.

Art Movements and Aesthetic Influences

Art movements have also played a crucial role in shaping the aesthetic aspects of escape rooms. From the surrealism of Salvador Dalí to the minimalist designs of modern art, the visual and conceptual elements of various art movements can be seen in the design of escape rooms. These artistic influences allow for a diverse range of themes and atmospheres, from the bizarre and abstract to the sleek and contemporary. The incorporation of art into escape rooms does more than just create visually appealing spaces; it adds layers of meaning and interpretation to the experience. Artistic elements can be used to convey themes, set the mood, and even serve as clues or components of puzzles. This fusion of art and entertainment in escape rooms not only enhances the aesthetic appeal but also provides a unique platform for artistic expression and appreciation.

The cultural and artistic influences on escape rooms are diverse and multifaceted. From literary inspirations to cinematic and theatrical elements, and art movements, these influences enrich the escape room experience, making it not just a game but a confluence of culture, art, and entertainment.

Video Games

Video games have long been an ideal platform for adapting the concept of escape rooms, leveraging interactivity, complex puzzle design, and atmospheric storytelling to create compelling experiences (Beatty, 2014). Among these, "The Room" series and "Escape Simulator" stand out as prime examples of how digital games can simulate the thrill and mental challenge of physical escape rooms.

"The Room" series, developed by Fireproof Games (https://www.fireproofgames.com/), is a critically acclaimed puzzle game series that began on mobile platforms and later expanded to PC and consoles. The game's premise revolves around solving complex puzzles and unlocking complex contraptions within beautifully rendered, detailed environments that evoke a strong sense of mystery. The puzzles are the core of the game, often involving mechanical and optical illusions that require logical thinking, keen observation, and trial-and-error to solve. The game's graphics are rich and detailed, often involving elements that are both gothic and steampunk in nature, which contribute to an immersive narrative experience. While the narrative is subtle, it weaves through cryptic notes and visual clues, compelling players to piece together a larger, mysterious story as they progress through the games. The Room series has been praised for its innovation in game design, particularly in how it translates the tactile nature of physical puzzles into the digital realm. Its success demonstrates the appeal of escape room mechanics when combined with story-driven gameplay and high production values.

"Escape Simulator" by Pine Studio (https://pinestudio.com/games/escape-simulator/) is another notable entry in the digital escape room genre. It stands out for its emphasis on realism and detail in both the environment and the puzzles. Unlike The Room, Escape Simulator is designed around more traditional escape room concepts, featuring rooms that could conceivably exist in real life. The game offers a variety of rooms, each with a unique theme and meticulously designed to ensure a lifelike escape room experience. Settings range from ancient tombs to modern offices, each filled with realistic objects and tools. One of the standout features of Escape Simulator is its cooperative multiplayer mode, which allows players to tackle puzzles together, replicating the social aspect of real-life escape rooms. The game includes a level editor, which allows players to create and share their own puzzle rooms, greatly enhancing replayability and community engagement. Escape Simulator's approach to realism and community involvement highlights the potential of video games to not only replicate but also expand on the traditional escape room experience. By enabling players to design their own puzzles, it fosters a creative outlet and a shared space for puzzle enthusiasts.

Both The Room and Escape Simulator exemplify how video games can elevate the escape room concept by making it more accessible and diverse. They also show how technology can enhance engagement through immersive graphics, complex interactions, and narrative depth. These games can be played from the comfort of one's home, making the escape room experience accessible to those who might not have physical rooms nearby or those with mobility constraints. Both games encourage critical thinking, problem-solving, and attention to detail, which are valuable educational skills. The popularity of these games reflects a broader cultural interest in puzzles and immersive stories, resonating with audiences who enjoy mental challenges (Table 3) and interactive storytelling.

Table 3. Common escape game challenges

Challenge Type	Example
Logic Puzzles	Sudoku, code-breaking
Spatial Challenges	Navigating mazes, arranging objects
Observation Tasks	Spotting hidden clues, finding patterns
Physical Challenges	Unlocking mechanisms, manipulating objects
Team Coordination	Delegating tasks, combining knowledge
Time Pressure	Completing tasks within a strict deadline

HISTORY AND EVOLUTION OF ESCAPE GAMES

Definition, Origins and Evolution of Escape Rooms

Definition of Escape Games

Escape games, a popular form of social entertainment, have evolved to become an engaging mix of adventure and puzzle-solving. These games, typically set in a locked room, require players to collaboratively solve puzzles and riddles within a specific time frame. The allure of escape games lies in their diverse nature, involving mental, physical, and emotional challenges, thus providing a rich and immersive experience (Krekhov et al., 2021). These activities are not just limited to physical spaces; the advent of augmented reality and VR technologies has led to the development of hybrid formats like Serious Escape Games, further enhancing the learning and engagement aspect of these games (Tzima et al., 2020).

Definition 1*An escape game, often referred to as an escape room or escape the room game, is an interactive and immersive adventure or puzzle-based entertainment experience in which participants are placed in a themed environment and tasked with solving a series of puzzles, riddles, and challenges in order to achieve a specific objective, typically to "escape" from the room or scenario within a set time limit.*

The appeal of escape games, particularly in educational settings, is multifaceted. They provide an adventurous environment where problem-solving, critical thinking, and teamwork are essential. This makes them an effective tool for enhancing learning and engagement in various disciplines, from nursing education to computer science and chemical engineering (Morrell and Ball, 2020; Monnot et al., 2020; Borrego et al., 2017). The design of these games, which involves searching for clues, solving puzzles, and completing tasks within a time limit, not only engages players but also adapts well to educational purposes, making them a versatile tool for both learning and entertainment (Benassi, 2019).

Origins and Early Development

The origins and early development of escape games can be traced back to the pre-digital era, rooted in physical escape rooms, and influenced by adventure and puzzle video games, leading to the advent of digital escape games. Physical escape rooms, a recent cultural phenomenon, emerged as collaborative live-action games requiring teamwork, problem-solving, and critical thinking (Morrell and Ball, 2020). These interactive experiences typically involved players working together to solve riddles and puzzles to escape from a locked room, often incorporating elements of physics and other educational content (Vörös and Sárközi, 2017).

The influence of adventure and puzzle video games on escape games is significant. These games, characterized by their narrative-driven puzzle-solving mechanics, laid the groundwork for the structure and dynamics of escape games. Video games in 3D environments, AR and VR technologies have further evolved the concept, creating immersive experiences that combine logic, pattern recognition, and spatial challenges (Bolivar et al., 2018; Zeng et al., 2020). The transition to digital escape games marks a significant evolution in the genre. The first digital escape games, such as those created in the Breakout EDU format, were designed for educational purposes, particularly in mathematics (Clune, 2019). These digital games, derived from their physical counterparts, maintained the core elements of teamwork and puzzle-solving while leveraging technology to enhance the experience. The development of these games was influenced by the early history of digital gaming, which saw the emergence of games from laboratories and research centers in the 1960s and 1970s (Lowood, 2009).

Definition 2*A Digital Escape Game is an interactive and immersive gaming experience that takes the concept of traditional escape rooms and adapts it to a digital format, often accessible through computers, smartphones, or gaming consoles.*

Escape games have evolved from physical rooms to digital platforms, driven by the foundational mechanics of adventure and puzzle video games. This evolution has expanded their application beyond entertainment, making them valuable tools in educational and training contexts.

Evolution of Game Mechanics

The evolution of game mechanics from text-based puzzles to interactive environments, bolstered by technological advancements, has dramatically transformed the gaming landscape. This transformation is evident in the shift from simple, linear gameplay to complex, multifaceted interactive experiences. Initially, games were primarily text-based, offering puzzles and challenges through descriptive narratives. However, with the advent of more sophisticated technology, these mechanics evolved into rich, interactive 3D environments, enabling a more immersive and engaging user experience (Kosmadoudi et al., 2013; Oktriono and Chong, 2015).

Definition 3*Game mechanics is the fundamental rules, systems, and interactions that govern the behavior and structure of a game.*

Technology has played a pivotal role in advancing game mechanics. The integration of artificial intelligence (AI), augmented reality (AR), and virtual reality (VR) has led to the creation of more dynamic and responsive gaming environments.

These technologies have facilitated the development of games that are not only entertaining but also serve educational and training purposes, such as in the fields of engineering and healthcare (Zook and Riedl, 2014; Ivory and Kalyanaraman, 2007).

Game design has become increasingly sophisticated, with greater variety and complexity in puzzle design. Modern games often incorporate complex strategies and multifaceted challenges, combining various elements like spatial puzzles, algorithmic thinking, and narrative-driven gameplay. This complexity not only enhances the entertainment value but also contributes to the cognitive and strategic development of players (McGuire and Jenkins, 2008; Costa et al., 2020). The evolution of game mechanics, driven by technological advancements, has led to a significant shift in how games are designed and experienced. This evolution reflects a trend towards more complex, interactive, and technologically sophisticated gaming experiences, providing both entertainment and educational value.

Thematic Diversification

Thematic diversification in modern escape games has significantly expanded beyond the traditional 'locked room' concept, incorporating a wide range of themes, including historical, cultural, futuristic, and fantasy elements (Table 4). Modern escape games have evolved to include themes that combine real-world environments with virtual world interactions, offering players the opportunity to interact with both physical and virtual objects (Tsai and Hsu, 2016; Krekhov et al., 2021).

Table 4. Common Genres in escape games

Genre	Examples of Themes
Mystery	Detective stories, murder investigations
Horror	Haunted houses, zombie apocalypses
Adventure	Treasure hunts, jungle explorations
Science Fiction	Space missions, futuristic labs
Historical	Ancient ruins, medieval castles
Fantasy	Magical quests, enchanted forests
Educational	STEM puzzles, cultural heritage themes

Historical and cultural influences have been increasingly integrated into game themes. Projects like "The Archivist's Dream" (Koenig et al., 2022) and "Escape The Countries" VR game (Yeasmin and Albabtain, 2020a) exemplify this trend, where players engage with rooms representing different cultures and historical contexts, solving puzzles based on mathematics, arrangement, and pattern recognition .

Futuristic and fantasy elements are becoming prevalent in modern escape games. The use of VR and AR technologies has enabled the creation of immersive and interactive game environments.

Games like the Escape Room VR and Educational Escape Games for Mixed Reality incorporate these elements to create unique and engaging experiences for players (David et al., 2019; Klamma et al., 2020). Modern escape games have diversified significantly, exploring various themes and integrating historical, cultural, futuristic, and fantasy elements. This thematic expansion enhances the overall gaming experience, making it more engaging and educational.

From Physical to Virtual Escape Games: The Digital Transition

The transition from physical to virtual escape games marks a significant evolution in the gaming landscape, characterized by an complex interplay of technological advancements, creative ingenuity, and the shifting preferences of a global gaming audience.

The shift from physical to virtual escape games has been a transformative journey. Initially rooted in the physical realm, escape games were synonymous with live-action experiences, involving tangible puzzles and collaborative problem-solving in real-world settings. The digital transition ushered in a new era, leveraging VR and AR technologies to create immersive, interactive environments. This shift has enabled the expansion of escape game experiences beyond physical constraints, allowing for more complex and varied puzzle designs and narratives (Yeasmin and Albabtain, 2020b; David et al., 2019; Kohn and Rank, 2016; Benassi, 2019; Monaghan and Nicholson, 2017; Mystakidis et al., 2019).

Several factors have propelled the digitalization of escape games. The increasing affordability and accessibility of VR devices, such as Google's Cardboard and Daydream SDK, have played a crucial role. This technological democratization has made immersive gaming experiences more achievable for a broader audience. The digital format's adaptability and sustainability in formal educational contexts have made it an attractive option for educators and game designers alike. The logistical ease of setting up digital escape rooms and their iterative nature have further contributed to their growing popularity in educational and recreational domains (Gupta et al., 2016; Videnovik et al., 2022; Huraj and Hudáková, 2022; Paton, 2022).

The puzzle design in escape games has undergone significant transformation during this digital shift from logical puzzles to highly interactive tasks combining digital and physical elements (Table 5). Digital puzzle gaming has brought new interactive interfacing techniques that positively impact learning, especially in early childhood education. The VR technology used in digital escape rooms has allowed for the regular replacement of puzzle elements, enhancing the versatility and complexity of puzzles. The integration of sound, light, and video in virtual environments has contributed to creating an original and engaging scenario for players. These advancements have facilitated a more profound and interactive gaming experience, aligning with modern players' expectations and preferences (Wang et al., 2023; Krekhov et al., 2021; David et al., 2019; Délécraz, 2023; Kahila et al., 2020).

Core Elements of Escape Games

Puzzles and Riddles

The evolution of puzzles in the context of escape rooms reflects a fascinating interplay between traditional problem-solving methods and the incorporation of modern technological and narrative elements (Table 5). Each type of puzzle may contribute uniquely to the overall escape room experience, combining intellectual engagement with immersive entertainment.

Logic puzzles, one of the foundational elements in escape rooms, trace their origins to classic brain teasers and games like Sudoku (Figure 4) or riddles. These puzzles challenge players to use deductive reasoning, often serving as the centerpiece for problem-solving tasks. Over time, logic puzzles in escape rooms have evolved from simple pen-and-paper formats to complex, multi-layered challenges that require players to connect abstract ideas or piece together disparate clues within a limited timeframe. This evolution aligns with the increasing demand for intellectually stimulating experiences.

Cipher puzzles, inspired by cryptography, have become a staple in escape rooms, reflecting the human fascination with secret messages and hidden codes. From ancient ciphers like the Caesar shift to modern techniques involving complex algorithms, these puzzles engage players by requiring them to decode symbols, patterns, or encrypted text. Their integration into escape rooms often combines historical or espionage-themed narratives, making them both educational and thrilling.

Figure 4. An example of a 6x6 Sudoku puzzle, a variation of the classic logic-based game where players must fill a 6x6 grid with digits so that each row, column, and 2x3 subgrid contains all numbers from 1 to 6 without repetition

2	6	1	4	3	
4				1	
		6		4	
	3	4			
	4				1
6	1		3	5	

The digitalization of escape rooms has further enhanced cipher puzzles by allowing dynamic code generation and interactive decryption tools.

Physical manipulation puzzles add a tactile dimension to the escape room experience. These puzzles, which involve interacting with objects such as locks, levers, or concealed compartments, draw inspiration from traditional mechanical puzzles like sliding tiles or lock-and-key challenges. In escape rooms, physical manipulation puzzles serve as gateways to progress, often demanding spatial reasoning and dexterity. The advent of hybrid and digital escape games has expanded these puzzles by incorporating AR and haptic feedback technologies, enabling a seamless blend of the physical and digital realms.

Search-and-find puzzles, rooted in classic scavenger hunts, capitalize on players' observational skills. These puzzles require participants to locate hidden objects or subtle clues within a room. Originally simple in design, these tasks have evolved into multi-step challenges where finding an object triggers a cascade of events or unlocks a deeper layer of the game. Search-and-find puzzles have gained complexity through digital innovations, such as hidden objects that reveal themselves only under certain conditions or through the use of specialized tools like UV lights.

Pattern recognition puzzles harness players' ability to discern recurring motifs, sequences, or relationships among elements. These puzzles often require players to draw connections between seemingly unrelated items, leveraging visual, auditory, or logical patterns. As escape rooms have advanced, pattern recognition tasks have incorporated multimedia components, such as soundtracks with encoded rhythms or visual cues embedded in digital projections, increasing both their complexity and their capacity for engagement.

Mathematical challenges have a long-standing association with puzzles, as they foster critical thinking and quantitative reasoning. Escape rooms use these challenges to varying degrees of difficulty, from simple arithmetic problems to advanced calculations embedded within the game's storyline. The inclusion of mathematical puzzles not only adds diversity to the gameplay but also encourages logical progression and teamwork, especially when combined with time constraints or layered clues.

Interactive tasks represent the most recent and innovative addition to escape room puzzles. These tasks often require players to combine physical and digital elements, creating a truly immersive experience. For instance, players might manipulate a physical object to trigger a digital response or solve a digital puzzle that unlocks a physical mechanism. This integration exemplifies the evolution of escape rooms into hybrid environments, where creativity and technology converge to redefine player engagement.

The Rise of Online and Mobile Escape Games

The rise of online and mobile escape games represents a significant shift in the gaming industry, driven by technological advancements and changing consumer preferences. The transition to online platforms has been influenced by factors such as the appeal of experiential motives like concentration, enjoyment, and escapism, and the ease of integrating educational content into digital game formats (Koo, 2009). These online escape games have successfully engaged students and improved learning outcomes, indicating their effectiveness in educational settings (Li et al., 2022).

Table 5. Types of puzzles in escape games

Puzzle Type	Description
Logic Puzzles	Require reasoning and deduction (e.g., Sudoku)
Cipher Puzzles	Decoding messages using cryptographic techniques
Physical Manipulation	Interacting with objects to reveal clues
Search-and-Find	Locating hidden items or details
Pattern Recognition	Identifying recurring patterns or sequences
Mathematical Challenges	Solving equations or numerical problems
Interactive Tasks	Combining digital and physical elements

Definition 4 *A Mobile Escape Game is an interactive and immersive puzzle-solving experience designed specifically for mobile devices, such as smartphones and tablets. In this digital format, players engage with a themed scenario or storyline, solving a series of challenging puzzles, riddles, and mysteries using touchscreen interfaces and virtual environments, with the ultimate goal of escaping from the in-game scenario or achieving a specific objective within a set time frame.*

Mobile gaming has contributed significantly to the accessibility and popularity of escape games. Mobile platforms have enabled the development of immersive and interactive game environments that can be accessed anytime and anywhere, enhancing the appeal of escape games. Mobile escape room games, for example, have been shown to promote greater information recall and engage users in a holistic experiential learning process, teaching skills like problem-solving, creative thinking, and teamwork (Stellmacher et al., 2022; Trinh, 2021).

Social media and online communities have also played a crucial role in the popularity of escape games. These platforms have facilitated social interaction among gamers, leading to increased in-game purchases and engagement. They provide an effective escape from traditional media limits, allowing businesses to engage with potential consumers and create a sense of community among players (Park et al., 2018).

The rise of online and mobile escape games can be attributed to the combined impact of digital platforms, mobile accessibility, and social media communities. These factors have not only enhanced the reach and appeal of escape games but also transformed them into effective tools for learning and social interaction.

Collaborative and Multiplayer Aspects

The collaborative and multiplayer aspects of escape games have significantly evolved, emphasizing teamwork, communication, and competitive elements. The shift to multiplayer experiences is driven by the demand for realism and affordability in VR games and the appeal of high-quality graphics and sound (Gupta et al., 2016; Prasad et al., 2017). Multiplayer experiences are enhanced by AR and deep learning technologies, offering immersive and interactive gameplay (Zeng et al., 2020). Collaborative puzzles and team dynamics are central to modern escape games. These games often involve teams working under pressure to solve puzzles, combining communication skills with gamification (Vergne et al., 2019; Sanders et al., 2020; Friedrich et al., 2018). Such collaborative activities are employed in various educational settings, from nursing and healthcare to engineering and computer science, enhancing learning and team-building skills (Abensur Vuillaume et al., 2021; Morrell and Ball, 2020; Monaghan and Nicholson, 2017). Competitive aspects in escape gaming, such as tournaments and challenges, add another dimension to these experiences. Competitive escape gaming activities in educational settings, like relay races and review games, motivate students and improve learning outcomes (Yeasmin and Albabtain, 2020a). These competitive elements not only enhance the engagement and excitement of the games but also foster a spirit of healthy competition among participants. The collaborative and multiplayer aspects of escape games, encompassing teamwork, communication, and competitive elements, have significantly contributed to their popularity and effectiveness as educational and recreational activities.

Cultural and Technological Impacts

Cultural Impact and Global Reach of Digital Escape Games

The cultural impact and global reach of digital escape games are profound, reflecting their popularity across diverse cultures and their influence on popular culture. The global popularity of escape games stems from their entertainment value, diverse application domains, and economic success. This popularity is further enhanced by the games' ability to engage players through teamwork-building aspects and overall enjoyment (Marchand and Hennig-Thurau, 2013; Kerr, 2003; Tzima et al., 2020; Kutzin et al., 2021; Najm et al., 2022; Clune, 2019; Menon and Romero, 2020; Mat Sin et al., 2014).

Cultural variations significantly influence game design and themes in digital escape games. The unique cultural contexts contribute to diverse gameplay experiences, reflecting different cultural values, social norms, and expectations (Pornsakulpaisal et al., 2023; Videnovik et al., 2022; Najm et al., 2022). Escape games have also influenced popular culture, becoming increasingly integrated into various educational and leisure contexts. Their role in language classes, educational settings, and as a form of social entertainment highlights their versatility and appeal across different age groups and educational backgrounds (Morrell and Ball, 2020; Tzima et al., 2020; Veldkamp et al., 2020; Menon and Romero, 2020; Benassi, 2019; Monaghan and Nicholson, 2017).

The Emerging Role of AI and Virtual Reality in Digital Escape Games

Artificial Intelligence (AI) and VR in digital escape games is transforming the gaming experience by enhancing game design, adding new dimensions to gameplay, and shaping the future of escape gaming. Incorporating AI in escape room design allows for more sophisticated and interactive experiences. AI-powered VR experiences, such as Room Escape VR, enable real-time interaction with the game environment and puzzle-solving, providing a more dynamic and engaging gameplay experience (Gupta et al., 2016; Kiruthika et al., 2022). The integration of AI in escape games also includes the use of knowledge to solve riddles and find hidden treasures, enriching the narrative and gameplay (Back et al., 2019).

Example 1 *"Room Escape VR" refers to a genre of VR games where players are placed in a digital room or series of rooms and must find their way out by solving puzzles and uncovering clues.*

VR escape games offer a new dimension of immersion and interaction. Advances in VR technologies have enabled the creation of shared, lifelike experiences in online virtual worlds (Sherstyuk et al., 2009). These VR escape games, such as those developed using the Unity3D game engine, reduce VR sickness and enhance player skills through more interactive and immersive experiences (Yeasmin and Albabtain, 2020a; David et al., 2019).

Definition 5*A virtual reality escape game is an immersive, interactive game experience where players use virtual reality technology to explore and solve puzzles in a digitally created environment.*

The future of AI and VR in escape gaming holds immense potential. AI and VR technologies are expected to continue evolving, leading to more realistic and engaging escape game experiences. These advancements will likely result in games that are not only more immersive but also more personalized and adaptive to individual players' skills and preferences. AI and VR technologies are significantly enhancing digital escape games, offering immersive and interactive experiences that are likely to shape the future of gaming.

Future Trends and Directions in Digital Escape Rooms

The future trends and directions in digital escape rooms are shaped by emerging technologies and their potential to transform the gaming experience. These trends are expected to redefine escape gaming and provide a long-term outlook for digital escape games.

Emerging technologies, especially VR and AR, are playing a pivotal role in the evolution of digital escape rooms. Educational digital escape rooms, for instance, are utilizing these technologies to increase media literacy among young people and adults (Frania, 2021). Escape rooms are becoming valuable tools for climate change education, providing immersive learning experiences that foster problem-solving and critical thinking skills (Ouariachi and Wim, 2020). The integration of VR in escape rooms, as highlighted in studies like "Escape The Countries," enhances players' skills through puzzle-solving challenges based on math, arrangement, and pattern recognition (Yeasmin and Albabtain, 2020a). The future of escape room gaming is likely to be influenced significantly by VR and AR technologies. For example, the application of VR in business models offers new possibilities and innovative ideas for escape rooms (Rankovic´ et al., 2020). Digital educational escape rooms (DEERs) are fostering learning outcomes for online learners and can be applied in various educational contexts (Makri et al., 2021). The development of scalable interactive mixed-reality escape room simulations for anatomy learning is another example of how these technologies are enhancing educational experiences (Guo et al., 2020). In the long term, digital escape rooms are expected to continue evolving, with a focus on enhancing educational and training applications. The integration of VR and AR technologies in games like "MillSecret" aims to reveal hidden local cultural heritage and enhance learning in outdoor settings (Tzima et al., 2020). The development of digital escape rooms for pilots' training in flight safety procedures demonstrates the potential of these technologies in specialized training contexts (Lekea et al., 2021). The future of digital escape rooms lies in harnessing emerging technologies to create more immersive, educational, and engaging gaming experiences. This evolution is likely to continue shaping the landscape of escape gaming, offering novel opportunities for learning, skill development, and entertainment.

THE PSYCHOLOGY OF PUZZLE-SOLVING AND ESCAPISM IN DIGITAL ESCAPE ROOMS

Introduction

Introduction to Puzzle-Solving Psychology

The psychology of puzzle-solving in digital escape rooms is an intriguing area of study that delves into the cognitive processes and intrinsic appeal of problem-solving. This subsection provides an insight into these aspects, exploring why problem-solving captivates us and the cognitive mechanisms at play during puzzle-solving. The intrigue of problem-solving in puzzle games is rooted in cognitive and psychological factors. The appeal of solving puzzles, such as in chess, is based on heuristic strategies rather than just memory or insight (Simon and Simon, 2007). Puzzle-solving in games like "Tangram" involves a wider search of the problem space, demonstrating a shift in constraints and representations (Nakano, 2017). This process is influenced by biases originating from individuals' decomposition patterns and categorization of the problem (Egidi, 2006). Solving puzzles aids in the development of logical thinking and problem-solving strategies (Huang et al., 2007), highlighting the educational potential of puzzle games.

The cognitive processes involved in puzzle-solving are multifaceted. In 3D puzzles, for instance, solving involves building a representation of the goal and formulating strategies to achieve it (Abbasi et al., 2009). Problem-solving encompasses reasoning, subgoaling, evaluating alternatives, and navigating large problem spaces (Langley and Cooper, 2017). These processes involve modifying decision patterns, identifying constraints, and reconstructing the spatio-temporal situation (Bégoin-Augereau and Caron-Pargue, 2010). The cognitive process of problem-solving in games interacts with other cognitive functions such as abstracting, searching, learning, decision-making, inference, analysis, and synthesis (Wang and Chiew, 2010). The psychology of puzzle-solving in digital escape rooms is an amalgamation of cognitive processes and intrinsic human curiosity for challenges. These aspects contribute significantly to the appeal and educational potential of puzzle-based games.

Historical Perspectives on Puzzles and the Human Mind

The historical perspectives on puzzles and the human mind encompass the evolution of puzzles in human culture and various psychological theories related to puzzle-solving. This subsection provides an insight into these aspects. Puzzles have evolved as significant elements in human culture, serving as psychological adaptations for prosocial behavior. This evolution is rooted in the interplay between social learning, culture, and social interactions in modern society (Schaller et al., 2006). The emergence of human culture itself required the coevolution of social learning ability and individual innovation (Mcelreath, 2010). Puzzles and games, with their inherent problem-solving challenges, reflect this coevolution and contribute to the development of complex social behaviors and cognitive abilities.

Psychological theories provide insight into the cognitive processes involved in puzzle-solving. One such theory is the SOAR cognitive architecture, which explains a range of cognitive phenomena including problem-solving and language (Vicente and Kirlik, 1992; LaPolla and Baars, 1992). Self-Determination Theory suggests that people engage in puzzle video games to fulfill psychological needs of autonomy, competence, and relatedness (Pusey et al., 2021; Deci and Ryan, 2000). The mental model theory offers a framework for understanding conditional reasoning in puzzles (Byrne and Johnson-Laird, 2009).

The historical perspectives on puzzles and the human mind reveal the significant role of puzzles in human culture and cognition, shaped by psychological adaptations and theories that explain our inclination towards problem-solving.

Core Concepts in Digital Escape Games

The Role of Challenges and Rewards in Escape Games

The role of challenges and rewards in escape games (Table 6), particularly in balancing difficulty and skill in puzzle design and understanding neurological responses to challenge and achievement, plays a pivotal role in the appeal and effectiveness of these games.

Table 6. Challenges and rewards in escape games

Challenges	Rewards
Complex Puzzles	Satisfaction of problem-solving
Time Pressure	Enhanced focus and time management skills
Team Coordination	Improved collaboration and communication
Immersive Environments	Increased engagement and escapism
Dynamic Scenarios	Development of adaptability and creativity

Balancing difficulty and skill in puzzle design is crucial for creating an engaging and rewarding escape game experience. This balance is achieved by incorporating mental, physical, and emotional challenges, ensuring puzzles are neither too easy nor too difficult for players. For instance, the escape room game "Escape the Countries" balances difficulty and skill by combining physical and virtual elements, smart lighting, and computational thinking puzzles (Kahila et al., 2020). Dynamic difficulty adjustment techniques are employed to tailor game difficulty to various player skill levels (Reis et al., 2021). Educational escape rooms also focus on developing valuable skills like teamwork and communication while ensuring puzzles are challenging yet achievable (Reuter et al., 2020).

Escapism and Immersion in Digital Environments

The concept of escapism and immersion in digital environments, particularly in digital escape rooms, is rooted in psychological foundations and manifests through various techniques and effects that enhance the user experience.

Escapism in digital environments, such as VR stores, is known to increase pleasure, vividness, and presence, positively impacting behavioral intentions. Calm background music in these settings can strengthen these effects, suggesting a deep interplay between sensory input and psychological responses (Loureiro et al., 2021). In multiplayer online role-playing games (MMORPGs), escapism is linked with decreased well-being, suggesting a complex relationship between digital immersion and psychological health (Kaczmarek and Dra̧zkowski,˙ 2014). Escapism can stem from intentions to prevent negative effects or promote positive effects, illustrating its role as a coping mechanism in digital spaces (Stenseng et al., 2012).

The role of escapism in education is multifaceted. In educational contexts, escapism through digital games can serve as a creative force, changing perceptions and promoting learning. For instance, humanistic psychology provides a foundation for understanding the role of escapism in digital learning environments, focusing on human experience, emotion, personal values, and self-realization (Hernández-Peña et al., 2022). This suggests that digital escapism can be harnessed positively in educational settings, allowing learners to engage with content in a more personal and immersive manner.

Creating immersive digital worlds involves various techniques that lead to different psychological and educational effects. Immersive virtual environments can develop a strong sense of presence and connection, inspiring personal and social transformation (Seo and Tindall, 2011). In learning environments, immersive digital worlds can create more relevant, effective, and efficient learning environments, engaging learners and increasing transparency and accountability (Kalantzis and Cope, 2004). The application of VR technology in the digital preservation of cultural heritage exemplifies how immersive digital worlds can make cultural and educational content more accessible and interactive (Zhong et al., 2021). Escapism and immersion in digital environments are complex phenomena with significant psychological underpinnings. They play a crucial role in both recreational and educational digital settings, shaping user experiences and learning outcomes.

Social and Educational Dynamics

Social Psychology in Multiplayer Scenarios of Digital Escape Rooms

The social psychology in multiplayer scenarios of digital escape rooms is deeply influenced by group dynamics, collaboration, communication, and role distribution within teams. These aspects significantly impact the experience and effectiveness of these collaborative games. In multiplayer digital escape rooms, group dynamics and collaboration play a crucial role. These games encourage teams to work together creatively to solve puzzles and challenges, which enhances the overall learning experience. For instance, in healthcare settings, escape rooms facilitate teamwork, critical thinking, and creative problem-solving among medical and paramedical teams (Sanders et al., 2020; Friedrich et al., 2018; Kaul et al., 2021). Cooperative gameplay can influence group collaboration in subsequent tasks, as demonstrated in computer games like Operation Sting (Nasir et al., 2013).

Communication and role distribution are integral to the success of team-based challenges in digital escape rooms. Effective communication and coordination are essential for achieving goals in multiplayer scenarios, as observed in games like World of Warcraft (Williams and Kirschner, 2012). The design of escape rooms can either facilitate or restrict collaboration and communication, impacting the overall team dynamics (Pan et al., 2017). Role distribution and team performance in online games are influenced by players' common interests and the size of the group (Huang et al., 2009). The social psychology of multiplayer digital escape rooms hinges on the complex interplay of group dynamics, collaboration, communication, and role distribution. These factors collectively shape the players' experiences and outcomes in these immersive digital environments.

Learning Principles of Digital Escape Rooms

The historical and philosophical evolution of escape rooms mirrors the Gee's principle of "situated cognition," where learners acquire knowledge in contexts that reflect real-world applications (Gee, 2003). From the symbolic journeys in ancient labyrinths to the immersive puzzles of Renaissance mazes, the educational value of such spaces has always stemmed from their ability to place participants in meaningful scenarios that demand intellectual engagement. This aligns with Gee's argument that learning is most effective when tied to experiences that simulate authentic challenges. The Gee's concept of "affinity spaces"—where learning communities form around shared goals and collaborative problem-solving—finds a parallel in the social dynamics of escape rooms. Modern escape rooms are designed to foster teamwork, mirroring the communal aspects of video game culture Gee highlighted. Participants in escape rooms, like players in cooperative video games, must rely on diverse perspectives and collective knowledge to succeed, which enhances their ability to engage with complex tasks. Digital escape rooms also reflect Gee's insights on multimodal literacy, blending visual, textual, and interactive elements to create an engaging learning experience. Digital escape rooms leverage technological advancements like AI and VR to enhance immersion, similar to how Gee notes that good video games integrate storytelling, gameplay mechanics, and feedback systems to teach implicitly.

Educational and Training Applications based on Digital Escape Games

The rise of educational and training applications based on digital escape games has marked progress in learning methodologies and corporate training approaches. These games blend entertainment with education, offering an innovative way to learn through play, enhance team-building, and address social causes. Educational escape games have become an effective tool for learning through play. Digital escape room-style games engage students in the learning process, achieving notable learning outcomes and fostering interest in various subjects. For instance, in STEM education, VR escape rooms have been appreciated by teachers for their value in learning, supporting playful learning, gameful learning, and serious games (Mystakidis et al., 2021; Clune, 2019; Videnovik et al., 2022; Karageorgiou et al., 2019). In the corporate sector, digital escape games have been effectively utilized for team-building exercises and training. These games teach teamwork in a practical and complete way, engaging employees, increasing awareness of effective teamwork, and enhancing problem-solving skills (López López and Lázaro, 2019; Jassawalla et al., 2022). Mixed reality escape room simulations and digital games like Asteroid Escape foster teamwork abilities in technology-enhanced learning environments (Guo et al., 2020; Pratticò et al., 2019).

Example 2 *"Asteroid Escape" (ast, 2024) is a sliding puzzle game designed for solo play. It challenges players to navigate a spaceship through an asteroid field. The game includes 60 missions, ranging from beginner to expert levels, and is suitable for ages 8 to adult. It focuses on developing cognitive skills such as flexible thinking, planning, problem solving, spatial insight, and visual perception.*

Serious games developed from digital escape games address social causes, providing educational experiences beyond mere entertainment. These games have been applied in various fields, including medicine, school education, and management. The integration of AR and VR in these games has further enhanced their potential for social and educational impact (Bakhsheshi, 2019; Itenge et al., 2021; Drosos et al., 2018).

Definition 6*A serious game is a digital or tabletop game designed with the primary intention of conveying educational, informational, or training content while maintaining an engaging and enjoyable gameplay experience.*

Digital escape games have expanded their reach to educational and training applications, offering innovative solutions for learning, skill development, and addressing social issues. Their interactive and engaging nature makes them a powerful tool in both educational and corporate settings.

The educational applications of digital escape games have become increasingly significant in modern pedagogy, owing to their ability to foster skills development and provide enriching learning opportunities. This evolution in educational methodology is underscored by various successful case studies that demonstrate the efficacy and versatility of these games in educational settings.

Digital escape games offer a unique blend of learning and skill development opportunities. They are particularly effective in motivating learners, enhancing adaptability, and developing responsive skills. Their integration into large classes or open house events has shown students engaged in an interactive and collaborative learning environment. For example, (Monnot et al., 2020) highlight how these games can be adapted to large audiences in chemical engineering education, demonstrating their versatility across different academic disciplines. Digital escape games have been found to reinforce and motivate learning in courses like chemical bonding (Ang et al., 2020), enhance mathematical content learning (Clune, 2019; Sidekerskiene˙ and Damaševicius,ˇ 2023), and engage students in the learning process to achieve specific outcomes (Videnovik et al., 2022).

Several case studies exemplify the successful implementation of digital escape games in educational contexts. For instance, (Videnovik et al., 2022) describe the successful migration of an escape room-style educational game to an online environment, with all students finding the game interesting for both playing and learning. (Offord et al., 2022) present an escape room game implemented in an undergraduate business course that promoted digital self-reliance and enhanced communication skills. In the context of STEM education, (Mystakidis et al., 2021) discuss the positive reactions of teachers to VR escape rooms, showcasing their eagerness to engage in professional development activities and embrace gameful learning methods. These examples underscore the potential of digital escape games as innovative pedagogical tools. Their ability to cultivate a range of skills—from problem-solving and critical thinking to collaboration and digital literacy—makes them a valuable asset in educational settings.

The educational and therapeutic applications of digital escape rooms extend from leveraging puzzle-solving as a learning tool to exploring the therapeutic benefits these environments offer. In educational contexts, puzzle-solving in digital escape rooms serves as a dynamic tool for enhancing various skills and competencies. For instance, the design of an escape room game aimed at fifth graders improves problem-solving and spatial reasoning skills, increasing their interest in STEM careers (Coffman-Wolph et al., 2018). Physics escape rooms are used as educational tools in classrooms for subjects not typically covered in the high-school curriculum (Vörös and Sárközi, 2017). In general chemistry courses, digital escape rooms like ChemEscape engage students, enhance discipline-specific skills, and apply problem-solving skills (Clapson et al., 2020). Digital escape rooms have also shown to increase motivation, engagement, and satisfaction among vocational high school students, leading to improved problem-solving skills and teamwork (Huraj and Hudáková, 2022).

Digital escape rooms also offer therapeutic benefits. For example, a digital-physical combined escape room significantly improves students' creative thinking and learning motivation, although it does not significantly affect science academic achievement (Kuo et al., 2022). Another study on computer network mechanic students shows that digital escape rooms increase cognitive, behavioral, and affective skills in an engaging way (Huraj et al., 2022). The implementation of digital escape rooms in science teaching in elementary schools led to higher learning motivation and problem-solving ability scores among fourth graders (Huang et al., 2020). These studies highlight the potential of digital escape rooms as innovative educational and therapeutic tools, offering unique opportunities for skill development, motivation, and therapeutic engagement.

Narrative and Cultural Aspects

Evolution of Narrative and Storytelling in Escape Games

The evolution of narrative and storytelling in escape games has been marked by the development of sophisticated storytelling techniques, immersive narratives enhancing player engagement, and the expansion of storytelling into cross-media platforms.

Definition 7*Game storytelling, also known as narrative design or interactive storytelling, is the art and process of crafting and delivering a coherent and engaging narrative within a game.*

The storytelling techniques in escape games have evolved to incorporate interactive story design, integrating narrative time and scenario verification (Cabioch et al., 2019), as well as distributed narrative and choice-based systems (Schäfer, 2018). These advancements have allowed for more complex and engaging storylines, utilizing genetic programming and co-evolution techniques to develop game strategies (Sun et al., 2009) and enhancing educational contexts (Rahim et al., 2022).

Immersive narratives in escape games increase player engagement through their intrinsically integrated fantasies, empathetic characters, and adaptiveness (Naul and Liu, 2019). These elements foster greater immersion and motivation, as well as repeat play sessions (Lu, 2015). The use of video conferencing platforms and tangible online narratives has expanded the scope of immersive storytelling in escape games (Harley et al., 2022).

Cross-media storytelling has further expanded the scope of escape games beyond gaming. This storytelling approach represents the basic elements of the media experience in each unique style of media (Cho, 2012) and facilitates the sharing of environments across different platforms. The integration of VR technology in escape room games (David et al., 2019) and the use of games as a narrative element in transmedia storytelling (Veugen, 2016) exemplify this expansion.

The evolution of narrative and storytelling in escape games has significantly enhanced the gaming experience, making these games more engaging, educational, and versatile. The impact of narrative and storytelling in digital escape rooms is profound, encompassing both the psychological power of storytelling in games and narrative design's role in enhancing engagement and emotional response. Narrative in digital escape room games is a powerful tool for intrinsic motivation, real-world exploration, and problem-solving strategies (Gómez-Cambronero et al., 2019). Educational digital escape rooms create environments of positive emotions and favorable opinions on knowledge acquisition tools (Ruiz et al., 2022). AR and deep learning technologies in digital escape rooms support more complex narratives, potentially achieving greater immersion (Zeng et al., 2020). Storytelling in digital escape rooms effectively forms high-demanded soft skills, improving learning quality (Vekua et al., 2020).

Narrative design in digital escape rooms plays a crucial role in enhancing engagement and emotional response. For instance, video conferencing platforms can facilitate collaborative storytelling using physical objects, focusing on the tangible and embodied characteristics of popular game formats like escape rooms (Harley et al., 2022). This blend of narrative and interactive elements in digital escape rooms not only captivates players but also supports learning across cognitive, affective, and psychomotor domains (Pornsakulpaisal et al., 2023). The incorporation of narrative elements in escape rooms enhances group learning experiences, requiring teams to solve problems collaboratively (Guo et al., 2020). The integration of narrative and storytelling in digital escape rooms significantly enhances the gaming experience, offering a rich, engaging, and emotionally resonant environment that supports both learning and personal development.

The Effect of Escape Game Design Elements on Player Psychology

The effect of escape game design elements on player psychology, particularly the impact of visual and auditory elements as well as feedback mechanisms (Kazimoglu et al., 2010), plays a significant role in shaping the player's experience. Visual and auditory elements in escape games significantly influence players' mood and cognition. The use of 3D sound in VR games like "Escape from Dark Jungle" effectively regulates emotions, making players feel more positive or negative, excited or calm (Gong et al., 2018). This demonstrates the profound impact of auditory elements on emotional responses. Visual elements also play a crucial role, as seen in action video gaming (Abrams and Gerber, 2021), which enhances encoding speed of visual information into visual short-term memory and generalizes into other activities (Wilms et al., 2013). Auditory distractions inhibit retrieval of implicit brand memory, while visual distractions have no significant effect, especially for familiar brands (Choi et al., 2013).

Feedback mechanisms in digital escape rooms have a psychological impact on players. These mechanisms can motivate learners, develop adaptability, and responsiveness skills, especially in large classes or open house events (Monnot et al., 2020). Video game players show enhanced abilities in distinguishing visual and auditory stimuli at the same moment and determining the temporal sequence of multisensory stimuli (Donohue et al., 2010). Intensive action video gaming improves cognitive flexibility, with players showing smaller switching costs and greater cognitive flexibility than individuals with little to no video game experience (Colzato et al., 2010). The design elements of digital escape rooms, including visual and auditory components and feedback mechanisms, have significant psychological effects on players. These elements enhance the gaming experience by influencing mood, cognition, and overall engagement.

Digital Escape Rooms and Culturally Sustaining Pedagogy

Digital escape rooms offer unique opportunities to align with culturally relevant and culturally affirming pedagogies (Macapugay and Nakamura, 2024), as well as the broader framework of Culturally Sustaining Pedagogy (CSP) (Paris, 2012). These pedagogical approaches emphasize the need to validate, sustain, and celebrate the cultural identities of learners while fostering meaningful engagement and academic success. The interactive and customizable nature of digital escape rooms positions them as effective tools to achieve these goals.

One of the core principles of culturally relevant pedagogy is the inclusion of learners' cultural narratives within the learning environment. Digital escape rooms, with their emphasis on storytelling and thematic immersion, provide a platform to integrate culturally significant narratives, symbols, and traditions. For example, escape room scenarios can be designed around historical events, folklore, or community stories relevant to the participants' cultural backgrounds. This approach not only validates students' identities but also promotes a deeper understanding of their heritage and its significance in contemporary contexts. CSP aims to sustain and expand learners' cultural competence while promoting cross-cultural understanding. Digital escape rooms can include puzzles and challenges that draw on diverse cultural practices, languages, or artistic expressions. For instance, a puzzle might require participants to decode a message using indigenous language patterns or solve a challenge inspired by traditional crafts or rituals. Such tasks encourage participants to appreciate and engage with cultural diversity, fostering an inclusive learning environment.

Culturally affirming pedagogies emphasize the communal aspects of learning, which resonate with the collaborative nature of escape rooms. Group-based challenges in escape rooms reflect the importance of community-oriented problem-solving, a value embedded in many cultures. By encouraging teamwork and dialogue, digital escape rooms can simulate real-world practices of collective knowledge-sharing and decision-making, aligning with CSP's focus on sustaining communal learning practices. Digital escape rooms inherently incorporate multimodal literacies—visual, textual, auditory, and interactive—which are central to CSP. By designing tasks that draw on diverse modes of communication and cultural expression, escape rooms cater to the varied learning styles and cultural preferences of participants. This multimodality sustains learners' cultural practices while equipping them with skills to navigate a digitally interconnected world.

REFERENCES

Abbasi, S., Burkhardt, J.-M., & Denis, M. (2009). Spatial problem solving: Assembling three-dimensional puzzles in real and virtual environments. *Cognitive Processing*, 10(S2), 179–181. DOI: 10.1007/s10339-009-0298-3 PMID: 19693560

Abensur Vuillaume, L., Laudren, G., Bosio, A., Thévenot, P., Pelaccia, T., & Chauvin, A. (2021). A didactic escape game for emergency medicine aimed at learning to work as a team and making diagnoses: Methodology for game development. *JMIR Serious Games*, 9(3), 9. DOI: 10.2196/27291 PMID: 34463628

Abrams, S. S., & Gerber, H. R. (2021). Videogames, feedback loops, and classroom practice. In *Teaching in the Game-Based Classroom* (1st ed., p. 9). Routledge.

Ang, J. W. J., Ng, Y. N. A., & Liew, R. S. (2020). Physical and digital educational escape room for teaching chemical bonding. *Journal of Chemical Education*, 97(9), 2849–2856. Advance online publication. DOI: 10.1021/acs.jchemed.0c00612

Asteroid escape. (2024). https://www.smartgames.eu/uk/one-player-games/asteroid-escape. Accessed: 2024-01-01.

Back, J., Back, S., Bexell, E., Stanisic, S., & Rosqvist, D. (2019). The quest: An escape room inspired interactive museum exhibition. In *Extended Abstracts of the Annual Symposium on Computer-Human Interaction in Play Companion Extended Abstracts*. ACM, DOI: DOI: 10.1145/3341215.3356987

Baeva, L. (2020). Existential risks and problem of escapism under conditions of e-culture. *Przeglad Wschodnioeuropejski*, 11(1), 299–313. DOI: 10.31648/pw.5991

Bakhsheshi, F. F. (2019). Serious games and serious gaming in escape rooms. In *2019 International Serious Games Symposium (ISGS)*, pages 42–47. IEEE, DOI: DOI: 10.1109/ISGS49501.2019.9047019

Baudrillard, J. (1981). *Simulacra and Simulations* (1st ed.). Routledge., DOI: 10.4324/9780367809195

Beatty, I. D. (2014). Gaming the system: Video games as a theoretical framework for instructional design. *ArXiv*. DOI: /arXiv.1401.6716DOI: 10.48550

Bégoin-Augereau, S., & Caron-Parque, J. (2010). Modified decision processes marked by linguistic forms in a problem solving task. *Cognitive Systems Research*, 11(3), 260–286. DOI: 10.1016/j.cogsys.2009.12.004

Benassi, A. (2019). Escape room a scuola: Ambienti fisici e virtuali per l'apprendimento. *Journal on Educational Technology*, 27, 174–185.

Bolivar, S., Ortega, F., Zock-Obregon, M., & Rishe, N. (2018). *3d spatial gaming interaction to broad cs participation. In 3D Spatial Gaming Interaction to Broad CS Participation*. Springer.

Borrego, C., Fernández, C., Blanes, I., & Robles, S. (2017). Room escape at class: Escape games activities to facilitate the motivation and learning in computer science. *Journal of Technology and Science Education*, 7(2), 162–171. DOI: 10.3926/jotse.247

Byrne, R. M., & Johnson-Laird, P. (2009). "if" and the problems of conditional reasoning. *Trends in Cognitive Sciences*, 13(7), 282–287. DOI: 10.1016/j.tics.2009.04.003 PMID: 19540792

Cabioch, T., Champagnat, R., Bosser, A.-G., Chiganne, J.-N., & Diéguez, M. (2019). Timing interactive narratives. *In2019 IEEE Conference on Games (CoG)*, pages 1–8. IEEE.

Cho, E. (2012). Meta-representation of video game through the cross-media storytelling: Focusing on the animated motion picture game over. *12*:25–36, DOI: .DOI: 10.7583/JKGS.2012.12.3.25

Choi, Y., Lee, S. M., & Li, H. (2013). Audio and visual distractions and implicit brand memory: A study of video game players. *Journal of Advertising*, 42(2-3), 219–227. DOI: 10.1080/00913367.2013.775798

Clapson, M. L., Gilbert, B. C. T., Mozol, V., Schechtel, S., Tran, J., & White, S. (2020). Chemescape: Educational battle box puzzle activities for engaging outreach and active learning in general chemistry. *Journal of Chemical Education*, 97(1), 125–131. Advance online publication. DOI: 10.1021/acs.jchemed.9b00612

Clune, M. (2019). Digital escape game. *Pacific Journal of Technology Enhanced Learning, 2.*

Coffman-Wolph, S., Gray, K., and Pool, M. (2018). Designing an escape room game to develop problem solving and spatial reasoning skills. DOI: DOI: 10.5703/1288284316855

Colzato, L., van Leeuwen, P. J. A., van den Wildenberg, W. P., & Hommel, B. (2010). Doom'd to switch: Superior cognitive flexibility in players of first person shooter games. *Frontiers in Psychology*, 1. Advance online publication. DOI: 10.3389/fpsyg.2010.00008 PMID: 21833191

Costa, D. M., Francisco, A. P., and Russo, L. M. S. (2020). Hardness of modern games. *ArXiv, abs/2005.10506.*

David, D., Edwin, , Arman, E., Hikari, , Chandra, N., & Nadia, N. (2019). Development of escape room game using vr technology. *Procedia Computer Science*, 157, 646–652. DOI: 10.1016/j.procs.2019.08.223

Deci, E. L., & Ryan, R. M. (2000). The "what" and "why" of goal pursuits: Human needs and the self-determination of behavior. *Psychological Inquiry*, 11(4), 227–268. DOI: 10.1207/S15327965PLI1104_01

Délécraz, C. (2023). Scoring the original soundtrack of an escape room. *Journal of Sound and Music in Games*, 4(1), 26–71. DOI: 10.1525/jsmg.2023.4.1.26

Donohue, S., Woldorff, M., & Mitroff, S. (2010). Video game players show more precise multisensory temporal processing abilities. *Attention, Perception & Psychophysics*, 72(4), 1120–1129. DOI: 10.3758/APP.72.4.1120 PMID: 20436205

Drosos, V., Alexakos, C., Alexandri, A., & Tsolis, D. (2018). Evaluating 3d serious games on cultural education. *In 2018 9th International Conference on Information, Intelligence, Systems and Applications (IISA)*, pages 1–5. IEEE, DOI: DOI: 10.1109/IISA.2018.8633689

Egidi, M. (2006). Chapter 1 decomposition patterns in problem solving. *Contributions to Economic Analysis*, 280:15–46, DOI: DOI: 10.1016/S0573-8555(06)80003-1

Fenwick, T. J. (2000). Expanding conceptions of experiential learning: A review of the five contemporary perspectives on cognition. *Adult Education Quarterly*, 50(4), 243–272. DOI: 10.1177/07417130022087035

Frania, M. (2021). *Educational e-escape room as an educational method of media literacy training for future teachers during the covid-19 pandemic. Media Education.* Mediaobrazovanie., DOI: 10.13187/me.2021.3.452

Friedrich, C., Teaford, H., Taubenheim, A., Boland, P., & Sick, B. (2018). Escaping the professional silo: An escape room implemented in an interprofessional education curriculum. *Journal of Interprofessional Care*, 33(5), 573–575. DOI: 10.1080/13561820.2018.1538941 PMID: 30362849

Gardner, M. (1960). Mathematical games. *Scientific American*, 202(3), 172–182. DOI: 10.1038/scientificamerican0360-172

Gee, J. P. (2003). *What Video Games Have to Teach Us About Learning and Literacy*. Palgrave Macmillan. DOI: 10.1145/950566.950595

Gómez-Cambronero, A., González-Pérez, A., Miralles, I., & Casteleyn, S. (2019). Mixed reality escape room video game to promote learning in indoor environments. *In EDULEARN19 Proceedings*, pages 6098–6104.

Gong, J., Shi, Y., Wang, J., Shi, D., and Xu, Y.-Q. (2018). Escape from the dark jungle: A 3d audio game for emotion regulation.

Guo, X., Rees, D., & Richards, M. (2020). A scalable interactive mixed reality escape room simulation for anatomy learning. *International Journal of Modeling and Optimization*, 10, 41–46. DOI: 10.7763/IJMO.2020.V10.744

Gupta, R., Nawani, R., & Talreja, V. (2016). Virtual reality content creation using unity 3d and blender. *International Journal of Computer Applications*, 156(3), 8–12. DOI: 10.5120/ijca2016912395

Harley, D., Grambart, S., Bergel, R. S., & Mazalek, A. (2022). Together alone: A tangible online narrative. *InSixteenth International Conference on Tangible, Embedded, and Embodied Interaction*. ACM. DOI: 10.1145/3490149.3501311

Hernández-Peña, H., Lagomarsino-Montoya, M., Aguirre-Martínez, G., Mansilla-Sepúlveda, J., Estay-Sepúlveda, J. G., & Ganga-Contreras, F. (2022). *On the digital age from humanistic psychology*. Environment and Social Psychology., DOI: 10.18063/esp.v7.i1.1401

Huang, O. W. S., Cheng, H. N. H., & Chan, T. (2007). Number jigsaw puzzle: A mathematical puzzle game for facilitating players' problem-solving strategies. *In2007 First IEEE International Workshop on Digital Game and Intelligent Toy Enhanced Learning (DIGITEL'07)*, pages 130–134. DOI: DOI: 10.1109/DIGITEL.2007.37

Huang, S., Kuo, Y.-H., & Chen, H.-C. (2020). Applying digital escape rooms infused with science teaching in elementary school: Learning performance, learning motivation, and problem-solving ability. *Thinking Skills and Creativity*, 35, 100635. DOI: 10.1016/j.tsc.2020.100681

Huang, Y., Zhu, M., Wang, J., Pathak, N., Shen, C., Keegan, B., Williams, D., & Contractor, N. (2009). The formation of task-oriented groups: Exploring combat activities in online games. *In2009 International Conference on Computational Science and Engineering,* volume 4, *pages* 122–127. *DOI:* DOI: 10.1109/CSE.2009.465

Huizinga, J. (2014). *Homo Ludens Ils 86*. Routledge., DOI: 10.4324/9781315824161

Huraj, L., Hrmo, R., & Sejutová Hudáková, M. (2022). The impact of a digital escape room focused on HTML and computer networks on vocational high school students. *Education Sciences*, 12(10), 682. DOI: 10.3390/educsci12100682

Huraj, L., & Hudáková, M. S. (2022). Digital escape room for computer network mechanic students. *In 2022 20th International Conference on Emerging eLearning Technologies and Applications (ICETA)*, DOI: DOI: 10.1109/ICETA54985.2022.983967

Itenge, H., Sedano, C. I., & Winschiers-Theophilus, H. (2021). Virtually escaping lockdown - co-designing a mixed reality escape room narrative with Namibian learners. *In3rd African Human-Computer Interaction Conference: Inclusiveness and Empowerment*, DOI: DOI: 10.1145/3448696.3448700

Ivory, J. D., & Kalyanaraman, S. (2007). The effects of technological advancement and violent content in video games on players' feelings of presence, involvement, physiological arousal, and aggression. *Journal of Communication*, 57(3), 532–555. DOI: 10.1111/j.1460-2466.2007.00356.x

Jassawalla, A., Sashtital, H., & Witt, A. (2022). Increasing students' felt need for teamwork: An experiential learning exercise that sets the stage. *In8th International Conference on Higher Education Advances (HEAd'22)*, DOI: DOI: 10.4995/HEAd22.2022.14590

Kaczmarczyk, K., & Salvoni, M. (2016). Hedge mazes and landscape gardens as cultural boundary objects. *Sign Systems Studies, 44*(1/2):53–68, ISSN: 1406-4243, DOI: DOI: 10.12697/SSS.2016.44.1-2.04

Kaczmarek, L. D., & Drążkowski, D. (2014). MMORPG escapism predicts decreased well-being: Examination of gaming time, game realism beliefs, and online social support for offline problems. *Cyberpsychology, Behavior, and Social Networking*, 17(5), 298–302. DOI: 10.1089/cyber.2013.0595 PMID: 24605951

Kahila, J., Parkki, T., Gröhn, A., Karvinen, A., Telimaa, E., Riikonen, P., Tiitta, R., Haantio, P., Keinänen, A., Kerkkänen, T., Jormanainen, I., Penttinen, S., & Tedre, M. (2020). Escape room game for CT learning activities in the primary school. *InProceedings of the 20th Koli Calling International Conference on Computing Education Research*. DOI: 10.1145/3428029.3428063

Kalantzis, M., & Cope, B. (2004). Designs for learning. *E-Learning and Digital Media*, 1(1), 38–93. DOI: 10.2304/elea.2004.1.1.7

Karageorgiou, Z., Mavrommati, I., & Fotaris, P. (2019). Escape room design as a game-based learning process for STEAM education. *InProceedings of the 12th European Conference on Game Based Learning*.

Kaul, V., Morris, A., Chae, J. M., Town, J., & Kelly, W. F. (2021). Delivering a novel medical education "escape room" at a national scientific conference: First live, then pivoting to remote learning because of COVID-19. *Chest*, 160(4), 1424–1432. Advance online publication. DOI: 10.1016/j.chest.2021.04.069 PMID: 34029564

Kazimoglu, C., Kiernan, M., & Bacon, L. (2010). Enchanting e-learning through the use of interactive-feedback loop in digital games. *In3rd International Conference on Human System Interaction*, DOI: DOI: 10.1109/HSI.2010.5514523

Kerr, A. (2003). Live life to the power of PS2: Locating the digital games industry in the new media environment. *Irish Communication Review, 9*(2).

Kiruthika, V., Jagadeeswari, M., Sneha Prabha, C., Vaishnavi, V., & Sreejaa, H. (2022). Virtual reality based escape room. *In2022 International Conference on Advanced Computing Technologies and Applications (ICACTA)*, pages 1–5. IEEE, DOI: DOI: 10.1109/ICACTA54488.2022.9753531

Klamma, R., Sous, D., Hensen, B., and Koren, I. (2020). Educational escape games for mixed reality. DOI: .DOI: 10.1007/978-3-030-57717-9_41

Koenig, N., Denk, N., Wimmer, S., & Prandstaetter, H. (2022). Creating an escape room for cultural mediation: Insights from "the archivist's dream". *European Conference on Games Based Learning, 16*(1):297–306, DOI: DOI: 10.34190/ecgbl.16.1.682

Kohn, J., & Rank, S. (2016). Evaluating physical movement as trigger for transitioning between environments in virtual reality. *InProceedings of the 2016 CHI Conference Extended Abstracts on Human Factors in Computing Systems.* DOI: 10.1145/2851581.2892506

Koo, D.-M. (2009). The moderating role of locus of control on the links between experiential motives and intention to play online games. *Computers in Human Behavior*, 25(2), 466–474. DOI: 10.1016/j.chb.2008.10.010

Kosmadoudi, Z., Lim, T., Ritchie, J., Louchart, S., Liu, Y., & Sung, R. (2013). Engineering design using game-enhanced CAD: The potential to augment the user experience with game elements. *Computer Aided Design*, 45(3), 777–795. DOI: 10.1016/j.cad.2012.08.001

Kotsonas, A. (2018). A cultural history of the Cretan labyrinth: Monument and memory from prehistory to the present. *American Journal of Archaeology*, 122(3), 367–396. DOI: 10.3764/aja.122.3.0367

Krekhov, A., Emmerich, K., Rotthaler, R., & Krueger, J. (2021). Puzzles unpuzzled. *Proceedings of the ACM on Human-Computer Interaction*, 5(CHI PLAY), 1–24. DOI: 10.1145/3474696

Kuo, H., Pan, A.-J., Lin, C.-S., & Chang, C.-Y. (2022). Let's escape! The impact of a digital-physical combined escape room on students' creative thinking, learning motivation, and science academic achievement. *Education Sciences*, 12(9), 615. Advance online publication. DOI: 10.3390/educsci12090615

Kutzin, J., Sanders, J., & Strother, C. (2021). Transitioning escape rooms to a virtual environment. *Simulation & Gaming*, 52(6), 796–806. DOI: 10.1177/10468781211035171

Lama, P., Payal, S., Bisht, S., & Gurung, A. (2021). *Escapism in the works of existentialist writers*. Webology., DOI: 10.29121/WEB/V18I2/3

Langley, P., & Cooper, R. (2017). Symposium on problem solving and goal-directed sequential activity. In: Gunzelmann, G. and Howes, A. and Tenbrink, T. and Davelaar, Eddy J. (eds.) Proceedings of the 39th Annual Conference of the Cognitive Science Society. Austin, U.S., pp. 29-30.

LaPolla, M., & Baars, B. (1992). A psychologically implausible architecture that is always conscious, always active. *Behavioral and Brain Sciences*, 15(3), 448–449. DOI: 10.1017/S0140525X00069582 PMID: 24924012

Lekea, D. I. K., Stamatelos, D., & Raptis, P. (2021). Learning how to escape the unthinkable with virtual reality: The case of pilots' training on emergency procedures. *InIOP Conference Series: Materials Science and Engineering,* volume 1024. DOI: DOI: 10.1088/1757-899X/1024/1/012098

Li, C.-T., Wang, C.-P., & Hou, H. (2022). Applying an escape room game in synchronous online learning activity: A preliminary analysis of learners' learning achievement, flow and technology acceptance. *In 2022 12th International Congress on Advanced Applied Informatics (IIAI-AAI)*, pages 659–660. IEEE.

López López, C. M. and Lázaro, M. (2019). Escape folroom: Propuesta de juego integrada en clase de formación y orientación laboral. *2:91–105, DOI*: .DOI: 10.24310/IJNE2.2.2019.7453

Loureiro, S., Guerreiro, J., & Japutra, A. (2021). How escapism leads to behavioral intention in a virtual reality store with background music. *Journal of Business Research*, 134, 288–300. DOI: 10.1016/j.jbusres.2021.05.035

Lowood, H. (2009). Videogames in computer space: The complex history of pong. *IEEE Annals of the History of Computing*, 31(3), 31. DOI: 10.1109/MAHC.2009.53

Lu, A. (2015). Narrative in exergames: Thoughts on procedure, mechanism, and others. *Games for Health Journal*, 4(1), 19–24. DOI: 10.1089/g4h.2014.0090 PMID: 26181676

Macapugay, K., & Nakamura, B. (2024). The student empowerment through narrative, storytelling, engagement, and identity framework for student and community empowerment: A culturally affirming pedagogy. *Genealogy*, 8(3), 94. DOI: 10.3390/genealogy8030094

Makri, A., Vlachopoulos, D., & Martina, R. (2021). Digital escape rooms as innovative pedagogical tools in education: A systematic literature review. *Sustainability (Basel)*, 13(8), 4587. DOI: 10.3390/su13084587

Marchand, A., & Hennig-Thurau, T. (2013). Value creation in the video game industry: Industry economics, consumer benefits, and research opportunities. *Journal of Interactive Marketing*, 27(3), 141–157. DOI: 10.1016/j.intmar.2013.05.001

Mat Sin, N., Talib, O., Norishah, T., Ishak, A., & Baki, R. (2014). Male students and digital game: Reason, motivation and feeling. *International Journal of Information and Education Technology (IJIET)*.

Mcelreath, R. (2010). The coevolution of genes, innovation, and culture in human evolution. In Kappeler, P., & Silk, J. (Eds.), *Mind the Gap*. Springer., DOI: 10.1007/978-3-642-02725-3_21

McGuire, M., & Jenkins, O. C. (2008). *Creating games - mechanics, content, and technology*. CRC Press. DOI: 10.1201/b10678

Menon, D., & Romero, M. (2020). Game mechanics supporting a learning and playful experience in educational escape games. In Global Perspectives on Gameful and Playful Teaching and Learning. DOI: DOI: 10.4018/978-1-7998-2015-4.ch007

Monaghan, S., & Nicholson, S. (2017). Bringing escape room concepts to pathophysiology case studies. *HAPS Educator*, 21(2), 49–65. DOI: 10.21692/haps.2017.015

Monnot, M., Laborie, S., Hébrard, G., & Dietrich, N. (2020). New approaches to adapt escape game activities to large audience in chemical engineering: Numeric supports and students' participation. *Education for Chemical Engineers*, 32, 50–58. DOI: 10.1016/j.ece.2020.05.007

Morrell, B., & Ball, H. M. (2020). Can you escape nursing school?: Educational escape room in nursing education. *Nursing Education Perspectives*, 41(3), 197–198. DOI: 10.1097/01.NEP.0000000000000441 PMID: 30614962

Morrison, T. (2021). *Labyrinth*. Springer International Publishing., DOI: 10.1007/978-3-319-57072-3_4

Mystakidis, S., Cachafeiro, E., & Hatzilygeroudis, I. (2019). Enter the serious e-scape room: A cost-effective serious game model for deep and meaningful e-learning. *In 2019 10th International Conference on Information, Intelligence, Systems and Applications (IISA)*.

Mystakidis, S., Papantzikos, G., & Stylios, C. (2021). Virtual reality escape rooms for stem education in industry 4.0: Greek teachers perspectives. *In 2021 6th South-East Europe Design Automation, Computer Engineering, Computer Networks and Social Media Conference (SEEDA-CECNSM)*, pages 1–5. IEEE, DOI: DOI: 10.1109/SEEDA-CECNSM53056.2021.9566265

Najm, A., Christofi, M., Hadjipanayi, C., Kyrlitsias, C., & Michael-Grigoriou, D. (2022). *The iterative development of an online multiplayer escape room game for improving social interaction through edutainment*. CSRN. DOI: 10.24132/CSRN.3201.8

Nakano, Y. (2017). Cognitive and attentional process in insight problem solving of the puzzle game "tangram.". *Cognitive Science*.

Nasir, M., Lyons, K. A., Leung, R., & Moradian, A. (2013). *Cooperative games and their effect on group collaboration.*, DOI: 10.1007/978-3-642-38827-9_43

Naul, E., & Liu, M. (2019). Why story matters: A review of narrative in serious games. *Journal of Educational Computing Research*, 58(3), 687–707. DOI: 10.1177/0735633119859904

Offord, M., Honeychurch, S., Quinn, N., Barr, M., and Mullen, H. (2022). Escape room: using a simple text-based game to promote business undergraduates' digital self-reliance. *Journal of Learning Development in Higher Education,* DOI: .DOI: 10.47408/jldhe.v123.855

Oktorino, K., & Chong, H. (2015). Developing a 3D action puzzle game application "the mechanic" using Unity. *In 2015 IEEE Conference on e-Learning, e-Management and e-Services (IC3e),* pages 120–124. IEEE.

Ouariachi, T., & Wim, E. J. L. (2020). Escape rooms as tools for climate change education: An exploration of initiatives. *Environmental Education Research*, 26(8), 1193–1206. DOI: 10.1080/13504622.2020.1753659

Pan, R., Lo, H., & Neustaedter, C. (2017). Collaboration, awareness, and communication in real-life escape rooms. *InProceedings of the 2017 Conference on Designing Interactive Systems,* DOI: DOI: 10.1145/3064663.3064767

Paris, D. (2012). Culturally sustaining pedagogy: A needed change in stance, terminology, and practice. *Educational Researcher*, 41(3), 93–97. DOI: 10.3102/0013189X12441244

Park, E., Rishika, R., Janakiraman, R., Houston, M., & Yoo, B. (2018). Social dollars in online communities: The effect of product, user, and network characteristics. *Journal of Marketing*, 82(1), 93–114. DOI: 10.1509/jm.16.0271

Paton, J. (2022). Place, space and object oriented learning methods with the library using ghosts and monsters to create safe spaces for learning. *The Journal of Play in Adulthood.*

Pornsakulpaisal, R., Ahmed, Z., Bok, H., de Carvalho Filho, M. A., Goka, S., Li, L., Patki, A., Salari, S., Sooknarine, V., Yap, S. W., & Moffett, J. (2023). Building digital escape rooms for learning: From theory to practice. *The Clinical Teacher*, 20(2), e13559. Advance online publication. DOI: 10.1111/tct.13559 PMID: 36691337

Prasad, B., Reddy, M., Reddy, V., & Kishore, R. (2017). Game development - bounty rescue step. *International Journal of Engineering*, 02, 25–31.

Prattico, F. G., Strada, F., Lamberti, F., & Bottino, A. (2019). Asteroid escape: A serious game to foster teamwork abilities. DOI: DOI: 10.2312/egs.2019013

Pusey, M., Wong, K. W., and Rappa, N. (2021). Using case studies to explore need satisfaction and frustration in puzzle video games. *Games and Culture, 17:*752–772, DOI: .DOI: 10.1177/15554120211056126

Rahim, A. S. A., Wahab, M. S. A., Ali, A. A., & Hanafiah, N. H. (2022). Educational escape rooms in pharmacy education: A narrative review. *Pharmacy Education*, 22(1), 540–557. Advance online publication. DOI: 10.46542/pe.2022.221.540557

Ranković, M., Dedić, V., Jovanović, M., Petrovic, D., & Ilić, M. (2020). Escape rooms in virtual reality, the concept and application in electronic business doing. *Independent Journal of Management & Production*, 11(4), 1247–1261. DOI: 10.14807/ijmp.v11i4.1117

Reis, S., Reis, L., & Lau, N. (2021). Game adaptation by using reinforcement learning over meta games. *Group Decision and Negotiation*, 30(2), 321–340. DOI: 10.1007/s10726-020-09652-8

Reuter, J., Dias, M. F., Amorim, M., Figueiredo, C., & Veloso, C. (2020). How to create educational escape rooms? strategies for creation and design. *InEighth International Conference on Technological Ecosystems for Enhancing Multiculturality.* DOI: 10.1145/3434780.3436664

Ruiz, L. M. S., López-Alfonso, S., Moll-López, S., Moraño-Fernández, J. M., & Vega-Fleitas, E. (2022). Educational digital escape rooms footprint on students' feelings: A case study within aerospace engineering. *Inf.*, 13(2), 92. DOI: 10.3390/info13020092

Salen, K., & Zimmerman, E. (2003). *Rules of Play: Game Design Fundamentals.* The MIT Press.

Sanders, J. E., Kutzin, J. M., & Strother, C. (2020). Escape the simulation room. *Simulation & Gaming*, 52(1), 62–71. DOI: 10.1177/1046878120963591

Schäfer, S. (2018). *Agency and Authorship in Ludic Narrative Environments*. Springer International Publishing., DOI: 10.1007/978-3-030-04028-4_19

Schaller, M., Simpson, J., & Kenrick, D. (2006). *Evolution and social psychology*. Psychology Press.

Seo, K., & Tindall, D. (2011). *Inspiring personal and social transformation through avatar role play in an online immersive virtual environment*., DOI: 10.4018/978-1-60960-046-4.ch021

Sherstyuk, A., Vincent, D., & Treskunov, A. (2009). Towards virtual reality games. *InProceedings of the 8th International Conference on Virtual Reality Continuum and its Applications in Industry, VRCAI '09,* page 315–316. ACM, DOI: DOI: 10.1145/1670252.1670322

Sidekerskienė, T., & Damaševičius, R. (2023). Out-of-the-box learning: Digital escape rooms as a metaphor for breaking down barriers in stem education. *Sustainability (Basel)*, 15(9), 7393. Advance online publication. DOI: 10.3390/su15097393

Simon, H. A., & Simon, P. A. (2007). Trial and error search in solving difficult problems: Evidence from the game of chess. *Systems Research and Behavioral Science*, 7, 425–429. DOI: 10.1002/sres.3850070507

Stellmacher, C., Ternieten, J., Soroko, D., & Schöning, J. (2022). Escaping the privacy paradox: Evaluating the learning effects of privacy policies with serious games. *Proceedings of the ACM on Human-Computer Interaction,* 6. DOI: 10.1145/3549495

Stenseng, F., Rise, J., & Kraft, P. (2012). Activity engagement as escape from self: The role of self-suppression and self-expansion. *Leisure Sciences*, 34(1), 19–38. DOI: 10.1080/01490400.2012.633849

Sun, K.-T., Lin, Y.-C., Wu, C.-Y., & Huang, Y.-M. (2009). An application of the genetic programming technique to strategy development. *Expert Systems with Applications*, 36(3), 5157–5161. DOI: 10.1016/j.eswa.2008.06.066

Trinh, M. P. (2021). Can you escape? using mobile app escape games to create a holistic experiential learning process online. *Management Teaching Review*, 7(3), 236–244. DOI: 10.1177/23792981211056921

Tsai, C.-Y., & Hsu, K.-H. (2016). An application of using bluetooth indoor positioning, image recognition and augmented reality. *In 2016 IEEE 13th International Conference on e-Business Engineering (ICEBE),* pages 276–281. IEEE.

Tzima, S., Styliaras, G., and Bassounas, A. (2020). Revealing hidden local cultural heritage through a serious escape game in outdoor settings. *Inf., 12*(10).

Vekua, N., Lubsky, A. A., Perevozchikova, M., & Folgerova, J. N. (2020). *Peculiarities of forming high-demanded soft skills in the educational space of the escape room*. Pedagogical Skills and Education. DOI: 10.32744/pse.2020.6.31

Veldkamp, A., van de Grint, L., Knippels, M. C. P. J., & van Joolingen, W. V. (2020). Escape education: A systematic review on escape rooms in education. *Educational Research Review*, 31, 100364. Advance online publication. DOI: 10.1016/j.edurev.2020.100364

Vergne, M., Simmons, J. D., & Bowen, R. S. (2019). Escape the lab: An interactive escape-room game as a laboratory experiment. *Journal of Chemical Education*, 96(5), 985–991. DOI: 10.1021/acs.jchemed.8b01023

Veugen, C. (2016). Assassin's creed and transmedia storytelling. *International Journal of Gaming and Computer-Mediated Simulations*, 8(2), 1–19. DOI: 10.4018/IJGCMS.2016040101

Vicente, K. J., & Kirlik, A. (1992). On putting the cart before the horse: Taking perception seriously in unified theories of cognition. *Behavioral and Brain Sciences*, 15(3), 461–462. DOI: 10.1017/S0140525X00069715 PMID: 24924025

Videnovik, M., Vold, T., Dimova, G., Kiønig, L., & Trajkovik, V. (2022). Migration of an escape room–style educational game to an online environment: Design thinking methodology. *JMIR Serious Games*, 10(3), 10. DOI: 10.2196/32095 PMID: 36155979

Vörös, A. I. V., & Sárközi, Z. (2017). Physics escape room as an educational tool. AIP Conference Proceedings, Volume 1916, Issue 1, id.050002. DOI: DOI: 10.1063/1.5017455

Wang, C., Lee, B., & Lo, T. (2023). The design of a novel digital puzzle gaming system for young children's learning by interactive multi-sensing and tangible user interfacing techniques. *Sustainability (Basel)*, 15(4), 3036. DOI: 10.3390/su15043036

Wang, Y., & Chiew, V. (2010). On the cognitive process of human problem solving. *Cognitive Systems Research*, 11(1), 81–92. DOI: 10.1016/j.cogsys.2008.08.003

Williams, J., & Kirschner, D. (2012). Coordinated action in the massively multiplayer online game world of warcraft. *Symbolic Interaction*, 35(3), 340–367. DOI: 10.1002/symb.22

Wilms, I., Petersen, A., & Vangkilde, S. (2013). Intensive video gaming improves encoding speed to visual short-term memory in young male adults. *Acta Psychologica*, 142(1), 108–118. DOI: 10.1016/j.actpsy.2012.11.003 PMID: 23261420

Yeasmin, S., & Albattain, L. A. (2020a). Escape the countries: A vr escape room game. *In 2020 3rd International Conference on Computer Applications & Information Security (ICCAIS),* pages 1–6. IEEE.

Yeasmin, S., & Albattain, L. A. (2020b). *Implementation of a virtual reality escape room game. 2020 IEEE Graphics and Multimedia.* GAME., DOI: 10.1109/GAME50158.2020.9315039

Zeng, H., He, X., & Pan, H. (2020). Implementation of escape room system based on augmented reality involving deep convolutional neural network. *Virtual Reality (Waltham Cross)*, 25(3), 585–596. DOI: 10.1007/s10055-020-00476-0

Zhong, H., Wang, L., & Zhang, H. (2021). The application of virtual reality technology in the digital preservation of cultural heritage. *Computer Science and Information Systems*, 18(2), 535–551. DOI: 10.2298/CSIS200208009Z

Zook, A., & Riedl, M. O. (2014). *Automatic game design via mechanic generation.* AAAI. DOI: 10.1609/aaai.v28i1.8788

KEY TERMS AND DEFINITIONS

Affinity Spaces: Social and collaborative environments where participants come together to engage in shared goals and problem-solving, reflecting communal aspects of learning and gaming.

Augmented Reality (AR): A technology that overlays virtual elements onto the physical world, creating interactive experiences that blend the real and digital environments.

Cognitive Flexibility: The mental ability to switch between different concepts, tasks, or problem-solving strategies, often enhanced through gaming experiences.

Collaborative Gameplay: A gaming mode emphasizing teamwork, communication, and collective problem-solving among participants to achieve shared goals.

Cultural Influence: The impact of diverse cultural elements, such as literature, cinema, and art, on the thematic and aesthetic design of escape games.

Digital Educational Escape Game: An interactive, immersive game designed specifically for educational purposes, blending the mechanics of traditional escape games with pedagogical objectives to engage learners in problem-solving, critical thinking, and collaborative tasks in a themed, digital environment.

Digital Escape Game: An interactive and immersive gaming experience that takes the concept of traditional escape rooms and adapts it to a digital format, often accessible through computers, smartphones, or gaming consoles.

Escape Game: An interactive and immersive adventure or puzzle-based entertainment experience in which participants are placed in a themed environment and tasked with solving a series of puzzles, riddles, and challenges to achieve a specific objective, typically to "escape" from the room or scenario within a set time limit.

Escapism: A psychological concept referring to the desire to retreat from reality, either to cope with stress or to explore alternative realities and perspectives, often associated with immersive gaming experiences.

Experiential Learning: A hands-on, immersive educational approach where participants actively engage with materials and scenarios, fostering practical skill development and critical thinking.

Game Mechanics: The fundamental rules, systems, and interactions that govern the behavior and structure of a game.

Game Storytelling: The art and process of crafting and delivering a coherent and engaging narrative within a game.

Immersion: The psychological state of being deeply engaged and absorbed in an experience, often enhanced by sensory-rich environments and compelling narratives.

Mobile Escape Game: An interactive and immersive puzzle-solving experience designed specifically for mobile devices, such as smartphones and tablets. Players engage with a themed scenario or storyline, solving puzzles using touchscreen interfaces and virtual environments.

Narrative Design: The process of creating and integrating engaging storylines into games to enhance immersion, player engagement, and emotional responses.

Puzzle-Solving Psychology: The cognitive processes and intrinsic appeal of problem-solving, highlighting why challenges captivate players and how they contribute to intellectual engagement and skill development.

Serious Game: A digital or tabletop game designed with the primary intention of conveying educational, informational, or training content while maintaining an engaging and enjoyable gameplay experience.

Simulacra: A concept from Jean Baudrillard's theory, describing representations or imitations of real-world scenarios that may bear little to no relation to reality, often seen in designed environments like escape rooms.

Virtual Reality Escape Game: An immersive, interactive game experience where players use virtual reality technology to explore and solve puzzles in a digitally created environment.

Chapter 2
Educational Theories and Digital Games

ABSTRACT

This chapter provides explores educational theories and their application in digital games and Digital Educational Escape Rooms (DEERs). Beginning with an overview of game-based learning, it discusses the definition, and key components that make this approach effective. The chapter introduces DEERs, examining their theoretical foundations, design principles, and impact on learning. Educational theories such as constructivism, behaviorism, and cognitivism are explored in the context of game design, along with their implications for DEER development. It focuses on gamification, explaining its principles, elements, and role in enhancing motivation and learning outcomes within DEERs. The cognitive benefits of puzzle games are highlighted, particularly their role in enhancing problem-solving, memory, and attention. Social learning and collaboration in virtual spaces are discussed through multiplayer environments, community-building, and social dynamics. The chapter concludes by examining methods for assessing learning outcomes in games.

EDUCATIONAL THEORIES IN DIGITAL GAMES

Game-Based Learning

Game-Based Learning (GBL) represents a transformative approach in education, integrating game elements into pedagogical practices to enhance learning outcomes (Bado, 2022). This innovative method leverages the principles of digital game design and mechanics to create learning experiences that are not only engaging but also effective in fostering cognitive, emotional, social, and skills-based development. The scope of GBL extends beyond traditional educational boundaries, utilizing the immersive and interactive nature of digital technologies to create environments

DOI: 10.4018/979-8-3693-4219-0.ch002

where learners can engage deeply with content in a manner that promotes active learning and retention. GBL is characterized by its emphasis on creating learning experiences that mirror the dynamics of gameplay, where challenges, feedback, and progression are central to the learning process (Shi and Shih, 2015). The definition of GBL underscores its role in not just imparting knowledge but also in cultivating critical thinking, problem-solving, and collaboration skills. By embedding educational objectives within the framework of a game, learners are motivated to engage with content more deeply, often without the traditional resistance associated with conventional educational methods.

Table 1 outlines key game design principles in educational games and their impact on learning. Engagement refers to elements that keep learners motivated and immersed, enhancing attention and learning time (Abdul Jabbar and Felicia, 2015), as seen in Minecraft Education and Kahoot!. Flow balances challenge and skill, maintaining optimal cognitive load and focus, exemplified by games like Portal and The Witness. Feedback, whether real-time or delayed, helps learners adjust and improve, with examples like Duolingo and Classcraft. Personalization tailors content to individual needs, ensuring relevance and engagement, as seen in DreamBox and Prodigy. Lastly, Progression, which involves gradually increasing difficulty and unlocking levels, fosters continuous learning and mastery, as demonstrated by Super Mario Odyssey and Lumosity.

The scope of GBL is broad, encompassing various disciplines and educational levels. It can be applied in formal education settings, such as schools and universities, as well as in informal contexts, including corporate training and lifelong learning programs. The adaptability of GBL makes it a powerful tool in addressing the diverse needs of learners, providing opportunities for differentiated instruction and personalized learning paths. Through the use of GBL, educators can create tailored learning experiences that cater to individual learning styles, thereby enhancing the effectiveness of the educational process.

Table 1. Key game design principles in educational games and platforms

Design Principle	Description	Impact on Learning	Example Games/ Platforms
Engagement	Elements that keep learners motivated and immersed	Increases attention, motivation, and time spent learning	Minecraft Education, Kahoot!
Flow	The balance between challenge and skill level	Maintains optimal cognitive load and helps learners stay focused	Portal, The Witness
Feedback	Real-time or delayed responses to player actions	Helps learners adjust, reflect, and improve their skills	Duolingo, Classcraft

continued on following page

Table 1. Continued

Design Principle	Description	Impact on Learning	Example Games/ Platforms
Personalization	Customizing content based on learner needs	Ensures content is relevant to each learner's abilities and interests	DreamBox, Prodigy
Progression	Gradually increasing difficulty and unlocking new levels	Encourages continuous learning and skill mastery	Super Mario Odyssey, Lumosity

The concept of Game-Based Learning (GBL) has undergone significant evolution, rooted in early educational practices and shaped by technological advancements over time (Ekin et al., 2023). The origins of GBL can be traced back to the use of educational toys and simulations, which were among the first tools designed to engage learners through play. These early forms of GBL focused primarily on physical activities that encouraged problem-solving and cognitive development in children. As educational theories advanced, the idea of integrating learning objectives with play became more structured, leading to the development of more sophisticated educational tools.

The late 20th century marked a pivotal shift in the evolution of GBL with the advent of digital technology. This era saw the emergence of computer-based educational games, which represented a new frontier in combining entertainment with learning. Early examples of such games include the iconic "Oregon Trail" and "Where in the World is Carmen Sandiego?". These games not only provided entertainment but also introduced educational content in a way that was engaging and memorable. "Oregon Trail" taught players about American history and survival, while "Where in the World is Carmen Sandiego?" focused on geography and problem-solving. These games demonstrated the potential of digital platforms to deliver educational content in a manner that was both enjoyable and effective.

The early 21st century witnessed an exponential increase in the research and application of GBL, driven by the growing accessibility of digital technology and a deeper understanding of its educational potential. The expansion of the internet, coupled with advancements in computer graphics and user interfaces, enabled the development of more interactive and immersive learning environments. This period also saw the rise of serious games—games specifically designed for educational purposes, rather than merely for entertainment. These games began to cover a wide range of subjects, from language learning and mathematics to science and social studies, reflecting the versatility and adaptability of GBL to different educational contexts.

The increasing interest in GBL during this period was supported by a growing body of research that highlighted its effectiveness in enhancing learning outcomes. Studies began to show that GBL could significantly improve student engagement, motivation, and retention of knowledge, particularly when compared to traditional educational methods. This research underscored the value of GBL as a tool for not only delivering content but also for fostering critical thinking, creativity, and collaboration among learners.

The effectiveness of GBL is fundamentally tied to several core components that must be carefully integrated to create a meaningful and engaging educational experience. These components serve as the building blocks for designing GBL environments that not only capture the interest of learners but also align with educational goals to facilitate deeper learning.

One of the most critical components of GBL is the establishment of clearly defined learning objectives. These objectives must be seamlessly aligned with the game mechanics to ensure that every aspect of the gameplay contributes to achieving the desired educational outcomes. This alignment is essential because it ensures that the activities within the game are not merely entertaining but also serve a pedagogical purpose. For example, a game designed to teach mathematical concepts might include puzzles that require the application of specific mathematical principles, thereby reinforcing the learner's understanding of these concepts through practice and application.

Engagement is another pivotal component in GBL. The use of narratives, challenges, and interactive feedback mechanisms is key to maintaining student interest throughout the learning experience. A well-crafted narrative can provide context and motivation, making the learning objectives more relatable and meaningful to the player. Challenges that are appropriately scaled in difficulty can keep learners in a state of flow, where they are sufficiently challenged to remain engaged but not so overwhelmed that they become frustrated. Feedback is crucial in this context, as it helps learners understand the consequences of their actions within the game, thereby encouraging them to reflect on their strategies and improve their performance.

Feedback and adaptation are closely related to engagement but warrant separate consideration due to their importance in personalizing the learning experience. Real-time feedback allows learners to see the immediate results of their decisions, which can be incredibly motivating and informative. Adaptive learning paths that respond to the learner's performance can help ensure that the game remains challenging yet achievable, thus supporting the learner's progression at a pace that suits their individual needs. This adaptability is particularly important in diverse learning environments where students may have varying levels of prior knowledge and skill.

Table 2 categorizes the different types of feedback in educational games, including immediate, delayed, formative, summative, peer, and automated feedback. Immediate feedback provides real-time responses to correct or reinforce behavior, promoting quick adjustments and engagement. Delayed feedback encourages reflection and deeper critical thinking by allowing learners to evaluate their own performance before receiving input. Formative feedback is ongoing and helps guide learners through the learning process, while summative feedback offers a final evaluation of overall performance. Peer feedback fosters collaboration and social learning, allowing learners to evaluate each other's work. Lastly, automated feedback is system-generated, providing consistent and scalable responses, particularly useful in large-scale or adaptive learning environments. Each feedback type impacts learning differently, from reinforcing behavior to fostering reflection, adjustment, and collaboration.

Figure 1 visualizes the impact of real-time feedback on learner performance over time in an educational game setting. It explains the importance of feedback for learner progression. Learners receiving immediate feedback show a rapid increase in performance, demonstrating how timely responses help them adjust their strategies more effectively. Learners without feedback show slower improvement in performance, as they lack the immediate information needed to adjust their actions and optimize learning.

Table 2. Types of feedback in educational games and their impact on learning

Feedback Type	Description	Impact on Learning	Examples in Games
Immediate Feedback	Provided instantly after a learner's action, giving real-time responses to correct or reinforce behavior.	Promotes immediate reinforcement of correct behavior, allowing quick adjustment and improved learning retention. Helps learners stay engaged and reduces frustration.	In puzzle games like *Portal*, learners receive immediate feedback when they solve puzzles correctly or incorrectly. In Duolingo, correct or incorrect answers are instantly acknowledged.
Delayed Feedback	Provided after a certain period or at the end of a task, giving learners time to reflect before receiving feedback.	Encourages deeper reflection and critical thinking, as learners must evaluate their own performance before receiving feedback. Can enhance long-term retention of knowledge.	In strategy games like *Civilization*, players often receive feedback about their choices after completing missions or tasks, allowing for strategic reflection.
Formative Feedback	Ongoing feedback provided throughout earning process to help guide learners towards improvement.	Supports continuous learning and adjustment, helping learners improve over time by identifying strengths and weaknesses. Builds a learning path with personalized insights.	In learning platforms like *Khan Academy*, learners receive formative feedback on problem-solving exercises, with hints and suggestions provided as they work

continued on following page

Table 2. Continued

Feedback Type	Description	Impact on Learning	Examples in Games
Summative Feedback	Given at the end of a task, level, or game, summarizing overall performance and providing an evaluation of success.	Allows learners to reflect on their overall achievement, understand their progress, and make adjustments for future tasks. Useful for final assessments or measuring progress.	In quiz games like *Kahoot!*, learners receive a final score or ranking at the end of the quiz, offering a summative evaluation of their performance.
Peer Feedback	Feedback provided by peers or other players in collaborative or multiplayer learning environments.	Encourages social learning and collaboration, helping learners to evaluate each other's work, build communication skills, and reflect on different perspectives.	In games like *Minecraft Education*, players provide feedback to each other during collaborative tasks, offering support or suggestions.
Automated Feedback	Feedback generated by the game system based on pre-defined criteria or algorithms, with little to no human intervention.	Provides consistent, scalable feedback for repetitive tasks. Useful for large-scale educational games where individual instructor feedback is impractical.	In adaptive learning platforms like *DreamBox*, feedback is automatically provided based on a learner's performance, adjusting the difficulty of tasks accordingly.

Figure 1. Expected impact of real-time feedback on learner performance

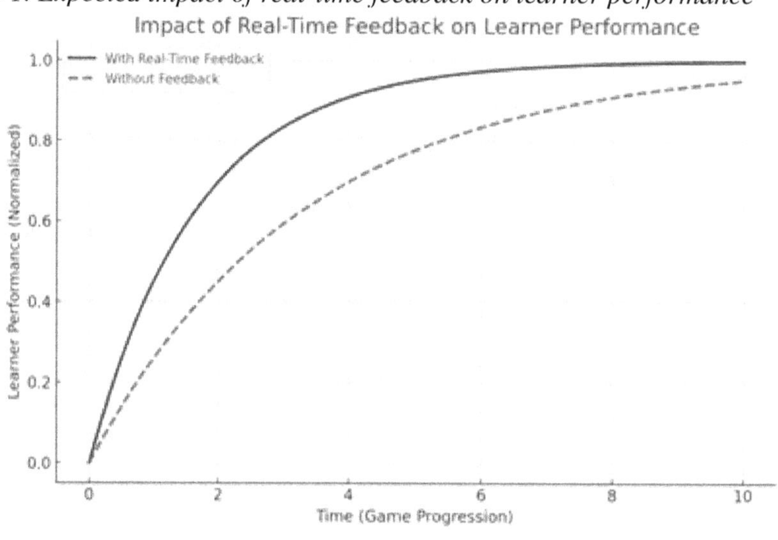

Assessment in GBL serves as the mechanism through which learning progress is measured. Integrated evaluation methods within the game allow for continuous assessment without disrupting the flow of the game. These assessments can take various forms, such as quizzes, performance metrics, or task completions, and are designed to gauge the learner's understanding and mastery of the content. By embedding assessment into the gameplay, educators can obtain a more holistic view of the learner's abilities and progress, allowing for timely interventions and support when necessary.

Each of these components—learning objectives, engagement, feedback and adaptation, and assessment—plays (Table 3) a crucial role in the success of GBL. Together, they create a learning environment that is not only immersive and enjoyable but also deeply educational, fostering the development of critical skills and knowledge in a way that traditional educational methods often cannot achieve.

Table 3. Key components of game-based learning

Component	Description
Learning Objectives	Clearly defined educational goals that are aligned with game mechanics to ensure that gameplay directly contributes to achieving desired learning outcomes.
Engagement	The use of narrative, challenges, and feedback mechanisms to maintain student interest and motivation throughout the learning experience.
Feedback and Adaptation	Real-time responses to student actions within the game, allowing for personalized learning paths that adjust to the learner's needs and abilities.
Assessment	Integrated evaluation methods embedded within the game to gauge learning progress and outcomes without disrupting the gameplay.

Digital Educational Escape Rooms (DEERs)

Digital Educational Escape Rooms (DEERs) represent an adaptation of the traditional escape room concept, specifically tailored for educational purposes and delivered through a digital medium (Veldkamp et al., 2020). These rooms are designed to offer immersive problem-solving experiences, where learners are challenged to complete a series of interconnected tasks or puzzles within a set timeframe. The ultimate aim is not merely to escape but to achieve predefined educational objectives. The digital format of DEERs allows for versatility and scalability, enabling their application across various educational contexts—from individualized learning modules to collaborative environments involving multiple participants. DEERs stand out due to their ability to simulate real-world scenarios that require the application of theoretical knowledge in practical, often complex, situations. This approach aligns closely with experiential learning theories, which emphasize learning through direct experience and reflection. By engaging in DEERs, learners are prompted to think critically, work collaboratively, and apply their knowledge to solve problems that are contextually embedded within the game's narrative. This method of learning is particularly effective in reinforcing concepts, as it encourages active participation and immediate application of knowledge.

The digital nature of DEERs introduces flexibility in design and implementation. Unlike physical escape rooms, DEERs can be customized to suit a wide range of educational needs and learning styles, making them accessible to diverse groups of learners (see the comparison between physical and digital escape rooms in Table 4). The digital platform allows for the integration of multimedia elements, such as videos, animations, and interactive simulations, which can enhance the learning experience by making it more engaging and visually stimulating. This adaptability makes DEERs a valuable tool in both formal education settings and informal learning environments.

DEERs are grounded in various educational theories, notably constructivism and experiential learning (Morris, 2020) (Figure 2). These theories emphasize the importance of active participation, problem-solving, and the construction of knowledge through experience. DEERs create a simulated environment where learners can apply theoretical knowledge in practical scenarios, thereby reinforcing their understanding through experience.

The design of DDEERs is guided by several fundamental principles that ensure their effectiveness as educational tools. One of the primary design principles is the integration of narrative and context. A compelling storyline serves as the backbone of DEERs, providing a framework that contextualizes the learning objectives and immerses learners in the experience. This narrative-driven approach not only captures the attention of learners but also makes the educational content more relatable and memorable. By embedding learning tasks within a coherent and engaging storyline, educators can create a more meaningful and impactful learning experience. Puzzles and challenges form the core of the DEER experience, acting as the primary mechanisms through which learning objectives are achieved. These tasks are carefully crafted to align with educational goals, requiring learners to engage in critical thinking, collaboration, and creativity. The design of these challenges is crucial; they must be neither too easy nor impossibly difficult, ensuring that learners remain in a state of flow, where they are constantly motivated to solve problems and advance through the game. The puzzles are designed to progressively build on one another, reinforcing previously learned concepts while introducing new ones, thus supporting a scaffolded learning approach.

Table 4. Comparison of physical and digital escape rooms

Aspect	Physical Escape Rooms	Digital Escape Rooms
Accessibility	Limited by physical location and availability. Participants need to visit a specific venue to participate. Typically more expensive due to operational costs.	Accessible from anywhere with an internet connection, allowing global participation. Often more affordable and can be played on various devices such as computers, tablets, and smartphones.
Scalability	Difficult to scale due to space limitations. The number of participants is often restricted by room size and physical setup.	Easily scalable. Digital environments can host many participants simultaneously without being limited by physical space. Game updates and new content can be rolled out easily.
Feedback	Typically real-time feedback provided by in-person game masters or facilitators. Limited in-game feedback mechanisms, often relying on hints provided by staff.	Immediate, automated feedback through digital systems. Integrated hint systems, scoring, and progress tracking are built into the game. Customizable feedback based on user input and performance.
Aspect	**Physical Escape Rooms**	**Digital Escape Rooms**
Immersion	High immersion through physical interaction with objects and environments. The tactile experience enhances the realism. Participants can physically move, touch, and manipulate items in real-world settings.	Immersion depends on the quality of the digital design, narrative, and user interface. Can simulate highly immersive environments with sound, visuals, and virtual interaction, but lacks physical tactile engagement.
Learning Outcomes	Promotes physical collaboration and teamwork through in-person communication. Supports kinesthetic learning and offers strong social interaction skills. Ideal for team-building exercises.	Supports a broader range of learning styles, including individual, collaborative, and distance learning. Data from digital games can be used to track performance, tailor feedback, and enhance learning outcomes. Provides flexible learning environments.
Flexibility	Limited flexibility. Rooms are often designed around specific puzzles and themes that are hard to modify without physical changes to the environment.	Highly flexible. Digital environments can be easily modified, updated, or repurposed to fit different educational objectives, themes, or audiences.
Social Interaction	Encourages face-to-face communication and collaboration in real-time. Builds in-person team dynamics and fosters direct social interactions.	Supports both collaborative and individual gameplay. Digital escape rooms can enable remote teamwork through online communication tools, though it may lack the depth of in-person interactions.

Figure 2. Theoretical foundations of game-based learning

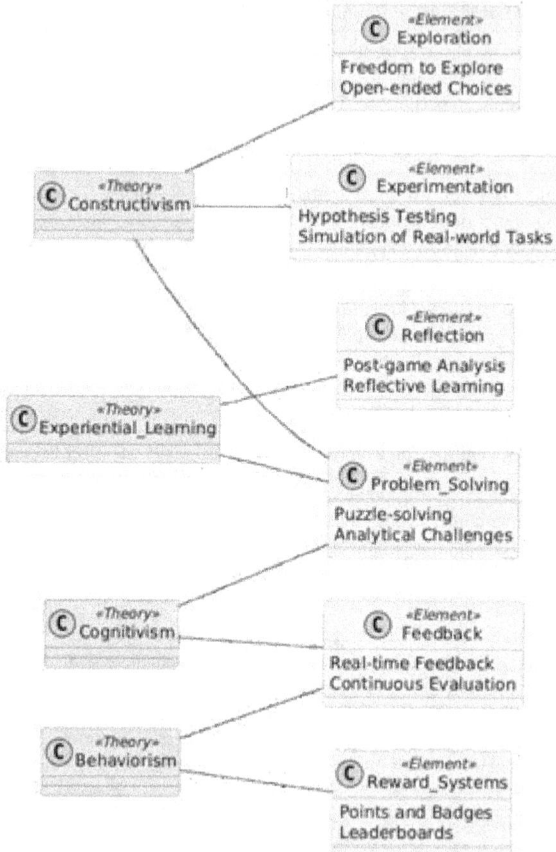

Another critical component of effective DEERs is the incorporation of feedback and debriefing mechanisms. Real-time feedback is essential in helping learners understand the consequences of their actions, guiding them towards the correct solutions and reinforcing learning. This feedback can take various forms, including hints, rewards, or corrective prompts, and is crucial for maintaining learner engagement and motivation. Debriefing sessions at the end of the game provide opportunities for reflection and discussion, allowing learners to consolidate their knowledge and receive formative feedback from educators. These sessions are vital for reinforcing the learning objectives and helping learners transfer the skills and knowledge gained from the game to real-world contexts.

The Components of Digital Educational Escape Rooms (DEERs) are summarized in Figure 3.

Figure 3. Components of Digital Educational Escape Rooms (DEERs)

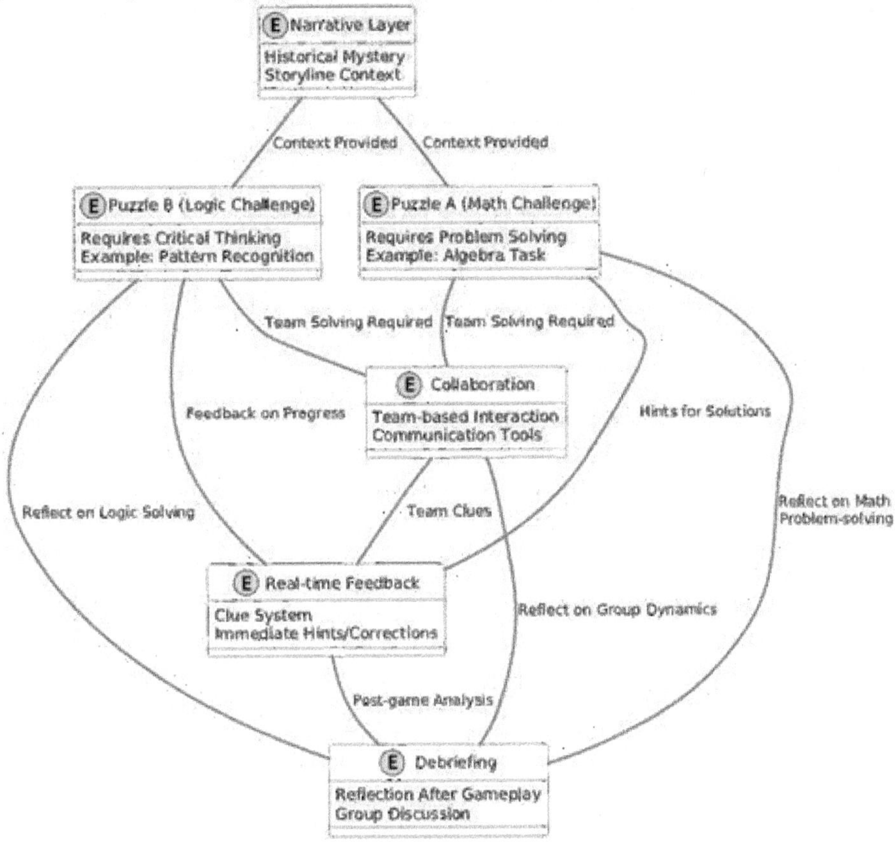

Implementing these design principles effectively requires careful planning and a deep understanding of the educational goals (see Table 5). The narrative must be well-integrated with the educational content, ensuring that every element of the DEER contributes to the learning objectives. Puzzles must be thoughtfully designed to challenge learners while remaining achievable, and feedback mechanisms must be seamlessly incorporated to guide and motivate learners throughout the experience. By adhering to these design principles, educators can create DEERs that are not only engaging and enjoyable but also powerful tools for learning.

Table 5. Key components of Digital Educational Escape Rooms (DEERs)

Component	Description	Example of Use	Learning Outcome
Narrative	A compelling story that contextualizes the learning objectives	A mystery to solve that incorporates historical events	Immersive engagement, contextual learning
Puzzles/ Challenges	Interactive tasks that require critical thinking to progress	Math problems to unlock codes	Problem-solving skills, subject mastery
Collaboration	Tasks designed for group interaction and teamwork	Group of students must work together to complete a puzzle	Communication, teamwork, leadership
Real-time Feedback	Immediate responses to learners' actions in the game	Hints or corrections provided when a puzzle is solved incorrectly	Self-regulation, iterative learning
Debriefing	Reflection sessions where learners discuss and analyze their gameplay	Post-game reflection on how knowledge was applied	Critical thinking, reflective learning

KEY THEORETICAL CONCEPTS IN DEER GAME DESIGN

Understanding Player Motivation and Engagement in Escape Rooms

Understanding player motivation and engagement in escape room games involves examining the theories of player motivation and the specific engagement mechanics used in digital escape rooms. Player motivation in escape room games is multifaceted. Students are motivated to engage in escape room games due to the satisfaction of playing, competing, and the possibility of winning (Mateos-Aparicio et al., 2020). Nine determinants of player motivation are associated with the model of goal-directed behavior (Lim Qian Pink et al., 2019).

The challenge-award system in mixed reality escape room video games promotes intrinsic motivation, real-world exploration, and the use of logic and problem-solving strategies (Gómez-Cambronero et al., 2019). Digital escape games motivate learners, develop adaptability, and responsiveness skills, especially in large groups (Monnot et al., 2020).

The engagement mechanics in digital escape rooms are designed to captivate players through various challenges and objectives. Escape rooms engage players by requiring them to search for clues, solve puzzles, and achieve pre-defined objectives, making them a popular educational game option (Benassi, 2019). Digital escape rooms use mental, physical, and emotional challenges to engage players (Krekhov et al., 2021). Digital escape room-style games engage students in the learning process and achieve learning outcomes, with all students finding the game interesting for both playing and learning (Videnovik et al., 2022).

These theories of player motivation and engagement mechanics are crucial in designing and implementing effective digital escape rooms.

Puzzle Design and Logical Frameworks in Escape Room Games

The design and logical frameworks of puzzles in escape room games, encompassing principles of puzzle creation and the balance between complexity and accessibility, are essential aspects of game design. Principles of puzzle creation in escape room games involve various elements that contribute to an engaging and challenging experience. One approach utilizes virtual reality technology to create interactive puzzles that can be regularly updated without altering the building's structure (David et al., 2019). In educational contexts, escape rooms involve students creating puzzles to foster creativity, teamwork, and demonstrate knowledge through non-traditional methods (Hong and Reiser, 2020). The construction of puzzles involves mental, physical, and emotional challenges, forming a common design framework and taxonomy necessary for analysis and building escape rooms (Krekhov et al., 2021). In physics education, escape rooms adapt puzzle-solving methods for multiple teams to work together in the same space (Vörös and Sárközi, 2017).

Balancing complexity and accessibility in puzzle design is crucial for engaging participants and ensuring the educational value of the game. For example, in general chemistry courses, ChemEscape engages students in problem-solving while enhancing discipline-specific skills (Clapson et al., 2020). The design of escape rooms for large audiences, such as in chemical engineering education, focuses on creating puzzles that motivate and engage students while being accessible to a broad range of participants (Monnot et al., 2020). The balance in puzzle design is pivotal for creating feelings of social presence and promoting curiosity (Shakeri et al., 2017).

The principles of puzzle design and the balance of complexity and accessibility are fundamental in creating effective and engaging escape room games. These aspects ensure that the games are challenging yet accessible, providing an enriching experience for players.

Narrative Design in Escape Room Games

The narrative design in escape room games, encompassing storytelling techniques and the integration of narrative and gameplay, is a crucial aspect of game design that enhances player experience and engagement. Storytelling techniques in escape room games utilize various creative methods. Escape rooms in STEAM education, for example, incorporate creative writing, theatrical elements, and technology such as Robotics, QR codes, and Virtual Reality to promote communication, teamwork, and peer learning (Karageorgiou et al., 2019). Augmented reality and deep learning technologies in escape rooms support more complicated narratives, achieving greater immersion (Zeng et al., 2020). Co-designing a mixed reality escape room narrative with learners involves identifying dominant emotions to guide the design of the narrative and puzzles (Itenge et al., 2021). Escape rooms like "The Archivist's Dream" combine archival practices with dream logic, allowing players to explore different levels of reality (Koenig et al., 2022).

Integrating narrative and gameplay is essential for creating an engaging escape room experience. Educational escape rooms recommend aligning game mechanics with pedagogical approaches to help players achieve game goals and educational objectives (Veldkamp et al., 2020). Escape room games are effective in forming high-demanded soft skills by considering game mechanics, professional aspirations, students' cognition style, and the correlation of formed skills with planned ones (Vekua et al., 2020). In nursing education, escape rooms involve problem-solving, critical thinking, and teamwork, promoting learning objectives (Morrell and Ball, 2020). The narrative design of escape room games set inside real buildings, connected to the virtual world, uses a challenge-award system to promote intrinsic motivation, real-world exploration, and logic and problem-solving strategies (Gómez-Cambronero et al., 2019). These insights reveal the significance of narrative design in enhancing the educational and entertainment value of escape room games, making them a popular choice for both learning and leisure activities.

The Role of Environment and Setting in Escape Room Games

The role of environment and setting in escape room games, particularly in creating immersive digital environments and integrating environmental storytelling and clues, is crucial for enhancing the player's experience. The creation of immersive digital environments in escape room games is achieved through various innovative approaches. Virtual reality technology is used to develop escape rooms that allow users to interact with the environment, solve puzzles, and improve creative thinking (Kiruthika et al., 2022). Passive haptics and motion tracking in VR escape rooms contribute to immersive environments by enabling players to physically interact and provide head and hand position cues to teammates (Hanus et al., 2019). Gamified Mixed Reality environments facilitate the quick and easy creation of educational escape rooms, thus enhancing the immersive digital environment (Klamma et al., 2020). 3D game development tools like Unity 3D and Blender are instrumental in creating virtual reality content that contributes to the immersive experience in escape rooms (Gupta et al., 2016).

Environmental storytelling and clue integration in escape rooms are essential for engaging players in the narrative. Escape rooms in STEAM education combine entertainment, creative writing, theater, and technology to promote communication, teamwork, and peer learning (Karageorgiou et al., 2019). Climate change education in escape rooms provides experiential and immersive learning, enhancing problem-solving and critical thinking skills, along with a sense of collaboration and urgency (Ouariachi and Wim, 2020). Educational digital escape rooms create an environment of positive emotions and favorable opinions on knowledge acquisition tools (Ruiz et al., 2022). A VR escape room game like “Escape The Countries” enhances player interaction and skills in math, arrangement, and pattern recognition (Yeasmin and Albabtain, 2020).

User Interface and User Experience (UI/UX) Design of Escape Rooms

The User Interface (UI) and User Experience (UX) design of escape room games, focusing on UI/UX principles in game design and tailoring UI/UX for escape room games, play a vital role in enhancing the overall player experience. UI/UX principles in game design, especially for escape room games, emphasize creating immersive and intuitive interfaces. Virtual reality technology in escape rooms allows for interaction with the environment and improves logic thinking, facilitating the regular replacement of puzzle elements without altering the building’s structure (David et al., 2019). Augmented reality combined with deep learning technology in escape rooms supports more complicated narratives, enhancing the gaming experience (Zeng et al., 2020). A high-level design framework and an atomic puzzle taxonomy, involving mental, physical, and emotional challenges, are crucial for the analysis and construction of escape rooms (Krekhov et al., 2021). Integrating STEAM education and problem-based learning principles into escape room game design can inspire problem-solving potential in students (Ma et al., 2018).

Tailoring UI/UX for escape room games involves aligning game mechanics with educational goals, thereby enhancing the gaming and learning experience. More alignment in game mechanics and pedagogical approaches is recommended to achieve both game goals and educational objectives (Veldkamp et al., 2020). Escape rooms use game design principles to augment knowledge acquisition in a fun, team-based learning experience (Kaul et al., 2021). Considering game mechanics, professional aspirations, students’ cognition style, and the correlation of formed skills with planned ones is key in forming high-demanded soft skills (Vekua et al., 2020). Escape rooms are valuable for acquiring professional competencies in higher education, increasing student motivation for learning (Ang et al., 2020). UI/UX design in escape room games is a critical component that requires a careful balance of immersive storytelling, interactive elements, and alignment with educational or entertainment objectives to provide an engaging and meaningful player experience.

Role of Technology and Innovation in Escape Room Games

The role of technology and innovation in escape room games, particularly concerning emerging technologies in game design and the impact of Virtual Reality (VR) and Augmented Reality (AR) on the escape room experience, is a dynamic and rapidly evolving aspect of modern gaming.

Emerging technologies have a significant impact on game design, particularly in escape room games. For example, the use of VR technology in escape rooms provides real-time, hands-on experience and enhances creative thinking, allowing for regular replacement of puzzle elements without altering the physical structure (Kiruthika et al., 2022). The implementation of AR in escape room systems, combining it with deep learning technology, creates more immersive and interactive experiences (Zeng et al., 2020). Escape rooms in STEAM education incorporate technologies like Robotics, QR codes, and VR to promote communication, teamwork, and peer learning (Karageorgiou et al., 2019). Serious games like Aventura Marina, inspired by escape room mechanics, aim to enhance visitor engagement in science museum exhibitions (Cordova-Rangel and Caro, 2021).

VR and AR have profoundly impacted the escape room experience, making it more immersive and engaging. VR escape rooms use headsets, sensor-equipped gloves, and hand controllers to create 3D artificial environments with puzzle-solving challenges (Yeasmin and Albabtain, 2020). AR technology in games like the Serious Escape Game MillSecret reveals hidden local cultural heritage, enhancing the learning and gaming experience (Tzima et al., 2020). The integration of VR in museum exhibits like 'The Quest,' inspired by escape room mechanics, aims to create deeper, more immersive engagement with the exhibit's subject (Back et al., 2019). VR technology in educational settings, like CySecEscape 2.0, has been transformed into physical settings without sacrificing player immersion for cybersecurity awareness training (Löffler et al., 2021).

Multiplayer and Social Interaction Design in Escape Room Games

The multiplayer and social interaction design in escape room games, particularly in designing for cooperative play and understanding social dynamics and communication, is a crucial component in enhancing player engagement and interaction.

Cooperative play is fostered through team-building and collaborative learning elements. Students, for instance, use analytical instruments in laboratory settings to solve puzzles collaboratively in an escape room game (Vergne et al., 2019). Edu-Escape Rooms incorporate cooperative gaming where players find clues, solve puzzles, and perform tasks within a limited time, enhancing student engagement and motivation (Grande-de Prado et al., 2020). Escape room games can develop perception, logical thinking, and cooperative work in children, with adult-child collaboration being essential (Melo, 2021). Escape rooms provide opportunities to practice collaboration skills, although the design may restrict such behaviors in real-world situations (Pan et al., 2017).

Social dynamics and communication are integral. Escape room games combine problem-solving, critical thinking, and teamwork, aligning with learning objectives (Morrell and Ball, 2020). Physics escape rooms adapt puzzle-solving methods for multiple teams to work collaboratively in the same space (Vörös and Sárközi, 2017). Aligning game mechanics with pedagogical approaches in educational escape rooms helps players achieve game goals and educational objectives (Veldkamp et al., 2020). AScapeD, a multiplayer virtual escape room, promotes equal cooperation and communication between children with ASD and their peers (Terlouw et al., 2021).

Feedback Loops and Player Learning in Escape Room Games

The implementation of feedback loops and facilitation of player learning and progression in escape room games are increasingly recognized for their effectiveness in enhancing the player experience and learning outcomes. Effective feedback mechanisms in escape room games involve various strategies to enhance player learning and progression. Immediate and ongoing constructive feedback in escape room exercises has been shown to decrease student anxiety and improve learning outcomes (Parker and Welch, 2021). Escape room games also use feedback loops to support exploratory learning and intrinsic motivation (Beatty, 2014). Interactive-feedback loops in digital games can effectively engage learners by providing timely and well-structured feedback (Kazimoglu et al., 2010). Adaptive feedback loops in escape room games must balance game difficulty accurately to be accepted by users and improve their overall game experience (Novak et al., 2014).

The design of escape room games facilitates player learning and progression through various educational approaches. Escape games in the classroom are found to increase student motivation and improve learning in computer science courses (Borrego et al., 2017). Feedback loop features in video game environments can engage adolescents in meaningful learning, both in and beyond the classroom (Abrams and Gerber, 2021). In chemical engineering education, escape game activities can motivate learners and develop adaptability and responsiveness skills (Monnot et al., 2020). Escape rooms have also been adapted for STEM teaching, demonstrating their potential in subject areas and for developing 21st-century skills.

Digital Educational Escape Rooms (DEERs) have shown a significant impact on various aspects of learning, primarily by enhancing learner engagement, motivation, and knowledge retention. The interactive and immersive nature of DEERs plays a crucial role in these outcomes. By involving learners in a series of well-designed, contextually relevant challenges, DEERs create an environment where students are not passive recipients of information but active participants in the learning process. This active participation is fundamental to deeper cognitive processing, which in turn leads to better retention and understanding of the material.

Research indicates that the engagement fostered by DEERs is one of the key factors contributing to their effectiveness. The immersive nature of these digital escape rooms, coupled with the pressure of time constraints, encourages students to remain focused and involved throughout the learning activity. This heightened engagement not only keeps learners interested but also reduces the cognitive load by breaking down complex information into manageable tasks, which are then solved progressively. As learners interact with the content in a hands-on manner, they are more likely to internalize the concepts being taught, leading to improved long-term retention of knowledge.

Figure 4 visualizes the Player Engagement Model in Digital Educational Games, following a funnel structure that narrows as players move from initial exposure to mastery:

- Awareness: At the top of the funnel, players become aware of the game's content and purpose. Game introductions and tutorials help capture initial attention.
- Interest: The game mechanics further engage players, building interest and encouraging them to explore deeper.
- Active Participation: This stage involves hands-on engagement where players interact with the game through active gameplay, developing skills and progressing in the game.

- Reflection: Players begin reflecting on their actions, strategies, and performance, often encouraged by feedback or in-game summaries.
- Mastery: At the narrowest point of the funnel, players achieve mastery through repeated play and learning, solidifying their understanding and skills within the game environment.

Each stage narrows toward mastery, representing how fewer players continue to deeper levels of engagement as learning and mastery are achieved.

Figure 4. Player engagement in digital educational games

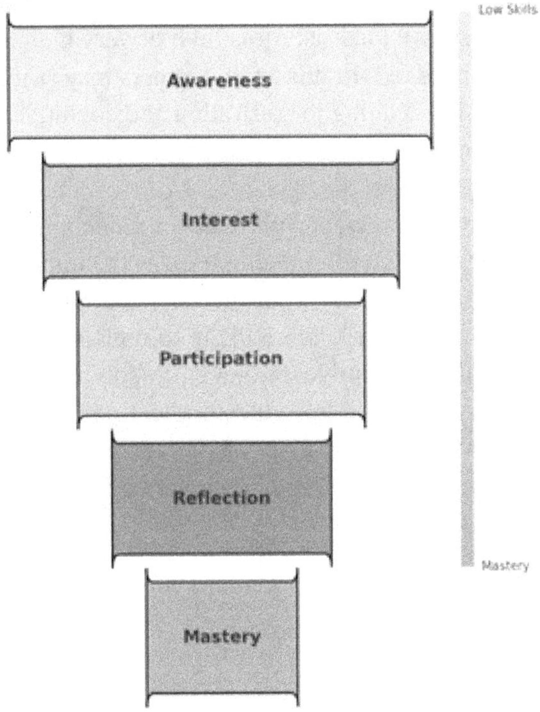

Figure 5 illustrates how cognitive load fluctuates during gameplay in educational settings, representing the relationship between task complexity and the cognitive effort required to engage with learning tasks. It provides a qualitative breakdown of how different levels of task complexity impact learners, ranging from underload to overload. The bell-shaped curve in the figure shows that cognitive load follows a predictable pattern during gameplay. When tasks are too easy (left side of the curve), the cognitive load is minimal, leading to an underload state, represented in light blue. In this zone, learners are likely to experience boredom or disengagement due to the lack of challenge, which hinders their motivation and learning outcomes. Tasks in the Low complexity range are unlikely to provide enough stimulation for meaningful cognitive development, as the load is "very low" or "low" on the Likert scale. As the complexity of the tasks increases to a medium level, cognitive load enters the Optimal Load zone, highlighted in light green. This zone represents the ideal balance between challenge and skill, where learners experience a moderate level of cognitive load that promotes effective learning. In this range, the task complexity is sufficient to engage learners without overwhelming them, creating the perfect environment for critical thinking, problem-solving, and knowledge retention. Here, learners are actively challenged but not discouraged, resulting in high levels of engagement and positive learning outcomes. However, as tasks become increasingly complex (right side of the curve), cognitive load rises to a high or very high level, represented by the Overload zone in light red. In this zone, learners may struggle to keep up with the demands of the game, leading to frustration and cognitive overload. This state occurs when task complexity exceeds the learner's capacity to process information effectively, causing them to feel overwhelmed. In educational settings, this overload can lead to disengagement, stress, and ultimately a reduction in the ability to absorb and apply new information. The figure emphasizes the importance of maintaining tasks within the Optimal Load zone for effective learning. When cognitive load is carefully balanced, learners are more likely to remain engaged, motivated, and capable of applying their cognitive resources to solving challenges. The visual representation serves as a guide for educators and game designers, highlighting the need to design learning tasks that are neither too easy nor too difficult, thereby maximizing the potential for meaningful cognitive development during gameplay.

Figure 5. Cognitive load during gameplay

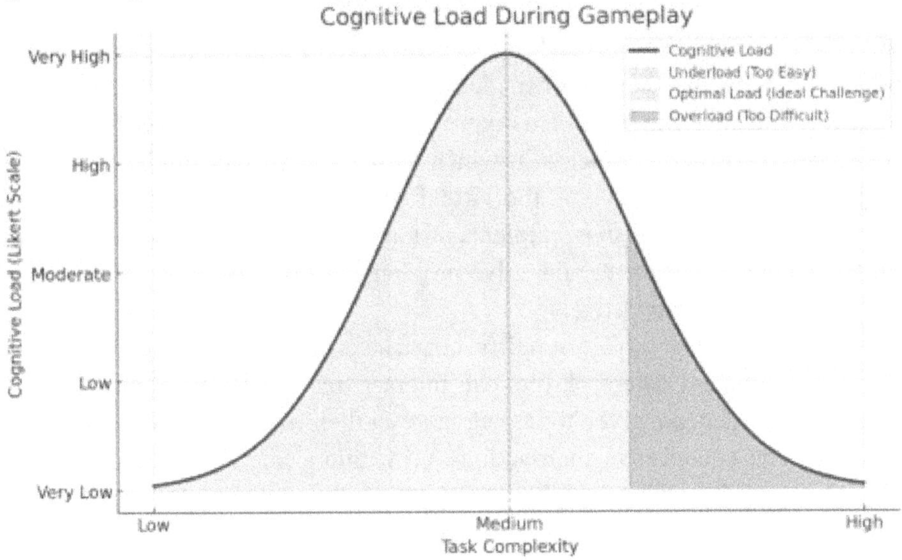

In addition to engagement, DEERs enhance motivation. The game-like structure of DEERs, which often includes elements such as scoring, competition, and rewards, taps into both intrinsic and extrinsic motivators. Intrinsically, students are motivated by the desire to solve puzzles and complete challenges, which are inherently rewarding. Extrinsically, the use of points, leaderboards, and badges can provide additional motivation, driving students to perform better and engage more deeply with the content. This combination of intrinsic and extrinsic motivation helps maintain high levels of participation and effort, which are crucial for effective learning.

DEERs offer a dynamic platform for the application of theoretical concepts, allowing students to transfer knowledge from abstract to practical contexts. This application of knowledge is particularly effective in developing critical thinking, problem-solving, and teamwork skills. As students work together to solve puzzles and overcome challenges, they are required to communicate, collaborate, and think critically, all of which are essential skills in both academic and professional settings.

The adaptability of DEERs to various educational contexts further amplifies their impact on learning. The digital format allows these escape rooms to be easily customized and scaled to different learning environments, making them accessible to a wide range of learners with diverse needs and abilities. Whether used in a classroom setting, in remote learning environments, or for corporate training, DEERs provide a versatile tool that can be tailored to specific educational objectives, ensuring that the learning experience is both relevant and effective.

Educational Theories in Educational Game Design

Constructivism plays a pivotal role in the design of educational games by promoting environments where learners actively construct knowledge through direct interaction with the game world. This approach is grounded in the belief that learning is an active, constructive process in which learners build their understanding by engaging with and manipulating the virtual environment. In constructivist game design, players are not passive recipients of information; instead, they are encouraged to explore, experiment, and solve problems, all of which contribute to the construction of new knowledge.

Games like "Minecraft" exemplify constructivist principles by providing a sandbox environment where players can interact freely with the game world. In such games, learners are given the freedom to set their own goals, make decisions, and see the consequences of their actions, all within a safe and controlled digital space. This type of learning is highly experiential, allowing players to form and test hypotheses, solve complex problems, and apply their learning in various contexts. The game environment thus becomes a laboratory for learning, where knowledge is built through experience rather than through the passive absorption of information.

Constructivist principles in game design emphasize the importance of context and experience in learning. Games designed with these principles often include scenarios that are closely aligned with real-world challenges, thereby making the learning experience more relevant and engaging. For example, in a game designed to teach principles of physics, players might be tasked with building structures that must withstand various forces, thereby applying theoretical knowledge in a practical, hands-on way. This not only enhances understanding but also helps in the retention of knowledge, as players are more likely to remember concepts that they have actively applied and experimented with. Constructivist games often encourage collaboration and social interaction, recognizing that learning is not just an individual process but a social one. By working together to solve problems or complete tasks, players can share their knowledge, learn from each other, and construct a deeper understanding of the subject matter. This collaborative aspect of constructivist game design is particularly powerful in educational settings, where peer learning can enhance the overall educational experience.

Constructivism in game design is about creating learning environments that are interactive, immersive, and aligned with real-world contexts. By allowing players to actively engage with the content, form and test hypotheses, and collaborate with others, these games promote a deeper and more meaningful learning experience. This approach not only fosters cognitive development but also equips learners with the critical thinking, problem-solving, and collaborative skills that are essential in today's world .

Behaviorism plays a critical role in the design of game mechanics, particularly in educational games, where the goal is to shape player behavior to reinforce learning. Grounded in the behaviorist theory, which posits that learning is a result of changes in overt behavior driven by the learner's response to stimuli, game mechanics inspired by behaviorism typically involve the use of reinforcement and punishment. These elements are used to encourage desired behaviors and discourage those that are undesirable, effectively guiding the player's actions toward achieving specific learning outcomes.

In the context of educational games, reinforcement is often implemented through rewards such as points, badges, or levels, which players earn by successfully completing tasks or challenges. This system of rewards serves to reinforce positive behavior, making it more likely that the player will repeat the behavior in the future. For example, in language learning games, players might receive points for correctly answering questions or completing exercises, which not only motivates them to continue playing but also reinforces the learning of new vocabulary or grammar rules. As players progress, the tasks typically become more complex, gradually increasing the level of difficulty and thus promoting the continuous development of skills.

Punishment, although less common in educational games, can also be employed to discourage behaviors that do not align with the learning objectives. For instance, a player might lose points or face in-game consequences for incorrect answers or failing to complete a task within a given time. This negative feedback helps players learn from their mistakes and adjust their strategies accordingly, promoting a more effective learning process. However, it is important that such punishments are carefully balanced to avoid discouraging players, which could negatively impact their motivation and engagement.

Behaviorist principles in game design are evident in many successful educational games, where the combination of reinforcement and punishment creates a structured environment that guides learning through incremental challenges. The repetitive nature of these mechanics helps solidify the learned material, making behaviorism a powerful tool in educational game design. By effectively using these principles, game designers can create experiences that are not only engaging but also highly effective in facilitating learning outcomes.

Cognitivism in game design is centered around understanding and enhancing the mental processes involved in learning, such as thinking, memory, knowing, and problem-solving. This educational theory emphasizes the importance of how information is received, organized, stored, and retrieved by the mind. In the context of game design, a cognitivist approach focuses on creating challenges that stimulate these cognitive processes, thereby fostering intellectual development and problem-solving skills in players. Games designed with a cognitivist approach often incorporate complex scenarios that require players to engage in higher-order thinking.

These games present tasks that challenge players to process information critically, recall knowledge, and make strategic decisions. For instance, puzzle-based games are a common example where players must recognize patterns, recall previous experiences, and apply logical reasoning to advance through levels. Such tasks are not just about memorization; they require players to understand underlying principles, make connections between different pieces of information, and adapt their strategies as new challenges arise.

A key aspect of cognitivist game design is the structuring of these cognitive challenges to gradually increase in complexity. This scaffolding approach ensures that as players progress through a game, they build upon their previous knowledge and skills, thus enhancing their cognitive development. For example, early levels might introduce basic concepts and simple puzzles, while later stages require the integration of multiple concepts to solve more complex problems. This progression mirrors the cognitive learning process, where learners move from acquiring basic knowledge to applying it in more sophisticated ways.

Cognitive challenges in games (see Figure 6) also often involve strategic planning and decision-making, which are essential components of cognitive development. Players are frequently required to evaluate various options, predict potential outcomes, and make decisions that will influence their progression in the game. This not only engages their critical thinking skills but also helps in developing executive functions such as problem-solving, planning, and adaptability. These skills are crucial for both academic success and real-world problem-solving.

Cognitivist game design emphasizes the importance of memory recall in learning. Many games use repetition and spaced recall techniques to reinforce learning. For instance, players might need to remember specific sequences, patterns, or strategies that they encountered earlier in the game. This repetition helps strengthen neural connections, leading to better retention and recall of information.

In summary, cognitivism in game design focuses on creating environments that challenge and develop the player's cognitive abilities. By structuring games to include tasks that require critical thinking, memory recall, and strategic decision-making, designers can create educational experiences that not only entertain but also significantly enhance cognitive development (Figure 6).

Figure 6. Stages of cognitive development through educational games

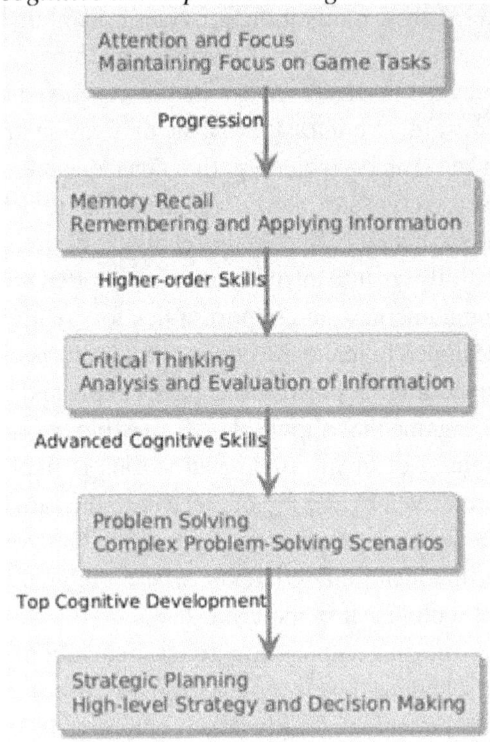

The discussed educational theories considering their application in game design are summarized in Table 6.

Table 6. Comparison of key educational theories in game design

Theory	Core Concepts	Application in Game Design	Example
Constructivism	Knowledge is actively constructed by learners through experience	Focus on exploration, experimentation, and problem-solving	Minecraft, SimCity
Behaviorism	Learning occurs through reinforcement of desired behaviors	Use of rewards (points, badges) to reinforce learning tasks	Duolingo, ClassDojo
Cognitivism	Learning depends on mental processes like memory, perception	Scaffolded puzzles and challenges that enhance cognitive load	Portal, Lumosity
Experiential Learning	Learning through reflection on doing	Hands-on tasks and reflective problem-solving	Oregon Trail, Breakout EDU
Motivation Theories	Intrinsic and extrinsic motivators drive engagement	Leaderboards, narratives, and achievement badges	Kahoot!, Quizizz

The exploration of motivation theories in gaming, particularly the distinction between intrinsic and extrinsic motivation, provides critical insights into how these factors influence player engagement and learning outcomes. Intrinsic motivation refers to the drive to engage in an activity for its inherent satisfaction or enjoyment, without any external reward. In contrast, extrinsic motivation involves participating in an activity to earn rewards or avoid negative consequences. Understanding the balance between these two types of motivation is crucial in designing educational games that are both engaging and effective in achieving desired learning outcomes.

Games that successfully tap into intrinsic motivation often provide rich, engaging narratives, challenging gameplay, and opportunities for creativity and exploration. These elements are designed to make the gaming experience enjoyable and rewarding in itself, thereby encouraging players to engage deeply and persistently with the content. For example, a game that allows players to explore a richly detailed world, make meaningful choices, or create their own stories is likely to foster intrinsic motivation. Players are driven by the joy of discovery, the satisfaction of overcoming challenges, or the pleasure of creative expression. This form of motivation is particularly effective in educational contexts, where the goal is to promote deep, sustained engagement with learning material.

On the other hand, games that emphasize extrinsic motivation typically use external rewards, such as points, badges, levels, or leaderboards, to encourage participation. These rewards serve as incentives for players to complete tasks or achieve certain milestones, thus driving their engagement through external reinforcement. While extrinsic motivators can be highly effective in increasing short-term engagement and providing clear goals for players, they may not always lead to long-term retention of knowledge or deep learning. However, when carefully balanced with intrinsic motivators, extrinsic rewards can enhance the gaming experience by adding a layer of recognition and achievement that complements the intrinsic satisfaction derived from the gameplay itself.

The interplay between intrinsic and extrinsic motivation is a central consideration in the design of educational games (see Table 7). An overreliance on extrinsic rewards can sometimes undermine intrinsic motivation, as players may become more focused on earning rewards than on the enjoyment or educational value of the activity itself. Conversely, games that rely solely on intrinsic motivation may struggle to maintain engagement if players do not find the content inherently interesting or rewarding. Therefore, successful educational game design often involves a careful blend of both types of motivation, ensuring that external rewards support rather than overshadow the intrinsic enjoyment of the game.

Table 7. Comparison of intrinsic vs. extrinsic motivation in gaming

Criteria	Intrinsic Motivation	Extrinsic Motivation
Source of Motivation	Derived from inherent enjoyment or interest in the activity itself	Driven by external rewards or avoidance of negative consequences
Examples in Gaming	Engaging narratives, challenging gameplay, creativity, exploration	Points, badges, levels, leaderboards, competition
Impact on Engagement	Leads to deep, sustained engagement due to personal satisfaction	Can lead to high initial engagement but may not sustain long-term interest
Effect on Learning	Promotes deeper understanding and long-term retention	Can motivate learning but may result in superficial engagement if overemphasized
Player Behavior	Encourages exploration, creativity, and intrinsic satisfaction	Focuses on task completion for rewards, potentially reducing creativity
Sustainability of Motivation	More sustainable as it is self-driven and does not rely on external factors	May diminish once rewards are no longer offered or perceived as valuable
Best Use in Educational Games	Ideal for fostering creativity, exploration, and intrinsic learning	Effective for setting clear goals, providing feedback, and encouraging achievement

The digital nature of games provides unique opportunities to harness both intrinsic and extrinsic motivators in ways that are dynamic and adaptable. For instance, games can adjust the difficulty of challenges to match the player's skill level, ensuring that the tasks remain engaging without being too easy or frustrating. Similarly, rewards can be tailored to individual players' preferences, offering a personalized gaming experience that maximizes both intrinsic enjoyment and extrinsic achievement.

In conclusion, the balance (Table 8) between intrinsic and extrinsic motivation is crucial in the design of educational games. By understanding and applying these motivational theories, game designers can create experiences that are not only engaging and enjoyable but also effective in promoting deep learning and knowledge retention.

Educational Theories in DEER Design

The integration of educational theories into the design of Digital Educational Escape Rooms (DEERs) has become increasingly crucial in creating effective learning experiences. Central to this integration is the application of constructivist principles, which emphasize learning as an active, constructive process. Constructivism aligns well with the interactive nature of DEERs, as these environments encourage learners to engage actively with content through problem-solving and exploration. For instance, in teaching chemical bonding, DEERs have been used to develop students' critical thinking and problem-solving skills by presenting challenges that require them to apply theoretical knowledge in practical contexts. This method resonates with constructivist ideals by allowing learners to build their understanding through direct interaction with the learning material.

Table 8. Balancing intrinsic vs. extrinsic motivation in game design

Motivation Type	Source of Motivation	Game Elements	Impact on Learning
Intrinsic Motivation	Inherent enjoyment or interest in the activity itself	Exploration, narrative immersion, creativity	Fosters deep engagement and long-term retention
Extrinsic Motivation	External rewards such as points, badges, or achievements	Points, badges, leaderboards, rewards	Increases short-term engagement, but may reduce long-term interest
Balanced Approach	Combination of intrinsic and extrinsic motivators	Engaging narrative plus rewards for achievements	Supports both engagement and achievement

In addition to constructivism, experiential learning theory plays a significant role in DEER design. This theory posits that learning occurs through reflection on doing, a principle that is particularly relevant in DEERs, where students engage in immersive, hands-on problem-solving tasks. These tasks are designed to simulate real-world challenges, enabling students to apply their knowledge in a controlled, yet dynamic environment. The experiential nature of DEERs fosters a deeper understanding of the subject matter as students not only solve puzzles but also reflect on their strategies and outcomes, reinforcing their learning through active participation and critical reflection.

DEERs provide a platform for the application of multiple learning theories, creating a rich, multifaceted educational experience (Table 9). By integrating constructivism, experiential learning, and other relevant theories, DEERs can address diverse learning styles and cognitive abilities, making them highly adaptable to various educational contexts. This theoretical grounding not only enhances the effectiveness of DEERs as educational tools but also contributes to their appeal as innovative learning environments that engage and motivate students.

In summary, the application of learning theories in DEER design is pivotal in creating engaging, effective, and educationally robust environments. By drawing on constructivist and experiential learning principles, DEERs can offer learners an active, immersive experience that not only enhances their understanding of complex concepts but also develops essential skills such as critical thinking and problem-solving. This integration of theory and practice in DEERs underscores their potential as powerful tools for modern education, capable of adapting to and meeting the needs of diverse learners in a variety of educational settings.

Table 9. Learning theories and game design elements matrix

Learning Theories	Reward Systems	Narrative	Collaboration	Problem-Solving
Constructivism	Indirect support, reward systems are less emphasized but can motivate through intrinsic rewards like discovering new levels or abilities.	Strongly supports narrative, as learners build knowledge through context and immersion in game stories.	Encourages collaboration, with peer learning and social interaction being key.	Critical in constructivist games where learners actively engage in solving problems and experimenting to build understanding.
Behaviorism	Strongly supports reward systems (e.g., points, badges) to reinforce desired behaviors and learning goals.	Not emphasized, though simple narratives may guide learners toward tasks.	Collaboration can exist but is typically secondary, with emphasis on individual reinforcement.	Problem-solving is task-focused with specific reinforcement for correct answers or actions.
Cognitivism	Moderate support for rewards that reinforce knowledge retention or recall tasks (e.g., completing memory tasks).	Narrative helps create meaningful contexts that support cognitive development and memory retention.	Collaboration supports cognitive strategies such as reflection and shared knowledge, promoting deeper understanding.	Central to cognitivist games, where players use critical thinking and strategy to solve complex challenges.
Experiential Learning	Rewards are less emphasized; the focus is on learning through experience, rather than external motivation.	Narratives create real-world scenarios that promote active, hands-on learning.	Strongly supports collaboration as players learn by doing, often in group settings.	Problem-solving is core to the learning experience, as players solve practical problems through trial and error.
Motivation Theories (e.g., Self-Determination Theory)	Rewards are important but need to support intrinsic motivation (autonomy, mastery, relatedness) rather than relying solely on extrinsic rewards.	Narrative supports intrinsic motivation, as it provides purpose and context for learner engagement.	Collaboration fosters relatedness, one of the key motivators in self-determination theory, helping learners feel connected.	Supports problem-solving as a means to foster mastery, where players derive motivation from overcoming challenges and improving skills.

While DEERs offer unique learning experiences, they also present challenges in design and implementation. One challenge is creating DEERs that are educationally effective across diverse learning styles and cognitive abilities. This requires careful consideration of how puzzles and tasks align with educational objectives, ensuring that they are both engaging and pedagogically sound. Technology integration, such as the use of virtual reality, presents both challenges and opportunities in DEER design. For example, the incorporation of VR in DEERs for STEM education has shown promise in enhancing learning experiences by supporting creative, student-centered teaching methods (Mystakidis et al., 2021). These technologies offer new avenues for innovation in DEER design, though they also require considerations of accessibility and technological expertise.

GAMIFICATION IN LEARNING ENVIRONMENTS

Principles of Gamification in Education

Gamification in education refers to the strategic integration of game design elements and principles into learning environments with the aim of enhancing engagement and motivation among learners. This approach leverages the engaging and interactive nature of games to make the learning process more immersive, enjoyable, and effective. At its core, gamification is about transforming traditional educational activities by infusing them with the dynamics and mechanics that make games compelling, such as goals, challenges, rewards, and feedback systems.

The concept of gamification is rooted in the idea that the intrinsic appeal of games—their ability to captivate and motivate—can be harnessed for educational purposes. In an educational context, gamification seeks to engage students more deeply in the learning process by making it feel less like a chore and more like an adventure or challenge. This approach is particularly effective in addressing the diverse learning styles and preferences of students (Table 10), as it provides multiple pathways for engagement and success.

Table 10. Learning styles and game features

Learning Style	Game Features	Description	Example Games
Visual Learners	Visual cues, graphics, animations	Engages learners through visual representations	Monument Valley, Minecraft
Auditory Learners	Narration, sound effects, audio feedback	Reinforces learning through spoken instructions or sounds	Duolingo, Kahoot!
Kinesthetic Learners	Physical manipulation, interaction with objects	Encourages learning through hands-on engagement and actions	LeapFrog Games, Osmo
Reading/ Writing Learners	Text-based instructions, narrative-driven content	Supports learners who prefer reading and writing-based activities	Text Adventure Games, RPGs

Key elements of gamification include points, badges, leaderboards, and progress bars, which serve to motivate learners by providing clear goals and immediate feedback on their performance. These elements not only foster a sense of accomplishment but also encourage healthy competition and collaboration among learners. By incorporating these game-like elements, educators can create a more dynamic and interactive learning experience that motivates students to participate actively and persist in their educational activities.

Gamification in education is not just about making learning fun; it is also about fostering a deeper connection with the content. By aligning game mechanics with educational objectives, educators can ensure that the gamified activities are not only engaging but also pedagogically sound. This alignment is crucial for ensuring that the use of gamification contributes to meaningful learning outcomes, rather than merely providing entertainment. Gamification in education relies on the culture of participation and the principle of intercreativity, fostering active involvement from students (Osuna-Acedo, 2021). Gamification processes have shown significant success in increasing student motivation, engagement, and participation, crucial for the training of citizens in the post-digital era.

The key elements of gamification in learning include points, badges, leaderboards, and challenges, designed to increase learner engagement and achievement. Nah et al. (2013) reviewed the literature on gamification and identify principles and system design elements for gamifying educational computer games. The framework proposed in their study encompasses principles of gamification, system design elements, and dimensions of user engagement. By implementing these elements, educators can create an environment that motivates students, which in turn increases learning achievement. The gamification of education is expected to foster a more engaging and interactive learning experience, thereby improving learning outcomes.

Impact of Gamified DEERs on Student Engagement in Classroom

Gamification in Digital Educational Escape Rooms (DEERs) has shown to be a powerful tool in enhancing student motivation and engagement in classroom settings. The incorporation of game elements like points, leaderboards, and timed challenges in DEERs creates an environment where students are not only learning but also actively participating and enjoying the process. A study (Benavente, 2022) discusses how gamification strategies, particularly through educational escape rooms, have led to active student engagement and yielded positive results in various educational strategies. This approach caters to the needs of contemporary learners by providing dynamic and interactive learning experiences. The use of gamified DEERs transforms traditional educational settings, making learning more appealing and effective for students.

The integration of gamification in DEERs also positively impacts learning outcomes. Gamified DEERs have been shown to improve not just engagement but also knowledge retention and understanding of complex concepts. For instance, an escape-room-inspired game for genetics review at Northwestern University illustrated how adding gaming elements to educational activities motivated students to study and helped organize and identify areas for further review (Brady and Andersen, 2019). This demonstrates the effectiveness of gamification in reinforcing learning objectives while making the educational process more enjoyable. Gamified DEERs, therefore, offer a balanced approach to education, merging entertainment with learning to achieve optimal educational outcomes.

Challenges and Opportunities in Learning Gamification using DEERs

Implementing gamification in the design of Digital Educational Escape Rooms (DEERs) presents several challenges. One major obstacle is ensuring that the gamified elements align well with the educational objectives and do not distract from the learning process. According to (Videnovik et al., 2022), implementing DEERs, especially in online environments, requires careful planning and understanding of students' engagement and motivation. This study presents how design thinking methodology was used to migrate an escape room style educational game online, maintaining student engagement in remote teaching and online learning. Another challenge is the technological aspect, where educators need to balance the use of technology in gamification without overwhelming the students or compromising the educational content.

The future of gamified learning, particularly in DEERs, is likely to see innovative trends and applications. The integration of advanced digital technologies, such as virtual and augmented reality, can enhance the immersive experience of DEERs, making them more engaging and effective for learning. Lyons et al. (2023) discuss the use of Google Forms to create virtual escape rooms during the COVID-19 pandemic, indicating a trend towards more accessible and adaptable gamified learning tools. These innovations suggest that the future of gamified learning in DEERs will continue to evolve, offering more sophisticated and varied educational experiences that cater to the diverse needs of learners.

Examples of Gamification Solutions in Real-World Applications

Gamification has been widely adopted in various fields, utilizing diverse strategies such as points, badges, leaderboards, narrative immersion, and collaborative gameplay to enhance engagement and learning outcomes. These approaches (summarized in Table 11) are applied across educational platforms, fitness applications, and even professional training programs to foster motivation and skill development.

Table 11. Gamification solutions and real-world examples

Gamification Solution	Real-World Examples
Points and Badges	• **Duolingo:** Users earn points and badges for completing language lessons. • **Khan Academy:** Rewards learners with badges for mastering topics.
Leaderboards and Competition	• **Kahoot!:** Displays leaderboards during quizzes to encourage competitive learning.
Interactive Puzzles and Problem Solving	• **Minecraft Education Edition**: Encourages students to solve spatial and logical challenges collaboratively. • **Portal**: Enhances physics-based problem-solving skills through interactive puzzles.
Scenario-Based Learning	• **SimCity:** Teaches urban planning and resource management. • **Digital Escape Rooms:** Customizable scenarios, such as solving chemistry puzzles for educational purposes.
Storytelling and Narrative Immersion	• **The Witness:** Uses a mysterious island narrative to drive puzzle-solving. • Assassin's Creed Discovery Tour: Provides historical context through immersive storytelling.
Progress Tracking and Adaptive Feedback	• **DreamBox:** Adapts difficulty based on user performance for math exercises. • Duolingo: Offers tailored lesson difficulty based on user progress.
Collaborative Gameplay	• **Minecraft Education Edition:** Encourages teamwork in building and exploration activities. • **Digital Educational Escape Rooms (DEERs):** Teams solve problems collaboratively in virtual settings.
Physical and Digital Integration	• **Osmo:** Combines physical objects with digital gameplay for interactive learning. • **LEGO Mindstorms:** Merges physical building with programming for STEM learning.

Points and Badges are among the most commonly used gamification elements. Platforms like Duolingo have perfected this strategy by awarding users with points and badges for completing language lessons and reaching milestones. This system encourages continuous learning by rewarding consistent effort and progress. Similarly, Khan Academy utilizes badges to incentivize learners, providing a sense of accomplishment as students master topics and advance through the curriculum.

Leaderboards and Competition take advantage of human competitiveness to drive engagement. Kahoot!, a popular educational platform, incorporates real-time leaderboards during quizzes, motivating students to compete for top rankings while reinforcing their learning. Fitness applications like Fitbit employ leaderboards in their step challenges, encouraging users to stay active by competing with friends to achieve the highest daily or weekly step count.

Interactive Puzzles and Problem-Solving are another effective gamification approach, often seen in games that challenge players to think critically. For instance, Minecraft Education Edition provides students with opportunities to collaborate on solving spatial and logical challenges, fostering teamwork and creativity. The physics-based game Portal pushes players to solve puzzles using innovative mechanics, enhancing their understanding of problem-solving and physics concepts.

Scenario-Based Learning engages users by placing them in simulated environments where they must apply knowledge to solve real-world problems. SimCity, for example, teaches urban planning and resource management by immersing players in the complexities of running a city. Similarly, Digital Educational Escape Rooms (DEERs) are versatile tools for education, allowing educators to design scenarios where students solve subject-specific puzzles, such as chemistry equations or historical mysteries.

Storytelling and Narrative Immersion add depth to gamified experiences, making them more engaging and memorable. The game The Witness uses an enigmatic island as its setting, compelling players to explore and solve puzzles driven by an overarching narrative. Meanwhile, Assassin's Creed Discovery Tour combines historical accuracy with immersive storytelling, enabling players to explore ancient civilizations and learn history in an engaging manner.

Progress Tracking and Adaptive Feedback help personalize the learning experience by responding to user performance. Platforms like DreamBox adapt their difficulty levels based on how well users perform in math exercises, ensuring that the tasks remain appropriately challenging. Similarly, Duolingo adjusts its lesson content dynamically, providing real-time feedback and tailored recommendations to improve user outcomes.

Collaborative Gameplay emphasizes teamwork and social interaction, crucial for learning and development. Minecraft Education Edition promotes collaborative projects, where students work together to build structures and solve challenges. Digital Educational Escape Rooms (DEERs) create virtual environments where groups must collaboratively solve interconnected puzzles, fostering communication and problem-solving skills.

Lastly, Physical and Digital Integration bridges the gap between tangible and virtual experiences. Osmo combines physical objects with digital interfaces, creating interactive learning games that engage kinesthetic learners. LEGO Mindstorms, on the other hand, merges the physical act of building robots with programming tasks, offering hands-on STEM education that builds both technical and creative skills.

These examples highlight how gamification strategies can be tailored to different contexts, making them versatile tools for education, fitness, and professional development. By tapping into intrinsic and extrinsic motivators, these solutions create engaging, meaningful experiences that drive learning, collaboration, and innovation.

COGNITIVE DEVELOPMENT THROUGH PUZZLE GAMES

Puzzle games offer a unique and engaging platform for fostering cognitive development in learners. These games require players to solve complex challenges, think critically, and apply strategic reasoning, making them ideal tools for enhancing mental processes such as problem-solving, memory, attention, and spatial reasoning. By presenting multi-step tasks and requiring players to recall information or manipulate objects, puzzle games stimulate various cognitive skills in an interactive and motivating environment.

Table 12 outlines design considerations for cognitive development in educational games by highlighting key cognitive processes and corresponding game design elements that enhance specific skills. Problem Solving is fostered through complex, multi-step challenges, which enhances critical thinking and strategic planning. Memory development is supported by requiring recall of learned information, improving working memory and long-term recall. Attention and Focus are targeted through time-bound tasks, which strengthen sustained attention. Lastly, Spatial Reasoning is enhanced by navigating 3D environments or building structures, such as in Minecraft, improving spatial awareness and creative problem-solving.

Enhancing Problem-Solving Skills through Puzzles

Puzzles play a pivotal role in developing critical thinking skills, a key component of cognitive development. They challenge learners to analyze, evaluate, and create solutions to complex problems. In the realm of education, puzzle-based learning has been recognized as an effective method for enhancing these skills. A study on teaching critical thinking and problem-solving in Information Systems (IS) education emphasized the value of puzzles in fostering these abilities (Matthee and Turpin, 2019). The research revealed that using puzzles as a teaching tool helps students understand the nature of a problem and enhances their analytical and evaluative capabilities. This approach aligns with the goals of modern education, which increasingly emphasizes the development of higher-order thinking skills.

Case studies in various educational settings have demonstrated the effectiveness of puzzles in improving problem-solving skills. For example, Asok et al. (2016) explored the use of active learning environments, including puzzles, to achieve higher-order thinking skills in engineering education. The study found that incorporating puzzles and similar activities in the curriculum significantly enhanced students' abilities to apply knowledge in solving real-world problems. This method of teaching not only engages students but also encourages them to think critically and creatively. Another study in nursing education (Bangun and Pragholapati, 2021) highlighted the role of educational strategies, such as puzzles, in developing critical thinking skills essential for the industrial revolution 4.0. These strategies, by facilitating the learning of complex judgmental skills, prepare students for the challenges of a technologically advanced society.

Table 12. Design considerations for cognitive development in educational games

Cognitive Process	Game Design Element	Example of Application in Games	Cognitive Skill Enhanced
Problem Solving	Complex, multi-step challenges	Puzzle-solving in Portal or The Witness	Critical thinking, strategic planning
Memory	Requiring recall of previously learned information	Memory sequences in Professor Layton games	Working memory, long-term recall
Attention and Focus	Time-bound tasks requiring sustained attention	Timed challenges in Lumosity or Brain Age	Focus, sustained attention
Spatial Reasoning	Navigating 3D environments or manipulating objects	Building structures in Minecraft	Spatial awareness, creative problem-solving

Puzzle Games and Their Impact on Memory and Attention

Puzzle games have been identified as effective tools for enhancing memory, particularly in the realms of working memory and episodic memory. The cognitive engagement required to solve puzzles can lead to improved memory retention and recall. For example, a study (Nouchi et al., 2013) found that playing brain training games, including puzzle games, boosted executive functions, working memory, and processing speed in young adults. These improvements were attributed to the cognitive demands of these games, which stimulate memory processing and consolidation. The results suggest that regular engagement with puzzle games can lead to significant improvements in memory functions, making them a valuable tool for cognitive development.

Puzzle games also play a significant role in enhancing attention and focus. The attentional demands of these games, which often require players to concentrate on complex problem-solving tasks, can improve overall attentional capacity and focus. A study (Ouankhamchan and Fujinami, 2018) demonstrated that playing a casual color-matching puzzle game improved spatial imaging, speed perception, and working memory, while also enhancing attentional focus. The study highlighted the potential of puzzle games to serve as cognitive training tools that can benefit various aspects of attention and cognitive flexibility.

Puzzle games, particularly those requiring complex problem-solving and critical thinking, have been shown to stimulate various areas of the brain, contributing to cognitive development. These games engage the prefrontal cortex, responsible for higher-order thinking skills such as planning, decision-making, and problem-solving. The concept of brain plasticity, or neuroplasticity, is central to understanding how puzzle games can lead to cognitive improvement. Neuroplasticity refers to the brain's ability to reorganize itself by forming new neural connections throughout life. Puzzle games are an effective way to promote this plasticity. They provide a form of mental exercise that can strengthen improve memory, attention and cognitive functions (Ouankhamchan and Fujinami, 2018).

Integrating Cognitive Skill Development in Escape Room Game Design

Integrating cognitive skill development into escape room game design involves creating experiences that not only entertain but also engage various cognitive processes such as problem-solving, critical thinking, and memory. A study (Vekua et al., 2020) highlights the importance of including game elements and digital services in students' cognitive activities as current trends in the development of didactic systems. The design of an interactive escape room as a form of cognition and learning technology is an effective way to gamify the digital educational space. The study emphasizes the significance of designing game scenarios that align with educational objectives, ensuring that the gameplay contributes to the development of high-demanded competencies necessary for graduates' professional success and socialization.

Effective cognitive skill integration in escape room game design can be seen in various educational settings. For instance, Vekua et al. (Vekua et al., 2020) involved students in the design of game areas for educational purposes, demonstrating how the escape room format can be used to develop over-professional competencies. The empirical value obtained in their experiment confirms that the improvement in learning quality after applying this approach is significant. This case exemplifies how escape rooms can be tailored to specific educational needs, creating immersive and interactive learning environments that actively engage students in developing their cognitive abilities and skills (Table 13).

Social Learning and Collaboration in Virtual Spaces

In the evolving landscape of education, social learning and collaboration in virtual spaces have gained increasing significance, particularly within the context of online learning games. As digital platforms become more ubiquitous and immersive, they offer unique opportunities for social interaction and collaborative problem-solving, essential components of 21st-century learning. These virtual spaces, particularly multiplayer games and online escape rooms, foster environments where learners are encouraged to engage, communicate, and work together to solve complex problems. The power of these collaborative dynamics lies not only in enhancing learning outcomes but also in building community and culture within virtual settings.

Table 13. Cognitive skills developed through game genres

Game Genre	Critical Thinking	Problem Solving	Spatial Reasoning	Memory	Collabo-ration	Creativity
Puzzle Games	High	High	Medium	High	Low	Low
Strategy Games	High	High	Low	Medium	Medium	Medium
Simulation Games	Medium	Medium	High	Medium	High	High
Role-Playing Games (RPGs)	High	Medium	Medium	High	High	High
Action Games	Low	Medium	High	Low	Low	Low
Educational Quiz Games	Medium	Medium	Low	High	Low	Low

Social interaction is fundamental to learning, and in virtual learning games, it plays an even more critical role by bridging the gap between isolation and engagement. Unlike traditional, solitary educational tools, multiplayer games encourage learners to interact with one another, share ideas, and collectively work toward a common goal. This interaction not only makes learning more engaging but also helps in reinforcing concepts through discussion and peer-to-peer teaching. In virtual spaces, learners can communicate in real time, exchange strategies, provide feedback, and learn from their peers' perspectives. This kind of interaction mimics real-world social learning environments but with the added benefit of digital tools that can track and enhance learning experiences.

Research into social learning theory, particularly Vygotsky's concept of the Zone of Proximal Development (ZPD), suggests that learners achieve better outcomes when they engage in tasks slightly beyond their current capabilities with the assistance of peers or mentors. In online games, this is often manifested through collaborative challenges where players must rely on each other's strengths to succeed. The interdependence fostered by such gameplay enhances cognitive development and deepens understanding, as learners must not only grasp the material themselves but also articulate and defend their reasoning to others. Social interaction in virtual learning spaces also allows for continuous feedback loops, where learners can adjust their approaches based on input from teammates, creating a dynamic learning environment that fosters both individual and collective growth.

One of the most compelling aspects of multiplayer game environments is their ability to simulate real-world problem-solving scenarios. These environments are designed to challenge players with tasks that require not only individual skill but also team collaboration. This is particularly true in educational games, where collaborative problem-solving becomes an essential part of the learning process. In multiplayer games, learners are often placed in situations where they must divide responsibilities, communicate effectively, and work towards a shared goal—skills that are crucial in both academic and professional contexts.

In these game environments, players must pool their cognitive resources to tackle challenges that are often too complex for one person to solve alone. For instance, in digital escape rooms, players are required to solve puzzles by interpreting clues, manipulating objects, or performing tasks in a coordinated manner. The need for collaboration becomes apparent as players must combine their individual perspectives and knowledge bases to progress through the game. This collaborative problem-solving promotes higher-order thinking skills, such as analysis, synthesis, and evaluation, as learners navigate through complex scenarios together. The competitive yet cooperative nature of these games helps learners practice essential interpersonal skills such as negotiation, leadership, and conflict resolution—skills that are vital for success in any team-based environment.

Beyond collaboration on immediate tasks, online learning games and multiplayer environments also serve as platforms for building community and culture. Virtual game spaces offer a unique opportunity to create a shared identity among players, fostering a sense of belonging and collaboration that extends beyond individual sessions. Through repeated interaction, learners in these spaces often form tight-knit communities characterized by their own norms, rituals, and cultural practices. This community-building aspect is crucial in virtual spaces, where learners may otherwise feel disconnected from their peers.

These virtual environments often reflect the culture of the physical classroom, but they also allow for the emergence of unique virtual subcultures, where players develop distinct ways of communicating, problem-solving, and interacting. For example, in multiplayer games with open-ended communication tools, learners often create their own jargon or in-game rituals, which help to solidify their community and improve collaboration. The shared challenges and experiences foster a collective identity, which not only strengthens teamwork within the game but also creates lasting bonds that can carry over into other learning environments.

Virtual spaces allow for the participation of diverse learners who might not have the same opportunities to engage in traditional face-to-face settings. The anonymity and flexibility of these spaces enable learners from different backgrounds, cultures, and skill levels to contribute equally, fostering inclusivity. This, in turn, enhances the richness of the learning experience, as learners are exposed to diverse perspectives and ways of thinking, which can improve problem-solving and creativity.

To illustrate the power of social learning and collaboration in virtual spaces, an analysis of popular online escape games provides insight into the dynamics at play. Games like "Breakout EDU" and "The Escape Game Digital" exemplify how multiplayer environments foster collaboration and community. These games require players to solve puzzles within a limited time frame, pushing them to communicate effectively and leverage each other's strengths. In these environments, players quickly form social hierarchies based on skills, leadership, and communication abilities. Strong communicators often emerge as leaders, guiding the team through complex puzzles by delegating tasks and synthesizing information. The necessity of time management and the pressure to succeed within a fixed time frame enhance the intensity of social interaction, requiring players to work together efficiently. These dynamics mirror real-world team challenges, making these games not only entertaining but also educational in teaching collaboration, time management, and problem-solving under pressure.

Studies of these online escape rooms (compared in Table 14) show that players often rate their experiences highly in terms of both enjoyment and educational value. Many players report improved relationships with teammates, increased confidence in problem-solving abilities, and a better understanding of the importance of clear communication in group settings. These games have been shown to improve engagement and motivation, as learners are more likely to participate when they feel connected to their peers and invested in a shared goal.

Assessing Learning Outcomes in Game Contexts

As the integration of digital games into education becomes more widespread, the need for robust methods to assess learning outcomes within game contexts has emerged as a critical consideration. Unlike traditional assessments, where learners are evaluated through tests and quizzes, educational games present unique challenges and opportunities for measuring learning. The interactive and immersive nature of games requires innovative approaches to assessment that can capture both cognitive and behavioral aspects of learning. In addition, balancing educational objectives with gameplay is essential to ensure that the game remains engaging while effectively promoting learning. Feedback mechanisms embedded within games can further enhance the assessment process, providing real-time data and personalized learning experiences. As technology advances, adaptive learning and personalized game experiences are poised to shape the future of educational game assessments, allowing for more dynamic and individualized evaluation methods (Table 15).

Table 14. Comparison of online escape room games

Game Name	Learning Objective	Social and Collaborative Dynamics	Assessment Strategy
Breakout EDU	Critical thinking and team collaboration	Teams must solve puzzles collaboratively in real-time	Performance-based, feedback integrated in gameplay
The Escape Game Digital	Problem-solving and application of subject-specific knowledge	Players collaborate via video chat to solve challenges	Game analytics track decision-making patterns
Escape Room EDU	STEM education through immersive, challenge-based gameplay	Players communicate to solve subject-specific puzzles	Embedded assessments through gameplay outcomes

Table 15. Evaluation methods for learning outcomes in Game-Based Learning (GBL)

Assessment Method	Description	Strengths	Challenges
Performance-based	Learners assessed based on their in-game actions and achievements	Measures application of knowledge, hands-on skills	May miss abstract learning; subjective scoring
Embedded Assessments	Assessments integrated into gameplay (e.g., completing a puzzle demonstrates learning)	Non-disruptive, continuous evaluation	Difficult to design; may oversimplify objectives
Game Analytics	Tracking player behavior, time on task, and decisions made	Provides granular data, detailed insight into progress	Requires technical expertise, data interpretation
Quizzes and Tests	Traditional assessments used at the end of gameplay	Easy to implement, measurable outcomes	Can disrupt game flow; focuses on knowledge recall
Reflection/ Debriefing	Learners reflect on their actions after gameplay	Enhances critical thinking, links experience to learning	Time-consuming, subjective evaluation

One of the key challenges in assessing learning outcomes in game-based learning is selecting appropriate methods that align with both the gameplay mechanics and educational objectives. Traditional assessment methods such as multiple-choice tests or written assignments may not fully capture the depth of learning that occurs in game environments, where problem-solving, creativity, and collaboration are often central components. As a result, alternative assessment strategies have been developed to measure learning in these contexts.

Performance-based assessments are a common method used in educational games, where learners are evaluated based on their actions, decisions, and achievements within the game. These assessments focus on how well players apply their knowledge and skills in the game's problem-solving scenarios, rather than merely recalling facts. For example, in a history-themed game, learners might be assessed on their ability to navigate historical events, make decisions that reflect an understanding of the period, and apply critical thinking to solve in-game challenges. This type of assessment reflects a learner's ability to transfer theoretical knowledge into practical application, providing a more holistic view of learning.

Game analytics offer a powerful tool for evaluating educational outcomes in digital games. By tracking player behavior—such as time spent on tasks, choices made, and strategies employed—educators can gather detailed data on learner engagement and progress. These analytics can reveal patterns that indicate a learner's strengths and weaknesses, allowing for more targeted feedback and support. Game analytics can track long-term progression, providing insights into how learners develop their skills over time.

Another emerging method is the use of embedded assessments, where assessment is seamlessly integrated into the gameplay experience. This approach eliminates the need for external tests and quizzes, as the game itself becomes the assessment tool. For instance, in a math game, players might be required to solve equations to progress through levels, with their success serving as a direct indicator of their understanding of mathematical concepts. Embedded assessments allow for continuous evaluation without disrupting the flow of the game, making the learning experience more natural and engaging for students.

One of the greatest challenges in educational game design is finding the right balance between educational objectives and gameplay. A game that is too focused on education risks becoming tedious and losing the engagement that makes games a powerful learning tool. Conversely, a game that prioritizes entertainment over education may fail to achieve the desired learning outcomes. Striking a balance between these two aspects is critical for creating effective educational games that are both enjoyable and instructive. To achieve this balance, it is essential to align game mechanics with educational goals from the outset of the design process. This involves integrating learning objectives into the core gameplay so that players acquire knowledge and skills as a natural part of the game experience. For instance, in a science-based game, players might engage in experiments, make observations, and test hypotheses as part of the game's progression. This alignment ensures that educational content is not an add-on but an intrinsic part of the gameplay.

Maintaining an appropriate level of challenge is crucial for balancing education and gameplay.

Games that are too easy may not provide the cognitive stimulation needed for effective learning, while games that are too difficult can lead to frustration and disengagement. Adaptive difficulty levels, where the game adjusts to the learner's ability, can help maintain an optimal balance, keeping players in a state of "flow"—a mental state where they are fully immersed and engaged in the task at hand. This adaptive approach ensures that learners are constantly challenged, but not overwhelmed, as they progress through the game (Figure 7).

Finally, narrative and storytelling play an important role in balancing educational content and gameplay. A well-crafted narrative can make learning objectives more relatable and meaningful by embedding them in a compelling story. For example, in a language learning game, players might take on the role of a traveler in a foreign country, using their language skills to navigate through challenges and complete missions. The narrative context not only enhances engagement but also provides a reason for learning, making the educational content feel relevant and necessary within the game world.

Feedback is a critical component of learning, and in digital games, it can take many forms—from simple notifications of success or failure to more complex systems that provide detailed insights into a player's performance. Immediate feedback is one of the key advantages of game-based learning, as it allows learners to see the consequences of their actions in real time, facilitating quicker adjustments and deeper understanding. In many educational games, feedback is provided through rewards, such as points, badges, or levels, which signal progress and achievement. These extrinsic motivators can be effective in encouraging players to persist through challenges and continue learning. However, more sophisticated feedback mechanisms focus on providing qualitative insights into a learner's performance. For instance, after completing a level, a game might offer a summary of the player's decisions, highlighting both strengths and areas for improvement. This type of feedback not only reinforces learning but also encourages reflection, prompting learners to think critically about their strategies and how they can improve.

Figure 7. Game flow and feedback loop in an educational escape room game

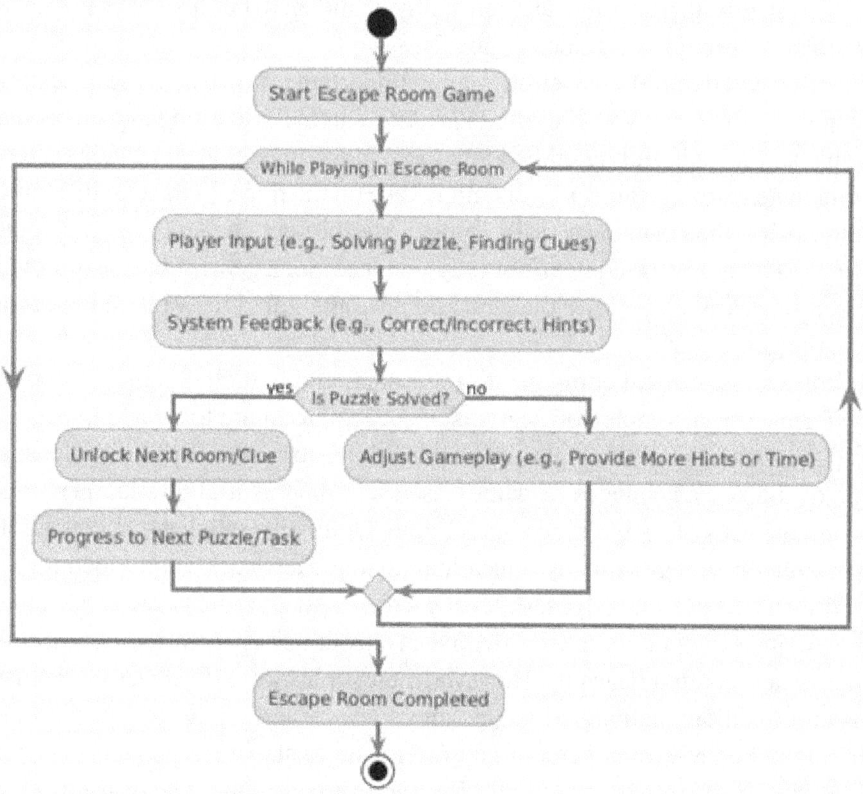

Assessment tools within digital games often include built-in analytics that track player behavior and performance over time. These tools can generate reports for educators, detailing each learner's progress and providing data-driven insights into their strengths and weaknesses. This information can be used to adjust the learning experience, offering more personalized feedback or interventions to support struggling learners.

Adaptive learning systems use player data to dynamically adjust the game's difficulty and content, tailoring the learning experience to each individual's needs (Table 16). For example, if a player is consistently struggling with a particular concept, the game might present additional practice problems or offer hints to guide them toward the correct solution. Conversely, if a player is excelling, the game may increase the difficulty to keep them challenged. This type of personalized feedback ensures that learners remain engaged and motivated while receiving the support they need to succeed.

Table 16. Adaptive learning elements in digital educational games

Adaptive Learning Element	Description	Impact on Player Experience	Example
Dynamic Difficulty Adjustment	Game adjusts difficulty based on player performance	Keeps players in a state of flow, prevents frustration	Portal adjusts puzzle complexity based on progression
Personalized Learning Pathways	Game content changes based on player's learning preferences or choices	Ensures content matches learner's needs and interests	Dreambox adapts math problems based on skill level
Real-time Feedback Customization	Feedback becomes more or less detailed based on player's performance	Tailors the learning experience to individual progress	Hints in Lumosity adjust based on performance
Content Unlocking	Learners unlock new content as they master previous material	Motivates learners to continue by providing incremental rewards	Duolingo unlocks more complex language lessons

CONCLUSION: INTEGRATING EDUCATIONAL THEORIES INTO DEER GAME DESIGN

Summarizing Key Educational Theories in DEER Games

Digital Educational Escape Rooms (DEERs) represent a convergence of various educational theories, offering a dynamic platform where theoretical concepts can be applied in practical, immersive environments. The theoretical foundations of DEERs are primarily rooted in constructivism and experiential learning. Constructivism emphasizes the active role of learners in constructing knowledge through experience, a principle that aligns well with the interactive nature of DEERs. These escape rooms engage learners in problem-solving activities that require the application of theoretical knowledge, thus reinforcing understanding through practical experience. Experiential learning further complements this approach by highlighting the importance of learning through reflection on doing. In DEERs, students actively engage in challenges that simulate real-world problems, allowing them to reflect on their actions and outcomes, thereby deepening their learning experience.

The theoretical frameworks guiding DEER game design are diverse and multifaceted. Constructivism, as mentioned, plays a crucial role by framing learning as an active, constructive process. This theory is complemented by behaviorism, which informs the use of rewards and feedback mechanisms within DEERs to reinforce learning behaviors. Cognitivism is another key framework, focusing on how games can be designed to enhance cognitive processes such as memory, problem-solving, and critical thinking. Cognitivist approaches are evident in DEERs that incorporate complex puzzles and challenges, requiring players to engage in higher-order thinking skills. Motivation theories, particularly the distinction between intrinsic and extrinsic motivation, are critical in designing DEERs that maintain student engagement.

The application of these educational theories within DEERs is evident in several design elements. For instance, constructivist principles are applied through the creation of narrative-driven puzzles that require learners to explore, hypothesize, and test their understanding in a contextualized environment. Behaviorist strategies are implemented through the use of immediate feedback, points, and rewards that encourage positive learning behaviors and enhance engagement. Cognitivism is applied by structuring DEERs in a way that challenges players' cognitive abilities, requiring them to recall information, apply knowledge in new situations, and solve complex problems. Motivation theories are integrated into game design by creating elements that tap into both intrinsic and extrinsic motivators, ensuring that learners remain engaged and motivated throughout the learning experience.

The integration of these theories not only makes DEERs effective educational tools but also demonstrates the potential of game-based learning environments to bridge the gap between theoretical knowledge and practical application. By designing DEERs that align with these educational theories, educators can create immersive learning experiences that are not only engaging but also deeply educational, fostering critical skills that are essential for students' academic and professional success.

Best Practices for Incorporating Learning Theories into DEER Game Design

Incorporating educational theories into Digital Educational Escape Rooms (DEERs) requires careful alignment of design strategies with educational objectives. One of the foundational strategies involves embedding learning objectives directly into the game mechanics. For instance, tasks within the DEERs should be designed to reflect and reinforce the educational goals, ensuring that each puzzle or challenge contributes to the learning outcomes. This involves integrating elements that promote critical thinking, problem-solving, and collaboration, which are often central to the educational objectives. The narrative context within DEERs should be constructed in a way that not only engages students but also situates the learning objectives within a meaningful storyline, thereby enhancing the relevance and retention of knowledge. By doing so, educators can create a cohesive learning experience where the gameplay is directly tied to the educational goals, making the learning process both immersive and effective .

Balancing educational content with engaging gameplay is crucial in the design of DEERs. While the primary goal is to educate, the effectiveness of DEERs lies in their ability to captivate and maintain student interest. To achieve this balance, it is essential to integrate game elements such as challenges, rewards, and feedback systems that are both engaging and educationally relevant. The use of gamification techniques, such as points, leaderboards, and timed challenges, can enhance motivation and encourage active participation. However, it is important that these elements do not overshadow the educational content; instead, they should serve to reinforce the learning objectives. The complexity of the gameplay should be calibrated to match the cognitive abilities of the learners, ensuring that the educational content remains accessible while still providing sufficient challenge to keep students engaged. By carefully balancing these elements, DEERs can effectively combine learning and fun, leading to improved educational outcomes.

Benefits and Challenges

Table 17 provides an overview of the benefits and challenges of using digital games in education. The benefits include high learner engagement and motivation, as games often capture attention and sustain interest. Games also allow for personalization, adapting to the needs and abilities of individual learners, and can enhance learning outcomes by improving retention and the application of skills through interactive learning. Collaboration in multiplayer games fosters teamwork and communication, making them ideal for developing social skills. On the other hand, challenges include the potential for distraction, as some learners may focus on the entertainment aspects rather than educational content. Designing and implementing high-quality digital games requires sophisticated technology and development, which can result in high initial costs. Assessing learning outcomes in game-based environments can be more difficult compared to traditional methods, especially in single-player modes where collaboration is limited.

Table 17. Benefits and challenges of using digital games in education

Aspect	Benefits	Challenges
Engagement	High engagement and motivation for learners	Some learners may become distracted by entertainment aspects
Personalization	Games can adapt to individual learner needs	Requires sophisticated technology and design
Learning Outcomes	Improved retention and application of skills	Assessing outcomes may be more difficult compared to traditional methods
Collaboration	Fosters teamwork and communication in multiplayer games	Limited collaboration in single-player modes
Cost	Can reduce costs for physical resources (e.g., textbooks)	High initial development costs for high-quality games

REFERENCES

Abdul Jabbar, A. I., & Felicia, P. (2015). Gameplay engagement and learning in game-based learning: A systematic review. *Review of Educational Research*, 85(4), 740–779. DOI: 10.3102/0034654315577210

Abrams, S. S., and Gerber, H. R. (2021). Videogames, libraries, and the feedback loop: Learning beyond the stacks. DOI: .DOI: 10.1108/9781800715059

Ang, J. W. J., Ng, Y. N. A., & Liew, R. S. (2020). Physical and digital educational escape room for teaching chemical bonding. *Journal of Chemical Education*, 97(9), 2849–2856. DOI: 10.1021/acs.jchemed.0c00612

Asok, D., Abirami, A., Angeline, N., & Lavanya, R. (2016). Active learning environment for achieving higher-order thinking skills in engineering education. In *2016 IEEE 4th International Conference on MOOCs, Innovation and Technology in Education (MITE)*, pages 47–53. DOI: DOI: 10.1109/MITE.2016.020

Back, J., Back, S., Bexell, E., Stanisic, S., & Rosqvist, D. (2019). The quest: An escape room inspired interactive museum exhibition. In *Extended Abstracts of the Annual Symposium on Computer-Human Interaction in Play Companion Extended Abstracts*. ACM, DOI: DOI: 10.1145/3341215.3356987

Bado, N. (2022). Game-based learning pedagogy: A review of the literature. *Interactive Learning Environments*, 30(5), 936–948. DOI: 10.1080/10494820.2019.1683587

Bangun, A. V., & Pragholapati, A. (2021). *Enhancing critical thinking skills in nursing higher education in preparation for the industrial revolution 4.0*. KnE Life Sciences., DOI: 10.18502/kls.v6i1.8756

Beatty, I. D. (2014). *Gaming the system: Video games as a theoretical framework for instructional design*. ArXiv.

Benassi, A. (2019). Escape room a scuola: Ambienti fisici e virtuali per l'apprendimento. *Journal on Educational Technology*, 27, 174–185.

Benavente, B. (2022). *Physical, Hybrid, and Digital Escape Rooms*. In Advances in Game-Based Learning (pp. 340–355). IGI Global. DOI: DOI: 10.4018/978-1-7998-8645-7.ch017

Borrego, C., Fernández, C., Blanes, I., & Robles, S. (2017). Room escape at class: Escape games activities to facilitate the motivation and learning in computer science. *Journal of Technology and Science Education*, 7(2), 162–171. DOI: 10.3926/jotse.247

Brady, S. C., & Andersen, E. (2019). An escape-room inspired game for genetics review. *Journal of Biological Education*, 55(4), 406–417. DOI: 10.1080/00219266.2019.1703784

Clapson, M. L., Gilbert, B. C. T., Mozol, V., Schechtel, S., Tran, J., & White, S. (2020). Chemescape: Educational battle box puzzle activities for engaging outreach and active learning in general chemistry. *Journal of Chemical Education*, 97(1), 125–131. Advance online publication. DOI: 10.1021/acs.jchemed.9b00612

Cordova-Rangel, J., & Caro, K. (2021). Designing a serious game to promote visitors' engagement in a science museum exhibition. In *Proceedings of the 8th Mexican Conference on Human-Computer Interaction*. DOI: DOI: 10.1145/3492724.3492732

David, D., Edwin, , Arman, E., Hikari, , Chandra, N., & Nadia, N. (2019). Development of escape room game using vr technology. *Procedia Computer Science*, 157, 646–652. DOI: 10.1016/j.procs.2019.08.223

Ekin, C. C., Polat, E., & Hopcan, S. (2023). Drawing the big picture of games in education: A topic modeling-based review of past 55 years. *Computers & Education*, 194, 104700. DOI: 10.1016/j.compedu.2022.104700

Gómez-Cambronero, Á., González-Pérez, A., Miralles, I., & Casteleyn, S. (2019). Mixed reality escape room video game to promote learning in indoor environments. In *EDULEARN19 Proceedings*, pages 6098–6104. DOI: 10.21125/edulearn.2019.1412

Grande-de Prado, M., García-Martín, S., Baelo, R., & Abella-García, V. (2020). Edu-escape rooms. In *Encyclopedia*. DOI: DOI: 10.3390/encyclopedia1010004

Gupta, R., Nawani, R., & Talreja, V. (2016). Virtual reality content creation using unity 3d and blender. *International Journal of Computer Applications*, 156(3), 8–12. DOI: 10.5120/ijca2016912395

Hanus, A., Hoover, M., Lim, A., & Miller, J. (2019). A collaborative virtual reality escape room with passive haptics. In *IEEE Conference on Virtual Reality and 3D User Interfaces (VR)*, pages 1413–1414. DOI: DOI: 10.1109/VR.2019.8798241

Hong, V., & Reiser, E. (2020). Make your escape: An active learning approach. In *Proceedings of the 51st ACM Technical Symposium on Computer Science Education*. DOI: DOI: 10.1145/3328778.3372557

Itenge, H., Sedano, C. I., & Winschiers-Theophilus, H. (2021). Virtually escaping lock down - co-designing a mixed reality escape room narrative with namibian learners. In *3rd African Human-Computer Interaction Conference: Inclusiveness and Empowerment*. DOI: DOI: 10.1145/3448696.3448700

Karageorgiou, Z., Mavrommati, I., & Fotaris, P. (2019). Escape room design as a game-based learning process for steam education. In *Proceedings of the 13th European Conference on Game Based Learning* (ECGBL 2019), Odense, Denmark.

Kaul, V., Morris, A., Chae, J. M., Town, J., & Kelly, W. F. (2021). Delivering a novel medical education "escape room" at a national scientific conference: First live, then pivoting to remote learning because of covid-19. *Chest*, 160(4), 1424–1432. Advance online publication. DOI: 10.1016/j.chest.2021.04.069 PMID: 34029564

Kazimoglu, C., Kiernan, M., & Bacon, L. (2010). Enchanting e-learning through the use of interactive-feedback loop in digital games. In *3rd International Conference on Human System Interaction*. DOI: DOI: 10.1109/HSI.2010.5514523

Kiruthika, V., Jagadeeswari, M., Sneha Prabha, C., Vaishnavi, V., & Sreejaa, H. (2022). Virtual reality based escape room. In *2022 International Conference on Advanced Computing Technologies and Applications (ICACTA)*, pages 1–5. IEEE, DOI: DOI: 10.1109/ICACTA54488.2022.9753531

Klamma, R., Sous, D., Hensen, B., & Koren, I. (2020). Educational escape games for mixed reality. In Alario-Hoyos, C., Rodríguez-Triana, M. J., Scheffel, M., Arnedillo-Sánchez, I., & Dennerlein, S. M. (Eds.), Lecture Notes in Computer Science: Vol. 12315. *Addressing Global Challenges and Quality Education. EC-TEL 2020*. Springer., DOI: 10.1007/978-3-030-57717-9_41

Koenig, N., Denk, N., Wimmer, S., & Prandstaetter, H. (2022). Creating an escape room for cultural mediation: Insights from "the archivist's dream". In *European Conference on Games Based Learning*. DOI: DOI: 10.34190/ecgbl.16.1.682

Krekhov, A., Emmerich, K., Rotthaler, R., & Krueger, J. (2021). Puzzles unpuzzled. *Proceedings of the ACM on Human-Computer Interaction*, 5(CHI PLAY), 1–24. DOI: 10.1145/3474696

Lim Qian Pink, M., Darun, M. R., & Nawani, G. (2019). *Decrypting the real life escape room experience*. KnE Social Sciences., DOI: 10.18502/kss.v3i22.5063

Löffler, E., Schneider, B., Zanwar, T., & Asprion, P. M. (2021). Cysecescape 2.0 - a virtual escape room to raise cybersecurity awareness. *International Journal of Serious Games*, 8(3), 51–69. DOI: 10.17083/ijsg.v8i1.413

Lyons, R., Sweeney, E. M., and Lynch, C. (2023). Applied gamification using virtual escape rooms: Case studies in entrepreneurial and anatomy classes. *Irish Journal of Technology Enhanced Learning*, DOI: .DOI: 10.22554/ijtel.v7i1.103

Ma, J.-P., Chuang, M.-H., & Lin, R. (2018). An innovated design of escape room game box through integrating steam education and PBL principle. In Rau, P. L. (Ed.), Lecture Notes in Computer Science: Vol. 10912. *Cross-Cultural Design. Applications in Cultural Heritage, Creativity and Social Development. CCD 2018.* Springer., DOI: 10.1007/978-3-319-92252-2_6

Mateos-Aparicio, I., Sánchez-Paniagua, M., García-Alonso, A., López-Ruiz, B., Molina, M., Moreno-Guzmán, M., Pérez-Rodríguez, M., Raggio, J., Rueda, C., Sevilla, P., Bodas-Folguera, C., & de la Peña-Armada, R. (2020). Student perception of an activity based on escape rooms for learning at university. In *Proceedings of ADVED 2020- 6th International Conference on Advances in Education.* DOI: 10.47696/adved.202018

Matthee, M., & Turpin, M. (2019). Invited paper: Teaching critical thinking, problem solving, and design thinking: Preparing its students for the future. *Journal of Information Systems Education*, 30, 242–252.

Melo, L. B. (2021). Playing with escape room games for children: An online experience at home in times of pandemic. In *2021 16th Iberian Conference on Information Systems and Technologies (CISTI).* DOI: DOI: 10.23919/CISTI52073.2021.9476536

Monnot, M., Laborie, S., Hébrard, G., & Dietrich, N. (2020). New approaches to adapt escape game activities to large audience in chemical engineering: Numeric supports and students' participation. *Education for Chemical Engineers*, 32, 50–58. DOI: 10.1016/j.ece.2020.05.007

Morrell, B., & Ball, H. M. (2020). Can you escape nursing school?: Educational escape room in nursing education. *Nursing Education Perspectives*, 41(3), 197–198. DOI: 10.1097/01.NEP.0000000000000441 PMID: 30614962

Morris, T. H. (2020). Experiential learning–a systematic review and revision of kolb's model. *Interactive Learning Environments*, 28(8), 1064–1077. DOI: 10.1080/10494820.2019.1570279

Mystakidis, S., Papantzikos, G., & Stylios, C. (2021). Virtual reality escape rooms for stem education in industry 4.0: Greek teachers perspectives. In *2021 6th South-East Europe Design Automation, Computer Engineering, Computer Networks and Social Media Conference (SEEDA-CECNSM)*, pages 1–5. DOI: DOI: 10.1109/SEEDA-CECNSM53056.2021.9566265

Nah, F., Telaprolu, V. R., Rallapalli, S., & Rallapalli Venkata, P. (2013). Gamification of education using computer games. In Lecture Notes in Computer Science, pages 99–107. DOI: DOI: 10.1007/978-3-642-39226-9_12

Nouchi, R., Taki, Y., Takeuchi, H., Hashizume, H., Nozawa, T., Kambara, T., Sekiguchi, A., Miyauchi, C., Kotozaki, Y., Nouchi, H., & Kawashima, R. (2013). Brain training game boosts executive functions, working memory and processing speed in young adults: A randomized controlled trial. *PLoS One*, 8(2), e55518. Advance online publication. DOI: 10.1371/journal.pone.0055518 PMID: 23405164

Novak, D., Nagle, A., & Riener, R. (2014). Linking recognition accuracy and user experience in an affective feedback loop. *IEEE Transactions on Affective Computing*, 5(2), 168–172. Advance online publication. DOI: 10.1109/TAFFC.2014.2326870

Osuna-Acedo, S. (2021). Gamification and MOOCs. In *MOOCs and the Participatory Challenge* (pp. 89–101). Springer International Publishing., DOI: 10.1007/978-3-030-67314-7_6

Ouankhamchan, P., and Fujinami, T. (2018). Effects of casual computer game on cognitive performance through hemodynamic signals. DOI: .DOI: 10.1007/978-3-030-31605-1_35

Ouariachi, T., & Wim, E. J. L. (2020). Escape rooms as tools for climate change education: An exploration of initiatives. *Environmental Education Research*, 26(8), 1193–1206. DOI: 10.1080/13504622.2020.1753659

Pan, R., Lo, H., & Neustaedter, C. (2017). Collaboration, awareness, and communication in real-life escape rooms. In *Proceedings of the 2017 Conference on Designing Interactive Systems*. DOI: DOI: 10.1145/3064663.3064767

Parker, K., & Welch, T. D. (2021). Innovative teaching strategies to escape the anxiety in simulation. *Teaching and Learning in Nursing*, 16(4), 414–417. Advance online publication. DOI: 10.1016/j.teln.2021.06.005

Ruiz, L. M. S., López-Alfonso, S., Moll-López, S., Moraño-Fernández, J. M., & Vega-Fleitas, E. (2022). Educational digital escape rooms footprint on students' feelings: A case study within aerospace engineering. *Inf.*, 13(2), 92. DOI: 10.3390/info13020092

Shakeri, H., Singhal, S., Pan, R., Neustaedter, C., & Tang, A. (2017). Escaping together: The design and evaluation of a distributed real-life escape room. In *Proceedings of the Annual Symposium on Computer-Human Interaction in Play*. DOI: DOI: 10.1145/3116595.3116601

Shi, Y.-R., & Shih, J.-L. (2015). Game factors and game-based learning design model. *International Journal of Computer Games Technology*, 2015, 1–11. Advance online publication. DOI: 10.1155/2015/549684

Terlouw, G., Kuipers, D., van 't Veer, J. T., Prins, J., & Pierie, J. (2021). The development of an escape room–based serious game to trigger social interaction and communication between high-functioning children with autism and their peers: Iterative design approach. *JMIR Serious Games*, 9(1), e19765. Advance online publication. DOI: 10.2196/19765 PMID: 33755023

Tzima, S., Styliaras, G., and Bassounas, A. (2020). Revealing hidden local cultural heritage through a serious escape game in outdoor settings. *Inf., 12*(10).

Vekua, N., Lubsky, A. A., Perevozchikova, M., & Folgerova, J. N. (2020). Peculiarities of forming high-demanded soft skills in the educational space of the escape room. *Pedagogical Skills and Education*, 48(6), 397–412. DOI: 10.32744/pse.2020.6.31

Veldkamp, A., van de Grint, L., Knippels, M. C. P. J., & van Joolingen, W. V. (2020). Escape education: A systematic review on escape rooms in education. *Educational Research Review*, 31, 100364. Advance online publication. DOI: 10.1016/j.edurev.2020.100364

Vergne, M., Simmons, J. D., & Bowen, R. S. (2019). Escape the lab: An interactive escape-room game as a laboratory experiment. *Journal of Chemical Education*, 96(5), 985–991. DOI: 10.1021/acs.jchemed.8b01023

Videnovik, M., Vold, T., Dimova, G., Kiønig, L., & Trajkovik, V. (2022). Migration of an escape room–style educational game to an online environment: Design thinking methodology. *JMIR Serious Games*, 10(3), 10. DOI: 10.2196/32095 PMID: 36155979

Vörös, A. I. V., & Sárközi, Z. (2017). Physics escape room as an educational tool. In *AIP Conference Proceedings*. DOI: DOI: 10.1063/1.5017455

Yeasmin, S., & Albabtain, L. A. (2020). Escape the countries: A vr escape room game. In *2020 3rd International Conference on Computer Applications & Information Security (ICCAIS)*, pages 1–6. IEEE. DOI: 10.1109/ICCAIS48893.2020.9096727

Zeng, H., He, X., & Pan, H. (2020). Implementation of escape room system based on augmented reality involving deep convolutional neural network. *Virtual Reality (Waltham Cross)*, 25(3), 585–596. DOI: 10.1007/s10055-020-00476-0

KEY TERMS AND DEFINITIONS

Adaptive Learning: A personalized learning approach where tasks adjust dynamically based on the learner's performance and needs.

Behaviorism: A theory of learning that focuses on observable behaviors, shaped by reinforcement and punishment to encourage desired actions or discourage undesired ones.

Cognitive Load: The mental effort required to process information during learning tasks, which needs to be balanced to optimize engagement and retention.

Cognitivism: An educational approach focusing on the mental processes involved in learning, such as memory, perception, and problem-solving, and their enhancement through structured and scaffolded challenges.

Constructivism: An educational theory emphasizing that learners construct knowledge actively through experience and interaction with their environment, making learning a personal and meaningful process.

Engagement: The degree of attention, curiosity, and interest that learners exhibit, which fosters a more immersive and effective learning experience.

Experiential Learning: A learning theory that emphasizes knowledge acquisition through active participation and reflection on doing.

Extrinsic Motivation: The external drive to participate in activities for rewards or to avoid negative consequences.

Feedback: Information provided during or after an activity to guide learners, reinforce correct behavior, or correct errors, essential in fostering learning and improvement.

Game-Based Learning (GBL): An educational approach that integrates game elements into pedagogical practices to enhance cognitive, emotional, and skills-based development through engaging and interactive digital environments.

Gamification: The application of game design elements such as points, badges, and leaderboards in non-game contexts, particularly in education, to enhance engagement and motivation.

Intrinsic Motivation: The internal drive to engage in activities for inherent satisfaction and enjoyment, without external rewards.

Narrative Immersion: The integration of storytelling into games to provide context and enhance engagement, making educational content more relatable and memorable.

Problem-Solving: A cognitive process involving analysis, evaluation, and the creation of solutions to complex challenges, often emphasized in educational and puzzle games.

Scaffolding: An instructional technique that provides successive levels of support to help learners achieve deeper understanding and greater independence.

Social Learning: Learning through collaboration and interaction with others, often facilitated in multiplayer or team-based educational games.

APPENDIX I: LINKS TO TOOLS, PLATFORMS AND GAMES

Table 18 provides a comprehensive list of tools, platforms, and games referenced in this chapter. These resources exemplify the integration of educational theories, gamification principles, and cognitive development strategies in digital environments. By leveraging these platforms, educators and learners can explore innovative methods to enhance engagement, motivation, and learning outcomes. Each entry includes a URL link for easy access to further information and utilization in educational or training contexts. This curated selection underscores the diverse applications of game-based learning in fostering effective and immersive educational experiences.

Table 18. Tools, platforms, and games mentioned in the chapter

Tool/Platform/Game	URL
Assassin's Creed Discovery Tour	https://www.ubisoft.com/en-us/game/assassins-creed/discovery-tour
Breakout EDU	https://www.breakoutedu.com
Classcraft	https://www.classcraft.com
DreamBox Learning	https://www.dreambox.com
Duolingo	https://www.duolingo.com
Kahoot!	https://kahoot.com
Khan Academy	https://www.khanacademy.org
LEGO Mindstorms	https://www.lego.com/mindstorms
Lumosity	https://www.lumosity.com
Lumosity Brain Games	https://app.lumosity.com/landing
Minecraft Education Edition	https://education.minecraft.net
Osmo	https://www.playosmo.com
Portal	https://www.thinkwithportals.com
Prodigy Math Game	https://www.prodigygame.com
Professor Layton Series	https://www.laytonseries.com/naen/
Quizizz	https://quizizz.com
SimCity	https://www.ea.com/games/simcity
The Witness	https://www.gog.com/en/game/the_witness

Chapter 3
Technology in Digital Escape Rooms

ABSTRACT

This chapter explores the key aspects of digital escape room design, focusing on player motivation, engagement, puzzle design, and storytelling. It examines how game mechanics, such as mental and physical challenges, balance complexity and accessibility to create an engaging player experience. The chapter highlights the role of immersive environments and effective user interface (UI) and user experience (UX) design in enhancing gameplay. It discusses emerging technologies like virtual reality (VR), augmented reality (AR), artificial intelligence (AI), and blockchain, showcasing their potential to transform digital escape rooms. Additionally, it covers multiplayer design, social interaction, and feedback loops to enhance learning and collaboration. The chapter concludes with an overview of essential tools and software for digital escape room development, emphasizing the importance of technology and innovation in creating adaptive, personalized, and immersive escape room experiences.

TECHNOLOGIES IN DIGITAL ESCAPE ROOMS

Metaverse, Virtual Reality and Augmented Reality

Digital escape rooms have emerged as a significant innovation, blending the thrill of problem-solving with immersive storytelling (Usmani et al., 2022). As we look into the future, the integration of advanced technologies such as the Metaverse, Virtual Reality (VR), and Augmented Reality (AR) is set to revolutionize these experiences, creating environments that transcend the physical limitations of tradi-

DOI: 10.4018/979-8-3693-4219-0.ch003

tional escape rooms and transport participants into realms previously confined to the imagination (Iqbal and Campbell, 2023).

The Metaverse, often described as the next frontier of digital interaction, offers a vast, interconnected virtual universe where users can interact with each other and their environment in real-time (Uddin et al., 2023). In the context of digital escape rooms, the Metaverse serves as the ultimate platform, allowing players to step into fully realized worlds where every element is intricately designed to challenge, engage, and captivate (Bhugaonkar et al., 2022). Imagine an escape room set in a sprawling cyberpunk city, where every street, building, and hidden alley is accessible, filled with puzzles that require not just logic but exploration and interaction with other players' avatars (Liu et al., 2023). The Metaverse's potential to host such expansive environments makes it the perfect medium for creating escape rooms that are not just games but immersive, social experiences (To et al., 2023).

Virtual Reality takes this immersion a step further by placing the player directly inside these digital worlds (Uddin et al., 2023). Unlike traditional gaming, where players interact with a world viewed on a screen, VR envelops the senses, providing a 360-degree perspective that makes the virtual feel almost tangible (Iqbal and Campbell, 2023). In a VR-enhanced digital escape room, participants wear headsets that transport them into environments that are visually stunning and meticulously detailed (Bhugaonkar et al., 2022). They can reach out to grab objects, physically move to avoid traps, or solve puzzles that require real-world actions translated into the virtual space (Liu et al., 2023). The sense of presence that VR brings is unparalleled, making every success and failure in the escape room feel deeply personal and exhilarating (Hajj, 2023).

Augmented Reality blends the virtual with the physical, overlaying digital elements onto the real world through devices like smartphones or AR glasses (Efe, 2022). This technology has the unique ability to extend the escape room beyond a single space, turning an entire building, city, or even natural environment into a playground of puzzles (Kim, 2021). For instance, a digital escape room could start in a participant's living room, with clues hidden in plain sight but visible only through an AR interface (Bhugaonkar et al., 2022). As the game progresses, it could lead players outside, where real-world locations are transformed into crucial elements of the narrative (To et al., 2023). AR's strength lies in its ability to make the ordinary extraordinary, using the familiar as a canvas for complex and layered storytelling (Tan et al., 2022).

The convergence of these technologies within digital escape rooms is not just about enhancing gameplay but about redefining what it means to escape (Mustafa, 2022). Traditional escape rooms are bound by physical space and time, but in the Metaverse, limitations dissolve (Durmaz and Gokalp, 2023). A single game can span galaxies, time periods, or entirely fictional universes, with players hopping between them seamlessly (Chen and Cheng, 2022). The social aspect of the Metaverse also introduces new dynamics, where players can collaborate with others from around the world, forming teams that are as diverse as they are geographically dispersed (Cheng et al., 2022).

VR's contribution is in its capacity to create deeply personal and tactile experiences (Rostami and Maier, 2022). The isolation of wearing a VR headset paradoxically intensifies the shared experience when multiple players are involved (Usmani et al., 2022). Every sound, sight, and interaction feels immediate and real, which makes the eventual escape—or failure—an emotionally charged event (Cheng et al., 2022). The technology's ability to track physical movements also opens the door to more physically interactive puzzles, where solving them requires not just mental acuity but physical dexterity (Nair et al., 2022).

AR, meanwhile, brings a level of accessibility that neither the Metaverse nor VR can match. It allows for escape rooms to be integrated into daily life, where a simple walk through a park could turn into the beginning of an elaborate game (Xu, 2022). AR escape rooms are also inherently flexible, easily adapting to different spaces and scales, making them perfect for both casual play and large-scale events (Fernandes and Werner, 2022). AR's integration into mobile devices makes these experiences accessible to a broader audience, requiring nothing more than a smartphone to participate (Xu, 2022).

Artificial Intelligence and Deep Learning

As digital escape rooms evolve, the integration of Artificial Intelligence (AI) and Deep Learning is set to transform them into more dynamic, responsive, and personalized experiences (Zeng et al., 2020). AI, with its ability to analyze and learn from data, brings a level of sophistication to escape rooms that was previously unimaginable (Montebello, 2021). Deep Learning, a subset of AI, further enhances this by allowing systems to understand complex patterns, behaviors, and preferences, enabling escape rooms to adapt in real-time to the players' actions and decisions (Horváth, 2021). Together, these technologies are pushing the boundaries of what digital escape rooms can offer, creating experiences that are not only challenging but also deeply immersive and tailored to each player (Tricarico, 2021).

AI's most significant contribution to digital escape rooms lies in its capacity to create intelligent, adaptive environments (Momenzad et al., 2018). Traditional escape rooms are static; once a puzzle is solved, it remains solved for every subsequent player. AI changes this by introducing dynamic puzzles that can adjust their difficulty based on the players' performance (Mystakidis et al., 2019). For instance, if a group is breezing through a series of challenges, the AI can escalate the difficulty, introducing more complex riddles or increasing the time pressure (Darejeh, 2023).

Conversely, if players are struggling, the AI can subtly offer hints or adjust the puzzle parameters to maintain engagement without making the game too frustrating (Kotzebue et al., 2022). This adaptability ensures that the escape room remains a compelling challenge, regardless of the players' skill level (Chen and Cheng, 2022).

AI can generate unique storylines and puzzles based on the players' preferences and past behaviors (Sun et al., 2023). Using natural language processing (NLP), AI systems can interpret players' inputs and interactions, allowing for more meaningful and context-sensitive dialogue within the game (Lauriola et al., 2021). Imagine an escape room where the narrative shifts based on the players' choices, leading to different outcomes and endings (Joo and Kim, 2019). This not only increases re-playability but also makes each session feel unique and personalized (Roohi et al., 2021). The AI can also track individual progress across multiple sessions, learning from each player's actions to provide a more tailored experience every time they play (Pfau et al., 2018).

Deep Learning takes this personalization even further by enabling the escape room to understand and predict player behavior (Nimpattanavong et al., 2022). By analyzing vast amounts of data, such as the choices players make, the time they take to solve puzzles, and their interactions with the environment, deep learning algorithms can create detailed player profiles (Roohi et al., 2021). These profiles can then be used to anticipate how players might react to certain challenges, allowing the escape room to present puzzles that are both engaging and appropriately challenging (Ge, 2023). For example, if a player is known to excel at logical puzzles but struggles with pattern recognition, the AI can adjust the game to emphasize their strengths while gradually helping them improve in weaker areas (Oh et al., 2019).

In addition to personalizing gameplay, AI and Deep Learning can enhance the social aspects of digital escape rooms (Liu et al., 2020). In multiplayer scenarios, AI can analyze the dynamics between players, such as how they communicate and collaborate (Soares and Bulitko, 2019). It can then introduce challenges that encourage teamwork or, in competitive modes, create scenarios that play to the strengths of different team members (Kopel and Hajas, 2018). AI-driven non-player characters (NPCs) can also be introduced, serving as guides, adversaries, or even fellow players. These NPCs, powered by advanced AI, can interact with human players in ways that feel natural and responsive, making the digital escape room environment more immersive and lively (Gomes et al., 2021).

One of the most exciting applications of AI in digital escape rooms is in the creation of procedurally generated content (Risi and Togelius, 2019). Rather than relying on a set of pre-designed puzzles and storylines, AI can generate new content on the fly, ensuring that no two games are the same (Magdy et al., 2021). This procedural generation can extend to every aspect of the escape room, from the layout of the environment to the nature of the puzzles and even the storyline itself (Ge, 2023).

By analyzing the players' previous experiences and preferences, the AI can craft a bespoke escape room experience that feels fresh and exciting each time (Ayas et al., 2023). This capability not only increases the longevity of digital escape rooms but also allows for an almost infinite variety of scenarios, making each playthrough a unique adventure (Liu et al., 2020).

AI can be used to enhance the realism of digital escape rooms by simulating complex environments and scenarios (Tang et al., 2020). For example, an escape room set in a post-apocalyptic world could use AI to simulate weather patterns, NPC behavior, and even the spread of a virtual disease (Soares and Bulitko, 2019). These simulations can react to the players' actions in real-time, creating a dynamic and ever-changing environment that requires constant adaptation (Gomes et al., 2021). This level of realism, combined with the immersive capabilities of VR and AR, can make digital escape rooms feel like living, breathing worlds, where every decision has consequences and every action matters (van der Boon et al., 2022).

Applications of AI Tools in DEER Design are summarized in Table 1.

Table 1. Applications of AI tools in DEER design

Design Stage	AI Tool Application	Examples in DEER Design
Content Creation	Generating narrative elements, brainstorming puzzles, drafting dialogues	Creating engaging storylines, generating riddle-like questions, designing character interactions
Gameplay Design	Testing scenario logic, automating clue creation, suggesting alternative solutions	Ensuring logical flow, generating adaptive puzzles, simulating player choices
Feedback Analysis	Summarizing player feedback, identifying trends, generating improvement suggestions	Analyzing beta tester feedback to refine puzzles, detecting usability issues
Language Translation	Translating game content, adapting to multilingual audiences	Localizing DEERs for global players, ensuring linguistic accuracy
Creative Assistance	Idea generation, brainstorming new features, resolving creative blocks	Suggesting innovative puzzle formats, enhancing storyline twists
Debugging and Testing	Identifying inconsistencies, simulating player actions, testing edge cases	Detecting logical gaps, validating puzzle complexity, ensuring error-free design
Educational Alignment	Suggesting educational content, aligning puzzles with learning objectives	Integrating curriculum-based challenges, tailoring content to educational goals

Table 2 summarizes specific AI tools, highlighting their key features and diverse applications in enhancing various aspects of Digital Educational Escape Room (DEER) design, from content creation and localization to adaptive gameplay and immersive visuals.

Table 2. Specific AI tools for DEER design

AI Tool	Key Features	Application in DEER Design
ChatGPT	Natural language processing, creative content generation, conversational AI	Writing storylines, generating puzzles, simulating NPC interactions
DALL·E	Image generation, customizable visual styles, text-to-image conversion	Creating game assets, designing immersive environments
DeepL Translator	High-quality language translation, multilingual support	Localizing DEER content for diverse audiences
GPT-4 Code Assistant	Code generation, debugging support, algorithm optimization	Writing and refining scripts for game logic and mechanics
Runway ML	Video editing, AI-based visual effects, text-to-video	Creating cinematic cutscenes, enhancing visual storytelling in escape rooms
Unity ML-Agents Toolkit	Reinforcement learning, training game agents, adaptive gameplay	Implementing intelligent NPC behavior, adapting gameplay challenges
Lumen5	AI-based video creation, multimedia integration, customizable templates	Designing introduction and tutorial videos for DEERs

- **ChatGPT** leverages natural language processing (NLP) to generate creative and coherent text, making it an invaluable tool for writing storylines, designing character dialogues, and generating puzzles for DEERs. It can simulate NPC interactions, ensuring engaging and dynamic conversations with players. ChatGPT can assist in brainstorming ideas, offering creative prompts for narrative twists or innovative game mechanics. Its flexibility allows developers to quickly draft, iterate, and refine content, streamlining the overall design process.
- **DALL·E** is a text-to-image generation tool capable of creating custom visuals in diverse styles. For DEERs, DALL·E can be used to design unique game assets, such as backgrounds, props, or clue illustrations, that enhance the immersive experience. Its ability to generate high-quality, context-specific images enables developers to quickly visualize and prototype game environments without relying solely on traditional graphic design, saving time and resources while maintaining creative control.
- **DeepL** Translator is renowned for its accuracy and fluency in translating text between multiple languages. It plays a critical role in localizing DEER content for diverse audiences, ensuring that puzzles, instructions, and narratives resonate culturally and linguistically. By providing high-quality translations, DeepL helps developers expand the reach of their games, making them accessible to global users and fostering inclusivity in educational gaming.
- **GPT-4 Code Assistant** is an AI-driven tool designed to help developers write, debug, and optimize code. In DEER design, it can assist in scripting

game logic, automating repetitive tasks, and resolving complex coding challenges. Its ability to generate efficient algorithms and identify potential issues in the code accelerates the development cycle, ensuring that the gameplay mechanics are seamless and error-free. Additionally, it can help developers integrate AI functionalities, such as adaptive puzzles or intelligent NPCs.

- **Runway ML** specializes in AI-based video editing, visual effects, and text-to-video generation, making it ideal for creating cinematic cutscenes or enhancing visual storytelling in DEERs. Its capabilities allow developers to produce professional-quality videos for game intros, tutorials, or narrative transitions, enriching the overall player experience. Runway ML's tools empower designers to incorporate dynamic and visually compelling elements that align with the game's themes and objectives.
- **Unity ML-Agents Toolkit** is a reinforcement learning platform that enables developers to train intelligent game agents. In DEER design, this toolkit can be used to implement NPCs with adaptive behavior, creating more realistic and challenging interactions. It also allows for the development of dynamic gameplay, where puzzles adjust to the player's skill level, ensuring an engaging and educational experience for diverse audiences.
- **Lumen5** is an AI-powered video creation platform that simplifies the process of producing multimedia content. In DEER design, it can be used to create introduction and tutorial videos that guide players through the rules, objectives, and storylines. Its customizable templates and multimedia integration allow developers to craft visually appealing and informative content that enhances the player's understanding and immersion. Lumen5's user-friendly interface makes it accessible to teams with varying levels of technical expertise.

Mobile Technologies and Accessibility

As digital escape rooms continue to evolve, the role of mobile technologies in enhancing accessibility and broadening the reach of these experiences cannot be overstated (dos Santos Nunes et al., 2018). Mobile devices, with their ubiquity and ever-increasing computational power, have become essential tools in bringing the excitement of escape rooms to a global audience. From smartphones to tablets, these devices offer a versatile platform that enables players to engage with escape rooms anytime, anywhere (Botelho, 2021). The integration of accessibility features within mobile technologies ensures that digital escape rooms are more inclusive, catering to a diverse range of players with varying needs and abilities (Žilak et al., 2022).

The portability of mobile devices is a significant advantage for digital escape rooms, as it allows players to participate in immersive experiences without being tethered to a stationary setup (Goh et al., 2019). Whether on a commute, during a break, or at home, players can seamlessly dive into the intricate puzzles and narratives of an escape room (Halloran and Minaeva, 2019). This level of convenience has contributed to the growing popularity of mobile-based escape rooms, where the gameplay is optimized for touchscreens and sensors native to mobile devices (dos Santos Nunes et al., 2018). These experiences can be as simple or complex as needed, from text-based adventures to richly animated and interactive environments that rival those found on more powerful gaming systems (Ahmetovic et al., 2021).

Mobile technologies also offer a unique opportunity to blend the digital and physical worlds through location-based and augmented reality (AR) features (Shang et al., 2020). By leveraging GPS, cameras, and sensors, mobile escape rooms can incorporate real-world locations into their puzzles, creating an experience that is both immersive and grounded in the player's surroundings (Tregel, 2021). For instance, an escape room could require players to visit specific landmarks, scan objects, or even interact with digital overlays that appear through the phone's camera (Alha et al., 2023). This fusion of the digital and physical realms not only enhances the realism of the escape room but also encourages players to explore their environment in new and engaging ways (Rocha and Lopes, 2020).

The accessibility of mobile devices plays a crucial role in making digital escape rooms more inclusive. Modern smartphones and tablets come equipped with a variety of accessibility features designed to support users with disabilities (Hassan, 2023). These include voice control, screen readers, haptic feedback, and customizable interfaces, all of which can be leveraged to ensure that escape rooms are accessible to players with different abilities (Garcez et al., 2020). For example, puzzles that rely on visual cues can be adapted for players with visual impairments through the use of audio descriptions or tactile feedback (Alkhathlan et al., 2022). Similarly, players with hearing impairments can benefit from captioned dialogues, visual alerts, or even vibrations that correspond to important game events (Teófilo et al., 2018).

The widespread use of mobile devices across different age groups and demographics means that digital escape rooms can be designed to cater to a broader audience (Trinh, 2021). Games can be tailored to different skill levels, from simple, beginner-friendly puzzles to complex, multi-layered challenges for seasoned players (Mystakidis et al., 2021). This adaptability ensures that everyone, regardless of their familiarity with escape rooms or their gaming experience, can enjoy and benefit from the unique mental and social challenges these games offer (Cairns et al., 2019). Mobile escape rooms can be localized into different languages, further expanding their accessibility to players around the world (Tregel, 2021).

One of the most exciting developments in mobile technologies is the advent of cloud gaming and streaming services, which allows even the most graphically intensive escape rooms to be played on devices with limited processing power (Zhang et al., 2019). By offloading the computational heavy lifting to cloud servers, players can enjoy high-quality graphics and seamless gameplay on their smartphones or tablets without the need for expensive hardware (Cai et al., 2018). This democratization of gaming technology ensures that digital escape rooms are accessible to a wider audience, regardless of the device they own (Chen et al., 2019).

Mobile technologies also facilitate social interaction and collaboration, which are key components of many escape room experiences (Singanamalla et al., 2018). Through mobile apps, players can easily connect with friends or join multiplayer games with others around the world (Ong et al., 2020). Features like in-game chat, video conferencing, and shared inventories allow teams to work together seamlessly, even when they are physically apart (Olin et al., 2020). This social connectivity is particularly valuable for digital escape rooms, as it preserves the collaborative spirit that is central to the genre, allowing players to share the thrill of solving puzzles and escaping together, no matter where they are (Revilla et al., 2021).

Mobile devices enable a level of personalization that enhances the overall experience of digital escape rooms (Tregel, 2021). Through the use of sensors and data analytics, mobile apps can track player behavior and preferences, adjusting the difficulty of puzzles or suggesting new challenges based on previous performance (Botev et al., 2019). Push notifications can remind players of unfinished games or alert them to new content, keeping them engaged over time (López-Faican and Martínez, 2020). This level of customization ensures that the escape room experience is not only accessible but also continuously engaging, adapting to the player's evolving skills and interests (Hanus et al., 2019).

TOOLS AND SOFTWARE FOR DIGITAL ESCAPE ROOM GAME DEVELOPMENT

Creating a digital escape room game can be an intricate process, requiring specific tools and software to ensure a captivating and interactive experience. The following extended outline offers a comprehensive guide to the tools and software essential for digital escape room game development, based on recent academic research:

Game Development Platforms

The landscape of game development platforms has grown remarkably diverse, offering a range of tools and environments tailored to different needs and skill levels. Among the most prominent in this field are Unity3D, Unreal Engine, and GDevelop, each bringing unique capabilities to the table, especially in the creation of immersive environments such as escape rooms.

- **Unity3D** has established itself as a versatile and user-friendly platform for game development. Its strengths lie in its robust scripting API, comprehensive asset library, and a supportive community. Unity's real-time 3D development capabilities make it particularly suited for creating immersive escape room experiences. Developers can leverage Unity's advanced physics engine, integrated audio tools, and intuitive user interface to craft detailed and responsive environments. Unity's compatibility with various platforms, including mobile devices and VR headsets, enables creators to deliver escape room experiences across a wide range of devices.
- **Unreal Engine**, known for its high-fidelity graphics and powerful rendering capabilities, is another popular choice for game developers. Its advanced visual scripting system, Blueprints, allows for rapid prototyping and development, which is crucial in the iterative design process of escape rooms. The engine's photorealistic rendering capabilities enable the creation of highly immersive and visually stunning environments, essential for the captivating and engaging nature of escape room games. Unreal Engine's robust multiplayer support can be utilized to create collaborative escape room experiences.
- **GDevelop** emerged as a unique platform, especially for those new to game development or with limited programming experience. It offers an intuitive event-based system that simplifies the game creation process. This accessibility makes GDevelop an appealing choice for educational purposes and for developers looking to rapidly prototype and test concepts. While it may not match the graphical prowess of Unity or Unreal, GDevelop is well-suited for creating simpler, 2D escape room games that focus more on puzzle mechanics and storytelling.
- **Godot Engine** is an open-source game development platform known for its lightweight structure and flexibility. Its node-based system allows developers to create games with a modular approach, while its built-in scripting language, GDScript, provides an intuitive way to implement game logic. Godot is particularly suitable for developing both 2D and 3D games, making it a versatile choice for escape room developers who require cost-effective tools. With its robust support for cross-platform deployment, Godot enables escape room games to be created for PC, mobile, and web platforms. Its open-source nature is especially appealing for educational environments, where cost and adaptability are critical. The engine is ideal for creating digital escape rooms with interactive puzzles, dynamic storytelling, and immersive environments, all while maintaining a low barrier to entry for developers.
- **CryEngine** is a powerful game development platform renowned for its photorealistic rendering, advanced AI systems, and sophisticated physics simulations. These features make it an excellent choice for creating highly

immersive and realistic digital escape rooms. With CryEngine, developers can craft environments that feel authentic, enhancing player engagement and the educational value of the game. Its integrated AI tools allow for the development of intelligent NPCs and adaptive puzzle mechanics, which can challenge players in innovative ways. CryEngine's strong emphasis on visual fidelity and simulation capabilities makes it particularly suitable for escape rooms that focus on experiential learning or storytelling in complex settings like historical reconstructions or futuristic landscapes.

- **Construct 3** is a browser-based game development platform with a drag-and-drop interface, making it accessible to developers with little or no programming experience. Its focus on simplicity and ease of use allows for the rapid prototyping and development of interactive 2D games. Construct 3 is particularly suited for creating digital escape rooms that rely on puzzle mechanics and linear narratives, such as text-based or image-driven challenges. The platform's HTML5 compatibility ensures that games can run seamlessly in web browsers, making it ideal for escape rooms designed for remote learning or online participation. Construct 3 is a great tool for educators and small development teams looking to create lightweight, interactive educational games.
- **RPG Maker** is a popular game development tool that specializes in creating narrative-driven games using pre-built templates, tile-based editing, and scripting capabilities. While originally designed for role-playing games, its features are highly adaptable for creating escape rooms with a strong emphasis on storytelling and character interaction. RPG Maker's user-friendly interface makes it accessible to beginners, while its scripting system allows more experienced developers to customize game mechanics. The platform excels at crafting immersive scenarios with branching narratives, allowing players to progress through puzzles and storylines that enhance educational themes. RPG Maker is especially effective for developing escape rooms that integrate rich, text-based storytelling with logical problem-solving tasks.
- **GameMaker Studio** is a beginner-friendly game development platform that supports both drag-and-drop functionality and scripting for more advanced customization. Known for its strong focus on 2D game development, GameMaker Studio is ideal for creating puzzle-heavy escape rooms with simple yet engaging mechanics. Its cross-platform support allows developers to create games for PC, mobile, and web platforms, ensuring wide accessibility. The platform's intuitive interface and rapid development capabilities make it suitable for educators, hobbyists, and small teams. GameMaker Studio is particularly well-suited for escape rooms that emphasize puzzles, mini-games, or interactive environments in a compact and efficient format.

Each of these platforms (Table 3) offers a unique set of tools and capabilities, making them well-suited for different aspects of escape room game development. Unity3D's flexibility and cross-platform support, Unreal Engine's graphical excellence, and GDevelop's accessibility provide a wide spectrum of options to developers looking to venture into this exciting genre of gaming.

Table 3. Game development platforms for digital educational escape rooms

Platform	Key Features	Educational Use Case
Unity3D	Versatile, scripting API, asset library, real-time 3D, multi-platform	Immersive experiences, responsive environments, mobile/VR compatibility
Unreal Engine	High-fidelity graphics, rendering, visual scripting, multiplayer	Visually stunning environments, collaborative rooms
GDevelop	Event-based system, simplified creation, no programming	Beginner-friendly, 2D games, puzzle mechanics, storytelling
Godot Engine	Open-source, lightweight, node-based system, built-in scripting	Cost-effective 2D and 3D game development for diverse educational scenarios
CryEngine	Photorealistic rendering, AI system, physics simulation	Developing immersive, realistic environments for advanced educational games
Construct 3	Drag-and-drop interface, HTML5 compatibility, browser-based	Simple interactive games for web, ideal for educational prototypes
RPG Maker	Pre-built templates, tile-based editing, built-in scripting	Creating narrative-focused puzzles and RPG-style educational games
GameMaker Studio	Beginner-friendly, drag-and-drop, cross-platform support	Quick development of 2D games, ideal for puzzle-based escape rooms

Interactive Design Tools

In the domain of game development, particularly for creating assets for escape room games, the utilization of interactive design tools like Adobe Creative Suite and Sketch plays a pivotal role. These tools not only facilitate the creation of visually appealing assets but also significantly contribute to the user interface (UI) and user experience (UX) design, which are crucial in enhancing the playability and immersive quality of escape room games.

- **Adobe Creative Suite**, a collection of software used for graphic design, video editing, web development, and photography, provides a comprehensive toolkit for game asset creation. Its components, such as Photoshop for image editing and Illustrator for vector graphics, are integral in developing detailed and engaging visuals for game environments. Photoshop's powerful editing capabilities allow designers to craft intricate textures and elements, vital for creating the ambiance and thematic consistency essential in escape room settings. Illustrator, on the other hand, is instrumental in designing clear and concise icons and UI elements, ensuring that players can navigate the game intuitively. The suite's ability to integrate different types of media and its compatibility with a myriad of formats make it a versatile choice for game developers looking to create a wide range of assets, from detailed backgrounds to interactive elements.

- **Sketch:** While Adobe Creative Suite excels in creating and manipulating raster and vector graphics, Sketch, known for its vector-based approach, is particularly beneficial in UI/UX design for games. It allows designers to create a prototype of the game interface quickly, enabling them to test and iterate designs efficiently. This rapid prototyping is crucial in escape room games, where the player's experience heavily relies on the intuitiveness and responsiveness of the UI. Sketch's extensive library of symbols and responsive design features ensures that UI elements are consistent and adaptable across different screen sizes and devices. This adaptability is essential in providing a seamless gaming experience, particularly in escape rooms that might be played on various platforms.
- **Figma** is a cloud-based design tool widely recognized for its collaborative capabilities and ease of use. It allows multiple team members to work on a project in real-time, making it ideal for designing interactive interfaces for digital escape rooms. With robust prototyping features, developers can create and test user interfaces, ensuring seamless navigation and immersive experiences. Figma's support for vector-based design ensures scalability, which is essential for adapting game assets across multiple devices. Its accessibility on both desktop and web platforms makes it a versatile choice for design teams working on educational escape room games.
- **Canva** is a user-friendly, drag-and-drop design tool with a vast library of templates and multimedia elements. Its intuitive interface makes it an excellent choice for educators and beginners creating engaging visuals and educational graphics for digital escape rooms. Canva's ability to integrate images, text, and animations enables the creation of aesthetically pleasing assets, such as posters, clue cards, or in-game visuals, with minimal effort. While it is less suited for intricate designs or complex animations, its simplicity and accessibility make it a valuable tool for quick and effective content creation.
- **Blender** is a powerful open-source tool for 3D modeling, animation, and rendering. It is particularly well-suited for designing immersive 3D environments and puzzles in digital escape rooms. Developers can use Blender to create realistic game elements, such as intricate props, rooms, or dynamic animations, adding depth and realism to the gaming experience. With its ex-

tensive feature set and active community, Blender is ideal for creating high-quality, interactive visuals that enhance the storytelling and puzzle-solving aspects of escape room games. While its complexity may require a learning curve, it offers unmatched flexibility for creating advanced assets.

- **InVision** is a prototyping and collaboration tool that excels in creating interactive workflows and testing interfaces. Its focus on interactivity makes it an excellent choice for designing and iterating on user interfaces for digital escape rooms. Developers can simulate gameplay navigation, ensuring that users experience intuitive and engaging interactions. The platform's collaboration features enable teams to gather feedback and make adjustments in real-time, streamlining the design process. InVision is particularly valuable for testing and refining game flows before full development, ensuring a polished final product.
- **Affinity Designer** is a high-performance vector graphics design tool that provides precision and flexibility in creating scalable game assets. Its intuitive interface and robust feature set make it suitable for both beginners and experienced designers. For digital escape rooms, Affinity Designer can be used to create detailed UI elements, icons, and game backgrounds that maintain their quality across various screen sizes. Its ability to handle complex designs efficiently makes it an excellent choice for developers seeking a cost-effective alternative to more established tools like Adobe Illustrator.
- **CorelDRAW Graphics Suite** is a comprehensive design toolkit that offers advanced layout and typography features. It is particularly suited for creating high-quality visuals for game settings, including intricate maps, signage, or thematic backgrounds for digital escape rooms. CorelDRAW's robust vector editing capabilities and support for a wide range of file formats make it versatile for integrating assets into game development pipelines. Its emphasis on creative freedom and precision allows developers to craft visually compelling and contextually rich elements for escape room games.
- **Gravit Designer** is a cloud-based vector graphics tool known for its cross-platform support and accessibility. Its lightweight nature makes it ideal for creating 2D assets for digital escape rooms, such as icons, clue graphics, and simple UI elements. Gravit Designer's user-friendly interface and range of design features cater to educators and hobbyists looking for an affordable, straightforward solution for game asset creation. While it lacks some advanced capabilities of professional tools, its ease of use and portability make it a practical choice for quick and efficient asset design.

The importance of user interface and user experience design in enhancing the playability of escape room games cannot be overstated. A well-designed UI not only guides players smoothly through the game but also enhances the storytelling aspect, immersing the player in the game's narrative. Effective UI/UX design in escape room games involves creating intuitive navigation systems, clear and contextually relevant icons, and interactive elements that are cohesive with the game's theme. This cohesive integration of UI elements with the game's environment contributes to a more immersive and engaging experience, a key aspect of successful escape room games.

User experience design in escape room games goes beyond the visual aspects. It encompasses the overall feel of the game, including the ease of interaction, the flow of the game, and the emotional response it elicits from the players. A well-designed UX ensures that the game is not only aesthetically pleasing but also functionally sound and emotionally engaging. This involves careful consideration of the game's pacing, the complexity of puzzles, feedback mechanisms, and the overall narrative structure. By focusing on these aspects, designers can create escape room games that are not only challenging and fun but also provide a memorable and satisfying experience to the players.

The utilization of interactive design tools (Table 4) such as Adobe Creative Suite and Sketch in the creation of game assets, coupled with a keen focus on UI/UX design, is paramount in developing escape room games that are visually striking, intuitive to navigate, and compelling to play. These tools enable designers to bridge the gap between artistic vision and functional design, resulting in games that not only look great but also provide an immersive and enjoyable experience for the player.

Table 4. Interactive design tools for digital educational escape rooms

Tool	Key Features	Educational Use Case
Adobe Creative Suite	Comprehensive toolkit, image editing, vector graphics	Creating detailed visuals, UI elements, and game assets
Sketch	Vector-based design, rapid prototyping, responsive design	UI/UX design, quick prototyping for game interfaces
Figma	Cloud-based design, collaboration features, prototyping	Real-time collaboration for designing interactive interfaces
Canva	Drag-and-drop design, templates, multimedia integration	Creating engaging visuals and educational graphics
Blender	3D modeling, animation, rendering	Designing immersive 3D environments and puzzles
InVision	Interactive prototyping, collaboration, feedback tools	Testing interactive game interfaces and workflows

continued on following page

Table 4. Continued

Tool	Key Features	Educational Use Case
Affinity Designer	High-performance vector graphics design	Creating scalable UI elements and detailed game assets
CorelDRAW Graphics Suite	Comprehensive design toolkit, layout and typography	Developing high-quality visuals for game settings
Gravit Designer	Cloud-based, vector graphics, cross-platform support	Accessible and lightweight tool for creating 2D assets

Augmented and Virtual Reality Integration

The integration of augmented reality (AR) and virtual reality (VR) in game development, especially for creating immersive escape room experiences, has been revolutionized by tools like ARKit, ARCore, Vuforia, Oculus Rift, and HTC Vive. Each of these tools (summarized in Table 5) brings unique capabilities and functionalities, pushing the boundaries of immersive gaming experiences.

Table 5. Tools for immersive experiences in digital educational escape rooms

Tool	Key Features	Educational Use Case
ARKit	AR framework for iOS, face tracking, environment texturing	Creating lifelike AR environments, integrating real-world elements
ARCore	AR development for Android, motion tracking, light estimation	Developing responsive AR escape rooms
Vuforia	3D object recognition, cross-platform AR experiences	Interactive AR games with complex object interactions
Meta Quest	Standalone VR headset, wireless setup, hand tracking	Developing immersive, accessible virtual escape rooms
HTC Vive	VR platform, room-scale tracking, handheld controllers	Creating detailed, interactive VR environments

- **ARKit**, developed by Apple, is a framework for building AR experiences on iOS devices. It offers advanced face tracking, realistic rendering, and environmental texturing, which are essential for creating lifelike AR environments. ARKit's ability to accurately track the real-world environment and overlay digital objects seamlessly makes it an excellent tool for developing AR-based escape rooms. These rooms can incorporate real-world elements, enhancing the immersion and interactivity of the game.

- **ARCore**, Google's counterpart to ARKit, enables the development of AR experiences for Android devices. It excels in motion tracking, environmental understanding, and light estimation. ARCore is pivotal in creating interactive escape room experiences that respond to the user's environment and movements, providing a more engaging and dynamic gaming experience.
- **Vuforia** is another prominent AR platform known for its wide range of features like 3D object recognition, ground plane detection, and VuMarks. It allows developers to create AR experiences across both iOS and Android platforms. Vuforia's robust tracking and real-time rendering capabilities make it ideal for escape room games that require precise interaction with complex objects and environments.
- **Meta Quest** is a standalone virtual reality (VR) headset that offers wireless functionality, high-resolution displays, and advanced hand-tracking capabilities, making it an excellent tool for developing immersive escape room experiences. Its untethered design allows players to move freely without being constrained by cables, enhancing the sense of immersion and making it ideal for room-scale VR environments. With built-in hand tracking and motion controllers, Meta Quest enables intuitive interactions with virtual objects, allowing players to solve puzzles and navigate environments naturally. The device's portability and ease of use make it accessible for a wide range of users, from educational institutions to gaming enthusiasts, ensuring an engaging and seamless escape room experience.

Puzzle and Scenario Development Software

In the intricate world of game development, particularly in crafting escape room games, the tools used for creating and implementing puzzles, such as RPG Maker MZ and Twine, play a fundamental role. These tools (see Table 6) not only facilitate the technical aspects of game development but also enrich the narrative and educational value of the games.

Table 6. Puzzle and scenario development software for DEERs

Software	Key Features	Educational Use Case
RPG Maker MZ	User-friendly, scripting system, custom assets	Creating intricate puzzles, immersive atmospheres, narrative integration
Twine	Interactive storytelling, branching narratives, text-centric design	Developing rich narratives, engaging scenarios, educational storytelling
Ink	Lightweight scripting, flexible integration, narrative-focused	Writing interactive scripts, branching narratives, game dialogue systems

continued on following page

Table 6. Continued

Software	Key Features	Educational Use Case
Ren'Py	Visual novel engine, Python-based scripting, multimedia integration	Creating story-driven puzzles, interactive visual scenarios, narrative-rich games
Quest	Text adventure creation, no programming required, cross-platform	Designing text-based escape rooms with puzzles and branching paths
Construct 3	Drag-and-drop, HTML5-based, fast prototyping	Creating interactive puzzles with quick development for web-based escape rooms
GameFroot	User-friendly, game templates, educational focus	Building accessible puzzle games and interactive learning tools

- **RPG Maker MZ** stands out as a versatile tool in the puzzle development landscape. Its user-friendly interface and powerful scripting system make it an ideal choice for developers who wish to create intricate puzzle mechanics without delving deeply into complex programming. RPG Maker MZ's tileset system and event management capabilities allow for the creation of diverse and interactive environments essential for puzzle-based games. Its ability to integrate custom graphics and sounds enables developers to craft unique and immersive atmospheres, enhancing the player's engagement and problem-solving experience. The software's emphasis on storytelling elements also allows for the seamless weaving of narratives into the game, providing a context and purpose to the puzzles, thereby enhancing the overall educational value.
- **Twine**, an open-source tool for telling interactive, nonlinear stories, offers a different approach to game and scenario development. Twine's simplicity in creating interactive stories makes it particularly suitable for developing engaging scenarios and narratives in educational games. It allows designers to focus on the branching paths of storytelling, enabling them to craft complex scenarios where player choices have significant impacts on the game's progression and outcome. This aspect of Twine is especially pertinent in escape room games, where narrative-driven puzzles are central to the gameplay. The software's text-centric approach encourages the creation of rich narratives, which can be instrumental in educational settings, allowing for the integration of learning objectives within the storyline of the game.
- **Ink** is a lightweight scripting language designed for creating interactive, branching narratives. It integrates seamlessly with game engines like Unity, making it highly versatile for digital escape rooms that require dynamic storytelling and decision-making. With its narrative-focused approach, Ink allows developers to craft intricate storylines where player choices influence

the game's progression. Its simplicity and flexibility make it ideal for educational escape rooms that incorporate complex dialogue trees or moral dilemmas, enabling players to engage deeply with the narrative while solving puzzles aligned with the story.

- **Ren'Py** is a visual engine that combines a simple Python-based scripting environment with robust multimedia support for creating rich, story-driven experiences. It is particularly suitable for designing puzzles and scenarios that rely on visual storytelling, dialogue, and branching choices. Developers can use Ren'Py to build interactive visual narratives where puzzles are integrated into the story, making it an excellent tool for educational escape rooms that aim to teach problem-solving through immersive storytelling. Its multimedia capabilities, including animations and sound effects, enhance engagement and provide a polished user experience.
- **Quest** is a user-friendly tool for creating text-based adventure games and interactive fiction, requiring no programming knowledge. Its drag-and-drop interface and robust support for branching narratives make it ideal for designing text-driven escape rooms with puzzles, riddles, and scenario-based learning. Quest allows developers to create accessible, story-rich environments where players progress through written prompts and solve logic-based challenges. This makes it particularly effective for educators seeking to create lightweight, narrative-focused digital escape rooms that emphasize reading comprehension and critical thinking.
- **Construct 3** is an HTML5-based game development platform with a drag-and-drop interface, enabling rapid prototyping and development of 2D games. It is ideal for creating interactive puzzles and mini-games for web-based escape rooms. Construct 3's ease of use allows educators and developers to focus on designing engaging puzzles without delving into complex coding. Its ability to export games to multiple platforms ensures accessibility, making it a great choice for educational escape rooms aimed at broad audiences, including remote learners.
- **GameFroot** is a browser-based platform tailored for creating educational games with an emphasis on accessibility and ease of use. It offers pre-designed templates and an intuitive interface, making it a practical choice for educators and novice developers. GameFroot is particularly suited for building simple puzzle games and interactive learning tools that align with curriculum objectives. Its focus on education enables the creation of escape room scenarios that promote teamwork, logical reasoning, and subject-specific knowledge, making it an excellent option for classroom-based digital escape rooms.

Incorporating narrative and educational elements into game scenarios enhances their educational value significantly. The use of storytelling in games serves not only as a medium for entertainment but also as a powerful tool for learning and engagement. By embedding educational content within the narrative, developers can create an immersive learning experience that captivates and educates simultaneously. This method of 'edutainment' proves effective in retaining the player's attention and interest, making the learning process more enjoyable and impactful.

The integration of software like RPG Maker MZ and Twine in game development reflects the evolving nature of educational technology. These tools bridge the gap between traditional learning methods and modern digital experiences, offering innovative ways to engage learners. As the field of educational game development continues to grow, the importance of such tools in creating compelling, educational, and enjoyable games becomes increasingly evident.

The use of RPG Maker MZ and Twine in developing puzzles and scenarios for escape room games highlights the convergence of technology, education, and entertainment. These tools not only facilitate the technical aspects of game development but also enrich the narrative and educational value, making them indispensable in the realm of educational game design.

Online and Cloud-Based Tools

The advent of online and cloud-based tools (Table 7) has revolutionized the way escape rooms are created, managed, and experienced. Platforms such as Google Forms and Genially have emerged as pivotal elements in this transformation, offering unique advantages in developing accessible and easy-to-manage digital escape rooms.

Table 7. Online and cloud-based tools for digital educational escape rooms

Tool	Key Features	Educational Use Case
Google Forms	Simple and versatile, real-time response collection, integrates with Google services	Creating puzzles and questionnaires, real-time feedback and analysis
Genially	Visually rich content, interactive multimedia, ease of use	Crafting immersive narratives with dynamic puzzles
Padlet	Collaborative boards, multimedia integration, real-time updates	Collaborative brainstorming, group-based puzzle solving, idea sharing
Miro	Digital whiteboards, visual collaboration, integration with other tools	Designing complex puzzle flows and collaborative planning
Kahoot!	Gamified quizzes, real-time participation, multimedia support	Engaging quiz-based challenges and gamified puzzles

continued on following page

Table 7. Continued

Tool	Key Features	Educational Use Case
ThingLink	Interactive image and video hotspots, immersive media creation	Creating interactive visual puzzles and explorative scenarios
Classcraft	Gamified learning, team collaboration, progress tracking	Designing team-based escape rooms with RPG-style elements

- **Google Forms**, known for its simplicity and versatility, has found a significant place in escape room development. Its utility lies in the ease of creating puzzles and questionnaires, which are essential components of any escape room. By leveraging Google Forms, developers can design a variety of puzzles, from multiple-choice questions to more complex logic-based challenges. The platform's seamless integration with other Google services, such as Sheets for data analysis and Drive for storage, further enhances its usability in escape room development. The real-time response collection and analysis feature of Google Forms allows for immediate feedback and adjustments, ensuring a smooth and engaging user experience.
- **Genially** takes the concept of interactive design a step further by providing a platform for creating visually rich and interactive content. Its strength lies in its ability to integrate multimedia elements like images, videos, and external links, making it an ideal tool for crafting immersive escape room narratives. Genially's interactivity features, including the ability to create hotspots and layered information, enable the development of dynamic puzzles that require exploration and discovery, key aspects of an engaging escape room experience. Genially's ease of use and flexibility in design make it accessible to educators and creators who may not have advanced technical skills, democratizing the process of digital escape room creation.
- **Padlet** is a collaborative online tool that enables the creation of shared boards where participants can post text, images, videos, or links in real-time. Its multimedia integration and intuitive interface make it a powerful tool for group-based activities in digital escape rooms. Educators can use Padlet to facilitate brainstorming sessions, where players collaboratively solve puzzles or piece together clues. Its real-time updates foster teamwork and allow for dynamic interactions, making it particularly suitable for escape rooms that emphasize collaboration and collective problem-solving.
- **Miro** is a digital whiteboard platform designed for visual collaboration and planning. It is ideal for designing complex puzzle flows or creating interactive maps for digital escape rooms. With its extensive template library, real-time collaboration features, and seamless integration with tools like Google

Drive and Slack, Miro is perfect for both planning and executing group-based activities. Educators can use Miro to structure escape room scenarios where participants visually interact with the game space, solve puzzles, or manage resources, enhancing engagement and strategy.

- **Kahoot!** is a gamified quiz platform that supports real-time participation and multimedia content. Its quiz-based approach makes it excellent for integrating trivia challenges or timed questions within digital escape rooms. Players can compete individually or in teams, creating an engaging and interactive atmosphere. Educators can use Kahoot! to craft quiz-based puzzles that test players' knowledge or problem-solving skills, incorporating visuals, videos, and music to enhance the immersive experience of the escape room.
- **ThingLink** enables the creation of interactive media by adding hotspots to images and videos, making it ideal for creating explorative scenarios and visual puzzles in digital escape rooms. Developers can craft environments where players discover clues by interacting with different elements of an image or video. Its immersive capabilities make it particularly effective for escape rooms that rely on visual storytelling or scavenger hunts. Educators can use ThingLink to design puzzles where players explore virtual environments and uncover hidden information critical to progressing through the game.
- **Classcraft** combines gamified learning with team collaboration and progress tracking, making it a unique tool for escape rooms with RPG-style elements. Players can work in teams to solve puzzles, earn rewards, and advance through levels, fostering cooperation and strategy. Classcraft's integration of storytelling and character development enhances immersion, making it ideal for escape rooms with rich narratives. Educators can use it to design team-based challenges where students collaborate to solve problems and achieve learning objectives, all while experiencing the thrill of a game-like environment.

The utilization of cloud-based tools like Google Forms and Genially offers significant advantages, particularly in the context of remote and collaborative game development. These platforms enable multiple creators to work simultaneously on a project, regardless of their geographical location. This collaborative aspect is crucial in the iterative process of escape room development, where puzzles and narratives often require input and refinement from a diverse team of creators. Cloud-based tools also ensure that updates and changes are synchronized in real time, providing a consistent and up-to-date version of the project to all team members.

The accessibility and ease of distribution of digital escape rooms created through these platforms make them an attractive option for educational and entertainment purposes. They can be easily shared with a broad audience through simple web links, eliminating the need for specialized software or hardware. This wide accessibility expands the reach of escape rooms, allowing them to be used as effective educational tools in classrooms, corporate training sessions, or remote team-building activities.

The integration of online and cloud-based tools like Google Forms and Genially in escape room development represents a significant leap forward in the field. These platforms not only streamline the development process but also enhance the creative possibilities, enabling the creation of more engaging, interactive, and accessible digital escape rooms. Their impact extends beyond mere convenience, fostering collaboration, innovation, and wider participation in the realm of digital game development.

For Digital Educational Escape Rooms (DEERs), the integration of analytics and feedback systems like Google Analytics and custom survey tools plays a critical role. These systems not only provide valuable insights into player behavior and preferences but also inform continuous improvements in the game's design and educational content.

Software tools such as Google Analytics offer a powerful means to track and analyze user interactions within the game. They provide data on various metrics, including user engagement, session duration, and the effectiveness of different game elements. This data is invaluable for understanding how players interact with the game, identifying parts that are particularly engaging or areas where players may be struggling. For example, Balaskas et al. (Balaskas et al., 2023) highlighted the importance of feedback in digital educational games and the integration of analytics to assess player engagement and learning outcomes. Such insights are critical for refining game design to enhance both educational impact and player satisfaction.

Custom survey tools also play a significant role in gathering qualitative feedback from players. These tools enable the collection of direct feedback on various aspects of the game, such as its educational content, entertainment value, and overall user experience. This feedback is essential for identifying areas of improvement from the player's perspective, ensuring that the game remains aligned with their needs and preferences.

The importance of collecting player data and feedback for continuous improvement cannot be overstated. This process allows developers to make data-driven decisions that enhance the game's educational and entertainment value. For instance, Liu and Hwang (Liu et al., 2023) discussed the effects of metalinguistic corrective feedback on digital game-based learning performances. Their study underscored the significance of feedback systems in adapting and improving educational content within digital games.

Examples of successful integration of analytics and feedback in DEERs further illustrate the importance of these systems. For instance, an educational escape room designed to teach history might use analytics to track which puzzles are most engaging and which historical topics capture players' interest the most. Coupled with feedback gathered from post-game surveys, developers can adjust the game narrative, puzzle difficulty, and educational content to better suit the players' learning needs and preferences.

The integration of analytics and feedback systems in DEERs is crucial for their success. By leveraging these tools, developers can gain a deeper understanding of player behavior and preferences, enabling them to continuously improve the game's educational efficacy and entertainment value. As DEERs continue to grow in popularity, the use of sophisticated analytics and feedback mechanisms will become increasingly important in ensuring their effectiveness as educational tools.

Accessibility and Inclusivity Features in DEERs

Ensuring accessibility and inclusivity in DEERs is paramount for creating games that are playable and enjoyable for a diverse audience. The integration of specialized tools such as screen readers, customizable difficulty settings, and other accessibility features is essential to accommodate players with various needs and preferences.

Screen readers and other assistive technologies are vital for making DEERs accessible to players with visual impairments. These tools convert text and interactive elements into speech or Braille, enabling visually impaired players to navigate and engage with the game effectively. Customizable difficulty settings are another crucial aspect of inclusivity. They allow players to adjust the game's challenge level to suit their individual skill sets and learning paces, thus creating a more personalized and accessible gaming experience.

The importance of developing DEERs that cater to a diverse audience cannot be overstated. Inclusive game design not only broadens the potential player base but also ensures that educational objectives are met for all players, regardless of their physical abilities or learning styles. For example, Zanto et al. (Zanto et al., 2024) highlights the potential of digital tools to provide accessible educational content. The study demonstrates how digital interventions can be tailored to diverse learning needs, emphasizing the importance of inclusive design in educational technologies.

In real-world applications, DEERs have been developed with a focus on inclusivity. For instance, an educational escape room designed for a museum might include audio descriptions and tactile elements for visually impaired visitors, ensuring that the narrative and puzzles are accessible to all. Another example could be an online DEER created for a school, featuring adjustable text sizes and contrast settings to accommodate students with dyslexia or other learning differences.

The integration of feedback systems in DEERs allows developers to continuously improve accessibility features based on player input. Collecting feedback from a diverse group of players ensures that the game design evolves to meet the needs of a wider audience. The incorporation of accessibility and inclusivity features in DEERs is crucial for creating educational experiences that are equitable and engaging for all players. By employing tools like screen readers, customizable difficulty settings, and designing with diverse needs in mind, DEER developers can create games that are not only fun and educational but also accessible to a broad spectrum of players.

Collaboration and Project Management Tools

The development of DEERs is a collaborative endeavor that requires effective coordination and communication among team members. Utilizing project management tools like Trello, Asana, and Slack is crucial in streamlining this process. These tools facilitate organization, task management, and real-time communication, which are essential for successful DEER development.

- **Trello**, with its intuitive board-based interface, allows teams to visualize project progress through a series of customizable lists and cards. This visual approach makes it easy to track different elements of the DEER development, from initial concept and narrative design to puzzle creation and testing.
- **Asana** provides a more structured platform for task management, offering features like timelines, task assignments, and progress tracking. This tool is particularly useful in managing the complex workflows and numerous tasks involved in DEER development.
- **Slack**, on the other hand, enhances team communication. It provides a centralized platform for discussion, file sharing, and integration with other tools like Trello and Asana. This integration ensures that communication is directly linked to tasks and project progress, facilitating a more cohesive and efficient development process.

Effective team collaboration and communication are vital during the DEER game development process. Strategies for achieving this include regular team meetings, clear role assignments, and the use of collaborative tools for brainstorming and feedback. Regular meetings, whether virtual or in-person, help ensure that all team members are aligned with the project goals and progress. Assigning clear roles and responsibilities to each team member fosters accountability and efficiency.

Examples of successful use of these tools in game development are numerous. A study (Almeida et al., 2023) highlighted how project management tools contribute to solution development and validation, ensuring the delivery of high-quality educational content. Another example is the development of a simulation game for teaching construction management, where tools like Trello and Asana could be used to manage the complex integration of educational content and game mechanics (Al-Jibouri and Mawdesley, 2001).

Project management tools like Trello, Asana, and Slack play a critical role in the development of DEERs. They facilitate effective collaboration and communication, ensuring that the development process is organized, efficient, and aligned with the project goals. By leveraging these tools (Table 8), DEER development teams can create engaging and educational escape rooms that meet the highest standards of quality and innovation.

Table 8. Collaboration and project management tools for DEERs

Tool	Key Features	Educational Use Case
Trello	Kanban boards, task tracking, intuitive interface	Organizing project tasks, monitoring progress, team collaboration
Asana	Project timelines, task dependencies, integrations	Managing complex projects, coordinating team efforts, ensuring timely delivery
Slack	Real-time messaging, file sharing, channel organization	Facilitating communication, sharing resources, quick decision-making

Procedural Content Generation Tools for DEERs

Procedural Content Generation (PCG) tools are instrumental in automating the creation of dynamic environments, puzzles, and game assets in Digital Educational Escape Rooms (DEERs). By leveraging algorithms and rule-based systems, these tools enable developers to design adaptable and engaging content that enhances replayability and reduces manual workload. The following are key tools and their applications in DEER design:

- **Houdini:** Houdini employs a node-based workflow for procedural modeling and dynamic simulations, making it ideal for creating intricate and evolving game environments. Developers can use it to generate diverse puzzles and modular assets that adapt to different scenarios.
- **Unity (Procedural Tools):** Unity's scripting capabilities and built-in procedural tools allow for the generation of randomized terrains, asset configurations, and puzzle elements. This ensures dynamic gameplay experiences tailored to players' progress and preferences.
- **Blender (Geometry Nodes):** Blender's geometry node editor supports procedural asset creation, enabling designers to craft modular escape room elements and layouts. Its open-source nature makes it accessible for developers with varied budgets and technical expertise.
- **Substance Designer:** Substance Designer specializes in procedural material authoring and texture generation, allowing developers to create dynamic tex-

tures and visual puzzle elements. These customizable materials enhance the aesthetic quality and interactivity of DEERs.

- **Unreal Engine (Blueprints):** Unreal Engine's visual scripting system, Blueprints, facilitates procedural terrain creation and asset generation. It is particularly useful for developing immersive environments and adaptive puzzles that respond to player actions.
- **CityEngine:** CityEngine provides tools for rule-based 3D city modeling and GIS integration, enabling the creation of urban-style escape room settings. Its procedural capabilities allow for dynamic layouts that enhance the storytelling aspect of DEERs.
- **PCG Algorithms (Custom):** Custom PCG algorithms enable developers to design tailored content generation systems. These algorithms can create adaptive puzzles and dynamic room configurations, ensuring a unique experience for each gameplay session.

By integrating these tools (summarized in Table 9) into the development process, DEER creators can achieve scalability, variety, and enhanced interactivity, offering players engaging and educationally enriching experiences.

Table 9. Procedural content generation tools for DEERs

Tool	Key Features	Application in DEERs
Houdini	Node-based workflow, procedural modeling, dynamic simulations	Creating complex, dynamic game environments, generating diverse puzzles
Unity (Procedural Tools)	C# scripting, asset randomization, terrain generation	Generating dynamic levels and randomized puzzle configurations
Blender (Geometry Nodes)	Open-source, geometry node editor, procedural asset creation	Designing modular escape room elements and layouts
Substance Designer	Material authoring, texture generation, parametric design	Creating dynamic textures and interactive visual puzzle elements
Unreal Engine (Blueprints)	Visual scripting, procedural terrain, asset generation	Developing procedural environments and adaptive puzzles
CityEngine	3D city modeling, rule-based procedural generation, integration with GIS	Designing urban-style escape room settings with dynamic layouts
PCG Algorithms (Custom)	Algorithmic content creation, customizable, AI-driven logic	Creating adaptive puzzles and dynamic room configurations

Audio and Voice Synthesis Tools for DEERs

Audio and voice synthesis tools play a critical role in enhancing the immersive experience of Digital Educational Escape Rooms (DEERs). These tools enable developers to create realistic sound effects, dynamic voiceovers, and atmospheric background music that enrich the gameplay environment. The following are key tools and their applications in DEER design:

- **Audacity:** Audacity is an open-source audio editing tool that supports multi-track editing and noise reduction. It is ideal for refining voiceovers, creating sound effects, and improving overall audio quality, making it a versatile choice for designing DEER audio assets.
- **Adobe Audition:** Adobe Audition offers professional-grade audio editing with features such as sound restoration, multi-channel support, and advanced mixing. This tool is perfect for producing high-quality NPC dialogue recordings and crafting immersive soundscapes for escape rooms.
- **Resemble AI:** Resemble AI uses AI-powered voice cloning and real-time synthesis to generate custom voices for NPCs. With multilingual support and customizable tones, it allows developers to create unique and engaging character voices tailored to specific narratives.
- **Descript:** Descript combines AI transcription, overdubbing, and audio editing to streamline the creation of dialogue and narration. It is particularly useful for refining voiceovers and integrating narration seamlessly into DEER gameplay.
- **Speechelo:** Speechelo is a text-to-speech platform that offers customizable voice styles and background music integration. It is ideal for creating voiceovers for NPCs, tutorials, and in-game announcements, enhancing the auditory experience for players
- **Voicemod:** Voicemod provides real-time voice modulation and soundboard integration, making it an excellent tool for adding dynamic voice effects. Developers can use it to create distinctive NPC voices or enable interactive player communication with voice effects.
- **Logic Pro X:** Logic Pro X is a professional audio workstation equipped with MIDI support and advanced mixing tools. It is well-suited for composing dynamic background music and designing complex audio environments that heighten the immersive quality of DEERs.

These tools (summarized in Table 10) empower developers to design rich auditory experiences, enabling players to become fully immersed in the escape room environment while also enhancing the educational and entertainment value of the game.

Table 10. Audio and voice synthesis tools for DEERs

Tool	Key Features	Application in DEERs
Audacity	Open-source audio editing, multi-track editing, noise reduction	Editing voiceovers, creating sound effects, enhancing audio quality
Adobe Audition	Professional audio editing, sound restoration, multi-channel support	Producing high-quality NPC dialogue recordings and immersive soundscapes
Resemble AI	AI-powered voice cloning, real-time synthesis, multilingual support	Generating NPC voices, customizing voice tones, supporting multilingual gameplay
Descript	AI transcription, overdubbing, audio editing	Refining dialogue, integrating narration, correcting voiceovers
Speechelo	Text-to-speech, customizable voice styles, background music integration	Creating voiceovers for NPCs and tutorials, enhancing immersion
Voicemod	Voice modulation, real-time effects, soundboard integration	Adding real-time voice effects for NPCs or player interactions
Logic Pro X	Professional audio workstation, MIDI support, advanced mixing tools	Composing dynamic background music, mixing audio for immersive environments

Educational Content Integration Tools

The integration of educational content into DEERs necessitates the use of specialized software that allows for seamless embedding of learning materials. Platforms like MOODLE and various e-learning platforms play a pivotal role in this process (Table 11). These tools facilitate the incorporation of educational content into the game environment, ensuring that learning objectives are met without compromising the engagement or interactivity of the game.

Table 11. Educational content integration tools for DEERs

Tool	Key Features	Educational Use Case
MOODLE	Open-source, customizable, multimedia integration	Embedding diverse educational content, tracking progress
E-learning Platforms	Interactive content, scenario-based learning, multimedia support	Developing narrative-driven puzzles, aligning with educational objectives
H5P	Interactive HTML5 content, open-source, easy embedding	Creating interactive videos, quizzes, and presentations for puzzles
Edmodo	Classroom management, content sharing, communication tools	Sharing learning materials, organizing escape room challenges
Nearpod	Interactive lessons, real-time feedback, multimedia integration	Designing multimedia puzzles and collaborative activities

continued on following page

Table 11. Continued

Tool	Key Features	Educational Use Case
BookWidgets	Customizable widgets, quizzes, and games, student tracking	Embedding dynamic activities into escape room scenarios
ClassFlow	Cloud-based, interactive content creation, student engagement	Delivering educational puzzles with live feedback and interaction

- **MOODLE**, as an open-source learning platform, offers a flexible and customizable environment for creating online courses and educational content. Its ability to integrate various types of media, including videos, quizzes, and interactive lessons, makes it an ideal tool for embedding educational content into DEERs. MOODLE's tracking and reporting features also enable game developers to monitor player progress and understand how effectively the educational content is being engaged with.
- **E-learning platforms**, in general, provide robust tools for creating interactive and engaging educational content. These platforms can be utilized to develop the narrative and puzzles in DEERs, ensuring that the educational elements are intricately woven into the gameplay. For example, e-learning platforms can be used to create scenario-based learning experiences within the game, where players need to apply knowledge or skills to progress.
- **H5P** is an open-source platform for creating interactive HTML5-based content, such as videos, quizzes, presentations, and interactive timelines. Its ability to easily embed interactive content into websites, e-learning platforms, or standalone applications makes it highly versatile for digital escape rooms. Educators can use H5P to create multimedia puzzles that challenge players through drag-and-drop activities, fill-in-the-blank questions, or branching scenarios. Its ease of use and compatibility with tools like MOODLE and WordPress make it a valuable addition for embedding engaging and educational tasks in escape room scenarios.
- **Edmodo** is a classroom management and communication platform designed to share learning materials and facilitate group activities. It allows educators to distribute tasks, manage student progress, and encourage collaboration. In digital escape rooms, Edmodo can be used to organize challenges, provide clues or hints, and track team interactions in real-time. Its communication tools enable players to collaborate effectively, making it especially useful for classroom-based escape room activities where teamwork and knowledge sharing are critical.
- **Nearpod** is an interactive lesson delivery platform that incorporates multimedia content, real-time feedback, and collaborative tools. Its ability to cre-

ate live or self-paced sessions makes it ideal for digital escape rooms where players need to solve puzzles in real-time or at their own pace. Nearpod's integration of videos, quizzes, and virtual tours enables educators to craft immersive challenges that align with educational objectives. The platform's real-time feedback feature allows instructors to monitor progress and guide players as they work through the puzzles, enhancing engagement and learning outcomes.

- **BookWidgets** is a tool for creating customizable widgets, such as quizzes, crossword puzzles, and interactive games, that can be embedded in escape room scenarios. Its intuitive design and robust tracking features make it an excellent choice for integrating dynamic activities into educational escape rooms. Educators can use BookWidgets to develop tailored challenges that test players' comprehension and problem-solving skills while providing instant feedback. This makes it ideal for scenarios where players interact with diverse content types, such as word games, sorting exercises, or visual puzzles.
- **ClassFlow** is a cloud-based platform designed for creating interactive lessons that foster student engagement through live participation. It allows educators to integrate multimedia content, quizzes, and collaborative activities into their lessons. For digital escape rooms, ClassFlow can be used to deliver puzzles and challenges that require real-time interaction and immediate feedback. Its ability to connect teachers and students in a virtual environment makes it an effective tool for facilitating educational escape rooms that emphasize group problem-solving and active participation.

Aligning game mechanics with educational objectives is a critical aspect of DEER development. Techniques for achieving this alignment include designing puzzles that require the application of specific knowledge or skills, incorporating educational content into the game's storyline, and using game-based assessments to gauge learning outcomes. This alignment ensures that players are not only entertained but also gain meaningful educational experiences from the game.

Examples of successful educational content integration in games are evident in various fields. For instance, the study Ito (2009) explores how games can be used as educational tools, embedding learning content within engaging gameplay. Similarly, (McIntyre et al., 2023) discusses the interconnectedness of games with education, emphasizing the potential of games as platforms for delivering educational content.

The use of software like MOODLE and e-learning platforms for embedding educational content in DEERs is crucial for creating experiences that are both educational and engaging. By effectively aligning game mechanics with educational objectives, developers can ensure that DEERs serve as powerful tools for learning, catering to a wide range of educational needs and preferences.

Social Interaction Tools for DEERs

Social interaction tools are essential for enabling effective communication and collaboration in Digital Educational Escape Rooms (DEERs), particularly in multiplayer or team-based scenarios. These tools facilitate real-time interaction, allowing players to share ideas, discuss strategies, and coordinate actions as they navigate through puzzles and challenges. The following discusses several key tools and their applications in DEER design:

- **Discord** is a versatile communication platform that supports real-time voice and text chat, custom channels, and bot integrations. For digital escape rooms, Discord is particularly useful for facilitating team collaboration and communication during gameplay. Players can use dedicated voice channels to discuss strategies, share clues, and solve puzzles in real-time. The ability to integrate bots enables developers to create automated hint systems or manage game-specific interactions, making Discord a robust choice for multiplayer escape room experiences.
- **Microsoft Teams** offers a combination of video conferencing, file sharing, and collaborative workspaces, making it ideal for teamwork in educational and professional settings. In digital escape rooms, Teams can be used to enable seamless resource sharing and group discussions. Breakout rooms allow developers to organize players into smaller teams for specific challenges, while its integration with Office tools provides additional flexibility for incorporating educational content into the game environment.
- **Slack** is a channel-based messaging platform that supports app integrations, file sharing, and real-time collaboration. For digital escape rooms, Slack allows developers to create structured channels for different teams or puzzles, enhancing coordination among players. The platform's integration capabilities enable interactive features such as automated clue distribution, progress tracking, or time-bound challenges, making it an excellent tool for team-based gameplay.
- **Zoom** is a widely-used video conferencing tool with features like breakout rooms, screen sharing, and live reactions. In digital escape rooms, Zoom can host virtual gameplay sessions, with breakout rooms enabling teams to work on puzzles independently before reconvening to discuss progress. Screen sharing is particularly useful for collaborative problem-solving, as players can share clues, visuals, or solutions with teammates in real-time.

- **Gather.Town** is an interactive virtual environment platform that combines proximity-based video chat with customizable virtual spaces. For digital escape rooms, Gather.Town allows developers to create immersive multiplayer experiences where players can navigate a virtual map, interact with objects, and communicate with teammates as they solve puzzles. Its gamified design enhances player engagement and fosters a sense of immersion, making it ideal for virtual escape rooms.
- **Miro** is a collaborative online whiteboard tool designed for real-time problem-solving and visual communication. In digital escape rooms, Miro can serve as a shared space where players collaborate on complex puzzles, map out strategies, or organize clues. The tool's templates and interactive features make it easy to integrate with other game elements, supporting teamwork and enhancing the puzzle-solving experience.
- **Google Meet** provides video conferencing, screen sharing, and live captioning features, making it a straightforward tool for enabling player interaction during digital escape rooms. It is particularly useful for hosting remote gameplay sessions, where players can discuss strategies and share visual information in real-time. The platform's simplicity and accessibility ensure that players can focus on the game without technical distractions.

These tools (summarized in Table 12) play a critical role in enhancing the collaborative and interactive aspects of DEERs, fostering teamwork and engagement while ensuring smooth communication during gameplay.

Table 12. Social interaction tools for DEERs

Tool	Key Features	Application in DEERs
Discord	Real-time voice and text chat, custom channels, bot integrations	Facilitating team communication during collaborative escape room gameplay
Microsoft Teams	Video conferencing, file sharing, collaborative workspaces	Enabling teamwork and resource sharing in multiplayer DEERs
Slack	Channel-based messaging, app integrations, file sharing	Supporting group coordination and discussion during puzzle-solving
Zoom	Video and audio meetings, breakout rooms, screen sharing	Hosting virtual escape rooms and facilitating team collaboration
Gather.Town	Interactive virtual spaces, proximity-based chat, customizable environments	Creating immersive, multiplayer virtual escape room experiences
Miro	Visual collaboration, interactive whiteboards, real-time updates	Enabling visual problem-solving and team coordination for DEER puzzles
Google Meet	Video calls, screen sharing, live captions	Supporting player interaction and team discussions

Testing and Quality Assurance Software

In the development of DEERs, ensuring a bug-free and smooth gameplay experience is crucial. To achieve this, developers rely on a variety of tools for game testing and quality assurance, including beta testing platforms and bug tracking software. These tools are integral to identifying and resolving issues before the game reaches the end user, thus ensuring a high-quality product.

- **TestFlight** is Apple's official beta testing platform for iOS applications, allowing developers to distribute pre-release versions of their apps to testers. It supports seamless installation and updates for beta testers, enabling developers to gather feedback on user experience and identify bugs. TestFlight's user-friendly interface makes it easy for testers to provide feedback directly within the app, and its integration with the Apple ecosystem ensures smooth compatibility for iOS-based Digital Educational Escape Rooms (DEERs). This platform is ideal for testing features, gameplay mechanics, and performance on a variety of Apple devices.
- **Beta Family** is a cross-platform beta testing service that connects developers with a global network of testers. It offers structured feedback collection, detailed analytics, and user interaction tracking, making it an effective tool for evaluating DEERs during development. Developers can target specific user demographics to ensure diverse testing scenarios, helping to identify bugs, usability issues, and improvement opportunities. Beta Family is particularly suitable for small to medium-sized development teams seeking detailed feedback from real users.
- **Google Play** Beta Testing is a built-in feature of the Google Play Store that facilitates the distribution of beta versions of Android applications. It allows developers to reach a wide audience by inviting testers directly through the store. With features like staged rollouts and user feedback integration, this platform enables developers to test DEERs in real-world conditions and gather insights on performance, stability, and user engagement. It's especially useful for teams looking to validate Android-based escape rooms across a diverse range of devices.
- **App Center** by Microsoft is a cross-platform beta testing and distribution service that supports iOS, Android, Windows, and macOS applications. It offers crash reporting, continuous integration, and real-time analytics, making it an excellent choice for DEER developers who need comprehensive testing tools. App Center enables developers to distribute builds efficiently, track bugs, and monitor gameplay performance across multiple platforms, ensuring a consistent and bug-free experience for players.

- **TestFairy** is a beta testing platform that focuses on detailed user behavior tracking and session recording. It provides developers with video recordings of user sessions, real-time crash reports, and performance monitoring, enabling them to identify and resolve issues effectively. For DEERs, TestFairy is especially valuable for understanding how users interact with the game, uncovering usability issues, and fine-tuning gameplay mechanics. Its support for both iOS and Android makes it versatile for testing cross-platform escape rooms.

Beta testing platforms (Table 13) are essential in the testing phase, allowing developers to gather real-world feedback from early users. These platforms enable controlled distribution of pre-release versions of the game to a select group of users, who then provide feedback on their experience. This feedback is crucial for identifying bugs, gameplay issues, and areas for improvement. For example, platforms like TestFlight or Beta Family offer a structured environment for managing beta tests, collecting feedback, and tracking user interactions.

Table 13. Beta testing platforms for DEERs

Platform	Key Features	Educational Use Case
TestFlight	iOS app testing, controlled distribution, feedback collection	Testing iOS-based DEERs, collecting feedback from early users
Beta Family	Cross-platform testing, structured feedback, user analytics	Testing game mechanics, identifying bugs, tracking user interactions
Google Play Beta Testing	Android app testing, easy distribution, user review insights	Gathering feedback from Android users, improving DEERs
App Center	Cross-platform support, crash reporting, continuous integration	Streamlining DEER testing for multiple platforms
TestFairy	Video session recordings, real-time crash reports, user behavior tracking	Analyzing user behavior, debugging gameplay interactions

Bug tracking software is another vital component in the quality assurance process. Tools such as JIRA or Bugzilla (Table 14) help developers keep track of reported issues, prioritize them, and manage their resolution. These systems enable efficient communication within the development team and ensure that no bug or issue is overlooked. By systematically tracking and addressing each issue, developers can significantly improve the overall quality and stability of the game.

Table 14. Bug tracking software for DEERs

Software	Key Features	Educational Use Case
JIRA	Comprehensive issue tracking, project management tools, customizable workflows	Tracking bugs and tasks during DEER development, prioritizing resolutions
Bugzilla	Open-source, advanced search features, email notifications	Organizing bug reports, facilitating communication among developers
Trello	Kanban-style boards, task management, simple bug tracking	Tracking minor bugs, visualizing testing progress for DEERs
MantisBT	Web-based, customizable fields, user-friendly interface	Simplifying bug reporting and resolution for small teams
Redmine	Open-source, project management, Gantt charts	Integrating bug tracking with project timelines and milestones
Zoho BugTracker	Cloud-based, issue tagging, time tracking	Tracking bugs in real-time, managing development schedules
YouTrack	Agile support, query-based search, integration capabilities	Tracking and managing bugs in Agile-based DEER projects

- **JIRA** is a comprehensive project management and bug tracking tool designed for Agile development environments. It offers customizable workflows, detailed issue tracking, and robust integration capabilities with other software development tools. In the context of Digital Educational Escape Rooms (DEERs), JIRA is particularly useful for organizing and prioritizing bug reports, assigning tasks to team members, and ensuring timely resolution of critical issues. Its ability to create detailed reports and track progress helps developers monitor the health of the project and maintain a high-quality user experience.
- **Bugzilla** is an open-source bug tracking system known for its simplicity and powerful search features. It provides email notifications, advanced filtering options, and the ability to link bugs with related issues, making it a reliable choice for tracking complex problems in DEERs. Bugzilla is ideal for development teams looking for an affordable yet efficient solution to manage bug reports and communicate effectively during the debugging process. Its straightforward interface allows developers to focus on resolving issues without being overwhelmed by excessive features.
- **Trello** is a visual task management tool that uses a Kanban-style board to organize workflows. While not a dedicated bug tracking system, Trello's simplicity and flexibility make it effective for managing minor bugs and tracking testing progress in DEERs. Developers can create separate boards or lists for different types of issues, prioritize tasks, and assign responsibilities. Trello is particularly useful for smaller teams or educational projects where stream-

lined communication and visual task management are key to resolving issues efficiently.

- **MantisBT (Mantis Bug Tracker)** is a web-based bug tracking system with an intuitive user interface and customizable fields. It simplifies bug reporting and resolution for small to medium-sized teams working on DEERs. MantisBT supports role-based access control, which ensures that each team member has appropriate permissions for their tasks. Its email notifications and issue tracking features help maintain transparency and accountability in the debugging process, making it an excellent choice for teams that need a straightforward tool without sacrificing functionality.
- **Redmine** is an open-source project management tool with integrated bug tracking and project planning capabilities. Its features, such as Gantt charts, time tracking, and milestone management, make it ideal for teams that want to align bug resolution with broader project goals in DEER development. Redmine's versatility allows developers to track bugs alongside other tasks, ensuring that quality assurance efforts are well-integrated into the overall project timeline. Its support for plugins further enhances its functionality for specialized needs.
- **Zoho BugTracker** is a cloud-based solution that offers issue tagging, customizable workflows, and real-time collaboration tools. Its clean interface and integration with Zoho's ecosystem make it particularly effective for teams managing DEER projects with tight deadlines. Developers can tag and categorize bugs, track resolution progress, and generate reports to identify recurring issues. The time tracking feature ensures that teams stay on schedule while resolving bugs, providing a seamless way to maintain quality control.
- **YouTrack** is an Agile-friendly bug tracking and project management tool known for its query-based search and integration capabilities. It supports custom workflows, dashboards, and detailed reporting, making it a powerful choice for managing DEER development projects. Its Agile board feature allows teams to track bugs and tasks in sprints, ensuring that issues are resolved systematically. YouTrack's ability to integrate with version control systems and CI/CD pipelines enhances its utility for complex DEER projects requiring iterative testing and frequent updates.

Best practices for ensuring a bug-free experience in DEERs include rigorous testing at various stages of development, continuous feedback collection from users, and a structured approach to bug tracking and resolution. Developers should engage in both automated and manual testing to cover a wide range of scenarios and use cases. Automated testing tools can quickly identify issues that might be overlooked in manual testing, while manual testing allows for a more nuanced understanding of the user experience.

Examples of effective use of testing and quality assurance tools in game development are numerous.

For instance, the development of educational games like "Duolingo" or "Khan Academy Kids" involves extensive testing to ensure that the educational content is delivered seamlessly within the game environment. These platforms use a combination of beta testing and bug tracking tools to refine their games, ensuring that they are both educational and engaging.

The integration of testing and quality assurance software in DEER development is crucial for delivering a high-quality educational gaming experience. By utilizing beta testing platforms and bug tracking tools, and following best practices for testing and feedback collection, developers can create DEERs that are not only immersive and engaging but also stable and free of bugs.

REFERENCES

Al-Jibouri, S., & Mawdesley, M. (2001). Design and experience with a computer game for teaching construction project planning and control. *Engineering, Construction, and Architectural Management*, 8(5/6), 418–427. DOI: 10.1108/eb021201

Alha, K., Loerke, D., Koskinen, E., & Paavilainen, J. (2023). Augmented play: An analysis of augmented reality features in location-based games. *Convergence (London)*, 29(2), 342–361. Advance online publication. DOI: 10.1177/13548565231156495

Alkhathlan, M., Tlachac, M. L., Harrison, L., & Rundensteiner, E. (2022). Improving image accessibility by combining haptic and auditory feedback. In *Proceedings of the 24th International ACM SIGACCESS Conference on Computers and* Accessibility. DOI: DOI: 10.1145/3517428.3550362

Almeida, R., Marques, D. F., Durán, R. L., Aguado, Y. B., Cid-Bartolomé, T., González-González, S., & Hatzikou, M. (2023). A game changer joint design journey of prems and proms in the hosmartai virtual coach study. *International Journal of Integrated Care*, 23(S1), 659. DOI: 10.5334/ijic.ICIC23586

Ayas, A. Y., Aydın, H., Çetinkaya, A., & Güney, Z. (2023). *Artificial intelligence (ai)-based self-deciding character development application in two-dimensional video games*. Bilgi ve Ileti¸sim Teknolojileri Dergisi., DOI: 10.53694/bited.1247338

Balaskas, A., Schueller, S. M., Cox, A. L., Rashleigh, C., & Doherty, G. (2023). Examining young adults daily perspectives on usage of anxiety apps: A user study. *PLOS Digital Health*, 2(1), e0000185. DOI: 10.1371/journal.pdig.0000185 PMID: 36812622

Bhugaonkar, K., Bhugaonkar, R., & Masne, N. (2022). The trend of metaverse and augmented & virtual reality extending to the healthcare system. *Cureus*, 14. Advance online publication. DOI: 10.7759/cureus.29071 PMID: 36258985

Botelho, F. H. F. (2021). Accessibility to digital technology: Virtual barriers, real opportunities. *Assistive Technology*, 33(sup1), 27–34. DOI: 10.1080/10400435.2021.1945705 PMID: 34951832

Botev, J., Mayer, J., & Rothkugel, S. (2019). Immersive mixed reality object interaction for collaborative context-aware mobile training and exploration. In *Proceedings of the 11th ACM Workshop on Immersive Mixed and Virtual Environment Systems*. DOI: DOI: 10.1145/3304113.3326117

Cai, W., Chi, Y., Zhou, C., Zhu, C., & Leung, V. C. M. (2018). Ubgaming: Ubiquitous cloud gaming system. *IEEE Systems Journal*, 12(3), 2483–2494. DOI: 10.1109/JSYST.2018.2797080

Cairns, P., Power, C., Barlet, M., Haynes, G., Kaufman, C., & Beeston, J. (2019). Enabled players: The value of accessible digital games. *Games and Culture*, 16(2), 262–282. DOI: 10.1177/1555412019893877

Chen, H., Zhang, X., Xu, Y., Ren, J., Fan, J., Ma, Z., & Zhang, W. (2019). T-gaming: A cost-efficient cloud gaming system at scale. *IEEE Transactions on Parallel and Distributed Systems*, 30(12), 2849–2865. DOI: 10.1109/TPDS.2019.2922205

Chen, Y., & Cheng, H. (2022). *The economics of the metaverse: A comparison with the real economy*. Metaverse., DOI: 10.54517/met.v3i1.1802

Cheng, R., Wu, N., Varvello, M., Chen, S., & Han, B. (2022). Are we ready for metaverse?: a measurement study of social virtual reality platforms. *Proceedings of the 22nd ACM Internet Measurement Conference, DOI:* DOI: 10.1145/3517745.3561417

Darejeh, A. (2023). Empowering education through eerp: A customizable educational vr escape room platform. In *2023 IEEE International Symposium on Mixed and Augmented Reality Adjunct (ISMAR-Adjunct), pages 764–766*. DOI: DOI: 10.1109/ISMAR-Adjunct60411.2023.00166

dos Santos Nunes, E. P., da Conceição Júnior, V. A., & Borges, L. C. L. F. (2018). Interaction techniques to promote accessibility in games for touchscreen mobile devices: A systematic review. In *Proceedings of the 17th International Conference on Human-Computer Interaction, pages* 178–191. DOI: DOI: 10.1007/978-3-319-92049-8_13

Durmaz, Y., & Gokalp, D. (2023). *Introduction of metaverse to our lives and unlimited services in the world of metaverse*. International Business & Economics Studies., DOI: 10.22158/ibes.v6n1p1

Efe, A. (2022). *Taking virtual reality and augmented reality to the next level: Artificial intelligence with mixed reality*. Uluslararası Yönetim Bilişim Sistemleri ve Bilgisayar Bilimleri Dergisi., DOI: 10.33461/uybisbbd.1190125

Fernandes, F., & Werner, C. (2022). Accessibility in the metaverse: Are we prepared? *Anais do XIII Workshop sobre Aspectos da Interação Humano-Computador na Web Social (WAIHCWS 2022), DOI:* DOI: 10.5753/waihcws.2022.226618

Garcez, L., Thiry, M., & Fernandes, A. M. R. (2020). Accessible features to support visually impaired people in game development: A systematic literature review of the last 15 years. In *2020 15th Iberian Conference on Information Systems and Technologies (CISTI), pages 1–6*. DOI: DOI: 10.23919/CISTI49556.2020.9140904

Ge, M. (2023). *Review on the application of deep learning algorithms in video game ai agent*. Applied and Computational Engineering., DOI: 10.54254/2755-2721/4/2023322

Goh, E. S., Sunar, M. S., & Ismail, A. (2019). 3d object manipulation techniques in handheld mobile augmented reality interface: A review. *IEEE Access : Practical Innovations, Open Solutions*, 7, 40581–40601. DOI: 10.1109/ACCESS.2019.2906394

Gomes, G., Vidal, C., Neto, J. B. C., & Nogueira, Y. L. (2021). Two level control of non-player characters for navigation in 3d games scenes: A deep reinforcement learning approach. In *2021 20th Brazilian Symposium on Computer Games and Digital Entertainment (SBGames), pages 182–190*. DOI: DOI: 10.1109/SBGames54170.2021.00030

Hajj, M. H. E. (2023). The future of digital arts. *International Journal of Education and Learning Research, DOI:*.DOI: 10.21608/ijelr.2023.211965.1002

Halloran, J., & Minerva, A. (2019). Touch and play? investigating the value of touchscreens for gamer experience. *Entertainment Computing*, 32, 100312. Advance online publication. DOI: 10.1016/j.entcom.2019.100312

Hanus, A., Hoover, M., Lim, A., & Miller, J. (2019). A collaborative virtual reality escape room with passive haptics. In *IEEE Conference on Virtual Reality and 3D User Interfaces (VR), pages 1413–1414*. DOI: DOI: 10.1109/VR.2019.8798241

Hassan, L. (2023). Accessibility of games and game-based applications: A systematic literature review and mapping of future directions. *New Media & Society*. Advance online publication. DOI: 10.1177/14614448231204020

Horváth, I. (2021). An analysis of personalized learning opportunities in 3d vr. *Frontiers of Computer Science*, 3, 673826. Advance online publication. DOI: 10.3389/fcomp.2021.673826

Iqbal, M. Z., & Campbell, A. G. (2023). *Metaverse as tech for good: Current progress and emerging opportunities*. Virtual Worlds., DOI: 10.3390/virtualworlds2040019

Ito, M. (2009). *Engineering Play: A Cultural History of Children's Software*. The MIT Press., DOI: 10.7551/mitpress/7939.001.0001

Joo, H.-T., & Kim, K.-J. (2019). Visualization of deep reinforcement learning using grad-cam: How ai plays atari games? In *2019 IEEE Conference on Games (CoG), pages* 1–2. DOI: DOI: 10.1109/CIG.2019.8847950

Kim, J. (2021). Advertising in the metaverse: Research agenda. *Journal of Interactive Advertising*, 21(3), 141–144. DOI: 10.1080/15252019.2021.2001273

Kopel, M., & Hajas, T. (2018). Implementing ai for non-player characters in 3d video games. In *Springer, pages 610–619*. DOI: DOI: 10.1007/978-3-319-75417-8_57

Kotzebue, L., Zumbach, J., & Brandlmayr, A. (2022). Digital escape rooms as game-based learning environments: A study in sex education. *Multimodal Technologies and Interaction*, 6(2), 8. DOI: 10.3390/mti6020008

Lauriola, I., Lavelli, A., & Aiolli, F. (2021). An introduction to deep learning in natural language processing: Models, techniques, and tools. *Neurocomputing*. Advance online publication. DOI: 10.1016/j.neucom.2021.05.103

Liu, J., Snodgrass, S., Khalifa, A., Risi, S., Yannakakis, G. N., & Togelius, J. (2020). Deep learning for procedural content generation. *Neural Computing & Applications*, 33(1), 19–37. DOI: 10.1007/s00521-020-05383-8

Liu, Z., Gong, S., Tan, Z., & Demian, P. (2023). Immersive technologies-driven building information modeling (bim) in the context of metaverse. *Buildings (Basel, Switzerland)*, 13(6), 1559. Advance online publication. DOI: 10.3390/buildings13061559

López-Faican, L., & Martínez, J. (2020). Emofindar: Evaluation of a mobile multiplayer augmented reality game for primary school children. *Computers & Education*, 149, 103814. DOI: 10.1016/j.compedu.2020.103814

Magdy, R., Rashad, S., Hany, S., Tarek, M., Hassan, M. A., & Mohammed, A. (2021). Deep reinforcement learning approach for augmented reality games. In *2021 International Mobile, Intelligent, and Ubiquitous Computing Conference (MIUCC), pages 330–336*. DOI: DOI: 10.1109/MIUCC52538.2021.9447671

McIntyre, P., Kerrigan, S., Fulton, J., King, E., & Williams, C. (2023). *Electronic Games and Interactive Content.In Palgrave Studies in Creativity and Culture, page 349–371*. Springer International Publishing, DOI: DOI: 10.1007/978-3-031-45972-6_14

Momenzad, M., Majidi, B., & Eshghi, M. (2018). Deep summarization of academic textbooks for adaptive gamified virtual learning environments. In *2018 2nd National and 1st International Digital Games Research Conference: Trends, Technologies, and Applications (DGRC), pages 88–94*. DOI: DOI: 10.1109/DGRC.2018.8712065

Montebello, M. (2021). Personalized learning environments. In *2021 International Symposium on Educational Technology (ISET), pages* 134–138. DOI: DOI: 10.1109/ISET52350.2021.00036

Mustafa, B. (2022). *Analyzing education based on metaverse technology*. Technium Social Sciences Journal., DOI: 10.47577/tssj.v32i1.6742

Mystakidis, S., Cachaferio, E., & Hatzilygeroudis, I. (2019). Enter the serious e-scape room: A cost-effective serious game model for deep and meaningful e-learning. In *2019 10th International Conference on Information, Intelligence, Systems and Applications (IISA), pages 1–6*. DOI: DOI: 10.1109/IISA.2019.8900673

Mystakidis, S., Papantzikos, G., & Stylios, C. (2021). Virtual reality escape rooms for stem education in industry 4.0: Greek teachers perspectives. In *2021 6th South-East Europe Design Automation, Computer Engineering,Computer Networks and Social Media Conference (SEEDA-CECNSM), pages1–5*. DOI: DOI: 10.1109/SEEDA-CECNSM53056.2021.9566265

Nair, V., Garrido, G. M., and Song, D. (2022). Exploring the privacy risks of adversarial vr game design. *ArXiv, abs/2207.13176*, DOI: .DOI: 10.48550/arXiv.2207.13176

Nimpattanavong, C., Choensawat, W., & Sookhanaphibarn, K. (2022). Action prediction of ai bot in fightingice by using deep learning model. In *2022 IEEE 4th Global Conference on Life Sciences and Technologies (LifeTech), pages 259–262*. DOI: DOI: 10.1109/LifeTech53646.2022.9754922

Oh, I.-S., Rho, S., Moon, S., Son, S., Lee, H., & Chung, J. (2019). Creating pro-level ai for a real-time fighting game using deep reinforcement learning. *IEEE Transactions on Games*, 14(2), 212–220. DOI: 10.1109/TG.2021.3049539

Olin, P. A., Issa, A. M., Feuchtner, T. M., & Grønbæk, K. (2020). Designing for heterogeneous cross-device collaboration and social interaction in virtual reality. In *Proceedings of the 32nd Australian Conference on Human-Computer Interaction*. DOI: DOI: 10.1145/3441000.3441070

Ong, Y., Ismail, A., Iahad, N. A., Rameli, M., & Dollah, R. (2020). A mobile game sdk for remote collaborative between two users in augmented and virtual reality. In *IOP Conference Series: Materials Science and Engineering,* volume 979. DOI: DOI: 10.1088/1757-899X/979/1/012003

Pfau, J., Smeddinck, J. D., & Malaka, R. (2018). Towards deep player behavior models in mmorpgs. In *Proceedings of the 2018 Annual Symposium on Computer-Human Interaction in Play*. DOI: DOI: 10.1145/3242671.3242706

Revilla, A., Zamarvide, S., Lacosta, I., Pérez, F., Lajara, J., Kevelham, B., Juillard, V., Rochat, B., Drocco, M., Devaud, N., Barbeau, O., Charbonnier, C., Lange, P., Li, J., Mei, Y., Lawicka, K., Jansen, J., Reimat, N., Subramanyam, S., & César, P. (2021). A collaborative vr murder mystery using photorealistic user representations. In *2021 IEEE Conference on Virtual Reality and 3D User Interfaces Abstracts and Workshops (VRW), pages 766–766*. DOI: DOI: 10.1109/VRW52623.2021.00266

Risi, S. and Togelius, J. (2019). Procedural content generation: From automatically generating game levels to increasing generality in machine learning. *ArXiv, abs/1911.13071*.

Rocha, S., & Lopes, A. (2020). Navigation based application with augmented reality and accessibility. In *Extended Abstracts of the 2020 CHI Conference on Human Factors in Computing Systems*. DOI: DOI: 10.1145/3334480.3383004

Roohi, S., Guckelsberger, C., Relas, A., Heiskanen, H., Takatalo, J., & Hämäläinen, P. (2021). Predicting game difficulty and engagement using ai players. *Proceedings of the ACM on Human-Computer Interaction*, 5(CHI PLAY), 1–17. DOI: 10.1145/3474658

Rostami, S., & Maier, M. (2022). The metaverse and beyond: Implementing advanced multiverse realms with smart wearables. *IEEE Access : Practical Innovations, Open Solutions*, 10, 110796–110806. DOI: 10.1109/ACCESS.2022.3215736

Shang, J., Chen, S., Wu, J., & Yin, S. (2020). Arspy: Breaking location-based multiplayer augmented reality application for user location tracking. *IEEE Transactions on Mobile Computing*, 21(2), 433–447. DOI: 10.1109/TMC.2020.3007740

Singanamalla, S., Thies, W., & Scott, C. (2018). Avatar: Enabling immersive collaboration via live mobile video. In *Proceedings of the 3rd International Workshop on Multimedia Alternate Realities*. DOI: DOI: 10.1145/3268998.3269003

Soares, E. S., & Bulitko, V. (2019). Deep variational autoencoders for npc behaviour classification. In *2019 IEEE Conference on Games (CoG), pages* 1–4. DOI: DOI: 10.1109/CIG.2019.8848095

Sun, Y., Yuan, B., Xiang, Q., Zhou, J., Yu, J., Dai, D., & Zhou, X. (2023). Intelligent decision-making and human language communication based on deep reinforcement learning in a wargame environment. *IEEE Transactions on Human-Machine Systems*, 53(1), 201–214. DOI: 10.1109/THMS.2022.3225867

Tan, T. F., Li, Y., Lim, J. S., Gunasekeran, D., Teo, Z. L., Ng, W. Y., & Ting, D. (2022). Metaverse and virtual health care in ophthalmology: Opportunities and challenges. *Asia-Pacific Journal of Ophthalmology*, 11(3), 237–246. DOI: 10.1097/APO.0000000000000537 PMID: 35772084

Tang, C., Wang, Z., Sima, X., & Zhang, L. (2020). Research on artificial intelligence algorithm and its application in games. In *2020 2nd International Conference on Artificial Intelligence and Advanced Manufacture (AIAM), pages 386–389*. DOI: DOI: 10.1109/AIAM50918.2020.00085

Teófilo, M., Lucena, V., Nascimento, J., Miyagawa, T., & Maciel, F. (2018). Evaluating accessibility features designed for virtual reality context. In *2018 IEEE International Conference on Consumer Electronics (ICCE), pages* 1–6. DOI: DOI: 10.1109/ICCE.2018.8326167

To, W. M., Yu, B. T. W., Chung, A., & Chung, D. W. K. (2023). Metaverse: Trend, emerging themes, and future directions. *Transactions on Emerging Telecommunications Technologies*. Advance online publication. DOI: 10.1002/ett.4912

Tregel, T. (2021). Online content generation in mobile applications - adaptation and personalization for location-based game systems. *Technische Universität Darmstadt, DOI:* DOI: 10.26083/TUPRINTS-00017499

Tricariaco, M. (2021). The educational value of the escape room in virtual environments. In *Encyclopedia of Information Science and Technology, Fifth Edition, pages 341–372*. DOI: DOI: 10.4018/978-1-7998-7638-0.ch015

Trinh, M. P. (2021). Can you escape? using mobile app escape games to create a holistic experiential learning process online. *Management Teaching Review*, 7(3), 236–244. DOI: 10.1177/23792981211056921

Uddin, M., Manickam, S., Ullah, H., Obaidat, M. A., & Dandoush, A. (2023). Unveiling the metaverse: Exploring emerging trends, multifaceted perspectives, and future challenges. *IEEE Access : Practical Innovations, Open Solutions*, 11, 87087–87103. DOI: 10.1109/ACCESS.2023.3281303

Usmani, S., Sharath, M., & Mehendale, M. (2022). Future of mental health in the metaverse. *General Psychiatry*, 35(4), e100825. Advance online publication. DOI: 10.1136/gpsych-2022-100825 PMID: 36189180

van der Boon, M., Fermoselle, L., Haar, F. T., Dijkstra-Soudarissanane, S., & Niamut, O. (2022). Deep learning augmented realistic avatars for social vr human representation. In *ACM International Conference on Interactive Media Experiences*. DOI: DOI: 10.1145/3505284.3532976

Xu, J. (2022). From augmented reality location-based games to the real-world metaverse. *Extended Abstracts of the 2022 Annual Symposium on Computer-Human Interaction in Play, DOI:* DOI: 10.1145/3505270.3558363

Zanto, T. P., Giannakopoulou, A., Gallen, C. L., Ostrand, A. E., Younger, J. W., Anguera-Singla, R., Anguera, J. A., & Gazzaley, A. (2024). Digital rhythm training improves reading fluency in children. *Developmental Science*, 27(3), e13473. Advance online publication. DOI: 10.1111/desc.13473 PMID: 38193394

Zeng, H., He, X., & Pan, H. (2020). Implementation of escape room system based on augmented reality involving deep convolutional neural network. *Virtual Reality (Waltham Cross)*, 25(3), 585–596. DOI: 10.1007/s10055-020-00476-0

Zhang, X., Chen, H., Zhao, Y., Ma, Z., Xu, Y., Huang, H., Yin, H., & Wu, D. O. (2019). Improving cloud gaming experience through mobile edge computing. *IEEE Wireless Communications*, 26(4), 178–183. DOI: 10.1109/MWC.2019.1800440

Žilak, M., Car, Z., & Čuljak, I. (2022). A systematic literature review of handheld augmented reality solutions for people with disabilities. *Sensors (Basel)*, 22(20), 7719. Advance online publication. DOI: 10.3390/s22207719 PMID: 36298070

KEY TERMS AND DEFINITIONS

Accessibility: Features integrated into digital escape rooms to accommodate players with various needs, such as screen readers or customizable difficulty settings.

Analytics and Feedback Systems: Tools like Google Analytics that track player behavior and provide insights for improving gameplay and educational content.

Artificial Intelligence (AI): Computational systems that analyze data and enable dynamic, adaptive environments, such as real-time puzzle adjustments in escape rooms.

Augmented Reality (AR): A technology that overlays digital content onto the real world using devices like smartphones, enabling location-based or interactive escape room experiences.

Cloud-based Tools: Online platforms like Google Forms or Genially that simplify the creation, sharing, and management of escape room scenarios.

Collaborative Tools: Platforms like Trello, Slack, and Miro that facilitate communication and project management during the escape room development process.

Digital Educational Escape Rooms (DEERs): Innovative gamified environments integrating storytelling, puzzles, and educational content designed to foster engagement and deep learning.

Educational Content Integration Tools: Platforms such as MOODLE or H5P that embed learning materials into escape rooms, aligning gameplay with educational objectives.

Game Development Platforms: Software used to design, develop, and implement digital escape rooms.

Gamification: The application of game-design elements and principles in non-gaming contexts to enhance engagement and motivation.

Interactive Design Tools: Software such as Adobe Creative Suite and Blender, used for creating assets like visuals, UI elements, and animations in escape rooms.

Metaverse: A virtual universe enabling real-time interaction between users and their environment, offering expansive and immersive digital escape room experiences.

Non-Player Characters (NPCs): AI-driven virtual entities in escape rooms that interact with players, enhancing the narrative and gameplay through dialogue or assistance.

Procedural Content Generation (PCG): Techniques and tools that algorithmically create game assets, dynamic puzzles, and environments to ensure unique and replayable escape room experiences.

Social Interaction Tools: Platforms such as Discord and Zoom that enable real-time communication and collaboration among escape room players.

User Experience (UX): The overall experience of a player, encompassing usability, game flow, and emotional engagement within an escape room.

Virtual Reality (VR): A technology that immerses users into a fully digital environment through headsets, providing a 360-degree experience for solving puzzles and navigating escape rooms.

APPENDIX I: TOOLS

This table provides a comprehensive overview of the tools and platforms mentioned throughout the chapter, along with their respective URL links. These tools encompass a wide range of applications, including game development, design, collaboration, and educational content integration, making them invaluable for creating and managing Digital Educational Escape Rooms (DEERs). Readers can explore these resources to further enhance their understanding and implementation of DEERs.

Table 15. Resources for further understanding of DEERs

Tool	URL
Adobe Audition	https://www.adobe.com/products/audition.html
Adobe Creative Suite	https://www.adobe.com/
Affinity Designer	https://affinity.serif.com/en-us/designer/
ARCore	https://developers.google.com/ar
ARKit	https://developer.apple.com/augmented-reality/arkit/
Audacity	https://www.audacityteam.org/
Blender	https://www.blender.org/
BookWidgets	https://www.bookwidgets.com/
Canva	https://www.canva.com/
CityEngine	https://www.esri.com/en-us/arcgis/products/arcgis-cityengine/overview
ClassFlow	https://classflow.com/
Construct 3	https://www.construct.net/
CorelDRAW Graphics Suite	https://www.coreldraw.com/
CryEngine	https://www.cryengine.com/
Descript	https://www.descript.com/
Discord	https://discord.com/
Edmodo	https://www.edmodo.com/
Figma	https://www.figma.com/
wGameMaker Studio	https://gamemaker.io/
GDevelop	https://gdevelop.io/
Godot Engine	https://godotengine.org/
Gravit Designer	https://www.designer.io/
H5P	https://h5p.org/
Houdini	https://www.sidefx.com/products/houdini/
InVision	https://www.invisionapp.com/
Logic Pro X	https://www.apple.com/logic-pro/

continued on following page

Table 15. Continued

Tool	URL
Meta Quest	https://www.meta.com/quest/
Microsoft Teams	https://www.microsoft.com/en-us/microsoft-teams/group-chat-software
MOODLE	https://moodle.org/
Nearpod	https://nearpod.com/
Resemble AI	https://www.resemble.ai/
RPG Maker	https://www.rpgmakerweb.com/
Sketch	https://www.sketch.com/
Slack	https://slack.com/
Speechelo	https://speechelo.com/
Substance Designer	https://www.adobe.com/products/substance3d-designer.html
Unity3D	https://unity.com/
Unreal Engine	https://www.unrealengine.com/
Voicemod	https://www.voicemod.net/
Vuforia	https://developer.vuforia.com/
Zoom	https://zoom.us/

Chapter 4
Designing a Digital Educational Escape Room

ABSTRACT

The chapter discusses the key principles and mechanics of designing Digital Educational Escape Rooms (DEERs) to enhance learning experiences. It emphasizes aligning educational goals with immersive game narratives and activities, focusing on cohesive storytelling to make learning content engaging and memorable. Important game mechanics like time constraints and puzzle integration are highlighted, along with how DEERs promote cognitive skills such as critical thinking and problem-solving. Real-world examples demonstrate these principles in practice, and the abstract also explores how gamification strategies maintain student motivation and engagement. Finally, it covers the role of narrative development and storyboarding in creating immersive educational experiences.

DEER DESIGN PRINCIPLES AND GAME MECHANICS

Creating DEERs involves a combination of sound design principles and effective game mechanics. The following subsections provide a comprehensive guide to the key aspects of DEER design, drawing on insights from recent academic research.

Foundational DEER Design Principles

The design of DEERs necessitates a deep understanding of educational objectives and their effective integration into the game's activities (Hakshurian, 2023). Foundational principles of DEER design (Table 1) include aligning educational goals with game mechanics and ensuring that the narrative and theme are cohesive and contribute meaningfully to the learning experience (Moffett et al., 2023a). Understanding and aligning educational objectives with DEER activities is crucial (Veldkamp et al.,

DOI: 10.4018/979-8-3693-4219-0.ch004

2020b). The design process should begin with a clear definition of what learners are expected to achieve by the end of the game. These objectives could range from acquiring specific knowledge, developing skills, or fostering attitudes. For example, if the objective is to teach historical facts, the game could involve solving puzzles that require understanding historical events or figures. This alignment ensures that every aspect of the DEER, from the storyline to the challenges, contributes towards the educational goals (Moffett et al., 2023a).

The narrative and theme of the DEER must be cohesive and integral to the learning experience (Hakshurian, 2023). A well-crafted narrative not only engages players but also provides context to the educational content, making it more relatable and memorable. For instance, a DEER themed around a science-fiction story could be used to teach scientific concepts, where players solve puzzles based on scientific principles to progress in the story. The theme should be appealing to the target audience and should be used to create an immersive environment that enhances learning (Hakshurian, 2023). Examples of DEERs that successfully implement these foundational design principles are evident in various educational settings. One such example is a DEER developed for medical training, where healthcare professionals solve clinically relevant puzzles in a simulated environment. This approach allows them to apply medical knowledge in a practical, engaging context, thereby enhancing their learning experience (Moffett et al., 2023a). Another example is found in language learning DEERs, where players engage in conversations or solve language-based puzzles, integrating language skills development into an exciting game format (Hakshurian, 2023).

Table 1. Foundational DEER design principles

Principle	Description
Educational Alignment	Align educational objectives with game mechanics to ensure that every aspect of the DEER contributes to learning outcomes. This involves clear definition of learning goals, such as knowledge acquisition, skill development, or attitude fostering.
Cohesive Narrative and Theme	Ensure the narrative and theme are cohesive and integral to the learning experience. A well-crafted story engages players and provides context for educational content, making it more relatable and memorable.
Immersive Experience	Design an immersive environment where the narrative, puzzles, and learning content are seamlessly integrated, enhancing engagement and making the educational content more impactful.

The foundational principles of DEER design play a critical role in creating effective and engaging educational experiences. By understanding and aligning educational objectives with game activities and ensuring that the narrative and theme are cohesive and contribute to the learning experience, DEERs can become powerful tools for education, combining the excitement of gaming with the rigor of learning.

Effective Game Mechanics in DEERs

In the domain of DEERs, the implementation of effective game mechanics is pivotal in enhancing engagement and ensuring the educational efficacy of the game (Repetto et al., 2023). Utilizing elements like time constraints, puzzles, and narrative, as well as designing challenges suitable for the target audience's skill level, are key strategies in this process (Huraj et al., 2022).

Time constraints in DEERs add a sense of urgency and excitement, compelling players to engage more deeply with the content (Hermanns et al., 2017). This mechanic can be used to simulate real-world scenarios where quick thinking and decision-making are essential. For instance, setting a time limit for solving a puzzle not only enhances engagement but also mirrors real-life situations where timely responses are crucial (Huraj et al., 2022).

Puzzles are the core of any escape room experience. In DEERs, puzzles must be intricately woven into the narrative and educational objectives (Repetto et al., 2023). Designing puzzles that require the application of specific knowledge or skills ensures that players are learning while playing (Makri et al., 2021). For example, a math-based DEER might include puzzles that require arithmetic or algebraic calculations to proceed, thereby reinforcing mathematical concepts in a fun and interactive way (Ross and Bennett, 2022).

Narrative is a powerful tool in DEERs to drive player engagement and contextualize learning (Huynh et al., 2020). A compelling story not only makes the game more interesting but also provides a framework within which educational content can be delivered (Verswijvelen et al., 2020). The narrative should be relevant to the educational objectives and resonate with the target audience, thereby enhancing the overall learning experience (Min et al., 2020). When designing challenges and tasks, it's crucial to align them with the target audience's skill level (Recke and Perna, 2021). This alignment ensures that the game is neither too easy nor too challenging, maintaining player interest and motivation (Omarali, 2022). For instance, a DEER designed for young children should include simpler puzzles and a narrative that is easy to follow, while one targeted at adults can incorporate more complex challenges and sophisticated storylines (Huynh et al., 2020). Examples of effective game mechanics in DEERs can be seen in educational settings. For example, a study (De Castell et al., 2023) discusses the use of game-based interactions to enhance language learning, emphasizing the role of game mechanics in improving understanding and engagement (Malegiannaki et al., 2020). Another example is the use of the falling-puzzle game "Puyo Puyo" in programming education for young people, as explored by (Tian et al., 2024). This study demonstrates how consumer digital game rules can be effectively applied in an educational context (Suter et al., 2021). The effective use of game mechanics such as time constraints, puzzles, narrative, and tailored challenges (Table 2), is crucial in the design and development of DEERs. These elements not only enhance engagement and enjoyment but also ensure that the educational goals of the game are met, making DEERs a valuable tool in the field of educational technology (Pecheranskyi, 2023).

Table 2. Effective game mechanics in DEERs

Mechanic	Description
Time Constraints	Adding urgency to the gameplay, compelling players to engage deeply with the content, simulating real-world scenarios where quick thinking is essential.
Puzzles	Core component of escape rooms, intricately woven into the narrative and educational objectives, requiring the application of specific knowledge or skills to progress.
Narrative	A compelling story that drives player engagement and contextualizes learning, enhancing the overall educational experience. The narrative should align with the educational objectives and resonate with the target audience.
Tailored Challenges	Designing puzzles and tasks that align with the target audience's skill level ensures the game is appropriately challenging, maintaining player interest and motivation.

Cognitive Skill Development in DEERs

The design of DEERs offers a unique opportunity to foster cognitive skill development, such as critical thinking and problem-solving (Tang et al., 2020). By incorporating thoughtfully designed puzzles and tasks, DEERs become effective tools for enhancing these essential skills (Makri et al., 2021). They can be tailored to develop over-professional competencies that are crucial for graduates in the modern world (Vodă et al., 2022). Incorporating puzzles and tasks that foster critical thinking and problem-solving skills involves creating challenges that require learners to apply knowledge, analyze information, and make reasoned decisions (Shanta and Wells, 2020). For instance, a DEER designed for a science curriculum might include puzzles that require learners to apply scientific principles to solve a mystery or to conduct virtual experiments to deduce a solution (Hakshurian, 2023). These activities not only reinforce the educational content but also encourage students to think critically and solve problems creatively (Tang et al., 2020). Using DEERs as a tool to develop over-professional competencies involves aligning the game's objectives with skills necessary for graduates, such as teamwork, communication, digital literacy, and adaptability (Makri et al., 2021). For example, a DEER can be designed where students need to collaborate to solve puzzles, thereby enhancing their teamwork and communication skills (Stork, 2020). Similarly, puzzles that require navigating digital platforms or interpreting data can improve digital literacy and adaptability (Stokes and Pike, 2022). Examples of DEERs that effectively incorporate cognitive skill development include:

- An educational Escape Room implemented to enhance analytical and sustainable engagement on social media, as discussed by (Gil Ruiz, 2024). This DEER focuses on developing critical and digital competencies through interactive and engaging puzzles.
- A digital escape room game designed by Sahin (Sahin, 2023) in basic education settings. This DEER provides a scenario different from traditional games, aiming to develop students' cognitive skills along with the educational content.
- An example by Abd Rahman et al. (Rahman et al., 2023), where STEM modules using gamification were used as induction teaching aids. This approach incorporates cognitive and psychomotor skills in non-digital games, making it an interesting teaching and learning aid.

DEERs present an innovative and engaging approach to cognitive skill development. By incorporating challenging puzzles and tasks, aligning the game with over-professional competencies, and using real-world examples as inspiration, DEER developers can create experiences that are not only educational but also instrumental in preparing students for professional success.

Supporting Student Engagement and Motivation in DEERs

In DEERs, supporting student engagement and motivation is key to the success of the educational experience (Bharti, 2023). Applying gamification strategies and balancing educational content with entertaining elements are effective ways to maintain student interest and participation (Zahedi et al., 2021).

Applying gamification strategies involves incorporating elements commonly found in games, such as point scoring, competition, and rules of play, into the educational content (Zahedi et al., 2021). This approach makes learning more engaging and can significantly increase student motivation (Sánchez, 2023). For example, a DEER designed for teaching mathematics might include a point system where students earn points for solving math puzzles correctly, fostering a sense of achievement and competition (Bharti, 2023).

Balancing educational content with entertaining elements is crucial to keep students engaged (Zahedi et al., 2021). While the primary goal of DEERs is educational, the incorporation of fun and interactive elements ensures that students remain interested and motivated throughout the learning process (Panchal and Sharma, 2022). This balance can be achieved by integrating puzzles and tasks that are not only educational but also creatively stimulating and enjoyable (Zahedi et al., 2021).

Examples of DEERs that effectively support student engagement and motivation include:

- Gamification in social media education: As explored by Gil Ruiz (2024), an educational Escape Room was implemented to enhance analytical and sustainable engagement on social media. This DEER used gamification strategies to engage students in critical and digital competencies through interactive and engaging puzzles.
- STEM modules with gamification for non-academic students: Abd Rahman et al. (2023) developed STEM modules using gamification as induction teaching aids. These modules incorporated cognitive and psychomotor skills in non-digital games, making them an interesting teaching and learning aid.
- Immersive learning in physiology through DEERs: Surapaneni (2023) created a digital escape room in physiology education, offering an immersive learning experience. This DEER highlighted the potential benefits of using educational escape room strategies in physiology education, combining educational content with engaging game mechanics.

Supporting student engagement and motivation in DEERs requires a careful blend of gamification strategies and a balance between educational and entertaining elements. By doing so, DEER developers can create experiences that are not only educational but also engaging and enjoyable for students.

CREATING ENGAGING PUZZLES AND CHALLENGES

Designing engaging puzzles for DEERs is crucial for creating an effective learning experience. This section outlines the key aspects and challenges in developing captivating puzzles for DEERs.

Cognitive and Educational Aspects in DEERs

In DEERs, the integration of cognitive and educational aspects plays a pivotal role in enhancing the learning experience (Hakshurian, 2023). This integration involves designing puzzles that not only engage and entertain but also promote cognitive skills such as problem-solving and critical thinking (Shmelev, 2021). These puzzles are infused with educational content to provide a meaningful and enriching experience (Pratiwi et al., 2020).

Designing puzzles to enhance cognitive skills like problem-solving and critical thinking involves creating challenges that require learners to apply knowledge, analyze situations, and devise strategies (Kazem, 2022). For instance, a puzzle in a DEER might involve deciphering codes based on historical facts, requiring students to use their understanding of history and logic skills (Akcaoglu et al., 2021). Such puzzles engage learners in active thinking processes, enhancing their cognitive abilities in a fun and interactive way (Gyaurov et al., 2022). Integrating educational content seamlessly into the puzzles is crucial for ensuring that the DEER is not only entertaining but also educational (Mulyana and Nurcahyani, 2022). This can be achieved by designing puzzles that directly relate to the learning material. For example, in a DEER designed for a science class, puzzles might involve performing virtual experiments or solving problems using scientific methods (Chen et al., 2021). This approach ensures that the puzzles serve as an extension of the learning experience, making the educational content more engaging and memorable (Geremias et al., 2023).

Balancing the educational and entertainment value in puzzle design is essential to maintain learner interest and engagement (Liu et al., 2020). While the primary aim of the DEER is educational, the incorporation of engaging and interactive elements ensures that the experience remains enjoyable (Wells et al., 2021). This balance can be achieved by embedding educational tasks within an intriguing narrative and presenting challenges that are intellectually stimulating yet accessible (Pratiwi et al., 2020). Examples of DEERs where cognitive and educational aspects are effectively integrated include:

- A DEER designed to teach STEM subjects involved puzzles that required applying mathematical equations and scientific principles to solve challenges. This DEER not only provided an interactive learning experience but also enhanced students' problem-solving skills and scientific knowledge (Lathwesen and Belova, 2021).
- In a DEER designed for language learning, puzzles involved translating phrases or constructing sentences in the target language. This approach made learning a new language more interactive and engaging, encouraging students to apply their language skills in a practical scenario (Voreopoulou et al., 2024).

The integration of cognitive and educational aspects in DEERs is crucial for creating an effective learning tool. By designing puzzles that enhance cognitive skills and seamlessly integrate educational content, while balancing educational and entertainment values, DEER developers can create experiences that are not only enjoyable but also highly beneficial for learners.

Collaborative and Team-Based Puzzles in DEERs

Incorporating collaborative and team-based puzzles in DEERs is a strategic approach that not only enhances the learning experience but also fosters essential skills like teamwork, communication, and joint problem-solving (Eukel et al., 2020). These types of puzzles require participants to work together, sharing knowledge and strategies to achieve a common goal (Mateos-Aparicio et al., 2020). Creating puzzles that encourage teamwork and communication involves designing challenges that cannot be solved by a single individual but require the collective effort of a group (Manojlovic, 2022). For instance, a puzzle in a DEER might require each team member to solve a different part of a riddle, with the solution only becoming apparent when all parts are combined. This approach encourages participants to communicate effectively and leverage each other's strengths (Vuillaume et al., 2021).

Designing challenges that require collaborative problem-solving means creating scenarios where the success of the team depends on the cooperation and coordination of its members (Garwood, 2020). A real-world example of this can be seen in a study titled "The Great Escape: A Novel Approach to Collaborative Learning" (Carpenter et al., 2023). In this study, healthcare students participated in an educational escape room where they needed to collaboratively solve puzzles related to their field of study. This experience not only reinforced their knowledge but also highlighted the importance of teamwork in healthcare settings (Carpenter et al., 2023).

Another real-world example is illustrated in the study "Teaching Science, Sustainability, and Technology Through Gamification: The Approach of Escape Rooms" by Alexandropoulou et al. (Alexandropoulou and Tsezou, 2023). This study combined team-based learning methods and project-based methods in digital escape rooms, where students had to solve puzzles related to science and sustainability. This approach not only made the learning process more engaging but also encouraged students to work collaboratively, fostering a deeper understanding of the subject matter (Sarage et al., 2020).

Collaborative and team-based puzzles in DEERs play a crucial role in enhancing the educational value of these interactive experiences. By requiring teamwork and cooperative problem-solving, these puzzles not only make the learning process more engaging but also equip learners with vital collaborative skills (Friedrich et al., 2020). Examples in educational settings demonstrate the effectiveness of this approach in fostering an interactive and collaborative learning environment (Turner et al., 2021).

PRACTICAL NARRATIVE DEVELOPMENT AND STORYBOARDING

Fundamentals of Narrative Development for DEERs

Narrative development in DEERs is a critical aspect that profoundly influences the effectiveness and engagement of the game (Huynh et al., 2020). A well-crafted narrative serves as the backbone of the DEER, providing context and motivation for players to engage with the educational content (Luhova, 2020). Understanding the role of storytelling in DEERs is essential (Verswijvelen et al., 2020). A compelling narrative can transform educational content from abstract and theoretical to concrete and experiential (Teixeira et al., 2023). It provides a framework within which educational objectives are embedded, making learning more meaningful and enjoyable (Verswijvelen et al., 2020). For instance, a DEER designed to teach history might use a narrative that transports players back in time, allowing them to experience historical events first-hand (Malegiannaki et al., 2020). Crafting an engaging and coherent narrative that aligns with educational goals is a delicate balancing act (Maksimenko, 2023). The narrative should be intriguing enough to captivate players, yet clear and structured to facilitate learning (Huynh et al., 2020). This involves creating a storyline that seamlessly integrates educational content without disrupting the flow of the game (Teixeira et al., 2023). For example, in a DEER focused on science education, puzzles might involve solving scientific problems or conducting experiments that are integral to the storyline (Kanareiko, 2023).

Techniques for creating immersive and interactive storylines include character development, branching narratives, and integrating multimedia elements (Lima et al., 2022). Characters in the DEER can serve as guides, adversaries, or companions, adding depth to the story and facilitating player engagement (Shoulson, 2021). Branching narratives, where players' choices influence the direction and outcome of the story, enhance interactivity and replayability (Lima et al., 2021). Incorporating multimedia elements like audio, video, and animations can bring the story to life, making the DEER more dynamic and engaging (Calvert and Hume, 2022). Examples of DEERs with effective narrative development include educational games that teach language skills through immersive storytelling (Malegiannaki et al., 2020). In such games, players might embark on adventures where they interact with characters in the target language, solving puzzles and challenges that require language comprehension and usage (Garrand, 2020). Another example is a DEER designed for corporate training, where the narrative involves solving a company crisis, requiring players to apply business concepts and teamwork skills (Mercado, 2023).

The narrative development in DEERs plays a pivotal role in enhancing the educational and entertainment value of the game (Huynh et al., 2020). By crafting engaging and coherent narratives that align with educational objectives and employing interactive storytelling techniques, DEER developers can create immersive and impactful learning experiences (Verswijvelen et al., 2020).

Elements of a Strong Narrative in DEERs

The crafting of a strong narrative in DEERs is fundamental to their success as educational tools (Verswijvelen et al., 2020). Key elements such as character development, engaging settings, coherent plots, thematic integration, and the balance between educational content and storytelling play pivotal roles in enhancing the learning experience (Luhova, 2020). Developing characters, settings, and plots that resonate with the students involves creating a narrative that is relatable and engaging (Saripudin et al., 2021). Characters should be well-defined with distinct personalities and roles that contribute to the storyline (Halimah et al., 2020). For instance, a character in a DEER designed for a biology course could be a scientist who guides students through a series of experiments (Lima et al., 2022). Settings should be immersive and relevant to the educational content. For example, a DEER teaching history could be set in a historically significant period or location, providing a rich backdrop for the learning activities (Malegiannaki et al., 2020). The plot should be compelling and logically structured, with a clear beginning, middle, and end, guiding students through the educational journey (Calvert and Hume, 2022).

Incorporating themes and messages that reinforce the learning objectives of DEERs is crucial (Russell, 2020). The narrative should be designed in a way that the central themes align with the educational goals (Friedmann, 2021). For instance, a DEER focused on environmental science might include themes of sustainability and conservation (Kanareiko, 2023). This thematic integration ensures that the storytelling serves not just as entertainment but as a vehicle for conveying important educational messages and concepts (Adwiah et al., 2023).

Balancing educational content with storytelling is essential to maintain learner interest (Lima et al., 2021). The narrative should not overshadow the educational content, nor should the educational elements feel forced or out of place within the story (Saripudin et al., 2021). Achieving this balance requires careful planning and creativity, ensuring that the educational objectives are woven seamlessly into the narrative (Palioura and Dimoulas, 2022). Interactive elements, such as decision-making points or problem-solving tasks, can be used to blend the educational content with the storyline effectively (Halimah et al., 2020).

The elements of a strong narrative in DEERs – character development, engaging settings, coherent plots, thematic integration, and the balance between educational content and storytelling (Table 3) – are essential for creating an effective and engaging learning experience. By including these elements, DEER developers can craft narratives that not only entertain but also educate, providing students with an immersive and impactful educational journey.

Storyboarding Techniques for DEERs

Storyboarding is an essential technique in the development of DEERs, as it helps in visualizing the narrative, mapping out puzzles and challenges, and planning the player's journey. This process involves creating a visual representation of the game's flow, which includes key scenes, puzzles, and interactions, thereby ensuring a cohesive and engaging experience for the player. The process of visualizing the DEER narrative through storyboarding starts with outlining the storyline (Figure 1). This could involve sketching scenes, dialogues, and key interactions that drive the narrative forward. The storyboard acts as a blueprint, helping designers and developers visualize how the story unfolds and how it is integrated with the educational content. For example, a storyboard for a history-themed DEER might include scenes depicting historical events, with puzzles that require players to interact with these events to proceed (Halimah et al., 2020).

Table 3. Elements of a strong narrative in DEERs

Element	Description
Character Development	Creating well-defined characters with distinct personalities that contribute to the storyline, making the narrative relatable and engaging.
Engaging Settings	Designing immersive environments relevant to the educational content, such as historical periods or scientifically significant locations, to enhance the learning experience.
Coherent Plots	Developing a compelling and logically structured storyline with a clear beginning, middle, and end that guides students through the educational journey.
Thematic Integration	Ensuring the narrative themes align with the educational objectives, such as sustainability or historical accuracy, to reinforce learning goals.
Balance Between Education and Storytelling	Blending educational content seamlessly with the narrative, ensuring that neither overshadows the other, to maintain learner interest and achieve educational outcomes.

Mapping out the sequence of puzzles and challenges within the storyline is another crucial aspect of storyboarding. This involves deciding where and how puzzles are introduced in the narrative, ensuring they are logically placed and contribute to the progression of the story. The puzzles should be designed to gradually increase in complexity, aligning with the rising action in the storyline (Halimah et al., 2020). This approach ensures that the player remains engaged and challenged throughout the game (Arnold et al., 2013). Using storyboarding to plan the player's journey and interactions in the DEER is vital for creating an immersive experience (Arnold et al., 2013). It allows designers to anticipate player actions and reactions, and plan accordingly. For instance, the storyboard might include branching paths based on player choices, with different outcomes and educational points. This level of planning ensures a dynamic and interactive learning experience, where the player's decisions impact their journey through the game (Halimah et al., 2020). Examples of effective storyboarding in DEERs include educational games that teach scientific concepts through an adventure storyline (Arnold et al., 2013). In such a game, the storyboard might depict a journey through different ecosystems, with puzzles related to environmental science. Another example could be a language-learning DEER, where the storyboard outlines various social interactions and dialogues, requiring the player to use language skills to navigate through the game (Halimah et al., 2020).

Storyboarding is a critical technique in game development (P. Jantke and Spundflasch, 2013). It enables the effective visualization of the narrative, organization of puzzles and challenges, and planning of the player's journey (Table 4). By employing storyboarding, developers can create DEERs that are not only educational but also narratively rich and engaging, offering players an immersive learning experience (Arnold et al., 2013).

Figure 1. Storyboarding techniques for DEERs

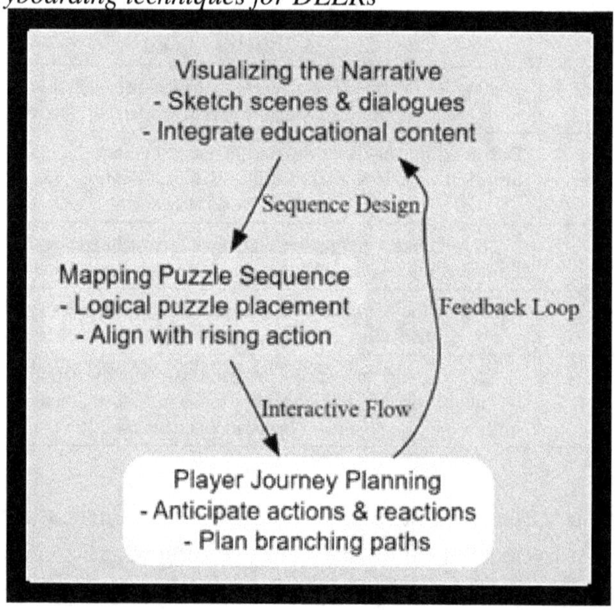

Table 4. Storyboarding techniques for DEERs

Technique	Description
Visualizing the Narrative	Outlining the storyline through sketches, dialogues, and key interactions to help visualize how the narrative unfolds and integrates with educational content.
Mapping Puzzle Sequence	Planning the placement and progression of puzzles within the narrative, ensuring they align with the storyline's rising action and maintain player engagement.
Player Journey Planning	Using storyboards to anticipate player actions, reactions, and choices, ensuring a dynamic and interactive learning experience. This includes planning branching paths and different outcomes based on player decisions.

Integrating Story and Puzzles in DEER Coherently

The integration of story and puzzles in DEERs is essential for creating an immersive and coherent learning experience. This integration ensures that puzzles and tasks not only challenge the player but also contribute to the narrative, making the educational content more engaging and memorable. Ensuring puzzles and tasks naturally fit within the narrative of DEERs requires careful planning. Each puzzle should feel like a natural part of the story, contributing to the progression of the narrative (Breien and Wasson, 2020). For example, in a DEER themed around a historical adventure, puzzles could involve deciphering ancient scripts or solving riddles based on historical facts (Foster et al., 2024). This approach helps maintain the thematic consistency of the DEER and ensures that the educational content is integrated seamlessly into the gameplay (Knox, 2023).

Using story elements to guide puzzle design and flow is another critical aspect. The storyline can provide context for the puzzles, setting up scenarios where the player must use specific knowledge or skills to progress. This method also helps in pacing the game, ensuring that puzzles are introduced at appropriate points in the story to maintain engagement and interest (Knox, 2023). For instance, in a DEER designed to teach scientific concepts, the story could lead players to a laboratory where they must conduct experiments to uncover the next part of the narrative (Schimpf et al., 2023).

Techniques for making the narrative and puzzles interdependent involve intertwining the storyline and the challenges so that one drives the other (Breien and Wasson, 2020). Narrative-driven puzzles, where the story unfolds as players solve the puzzles, can create a sense of accomplishment and curiosity (Knox, 2023). Conversely, puzzle-driven narratives, where the story is shaped by the outcomes of the puzzles, can provide a dynamic and interactive storytelling experience (Polydorou, 2024). Examples of DEERs that effectively integrate story and puzzles include educational games developed for sustainable development education (Lathwesen and Eilks, 2024; Vlachopoulos, 2021; Ouariachi and Wim, 2020). These puzzles are embedded within a narrative that educates students about sustainable practices, making the learning process engaging and contextually rich. The coherent integration of story and puzzles in DEERs is crucial for creating an engaging and effective educational experience. By ensuring that puzzles fit naturally within the narrative, using story elements to guide puzzle design, and making narrative and puzzles interdependent, DEER developers can craft games that are not only entertaining but also educational (Knox, 2023).

Leveraging Technology in Narrative Development of DEERs

In DEERs, leveraging technology in narrative development has become increasingly significant (Khundam, 2020). This integration encompasses a range of digital tools and software for storyboarding and narrative design, the implementation of augmented reality (AR) and virtual reality (VR) elements, and the exploration of interactive storytelling platforms (Huerta-Cardoso et al., 2023). These technological advancements contribute to creating more immersive, engaging, and educationally effective DEER experiences (Ticala et al., 2020).

Utilizing digital tools and software for storyboarding and narrative design in DEERs enables creators to visualize and plan the storyline, character arcs, and puzzle integration in a more organized and efficient manner (Talbot et al., 2020). Software like Storyboard That or Adobe Creative Cloud tools can be used to create visual representations of the DEER narrative, allowing designers to map out the sequence of events and how puzzles fit into the story (Huerta-Cardoso et al., 2023). This approach helps in ensuring that the narrative flows logically and that educational content is effectively integrated into the storyline (Talbot et al., 2020).

Implementing AR and VR elements to enhance the storytelling experience in DEERs opens up new dimensions of interactivity and immersion (Skult and Smed, 2020). AR technology can be used to overlay digital information onto the real world, providing an engaging way to present educational content and puzzles (Reeves et al., 2021). For instance, an AR-powered DEER might involve students scanning real-world objects with their devices to reveal hidden information or puzzles (Khundam, 2020). VR takes this a step further by immersing players in a completely digital environment, offering a more intense and engaging experience (Vemula, 2021). VR-driven DEERs can transport players to different settings, such as historical sites or scientific labs, where they interact with the environment to solve puzzles and learn (Zhang et al., 2021).

Exploring interactive storytelling platforms for DEER development is another avenue where technology plays a crucial role (Skult and Smed, 2020). Platforms like Twine or Inklewriter allow for the creation of branching narratives, where player choices influence the direction and outcome of the story (Khundam, 2020). This interactive element adds a layer of engagement and personalization to the learning experience, as students actively participate in shaping the narrative (Bruckner et al., 2022). Examples of leveraging technology in DEER narrative development include projects like "The Limits of Immersion: Case Study of Desolatium," a VR graphic adventure, which explores specific technical and narrative strategies to enhance player immersion in a digital environment (Parra-Martínez, 2023).

Leveraging technology in the narrative development of DEERs is essential for creating rich, immersive educational experiences (Khundam, 2020). By using digital tools for storyboarding, implementing AR and VR elements, and exploring interactive storytelling platforms, DEER developers can craft narratives that are not only engaging but also effective in achieving educational objectives (Reeves et al., 2021).

Collaborative Story Development for DEERs

Collaborative story development is a crucial aspect in the creation of DEERs (Figure 2). Involving a diverse team of educators, game designers, and students in the narrative development process ensures a rich and multifaceted storyline that aligns with educational goals and resonates with the target audience (Puga, 2022). Involving a team in the narrative development process brings together different perspectives and expertise (Grabska-Gradzinska et al., 2020). Educators bring an understanding of curriculum and learning objectives, ensuring that the narrative aligns with educational goals (Coughlan and Lister, 2022). Game designers contribute expertise in game mechanics and player engagement, ensuring the story is engaging and interactive (Cardoso et al., 2020). Including students in the process provides insights into what is appealing and relevant to the target audience, making the narrative more relatable and effective (Puga, 2022).

Gathering input from educators, game designers, and students is a collaborative effort that requires open communication and respect for each member's expertise (Grabska-Gradzinska et al., 2020). Educators can provide feedback on the educational content, ensuring it is accurate and appropriate for the learning outcomes (Malegiannaki et al., 2020). Game designers can suggest ways to integrate this content into the game mechanics seamlessly (Recke and Perna, 2021). Students can offer insights into what makes a story engaging and suggest ideas that could enhance the narrative (Puga, 2022).

Techniques for collaborative brainstorming and ideation are essential for successful story development (Grabska-Gradzinska et al., 2020). Techniques like mind mapping, where ideas are visually organized, can help in generating and connecting concepts (Kilanioti, 2022). Workshops or creative sessions, where team members can freely share and develop ideas, are also beneficial (Dimitriadou et al., 2020). These techniques encourage creative thinking and allow for the exploration of various narrative possibilities (Recke and Perna, 2021). Examples of DEERs that have successfully employed collaborative story development include projects where teachers, game developers, and students have worked together to create an educational escape room (Grabska-Gradzinska et al., 2020). For instance, a DEER teaching environmental science might involve a collaborative effort where educators provide content expertise, game designers develop engaging puzzles, and students suggest scenarios that make the story more engaging and relevant to their peers (Caeiro et al., 2021).

Collaborative story development is a key factor in creating effective and engaging DEERs (Puga, 2022). By involving a diverse team of educators, game designers, and students in the narrative development process and employing collaborative brainstorming and ideation techniques, DEER developers can create rich, multifaceted narratives that effectively integrate educational content and resonate with the target audience (Grabska-Gradzinska et al., 2020).

Adapting Narratives for Diverse Learners of DEERs

DEERs have become a popular tool for immersive learning experiences. However, creating narratives that resonate with diverse audiences across different cultural and educational contexts presents unique challenges (Malegiannaki et al., 2020). Adapting DEER narratives requires a nuanced understanding of cultural sensitivities, language use, thematic appropriateness, and the localization of content for global audiences (Dutra et al., 2023).

One of the essential aspects of adapting DEER narratives for diverse audiences is cultural sensitivity and relevance (Figure 3). It involves tailoring the story to suit various cultural backgrounds while ensuring that it remains engaging and educational (Guo, 2021). This approach was exemplified in a study Pirrone and D'Ulizia (2024), which explored the strategies of localization used by video game creators to promote their products worldwide. They emphasized the importance of cultural adaptation in storytelling, ensuring that narratives are relevant and respectful to different cultural backgrounds (Dutra et al., 2023).

Figure 2. Collaborative story development

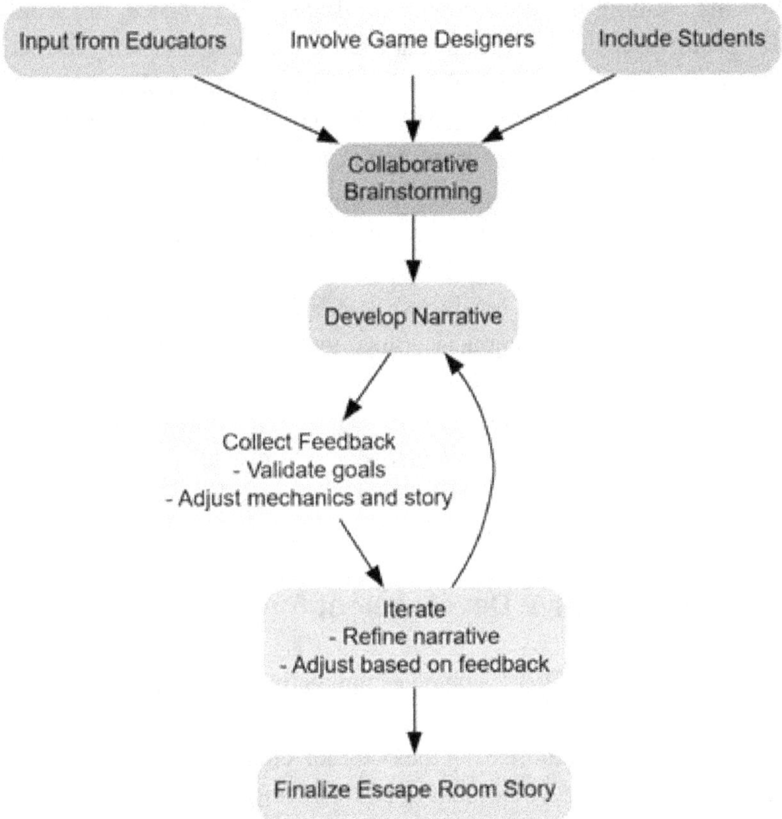

Language plays a crucial role in making DEERs inclusive and accessible (Chao, 2023). Adjusting the language to suit the audience's proficiency level and using clear, concise, and culturally appropriate terms can enhance the overall experience (Dutra et al., 2023). For instance, research on translating intertextuality in "The Witcher" game (Makuch, 2023) highlighted the significance of language adaptation in engaging players with immersive narratives and rich settings. This adaptation is critical as video games cross borders and reach various audiences, where the choice of language and its translation can impact the narrative's reception (Dutra et al., 2023).

In terms of thematic appropriateness, DEERs should offer content that is universally relatable yet adaptable to local narratives. This consideration is crucial in educational contexts, where the goal is to provide a learning experience that is both informative and engaging (Malegiannaki et al., 2020). Cesário et al.'s study Cesário et al. (2023) on "The Witcher III: Wild Hunt" is a pertinent example. It explored the intersection of storytelling, localization, and immersion, suggesting the need to explore the impact of localization on story and thematic elements. The study underscored the importance of adapting themes to fit different cultural contexts while maintaining narrative coherence. Examples of successful localization in DEER narratives include projects that have effectively adapted their stories to suit diverse educational and cultural settings (Dutra et al., 2023). For instance, a project that uses a popular local legend as the basis for an escape room narrative in one country could adapt the story to feature a well-known historical event or a universally recognized scientific concept in another. This approach ensures that the DEER is culturally relevant and educationally valuable, regardless of the audience's geographic location (Dutra et al., 2023).

Adapting DEER narratives for diverse audiences is a multifaceted process involving cultural sensitivity, language adjustment, thematic appropriateness, and localization (Guo, 2021). It requires a deep understanding of the target audience's cultural and educational background to create narratives that are not only engaging and informative but also respectful and inclusive (Dutra et al., 2023). By considering these factors, DEER developers can create experiences that resonate with a global audience, making learning both enjoyable and impactful (Malegiannaki et al., 2020).

Challenges in Narrative Development for DEERs

Creating narratives for DEERs involves navigating a complex landscape where educational objectives, entertainment value, and technical constraints intersect (Veldkamp et al., 2020c). Addressing the common challenges in crafting these narratives is essential for the development of effective and engaging DEERs (Klamma et al., 2020).

One of the primary challenges in DEER narrative development is balancing educational aims with entertainment value (Klamma et al., 2020). The narrative must be engaging enough to captivate participants while simultaneously delivering educational content (Borrás-Gené et al., 2022). This balancing act can be difficult, as too much focus on educational aspects may reduce the entertainment value, while an overly entertaining narrative might dilute the educational impact (Ross and Bennett, 2022). Meletiadou's study on applying digital escape rooms infused with cross-cultural management theories in higher education highlights this challenge. It underscores the need to blend storytelling with educational content in a manner that maintains student engagement and enhances academic performance (Meletiadou, 2023).

Figure 3. Adapting narratives for diverse learners

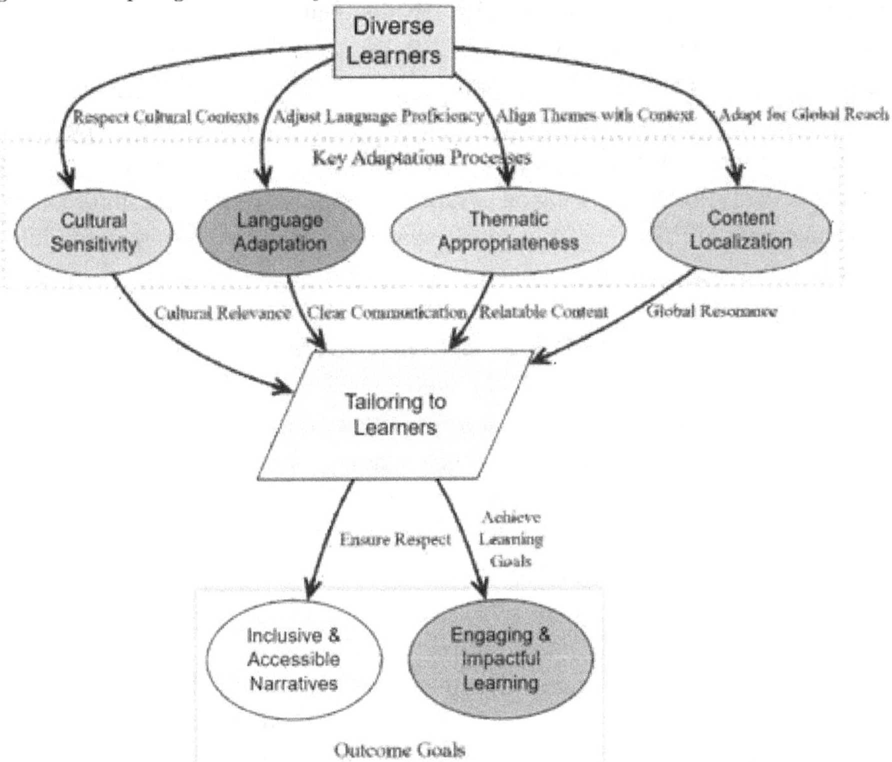

Another significant challenge is overcoming technical limitations in narrative execution (Videnovik et al., 2022). The use of digital platforms imposes certain constraints on how narratives can be presented and interacted with (Veldkamp et al., 2020c). These limitations might include the complexity of puzzles, the extent of interactivity, or the integration of multimedia elements (Kotzebue et al., 2022). (Minor et al., 2024) discuss the technical challenges of creating 3D models and location-based games for an escape room that teaches ancient Egyptian religious beliefs. The study illustrates the importance of balancing technical feasibility with narrative and educational goals.

To address these challenges, developers of DEERs can draw lessons from real-world examples where successful balance and technical adaptation have been achieved (Moffett and Cassidy, 2023). For instance, a DEER designed for a museum setting might focus on creating a narrative that is historically accurate and educational but also incorporates elements of mystery and exploration to keep visitors engaged (Videnovik et al., 2022). Similarly, in an academic setting, a DEER could use a fictional storyline relevant to the course content, ensuring that the puzzles and tasks are directly tied to the learning objectives (Najm et al., 2022).

The development of narratives for DEERs presents several challenges, notably balancing educational content with entertainment and overcoming technical limitations (Veldkamp et al., 2020c). Successful DEERs address these challenges by crafting narratives that are engaging, educationally relevant, and technically feasible (Videnovik et al., 2022). Drawing on best practices and real-world examples, developers can create DEERs that offer immersive and impactful learning experiences (Moffett and Cassidy, 2023).

Case Studies: Successful DEER Narratives

The concept of DEERs has been gaining traction in educational settings, offering an innovative approach to learning through engaging, narrative-driven experiences. Analyzing successful DEER narratives provides valuable insights into the effective integration of storytelling and educational content. This analysis focuses on various case studies that highlight the successful application of DEERs in different educational contexts.

One notable example is the use of DEERs in teaching Ancient Egypt in museums. In "From Glyphs to Bytes: Ancient Egypt and the Future of Digital Humanities in Museum-Based Learning" (Minor et al., 2024), the authors discuss the application of digital humanities best practices in creating immersive interactives. This includes a DEER that builds stories around artifacts, allowing visitors to engage with history in a more interactive and meaningful way. This approach enhances the museum experience by blending education with entertainment, making learning more engaging and memorable.

Another case study is the integration of DEERs in higher education, as explored in "Crossword Puzzle Games, Short Stories, and Mind Maps Assignments as Innovative Online Teaching Methods" (Yousof, 2023). This study highlights the use of DEERs as part of innovative online teaching methods, where students engage in interactive learning activities. The DEER format encourages students to apply critical thinking and problem-solving skills, thereby enhancing their learning outcomes (Yousof, 2023).

A microlearning approach using DEERs is detailed in "A MICROLEARNING APPROACH BY THE #ESTUDOEMCASA APOIA PROJECT" (Torres et al., 2023). This approach involves using a narrative (storytelling) that allows students to study learning sequences through sequential digital resources. Such an approach demonstrates the effectiveness of using storytelling and microlearning strategies in educational escape rooms, providing a compact yet impactful learning experience.

"Maruxaina: The Sea Nymph Who Will Be Judged" (Lopez-Bermudez and Martinez, 2023) presented a DEER based on a popular legend, designed to enhance student engagement. The success of this escape room among students highlights the effectiveness of using culturally relevant stories to create immersive and educational experiences. This case study emphasizes the importance of cultural relevance in educational narratives, which can significantly increase student engagement and interest.

From these case studies, several lessons can be learned about effective storytelling in educational escape rooms:

- Integration of educational content with narrative: Successful DEERs seamlessly integrate educational content within the narrative. This integration ensures that the learning objectives are met without compromising the entertainment value of the experience.
- Cultural relevance and engagement: Using culturally relevant narratives, as seen in the Maruxaina case study, can significantly enhance engagement and interest among students, making the learning experience more relatable and impactful.
- Interactivity and immersion: The use of interactive elements, as demonstrated in the Ancient Egypt case study, plays a crucial role in immersion. This approach fosters a deeper connection between the learner and the content.
- Adaptability in various educational settings: DEERs have shown adaptability across different educational settings, from museums to higher education, indicating their versatility as a teaching tool.
- Use of Microlearning and sequential learning: The incorporation of microlearning techniques (Damaševičius and Sidekerskienė, 2023), as seen in the #ESTUDOEMCASA APOIA project, highlights the effectiveness of DEERs in delivering compact, focused educational experiences.

Successful DEER narratives demonstrate the power of storytelling in creating engaging and effective educational experiences. By blending educational content with interactive storytelling, DEERs offer a unique and impactful way to enhance learning across various settings.

Future Trends in DEER Storytelling

The future of DEERs storytelling is poised for significant transformation, driven by technological advancements and evolving pedagogical approaches. One of the emerging trends is the integration of adaptive narratives, where the storyline dynamically adjusts based on the player's decisions and progress (Horváth, 2021).

This creates a more personalized learning experience, allowing educators to cater to individual learning paces and styles. The use of artificial intelligence (AI) in narrative development is anticipated to grow, with AI-driven characters and scenarios that respond to player input in real-time, enhancing the interactivity and immersion of the DEER experience (Yin, 2022).

Another trend is the increased use of augmented reality (AR) and virtual reality (VR) to create more immersive storytelling environments. These technologies allow for the creation of rich, multi-sensory experiences that can transport players to different worlds or historical settings, making the learning experience more engaging and memorable (Xanthidis et al., 2020). As these technologies become more accessible, they are expected to play a central role in the design of DEERs, enabling more complex and visually stunning narratives (Mystakidis et al., 2021).

There is a growing emphasis on the development of collaborative and multi-user narratives in DEERs. These narratives enable multiple players to interact and work together to solve challenges, promoting teamwork and communication skills (Sereno et al., 2020). The integration of social and collaborative elements into the narrative structure of DEERs reflects the broader trend towards more interactive and community-oriented learning experiences (Harley et al., 2022).

Finally, the importance of culturally relevant storytelling is becoming increasingly recognized in the design of DEERs. As educational tools are used in diverse global contexts, there is a need for narratives that resonate with different cultural backgrounds and experiences (Makri et al., 2021). This involves not only localizing content but also incorporating themes and stories that are meaningful and relevant to diverse audiences. The future of DEER storytelling will likely see a greater emphasis on inclusivity and cultural sensitivity, ensuring that the educational experiences are accessible and engaging for all learners (Lopez-Bermudez and Martinez, 2023).

PLAYTESTING AND ITERATIVE DESIGN OF DEERS

Understanding the Importance of Playtesting in DEER Design

Playtesting is a critical component in the development of DEERs, serving as a vital process for refining and enhancing the game before its final release. The primary role of playtesting in DEER design is to identify areas for improvement, ensuring that the game is both engaging and educationally effective (Moffett et al., 2023a). Through playtesting, developers can observe how real users interact with the game, revealing any issues related to gameplay mechanics, narrative flow, or the integration of educational content (Kotzebue et al., 2022). This feedback is invaluable, as it allows developers to make informed adjustments that enhance the overall user experience and ensure that the educational objectives are met (Terlouw et al., 2021).

Effective playtesting involves a variety of methods that provide comprehensive insights into how a DEER performs. One common approach is to conduct iterative playtesting sessions, where the game is tested multiple times throughout its development (Najm et al., 2022). Each session allows for incremental improvements based on player feedback, gradually refining the game (Videnovik et al., 2022). These sessions can range from informal in-house testing with team members to more formal sessions with external participants who represent the target audience (Shin et al., 2020).

Another important method for conducting effective playtesting is the use of diverse testing environments. By testing the DEER in different settings—such as classrooms, online platforms, or controlled environments—developers can observe how the game performs under various conditions (Reed and Ferdig, 2021). This helps to identify any technical issues that may arise, such as platform compatibility or network reliability, and ensures that the game is accessible and functional in the intended deployment context (Huraj et al., 2022).

It is crucial to gather qualitative and quantitative data during playtesting. Qualitative data, such as player feedback through interviews or surveys, provides insights into the players' emotional responses, engagement levels, and perceived educational value (Borrás-Gené et al., 2022). Quantitative data, on the other hand, involves metrics such as time spent on tasks, completion rates, and error frequencies, which offer a more objective view of how well the game functions (Mijal et al., 2020). Combining these data types gives a holistic understanding of the DEER's effectiveness (Kuo et al., 2022).

Ultimately, playtesting is an iterative and feedback-driven process that is essential for the success of DEERs. By rigorously testing the game in various stages of development and across different environments, developers can ensure that the final product is both educationally robust and engaging, providing an optimal learning experience for users (Sahin, 2023).

Setting Up Playtesting Environments

Creating an effective playtesting environment is a crucial step in the development of DEERs. The environment in which playtesting occurs can significantly impact the quality and relevance of the feedback received, thus influencing the overall success of the game design (Terlouw et al., 2021). Whether conducted in-person or online, the playtesting environment must be carefully designed to simulate real-world conditions as closely as possible, ensuring that the feedback is reflective of the actual user experience (Huraj et al., 2022).

For in-person playtesting, it is important to set up a controlled environment that minimizes distractions and allows participants to focus fully on the DEER (Reed and Ferdig, 2021). This environment should be equipped with all the necessary tools and resources, such as computers, VR headsets, or other relevant hardware, to mirror the intended deployment scenario (Kotzebue et al., 2022). In-person playtesting also benefits from the presence of facilitators who can observe participants directly, noting their interactions, behaviors, and any difficulties they encounter. This direct observation can provide valuable insights that might not be captured through other means (Moffett et al., 2023a).

Online playtesting, on the other hand, requires a different approach. Since many DEERs are designed for digital platforms, testing in an online environment is essential to understand how the game performs across different devices and network conditions (Videnovik et al., 2022). Online playtesting environments should be set up to accommodate a variety of platforms, including desktop computers, tablets, and smartphones, as well as different internet speeds and connection qualities (Najm et al., 2022). This diversity ensures that the DEER is accessible and functional for all users, regardless of their technical setup (Borrás-Gené et al., 2022).

Selecting appropriate playtesters is another critical aspect of setting up a successful playtesting environment (Sahin, 2023). The playtesters should represent the target audience as closely as possible, encompassing a range of demographics, skill levels, and learning styles (Kuo et al., 2022). For educational escape rooms, this often means recruiting participants who are similar to the intended learners in terms of age, educational background, and familiarity with digital games. This alignment ensures that the feedback is relevant and that the DEER is tailored to meet the needs of its actual users (Mijal et al., 2020).

In addition to demographic considerations, it is important to select playtesters who can provide constructive feedback (Terlouw et al., 2021). Experienced gamers may offer insights into the game mechanics and usability, while educators can provide feedback on the educational content and its integration into the gameplay (Moffett et al., 2023a). By involving a diverse group of playtesters, developers can gather a wide range of perspectives, helping to identify both strengths and areas for improvement (Borrás-Gené et al., 2022).

Collecting and Analyzing Feedback

Collecting and analyzing feedback is a critical step in the iterative design process of DEERs. Gathering meaningful feedback from playtesters requires a strategic approach, combining various techniques to ensure that the data collected is comprehensive and actionable. Effective feedback collection begins with clear objectives—developers must decide what specific aspects of the game they want to evaluate, whether it be user engagement, educational effectiveness, or the usability of the game interface (Grönberg et al., 2021).

One effective technique for gathering feedback is through direct observation during playtesting sessions. By watching how playtesters interact with the DEER, developers can identify any difficulties or frustrations that arise, such as challenges with navigation or confusion about instructions. This real-time observation provides immediate insights into potential problem areas that might not be captured through surveys or questionnaires alone (Blahova and Kostolny, 2023).

Surveys and questionnaires are also invaluable tools for collecting feedback, particularly when playtesting is conducted online or when large groups of participants are involved. These tools can be designed to capture both quantitative and qualitative data, allowing developers to measure aspects such as user satisfaction, perceived difficulty, and overall enjoyment. Open-ended questions in surveys can elicit detailed responses from playtesters, providing deeper insights into their experiences and suggestions for improvement (Huang et al., 2020).

Focus groups and interviews are another technique that can be used to gather more nuanced feedback. These methods allow developers to engage in in-depth discussions with playtesters, exploring their thoughts and feelings about the game in a more conversational setting. This can reveal insights into the emotional and cognitive responses of players, which are often key to understanding how well the educational objectives are being met (Videnovik et al., 2022).

Once the feedback is collected, the next crucial step is analyzing it to identify patterns and areas requiring modification. Quantitative data, such as scores and completion rates, can be analyzed statistically to determine trends and common issues. For example, if a large percentage of playtesters consistently struggle with a particular puzzle, this might indicate that the puzzle is too difficult or unclear, requiring adjustment (Grönberg et al., 2020). On the other hand, if players spend more time than expected on certain tasks, this might suggest that these areas are particularly engaging or challenging (Makri et al., 2021).

Qualitative data, such as comments from surveys or insights from interviews, should be carefully reviewed to identify recurring themes or concerns. Developers can categorize this feedback to highlight the most frequently mentioned issues or suggestions. For example, if multiple playtesters mention that a certain aspect of the narrative is confusing or that the educational content is not well integrated with the gameplay, these are clear indicators that these areas need refinement (Ferrari et al., 2023). By systematically collecting and analyzing feedback, developers can gain a comprehensive understanding of how their DEER performs in practice. This process enables them to make targeted improvements that enhance the overall quality and effectiveness of the game. Ultimately, the goal of this feedback-driven approach is to create a DEER that is not only educationally robust but also engaging and enjoyable for its intended audience (Videnovik et al., 2022).

Iterative Design Process in DEERs

The iterative design process is a cornerstone of developing effective and engaging DEERs (Figure 4). At its core, iterative design involves the continuous cycle of designing, testing, analyzing, and refining a product. This approach is particularly well-suited to DEER development, where the dual goals of educational effectiveness and user engagement must be carefully balanced. The principles of iterative design emphasize the importance of flexibility, responsiveness to feedback, and gradual improvement, ensuring that the final product meets both educational objectives and user expectations (Terlouw et al., 2021).

Figure 4. Iterative design process in DEERs

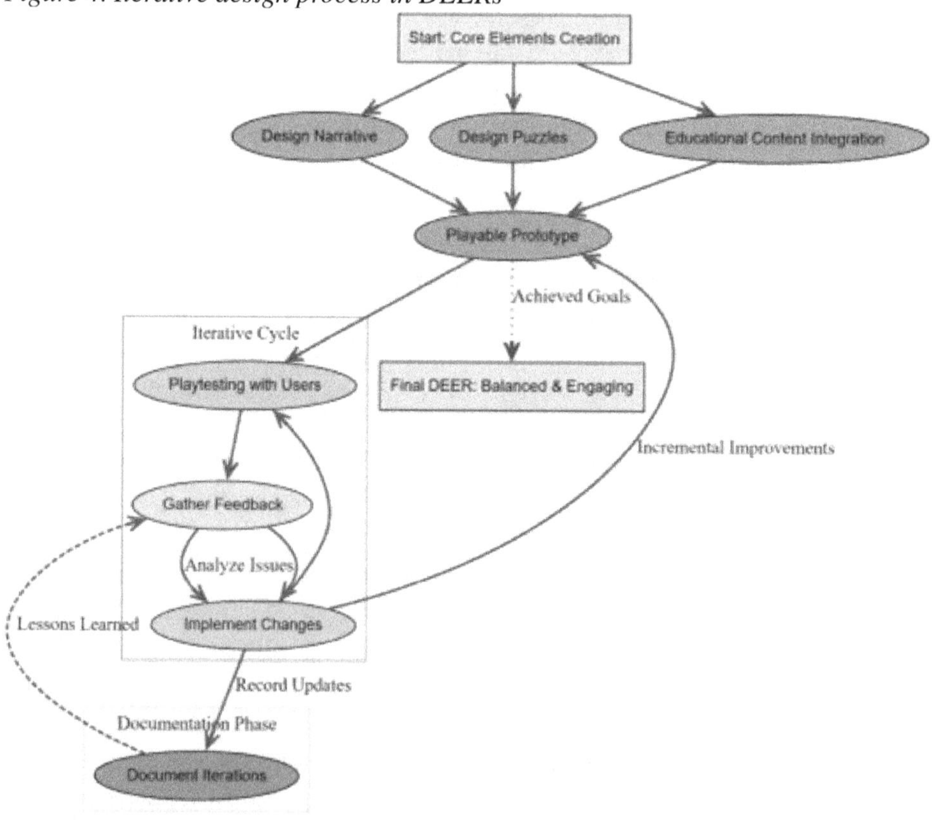

The iterative design process begins with the initial creation of the game's core elements, such as the narrative, puzzles, and educational content. These elements are then integrated into a playable prototype, which serves as the first version of the DEER. This prototype is subjected to playtesting, where real users interact with the game and provide feedback on their experience. The feedback gathered during this phase is critical, as it highlights areas where the game may not be functioning as intended or where the educational content is not effectively conveyed (Repetto et al., 2023). Based on the feedback from playtesting, developers proceed to the next step of the iterative design process: implementing changes. This may involve tweaking the difficulty of puzzles, refining the narrative to make it more engaging, or adjusting the way educational content is presented. The key to successful iterative design is to focus on incremental changes that address specific issues identified during playtesting. By making small, targeted modifications, developers can ensure that each iteration of the DEER is an improvement over the last (Eukel and Morrell, 2020).

After implementing these changes, the updated version of the DEER is subjected to further playtesting. This cycle of testing, feedback, and refinement is repeated multiple times, with each iteration bringing the game closer to its final form. One of the benefits of this approach is that it allows developers to remain agile and responsive to user needs. Instead of trying to create a perfect game from the outset, the iterative process embraces the idea that the best solutions often emerge through experimentation and continuous improvement (Viudes-Carbonell et al., 2021).

An essential aspect of the iterative design process in DEERs is the documentation of each iteration. Keeping detailed records of the changes made, the rationale behind them, and the outcomes of subsequent testing helps ensure that the development process is transparent and that lessons learned from each iteration can be applied to future projects. This documentation also facilitates communication among team members and stakeholders, ensuring that everyone involved in the project is aligned with the design goals and aware of the progress being made (Moffett and Cassidy, 2023).

Ultimately, the iterative design process in DEER development is about refining the game to achieve a balance between educational content and user engagement. By continuously testing and improving the DEER, developers can create a final product that is not only effective as a learning tool but also enjoyable and compelling for players. The iterative approach ensures that the DEER evolves in response to real user feedback, leading to a game that truly meets the needs of its intended audience (Repetto et al., 2023).

Balancing Educational Objectives and Engagement

Balancing educational objectives and user engagement is one of the most critical challenges in the development of DEERs. The primary goal of a DEER is to deliver educational content in a way that is both effective and enjoyable, but achieving this balance requires careful design and thoughtful integration of learning materials with gameplay (Rawlinson and Whitton, 2024). Ensuring that modifications enhance both the educational value and the user engagement of the DEER is essential for creating a product that fulfills its educational mission while also providing an immersive and enjoyable experience for players.

To maintain this balance, developers must first clearly define the educational objectives of the DEER. These objectives could range from teaching specific content knowledge, such as historical facts or scientific principles, to developing broader skills like critical thinking, problem-solving, and collaboration (Videnovik et al., 2022). Once these objectives are established, the challenge is to weave them into the game's narrative, puzzles, and activities in a way that feels natural and engaging rather than forced or didactic. Every modification made during the iterative design process should be evaluated against these objectives to ensure that the educational goals remain at the forefront of the design.

One effective strategy for maintaining the balance between learning and gameplay is to integrate educational content directly into the game mechanics (Mozelius et al., 2023). For example, puzzles can be designed to require the application of specific knowledge or skills, making learning an inherent part of the gameplay experience. Instead of presenting educational content separately from the game, successful DEERs embed it within the challenges players must overcome, ensuring that learning is both contextual and interactive. This approach not only reinforces the educational objectives but also enhances engagement by making the content relevant to the player's progress within the game (Rawlinson and Whitton, 2024).

Another strategy is to use the narrative as a vehicle for delivering educational content (Ross et al., 2022). A compelling story can motivate players to engage deeply with the material, providing context and purpose to the challenges they face. By carefully crafting a narrative that aligns with the educational objectives, developers can create a more immersive experience that encourages players to explore and learn. For instance, a DEER designed to teach environmental science might place players in a scenario where they must solve puzzles to prevent an ecological disaster, thereby teaching key concepts through the storyline (Veldkamp et al., 2020c).

It is important to consider the pacing of the game to maintain both engagement and educational effectiveness (Veldkamp et al., 2020c). If the game is too focused on educational content, it may become tedious, leading to disengagement. If the game is too focused on entertainment, the educational value may be lost. Balancing the pacing involves ensuring that moments of intense learning are interspersed with periods of lighter, more relaxed gameplay (Videnovik et al., 2022). This helps maintain the player's interest and prevents cognitive overload, making the educational experience more enjoyable and sustainable over time.

Finally, developers should continually seek feedback from playtesters to assess how well the game is balancing educational objectives with engagement (Moffett et al., 2023b). This feedback can highlight areas where the game may be leaning too heavily in one direction and provide insights into how to adjust the gameplay or content delivery to achieve a better balance (Videnovik et al., 2022). Iterative testing and refinement are key to ensuring that the DEER remains both educationally valuable and engaging throughout its development (Veldkamp et al., 2020c).

Balancing educational objectives and user engagement in DEERs requires a careful integration of learning content with game mechanics and narrative (Videnovik et al., 2022). By embedding educational objectives into the gameplay, crafting compelling narratives, pacing the game appropriately, and continually refining based on feedback, developers can create DEERs that are not only effective learning tools but also immersive and enjoyable experiences for players.

Technological Adaptations and Upgrades

Continually updating DEERs with the latest technologies and tools is crucial for maintaining their relevance and effectiveness. As playtesting feedback provides insights into areas where the DEER can be improved, developers must be proactive in incorporating technological adaptations that enhance the user experience and educational value of the game. These updates can range from integrating new software and hardware advancements to refining existing systems to better meet the needs of the users. One of the primary drivers for technological upgrades in DEERs is the feedback gathered during playtesting sessions. Playtesters often reveal practical issues related to the usability, functionality, and overall experience of the DEER. For instance, if playtesters report that certain puzzles are too difficult to navigate due to interface limitations, developers might consider updating the user interface with more intuitive design elements or incorporating new input methods that make interactions smoother. Similarly, if players encounter performance issues, such as lag or slow loading times, upgrading the underlying technology stack—such as switching to a more robust game engine or optimizing the code—can significantly improve the gameplay experience.

In addition to responding to specific feedback, developers must also stay abreast of emerging technologies that can be integrated into DEERs to enhance their educational impact. For example, the growing accessibility of augmented reality (AR) and virtual reality (VR) tools presents new opportunities for creating more immersive learning environments. By incorporating AR and VR into DEERs, developers can offer players richer, more interactive experiences that bring educational content to life in ways that were previously unimaginable. These technologies can be particularly effective in creating scenarios that simulate real-world environments or complex systems, allowing learners to explore and engage with the material in a hands-on manner.

As developers consider these technological upgrades, it is essential to keep accessibility and ease of use at the forefront of their decisions. New technologies, while potentially transformative, can also introduce barriers for some users, particularly those with disabilities or those who may not have access to high-end hardware. Ensuring that DEERs remain accessible to a broad audience involves adopting technologies that enhance the experience without excluding users. For example, while VR can offer incredible immersive experiences, not all users may have access to VR headsets. In such cases, providing alternative modes of interaction or ensuring that the DEER is fully functional on standard devices is crucial.

Ease of use is another critical factor when implementing technological upgrades. The goal of these adaptations should be to simplify and enhance the player's interaction with the DEER, not to complicate it. Developers must ensure that any new technologies are seamlessly integrated into the game and that they do not require users to undergo a steep learning curve. Clear instructions, intuitive design, and thoughtful implementation are key to making these upgrades successful.

The continuous adaptation and upgrading of DEERs with new technologies and tools are vital to keeping these educational platforms effective and engaging. By responding to playtesting feedback, integrating emerging technologies thoughtfully, and prioritizing accessibility and ease of use, developers can ensure that DEERs remain at the cutting edge of educational innovation while being inclusive and user-friendly for all learners.

Challenges in Playtesting and Iterative Design

The process of playtesting and iterative design is essential for the successful development of DEERs, but it is not without its challenges. Developers often encounter various obstacles that can impede the effectiveness of playtesting sessions and the overall iterative design process. Among the most common issues are limited resources, participant bias, scope creep, and time constraints. Overcoming these challenges requires careful planning, adaptability, and a clear focus on the project's educational and design goals.

One significant challenge in playtesting is the limitation of resources. Playtesting can be resource-intensive, requiring not only time and money but also access to appropriate participants, technology, and environments. For smaller teams or projects with constrained budgets, it may be difficult to conduct as many playtesting sessions as needed or to reach a sufficiently diverse group of participants. To mitigate this, developers can prioritize the most critical aspects of the DEER for testing, focusing on areas where feedback will have the most significant impact. Leveraging online tools and platforms can help reach a broader audience at a lower cost, though care must be taken to ensure that the testing environment remains controlled and reflective of real-world conditions.

Participant bias is another common issue that can skew the results of playtesting. Biases can arise from participants' prior experiences with similar games, their familiarity with the content, or their relationship with the developers. To reduce bias, it is important to select a diverse group of participants that accurately represents the target audience of the DEER. This includes varying levels of experience with educational games, different age groups, and diverse backgrounds. Anonymizing feedback and encouraging honest, constructive criticism can help to ensure that the insights gained are as objective as possible.

In the context of iterative design, scope creep and time constraints are persistent challenges that can derail a project if not carefully managed. Scope creep occurs when additional features, content, or changes are introduced during the development process, often without proper consideration of their impact on the timeline and resources. While the iterative process naturally involves making improvements based on feedback, it is crucial to maintain a clear focus on the core objectives of the DEER. Developers should establish clear boundaries and priorities at the outset and resist the temptation to continually expand the scope of the project. Regularly revisiting the original goals and evaluating whether proposed changes align with these objectives can help to keep the project on track.

Time constraints are another significant challenge in iterative design, particularly when working with fixed deadlines, such as those imposed by educational calendars or funding requirements. The iterative process inherently involves multiple cycles of testing and refinement, which can be time-consuming. To manage time effectively, developers need to create a detailed project timeline that allocates sufficient time for each iteration while also allowing for unexpected delays. Prioritizing the most critical feedback for each iteration and setting realistic goals for what can be achieved within the available time are essential strategies for meeting deadlines without sacrificing quality.

The challenges of playtesting and iterative design in DEER development, such as limited resources, participant bias, scope creep, and time constraints, require careful management and strategic decision-making. By being mindful of these challenges and proactively addressing them, developers can ensure that the DEER remains focused on its educational objectives and is refined through a structured, efficient, and effective iterative process.

Documenting the Development Process

Documenting the development process in the creation of DEERs is a critical practice that ensures the project's success and longevity. The importance of thorough documentation cannot be overstated, as it serves multiple functions, including tracking progress, facilitating communication among team members, and providing a detailed record for future reference and continuous improvement.

Each iteration of the DEER development should be meticulously documented, capturing the changes made, the rationale behind those changes, and the outcomes of subsequent playtesting sessions. This documentation serves as a roadmap of the project's evolution, allowing developers to understand how and why the DEER has transformed over time. By recording the details of each modification—whether it involves adjustments to game mechanics, narrative elements, or educational content—developers can ensure that all team members are on the same page and that the project maintains a cohesive vision. This is particularly important in iterative design, where frequent adjustments are made based on user feedback. Without proper documentation, there is a risk of losing sight of the original goals or repeating past mistakes.

Documentation plays a crucial role in facilitating continuous improvement. By keeping a detailed record of what has been tried and tested, developers can learn from each iteration, understanding what worked well and what did not. This historical record becomes an invaluable resource when making decisions about future changes or when embarking on new DEER projects. For example, if a particular narrative structure proved effective in engaging users, that approach can be documented and replicated in future designs. Conversely, if certain puzzles were consistently problematic, the documentation can provide insights into how they were adjusted or why they were ultimately discarded.

Documentation also serves as a tool for knowledge transfer, particularly in projects involving large or dispersed teams. Team members may come and go, and having a well-documented development process ensures that new members can quickly get up to speed. It also allows for smoother collaboration, as detailed records of decisions and changes reduce the likelihood of misunderstandings or redundant work. Documented processes and outcomes can be used to create best practices for DEER development. Over time, as more projects are completed and documented, developers can compile a set of guidelines that outline effective strategies and common pitfalls. These best practices can streamline future projects, reduce development time, and improve the quality of DEERs from the outset. Documenting the development process in DEERs is essential for maintaining clarity, ensuring consistency, and promoting continuous improvement. By systematically recording each iteration, the changes made, and the lessons learned, developers can build a comprehensive knowledge base that supports both current and future projects. This practice not only enhances the quality of individual DEERs but also contributes to the overall advancement of educational game design.

Future Trends in Playtesting and Design

The future of playtesting and design in DEERs is set to evolve significantly as new technologies and educational paradigms emerge. Anticipating these developments is crucial for developers who aim to create DEERs that are not only engaging and effective but also aligned with the latest trends in education and technology.

One of the most promising future trends in playtesting methodologies is the increased use of artificial intelligence (AI). These technologies have the potential to revolutionize how feedback is collected and analyzed during playtesting. AI can be used to track player behavior in real-time, identifying patterns and issues that might not be immediately obvious through traditional playtesting methods. For example, AI could automatically detect when players are struggling with a particular puzzle or when they are losing interest, allowing developers to make data-driven adjustments to the game design. Machine learning algorithms can also analyze vast amounts of playtesting data, providing insights into player engagement and educational effectiveness that can guide future iterations.

The use of virtual and augmented reality (VR/AR) in playtesting is expected to become more prevalent. As these technologies become more accessible, they will allow developers to create more immersive playtesting environments that closely mimic the intended user experience. VR/AR can provide testers with a more realistic sense of the DEER's final environment, leading to more accurate feedback on how the game will perform in real-world settings. This will be particularly valuable for DEERs designed to teach skills that require spatial awareness or physical interaction, such as engineering concepts or medical procedures.

Another trend likely to shape the future of DEER design is the growing emphasis on personalized learning experiences. As educational tools increasingly cater to individual learning styles and needs, playtesting methodologies will need to adapt to assess how well DEERs support personalized learning paths. This may involve more sophisticated data collection techniques that can track how different types of learners interact with the game and how effectively the DEER adapts to their individual progress and preferences. Playtesting will need to focus not just on the average user experience but also on how the game performs for diverse learners, including those with different levels of prior knowledge, learning speeds, and accessibility requirements.

As educational needs continue to evolve, DEERs will also need to incorporate emerging pedagogical approaches, such as competency-based learning and social-emotional learning. These approaches emphasize the development of specific skills and the importance of students' emotional engagement with the material. Future playtesting will likely include assessments of how well DEERs foster these competencies and emotional connections, requiring new tools and metrics that go beyond traditional measures of engagement and learning outcomes.

Preparing for these technological and educational advancements will require developers to stay informed about the latest trends and to be proactive in incorporating new tools and methodologies into their design processes. This includes investing in AI and machine learning technologies, exploring the possibilities of VR/AR, and continuously updating playtesting protocols to reflect the changing landscape of education. By doing so, developers can ensure that their DEERs remain at the cutting edge of educational innovation, providing learners with experiences that are both effective and deeply engaging.

The future of playtesting and design in DEERs is poised for significant transformation as AI, VR/AR, and personalized learning technologies become more integrated into the development process. By anticipating these trends and preparing for evolving educational needs, developers can create DEERs that are not only relevant to today's learners but also adaptable to the educational challenges of the future.

STEP-BY-STEP GUIDE TO ESCAPE ROOM GAME CREATION

Understanding the Purpose and Setting Objectives

In the context of DEERs, understanding the purpose and setting clear objectives is fundamental to their design and effectiveness. This involves defining the educational goals, aligning game activities and puzzles with learning outcomes, and ensuring that the DEER serves as an engaging educational tool (Veldkamp et al., 2020c).

Defining the educational goals and objectives of the escape room requires a thorough understanding of what the DEER aims to achieve in terms of learning outcomes. These goals might range from acquiring specific knowledge, developing problem-solving skills, to fostering critical thinking or teamwork (Sidekerskiene and Damaševicius,ˇ 2023). For instance, a DEER focused on environmental science might aim to teach concepts of ecology and sustainability. The objectives should be specific, measurable, achievable, relevant, and time-bound (SMART), providing clear direction for the development of the DEER (Videnovik et al., 2022).

Figure 5. DEER design workflow

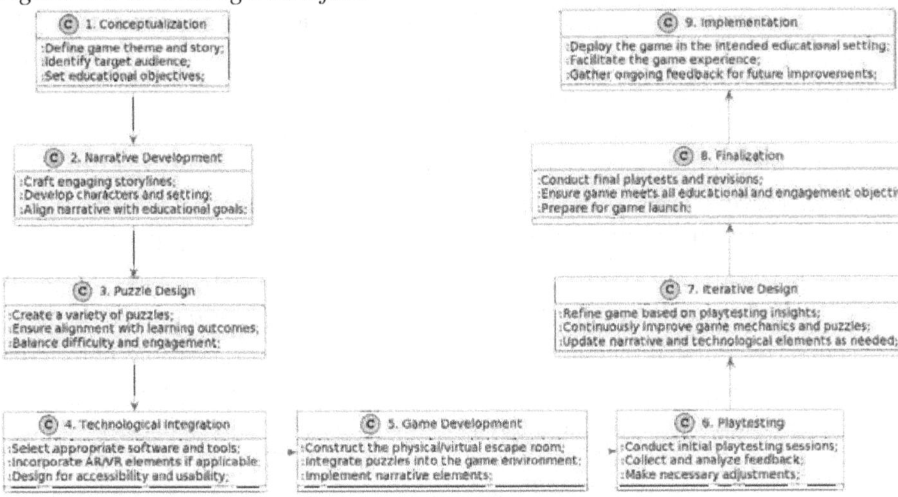

Aligning escape room game activities and puzzles with learning outcomes is crucial for ensuring that the DEER is educationally effective. Each puzzle or task within the escape room should contribute to the learning objectives (Hakshurian, 2023). This alignment means that solving puzzles not only progresses the game but also reinforces the educational content. For example, in a history-themed DEER, puzzles might involve analyzing historical sources or solving riddles based on historical events, thus promoting historical understanding and analytical skills (Mozelius et al., 2023). Examples of DEERs where educational goals are effectively integrated include a mental health nursing course where an escape room is used to teach specific nursing skills and concepts (Aragon, 2023). Another example is the design of an accessible VR-Escape Room for accessibility education, aiming to educate computer science students about digital accessibility, with puzzles and tasks aligned with this educational goal (Mateen et al., 2023).

Understanding the purpose and setting objectives in the development of DEERs are essential steps to ensure their success as educational tools. By defining clear educational goals and aligning game activities and puzzles with these objectives, DEER developers can create engaging, effective, and meaningful learning experiences for students.

Designing the Game Narrative and Theme of DEERs

The design of the game narrative and theme in DEERs plays a critical role in engaging learners and enhancing the educational experience (Fotaris and Mastoras, 2022). An effective DEER narrative not only captivates the participants but also complements and enriches the educational content (Repetto et al., 2023). Developing an engaging story or theme that captivates learners requires creativity and a deep understanding of the target audience (Moffett and Cassidy, 2023). The narrative should be compelling and immersive, drawing learners into the game world (Videnovik et al., 2022). This can be achieved by creating a storyline that resonates with the learners' interests or experiences (Ross and Bennett, 2022). For instance, a DEER designed for young learners might use a fantasy adventure theme, while one aimed at older students could involve solving a mystery in a historical setting (Kotzebue et al., 2022). The key is to create a narrative that is both appealing and relevant to the learners (Verswijvelen et al., 2020). Ensuring that the narrative complements and enhances the educational content involves integrating the learning objectives seamlessly into the storyline (Repetto et al., 2023). The narrative should not only serve as a backdrop but should be intertwined with the educational elements of the game (Veldkamp et al., 2020c). For example, in a DEER designed to teach environmental science, the storyline could revolve around solving an ecological crisis, with puzzles related to environmental concepts and sustainability. Examples of DEERs with effectively designed narratives and themes include:

- Mizuko Ito's Study on Children's Software: Mizuko Ito's research, highlighted in her book "Engineering Play: A Cultural History of Children's Software," (Ito, 2009) explores the intersection of young people, learning, and digital media. This research underscores the importance of narrative and thematic design in educational software, including DEERs, and how it influences children's engagement and learning experiences.
- A Historical Mystery DEER in Education: A real-world example of a DEER with an engaging narrative is one set in a historical context, where students play the role of detectives in outdoor setting. They unravel a mystery related to cultural heritage using their knowledge of history to solve puzzles and complete the game (Tzima et al., 2020). This thematic approach makes learning history interactive and exciting, enhancing students' understanding and retention of historical facts.

The design of the game narrative and theme in DEERs is a critical aspect that significantly impacts learners' engagement and the effectiveness of the educational experience. By developing captivating stories and themes that are seamlessly integrated with the educational content, DEER creators can provide learners with an immersive and enriching learning experience. These steps provide a comprehensive framework for educators and game designers to create escape room games that are not only fun and engaging but also serve educational purposes effectively. Each step emphasizes the importance of aligning game mechanics with learning objectives, ensuring player engagement, and continuously improving the game based on feedback and evaluation.

PRACTICAL RECOMMENDATIONS AND DESIGN GUIDELINES

Understanding the Educational Objectives

Clearly defining educational goals and aligning DEER activities with these objectives is fundamental for the effectiveness of these interactive learning tools. This alignment ensures that the DEER not only engages students but also meets specific educational outcomes (Veldkamp et al., 2020c). Defining educational goals involves identifying what students should learn or achieve through the DEER experience. These goals might include acquiring subject-specific knowledge, developing critical thinking skills, enhancing problem-solving abilities, or fostering collaborative skills (Borrás-Gené et al., 2022). For example, in a DEER designed for a science curriculum, the educational goals could include understanding scientific principles, applying them in problem-solving scenarios, and developing the ability to work effectively in teams (Videnovik et al., 2022). Aligning DEER activities and puzzles with learning outcomes requires integrating the educational content into the game mechanics seamlessly (Videnovik et al., 2022). This means that puzzles and challenges within the DEER should be directly related to the learning objectives (Veldkamp et al., 2020c). For instance, if the goal is to teach mathematical concepts, puzzles might involve solving equations or applying mathematical theories to progress in the game (Veldkamp et al., 2020a). This integration ensures that students are actively engaging with the educational material while navigating the DEER (Videnovik et al., 2022).

Specific real-world examples of DEERs where educational objectives are effectively integrated include:

- Accessible VR-Escape Room for Accessibility Education: aimed to educate computer science students about digital accessibility. The learning objectives focused on understanding the Web Content Accessibility Guidelines (WCAG), with puzzles based on these standards. This DEER effectively integrated educational goals into an immersive VR environment, enhancing the learning experience (Mateen et al., 2023).
- E-Learning Material for Computational Thinking: Kole's (Kole, 2023) work on developing e-learning materials for computational thinking in secondary education incorporated DEER elements. The puzzles in the escape room covered the learning objectives of computational thinking, aligning game activities with specific educational goals, like digital literacy and problem-solving.

The effective design of DEERs hinges on clearly defining educational objectives and aligning game activities with these goals. By ensuring that puzzles and tasks within the DEER contribute directly to the learning outcomes, educators can create engaging and educationally valuable experiences for students.

Incorporating Effective Game Mechanics in DEER

In the design of DEERs, incorporating effective game mechanics is essential for enhancing engagement and reinforcing learning. Utilizing elements like time constraints, narrative, feedback, puzzles, and challenges appropriately can significantly boost the educational impact of DEERs. Utilizing game elements like time constraints and narrative effectively in DEERs can heighten the sense of urgency and immersion. Time constraints add an element of challenge and excitement, prompting learners to engage more actively with the content. Meanwhile, a well-crafted narrative serves as the backbone of the DEER, providing context and motivation for players to engage with the educational content. For instance, a DEER themed around a historical adventure could use a narrative that transports players back in time, allowing them to experience historical events firsthand.

Designing puzzles and challenges that align with the target audience's skill level and educational background is crucial for ensuring that the DEER is both challenging and educationally effective. These puzzles should be tailored to the learning objectives and the capabilities of the learners. For example, a DEER designed for younger students might include more straightforward puzzles that focus on basic concepts, while one for higher education students could feature more complex problems requiring advanced critical thinking skills. Examples of DEERs where game mechanics are effectively integrated include projects like "Microgames and Language Learning: Performance Before Competence?" by de Castell et al. (De Castell et al., 2023), which discusses digital game-based interactions for language learning, emphasizing the role of game mechanics in improving understanding and engagement. Another example is the use of the falling-puzzle game "Puyo Puyo" in programming education for young people, as explored by (Tian et al., 2024). This study demonstrates how consumer digital game rules can be effectively applied in an educational context.

Incorporating effective game mechanics (Figure 6) in DEERs is vital for creating engaging and impactful learning experiences. By carefully utilizing time constraints, narrative, puzzles, and challenges, and aligning them with the learners' needs and educational objectives, DEER developers can craft experiences that are not only entertaining but also highly educational.

Figure 6. Effective game mechanics in DEER

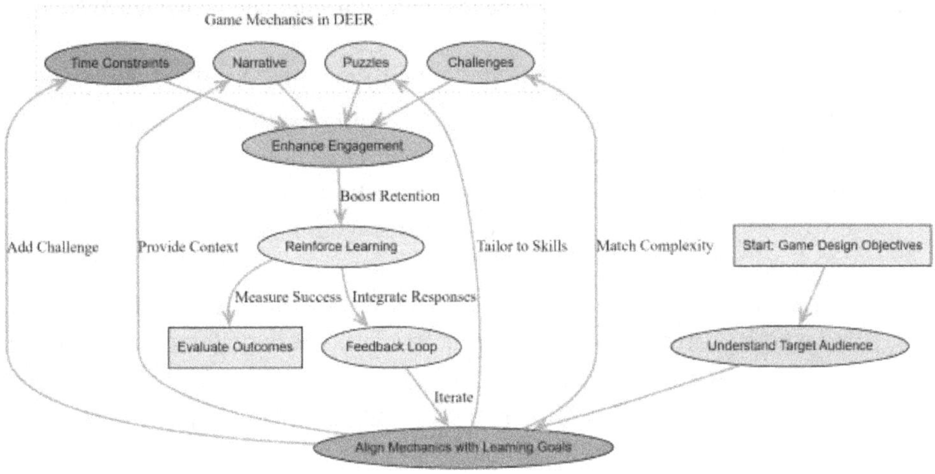

Balancing Entertainment and Education in DEERs

In DEERs, achieving a harmonious balance between entertainment and education is crucial. This balance ensures that while learners are engaged and entertained, they are also absorbing the educational content effectively (Veldkamp et al., 2020c). Striking this balance involves carefully designing the DEER experience to be fun and interactive while ensuring that the educational objectives are not overshadowed (Najm et al., 2022).

One key aspect of balancing entertainment and education is to integrate the educational content seamlessly into the narrative and mechanics of the escape room (Veldkamp et al., 2020c). The puzzles and tasks should be designed in a way that naturally incorporates learning objectives (Veldkamp et al., 2020c). For example, if the DEER aims to teach historical facts, the puzzles could involve deciphering codes or solving riddles based on historical events or figures (Veldkamp et al., 2020a). This approach allows learners to engage with the educational material in an enjoyable and interactive manner (Veldkamp et al., 2020c).

Avoiding overwhelming learners with too much information or overly complex puzzles is another important consideration (Veldkamp et al., 2020a). The challenge level of the puzzles should be appropriate for the target audience's age and educational background (Moffett and Cassidy, 2023). Too much complexity or information can lead to frustration, diminishing the enjoyment and educational value of the experience (Veldkamp et al., 2020c). Puzzles should be challenging enough to be engaging but not so difficult that they become a source of frustration (Veldkamp et al., 2020c). Examples of DEERs that successfully balance entertainment and education include:

- Science-Themed DEER for Middle School Students: A DEER designed to teach scientific concepts to middle school students utilized interactive puzzles related to basic physics and chemistry. The game was set in a fictional laboratory, where students had to apply scientific principles to escape, effectively blending educational content with an engaging storyline (Vergne et al., 2019).
- Historical Adventure in a Museum Setting: A museum created an AR based game where visitors explored different historical periods through a series of puzzles related to artifacts and historical events. This game combined the thrill of solving mysteries with learning about history, making the educational experience enjoyable and memorable (Paulauskas et al., 2023).

Balancing entertainment and education in DEERs is essential for maximizing their effectiveness as learning tools. By integrating educational content into engaging narratives and puzzles, and ensuring the challenges are appropriate for the learners, DEER developers can create experiences that are both educational and enjoyable.

Accessibility and Inclusivity

In the design and development of DEERs, accessibility and inclusivity are paramount to ensure that these educational tools are beneficial to a wide range of learners (Cezarotto and Armstrong, 2023). This involves creating DEERs that are accessible to individuals with various abilities and designing content that respects and represents diverse backgrounds and experiences (Salvador-Ullauri et al., 2020). Ensuring DEERs are accessible to a diverse range of learners, including those with different abilities, requires thoughtful design considerations (Terlouw et al., 2021). This includes implementing features like screen readers for visually impaired learners, subtitles for those with hearing impairments, and alternative control options for learners with physical disabilities (Cezarotto and Armstrong, 2023). For instance, a DEER designed for teaching language skills could include audio descriptions and sign language options to cater to visually and hearing-impaired students, respectively (Neto et al., 2019).

Designing inclusive content involves creating scenarios, characters, and narratives that reflect a wide range of cultures, experiences, and perspectives (Treglia et al., 2022). This inclusivity helps in fostering a sense of belonging and engagement among all learners (Veldkamp et al., 2020a). For example, a DEER focused on social studies could incorporate stories and characters from various cultural backgrounds, allowing students from different ethnicities to see themselves represented in the learning material (Sidekerskiene and Damaševičius, 2023). Examples of DEERs that have successfully incorporated accessibility and inclusivity include:

- A university developed a DEER to teach STEM subjects to students with disabilities. The DEER featured adaptive technologies, such as voice recognition software and adjustable difficulty levels, to ensure that students with various physical and cognitive abilities could participate and learn effectively (Martín-Gutiérrez et al., 2022).
- A museum created a history-themed DEER that allowed visitors to explore different historical periods and cultures. The DEER included narratives and puzzles related to diverse historical figures and events, ensuring that the content was inclusive and representative of various cultural backgrounds (The Quest (Back et al., 2019)).

Accessibility and inclusivity in DEERs are critical in creating educational experiences that are equitable, engaging, and effective for all learners. By focusing on these aspects, DEER developers can ensure that their creations are not only educational but also respectful and representative of the diverse society in which we live.

REFERENCES

Adwiah, A. R., Tantia, A. F., and Rantikassari, I. A. (2023). Implementation of storytelling method with folktales in instilling character values in children: A study at aba warungboto kindergarten. *Golden Age: Jurnal Ilmiah Tumbuh Kembang Anak Usia Dini*, DOI: .DOI: 10.14421/jga.2023.81-05

Akcaoglu, M., Jensen, L. J., and Gonzalez, D. (2021). Understanding children's problem-solving strategies in solving game-based logic problems. *International Journal of Technology in Education and Science*, DOI: .DOI: 10.46328/ijtes.98

Alexandropoulou, A., & Tsezou, A. 2023. Teaching science, sustainability and technology through gamification: The approach of escape rooms. DOI: DOI: 10.5194/egusphere-egu23-7397

Arnold, S., Fujima, J., & Jantke, P. K. (2013). Storyboarding Serious Games for Large-scale Training Applications. In Proceedings of the 5th International Conference on Computer Supported Education - CSEDU; SciTePress, pages 651-655. DOI: DOI: 10.5220/0004415606510655

Back, J., Back, S., Bexell, E., Stanisic, S., & Rosqvist, D. (2019). the quest: An escape room inspired interactive museum exhibition. In *Extended Abstracts of the Annual Symposium on Computer-Human Interaction in Play Companion Extended Abstracts, CHI PLAY '19*, page 81–86. ACM, DOI: DOI: 10.1145/3341215.3356987

Bharti, M. K. (2023). *Exploring the impact of gamification on students' motivation, and learning outcomes in secondary education*. International Journal For Multidisciplinary Research., DOI: 10.36948/ijfmr.2023.v05i05.7877

Blahova, L., & Kostolny, J. (2023). Development of a module for collecting students' feedback in an interactive manner. In *2023 21st International Conference on Emerging eLearning Technologies and Applications (ICETA)*, pages 62–67. DOI: DOI: 10.1109/ICETA61311.2023.10343671

Borrás-Gené, O., Diez, R., & Macías-Guillén, A. (2022). Digital educational escape room analysis using learning styles. *Information (Basel)*, 13(11), 522. DOI: 10.3390/info13110522

Breien, F. S., & Wasson, B. (2020). Narrative categorization in digital game-based learning: Engagement, motivation amp; learning. *British Journal of Educational Technology*, 52(1), 91–111. DOI: 10.1111/bjet.13004

Bruckner, F., Baumann, C., Husinsky, M., Jakl, A., Püringer, J., Suess, R. V., Vogt, G., & Wintersberger, M. (2022). Interactive storytelling in extended reality: Concepts for the design. In *Interactive Film & Media Journal*. DOI: DOI: 10.32920/ifmj.v2i4.1693

Caeiro, M., Vázquez, M. M., Nistal, M., Mikic-Fonte, F. A., Iglesias, M. J. F., Tsalapata, H., Heidmann, O., Carvalho, C. V., Jesmin, T., Terasmaa, J., & Tolstrup, L. (2021). A collaborative city-based game to support soft skills development in engineering and economics. In *2021 International Symposium on Computers in Education (SIIE)*, pages 1–4. DOI: DOI: 10.1109/SIIE53363.2021.9583639

Calvert, J., & Hume, M. (2022). Immersing learners in stories: A systematic literature review of educational narratives in virtual reality. *Australasian Journal of Educational Technology*, 55–71. Advance online publication. DOI: 10.14742/ajet.7032

Cardoso, P., Morgado, L., & Coelho, A. (2020). Authoring game-based learning activities that are manageable by teachers. *ERCIM News*, 2020, •••.

Carpenter, K. M., Cunningham, A. D., Smith, L., & Krigbaum, K. (2023). The great escape: A novel approach to collaborative learning (pilot). *Journal of Occupational Therapy Education*, 7(2). Advance online publication. DOI: 10.26681/jote.2023.070209

Cesário, V., Ribeiro, M., & Coelho, A. (2023). Design recommendations for improving immersion in role-playing video games. A focus on storytelling and localisation. [s]. *Interaction Design and Architecture*, (58), 207–225. DOI: 10.55612/s-5002-058-009)

Cezarotto, M., & Armstrong, A. L. (2023). *Designing inclusive educational games: accessibility rubric*. InfoDesign - Revista Brasileira de Design da Informação., DOI: 10.51358/id.v20i3.1064

Chao, T. (2023). The influence of cross-cultural adaptation on two language acquisition strategy. *Journal of Education and Educational Research*, 3(2), 2. DOI: 10.54097/jeer.v3i2.9038

Chen, S.-Y., Tsai, J.-C., Liu, S.-Y., & Chang, C.-Y. (2021). The effect of a scientific board game on improving creative problem solving skills. *Thinking Skills and Creativity*, 41, 100921. DOI: 10.1016/j.tsc.2021.100921

Coughlan, T., & Lister, K. (2022). *Creating stories of learning, for learning: Exploring the potential of self-narrative in education with 'our journey*. Creativity and Cognition., DOI: 10.1145/3527927.3535208

Damaševičius, R., & Sidekerskienė, T. (2023). Designing Metaverse Escape Rooms for Microlearning in STEM Education. In Advances in Higher Education and Professional Development (pp. 192–211). IGI Global. DOI: DOI: 10.4018/978-1-6684-8656-6.ch009

De Castell, S., Perry, N., Bailey, L., & Jenson, J. (2023). Microgames and language learning: Performance before competence? *European Conference on Games Based Learning*, 17(1):143–153, DOI: DOI: 10.34190/ecgbl.17.1.1766

Dimitriadou, A., Djafarova, N., Turetken, O., Verkuyl, M., & Ferworn, A. (2020). Challenges in serious game design and development: Educators' experiences. *Simulation & Gaming*, 52(2), 132–152. DOI: 10.1177/1046878120944197

Dutra, S. V. O., Chee, V., & Clochesy, J. (2023). Adapting an educational software internationally: Cultural and linguisticual adaptation. *Education Sciences*, 13(3), 237. DOI: 10.3390/educsci13030237

Eukel, H. N., Frenzel, J. E., Frazier, K., & Miller, M. (2020). Unlocking student engagement: Creation, adaptation, and application of an educational escape room across three pharmacy campuses. *Simulation & Gaming*, 51(2), 167–179. DOI: 10.1177/1046878119898509

Eukel, H. N., & Morrell, B. (2020). Ensuring educational escape-room success: The process of designing, piloting, evaluating, redesigning, and re-evaluating educational escape rooms. *Simulation & Gaming*, 52(1), 18–23. DOI: 10.1177/1046878120953453

Ferrari, A., Huichapa, T., Spoletini, P., Novielli, N., Fucci, D., & Girardi, D. (2023). Using voice and biofeedback to predict user engagement during product feedback interviews. *ACM Transactions on Software Engineering and Methodology*. Advance online publication. DOI: 10.1145/3635712

Foster, C., Urdl, E., Steller, C., Sharaf, S., Rößner, R., Mystakidis, S., Ingram, J., Ramamoorthy, H. K., & Mentzelopoulos, M. (2024). Work-in-progress—telecommunications history escape room puzzle game: Become a phyctoria archmaster. In *Academic Proceedings of the 10th International Conference of the Immersive Learning Research Network (iLRN2024)*, iLRN2024, page 90–95. The Immersive Learning Research Network, DOI: DOI: 10.56198/U6C0WURU6

Fotaris, P., & Mastoras, T. (2022). Room2educ8: A framework for creating educational escape rooms based on design thinking principles. *Education Sciences*, 12(11), 768. Advance online publication. DOI: 10.3390/educsci12110768

Friedmann, J. (2021). Storytelling for media: Introduction to the theory and practice of narrative design. ISBN: 9783838557649, DOI: DOI: 10.36198/9783838557649

Friedrich, C., Teaford, H., Taubenheim, A., & Sick, B. (2020). Interprofessional health care escape room for advanced learners. *The Journal of Nursing Education*, 59(1), 46–50. DOI: 10.3928/01484834-20191223-11 PMID: 31945176

Garrand, T. (2020). *Interactive multimedia narrative and linear narrative. Writing for Multimedia and the Web*. Elsevier., DOI: 10.4324/9780080474861

Garwood, J. (2020). Escape to learn! an innovative approach to engage students in learning. *The Journal of Nursing Education*, 59(5), 278–282. DOI: 10.3928/01484834-20200422-08 PMID: 32352543

Geremias, M. S., Dutra, T. C., Gasparini, I., & Maschio, E. (2023). An educational game to promote the development of computational thinking in children both neurotypical and with intellectual disabilities. *2023 IEEE International Conference on Advanced Learning Technologies (ICALT)*, DOI: DOI: 10.1109/ICALT58122.2023.00036

Gil Ruiz, P. (2024). Gamification as a methodology to enhance analytical and sustainable engagement on social media. *Discover Education*, 3(1), 4. Advance online publication. DOI: 10.1007/s44217-023-00074-7

Grabska-Gradzinska, I., Grabska, E., Nowak, L., & Palacz, W. (2020). Towards automatic generation of storyline aided by collaborative creative design. In *2020 International Conference on Interactive Collaborative Learning (ICL)*, pages 47–56. DOI: DOI: 10.1007/978-3-030-60816-3_6

Grönberg, N., Knutas, A., Hynninen, T., & Hujala, M. (2020). An online tool for analyzing written student feedback. *Proceedings of the 20th Koli Calling International Conference on Computing Education Research*, DOI: DOI: 10.1145/3428029.3428565

Grönberg, N., Knutas, A., Hynninen, T., & Hujala, M. (2021). Palaute: An online text mining tool for analyzing written student course feedback. *IEEE Access : Practical Innovations, Open Solutions*, 9, 134518–134529. DOI: 10.1109/ACCESS.2021.3116425

Guo, Y. (2021). Why are cultural adaptations such a prevalent strategy in the media business? *Scientific and Social Research*, 3(5), 1253. DOI: 10.36922/ssr.v3i5.1253

Gyaurov, D., Fabricatore, C., & Bottino, A. (2022). Features of entertainment digital games for learning and developing complex problem-solving skills: A protocol for a systemic review. *International Journal of Qualitative Methods*, 21, 16094069221128491. Advance online publication. DOI: 10.1177/16094069221128491

Hakshurian, Y. (2023). Teachers' competencies in developing digital educational escape rooms. *Educatia*, 21(25), 66–71. Advance online publication. DOI: 10.24193/ed21.2023.25.07

Halimah, L., Arifin, R. R. M., Yuliariatiningsih, M. S., Abdillah, F., & Sutini, A. (2020). Storytelling through "wayang golek" puppet show: Practical ways in incorporating character education in early childhood. *Cogent Education*, 7(1), 1794495. Advance online publication. DOI: 10.1080/2331186X.2020.1794495

Harley, D., Grambart, S., Bergel, R. S., & Mazalek, A. (2022). Together alone: A tangible online narrative. In *Sixteenth International Conference on Tangible, Embedded, and Embodied Interaction*. DOI: DOI: 10.1145/3490149.3501311

Hermanns, M., Deal, B., Campbell, A., Hillhouse, S., Opella, J., Faigle, C., & Campbell, R.IV. (2017). Using an "escape room" toolbox approach to enhance pharmacology education. *Journal of Nursing Education and Practice*, 8(4), 89. DOI: 10.5430/jnep.v8n4p89

Horváth, I. (2021). An analysis of personalized learning opportunities in 3d vr. *Frontiers of Computer Science*, 3, 673826. Advance online publication. DOI: 10.3389/fcomp.2021.673826

Huang, S., Kuo, Y.-H., & Chen, H.-C. (2020). Applying digital escape rooms infused with science teaching in elementary school: Learning performance, learning motivation, and problem-solving ability. *Thinking Skills and Creativity*, 37, 100681. DOI: 10.1016/j.tsc.2020.100681

Huerta-Cardoso, O., Unver, E., Kuş, A., & Arslan, R. (2023). Enhancing the pedagogic impact of augmented reality experiences: Storyboarding as a human-centred design approach. In *CAD'23*, pages 350–356. DOI: DOI: 10.14733/cadconfP.2023.350-356

Huraj, L., Hrmo, R., & Hudáková, M. S. S. (2022). The impact of a digital escape room focused on html and computer networks on vocational high school students. *Education Sciences*, 12(10), 682. Advance online publication. DOI: 10.3390/educsci12100682

Huynh, E., Nyhout, A., Ganea, P. A., & Chevalier, F. (2020). Designing narrative-focused role-playing games for visualization literacy in young children. *IEEE Transactions on Visualization and Computer Graphics*, 27(2), 924–934. DOI: 10.1109/TVCG.2020.3030464 PMID: 33048745

Ito, M. (2009). *Engineering Play: A Cultural History of Children's Software*. The MIT Press., DOI: 10.7551/mitpress/7939.001.0001

Jantke, K. P., & Spundflasch, S. (2013). Storyboarding pervasive learning games. In *Proceedings of the 2013 International Conference on Advanced ICT, ICAICTE-13*. Atlantis Press, DOI: DOI: 10.2991/icaicte.2013.9

Kanareiko, D. (2023). *Issue 3* (Vol. 2023). Narrative design in pedagogical scheming. *Ergodesign*. Bryansk State Technical University BSTU., DOI: 10.30987/2658-4026-2023-3-262-267

Kazem, M. M. A. (2022). The effects of educational games on the development of some cognitive skills in children aged (4-5) years. *Journal of STEPS for Humanities and Social Sciences*, DOI: .DOI: 10.55384/2790-4237.1147

Khundam, C. (2020). Storytelling platform for interactive digital content in virtual museum. *ECTI-CIT*, 15(1), 34–49. DOI: 10.37936/ecti-cit.2021151.239948

Kilanioti, I. (2022). Teaching a serious game for the sustainable development goals in the scratch programming tool. *European Journal of Engineering and Technology Research*, DOI: .DOI: 10.24018/ejeng.2022.1.CIE.2957

Klamma, R., Sous, D., Hensen, B., and Koren, I. (2020). Educational escape games for mixed reality. pages 437–442, DOI: .DOI: 10.1007/978-3-030-57717-9_41

Knox, A. (2023). *Game-Based Learning Design Optimized for Cognitive Load.* Springer International Publishing., DOI: 10.1007/978-3-031-25102-3_20

Kole, K. (2023). E-learning material for computational thinking in secondary education. Bachelor Thesis, University of Twente, The Netherlands.

Kotzebue, L., Zumbach, J., & Brandlmayr, A. (2022). Digital escape rooms as game-based learning environments: A study in sex education. *Multimodal Technologies and Interaction*, 6(2), 8. DOI: 10.3390/mti6020008

Kuo, H., Pan, A.-J., Lin, C.-S., & Chang, C.-Y. (2022). Let's escape! the impact of a digital-physical combined escape room on students' creative thinking, learning motivation, and science academic achievement. *Education Sciences*, 12(9), 615. Advance online publication. DOI: 10.3390/educsci12090615

Lathwesen, C., & Belova, N. (2021). Escape rooms in stem teaching and learning—Prospective field or declining trend? a literature review. *Education Sciences*, 11(6), 308. DOI: 10.3390/educsci11060308

Lathwesen, C., & Eilks, I. (2024). Can you make it back to earth? a digital educational escape room for secondary chemistry education to explore selected principles of green chemistry. *Journal of Chemical Education*, 101(8), 3193–3201. DOI: 10.1021/acs.jchemed.4c00149

Lima, E. S. D., Feijó, B., & Furtado, A. (2021). Adaptive branching quests based on automated planning and story arcs. In *2021 20th Brazilian Symposium on Computer Games and Digital Entertainment (SBGames)*, pages 9–18. DOI: DOI: 10.1109/SBGames54170.2021.00012

Lima, E. S. D., Feijó, B., & Furtado, A. (2022). A character-based model for interactive storytelling in games. In *2022 21st Brazilian Symposium on Computer Games and Digital Entertainment (SBGames)*, pages 1–6. DOI: DOI: 10.1109/SBGAMES56371.2022.9961071

Liu, Z.-Y., Shaikh, Z. A., & Gazizova, F. (2020). Using the concept of game-based learning in education. *International Journal of Emerging Technologies in Learning*, 15(14), 53–64. DOI: 10.3991/ijet.v15i14.14675

Lopez-Bermudez, V., & Martinez, A. (2023). Maruxaína: The sea nymph who will be judged. In *Proceedings of the 17th European Conference on Game-Based Learning*, pages 467–476. DOI: DOI: 10.4018/978-1-6684-6081-8.ch008

Luhova, T. (2020). Narrative and storytelling in the knowledge structure of the educational business video games as factors of the synergy of information technologies and spiritually-oriented pedagogy, pages 42–59, DOI: .DOI: 10.28925/2414-0325.2020.8.6

Makri, A., Vlachopoulos, D., & Martina, R. (2021). Digital escape rooms as innovative pedagogical tools in education: A systematic literature review. *Sustainability (Basel)*, 13(8), 4587. DOI: 10.3390/su13084587

Maksimenko, A. (2023). Experience of the narrative approach in foreign educational practice. *Vestnik Orenburgskogo gosudarstvennogo universiteta*, DOI: .DOI: 10.25198/1814-6457-238-33

Makuch, G. (2023). A linguistic analysis of translating intertextuality in "the witcher" game. BA thesis, University of Silesia in Katowice. DOI: DOI: 10.13140/RG.2.2.13032.03842

Malegiannaki, I., Daradoumis, T., & Retalis, S. (2020). Teaching cultural heritage through a narrative-based game. [JOCCH]. *Journal on Computing and Cultural Heritage*, 13(4), 1–28. DOI: 10.1145/3414833

Manojlovic, H. (2022). *Escape room as a teaching method.* Opus et Educatio., DOI: 10.3311/ope.504

Martín-Gutiérrez, A., Leal-Vega, L., Alcoceba-Herrero, I., Herrera-Medina, J., Arenillas-Lara, J. F., & Coco-Martín, M. B. (2022). *Educational Opportunities of Virtual Game-Based Initiatives for Students With Disabilities*. IGI Global., DOI: 10.4018/978-1-6684-4287-6.ch021

Mateen, S., Wiesemüller, P., & Voß-Nakkour, S. (2023). Design of an accessible vr-escape room for accessibility education. In *SMART ACCESSIBILITY 2023,The Eighth International Conference on Universal Accessibility in the Internet of Things and Smart Environments*, Venice, Italy.

Mateos-Aparicio, I., Sánchez-Paniagua, M., Garcia-Alonso, A., López-Ruíz, B., Molina, M., Moreno-Guzmán, M., Pérez-Rodríguez, M., Raggio, J., Rueda, C., Sevilla, P., Bodas-Folguera, C., and PEÑA-ARMADA, R. D. L. (2020). Students perceive the gamification through activities based on escape rooms very positively for learning at university. *IJAEDU- International E-Journal of Advances in Education*, 6:281–286, DOI: .DOI: 10.18768/ijaedu.819338

Meletiadou, E. (2023). Applying Digital Escape Rooms Infused With Cross-Cultural Management Theories in Higher Education: Academic Performance, Student Engagement, and Intercultural Intelligence Development. In *Handbook of Research on Redesigning Teaching*. Learning, and Assessment in the Digital Era., DOI: 10.4018/978-1-6684-8292-6.ch001

Mercado, M. A. T. (2023). Css quest: Gamifying computer system servicing module using gamified progress tracking and interactive storytelling algorithm for grade 12 students in calamba city. *International Journal of Advanced Research in Computer Science*, DOI: .DOI: 10.26483/ijarcs.v14i3.6996

Mijal, M., Ciesla, M., and Gromadzka, M. (2020). Escape room as an adult education tool. pages 163–178, DOI: .DOI: 10.14746/HL.2020.13.10

Min, W., Mott, B. W., Park, K., Taylor, S., Akram, B., Wiebe, E., Boyer, K., & Lester, J. C. (2020). Promoting computer science learning with block-based programming and narrative-centered gameplay. In *2020 IEEE Conference on Games (CoG)*, pages 654–657. DOI: DOI: 10.1109/CoG47356.2020.9231881

Minor, E., Oliver, K., & Fleming, K. (2024). From glyphs to bytes: Ancient egypt and the future of digital humanities in museum-based learning. In Minor, E. (Ed.), *Teaching Ancient Egypt in Museums* (1st ed.). Routledge. DOI: 10.4324/9781003348221-18

Moffett, J. and Cassidy, D. (2023). Building a digital educational escape room using an online design-thinking process. *Online Learning* (Vol. 27, Issue 2). The Online Learning Consortium. DOI: .DOI: 10.24059/olj.v27i2.3279

Moffett, J., Cassidy, D., Collins, N., Illing, J., de Carvalho Filho, M. A., & Bok, H. (2023a). Exploring medical students' learning around uncertainty management using a digital educational escape room: A design-based research approach. *Perspectives on Medical Education*, 12(1), 86–98. DOI: 10.5334/pme.844 PMID: 36969324

Moffett, S., Torres, C., & Brooks, C. (2023b). Building effective educational escape rooms: Insights from iterative design and user feedback. *British Journal of Educational Technology*, 54(1), 114–128.

Mozelius, P., Sällvin, L., and Humble, N. (2023). An educational escape room for computational thinking - defining the requirements. *Education and New Developments 2023 – Volume 2*, DOI: .DOI: 10.36315/2023v2end040

Mulyana, A., & Nurcahyani, N. (2022). *The effect of the puzzle playing method on improving the cognitive development of children aged 4-6 years*. KnE Life Sciences., DOI: 10.18502/kls.v7i2.10354

Mystakidis, S., Papantzikos, G., & Stylios, C. (2021). Virtual reality escape rooms for stem education in industry 4.0: Greek teachers perspectives. In *2021 6th South-East Europe Design Automation, Computer Engineering, Computer Networks and Social Media Conference (SEEDA-CECNSM)*, pages 1–5. DOI: DOI: 10.1109/SEEDA-CECNSM53056.2021.9566265

Najm, A., Christofi, M., Hadjipanayi, C., Kyrlitsias, C., & Michael-Grigoriou, D. (2022). *The iterative development of an online multiplayer escape room game for improving social interaction through edutainment*. CSRN., DOI: 10.24132/CSRN.3201.8

Neto, L. V., Junior, P. H. F., Bordini, R. A., & Otsuka, J. (2019). Design and implementation of an educational game considering issues for visually impaired people inclusion. *Smart Learning Environments*, 7(1), 4. Advance online publication. DOI: 10.1186/s40561-019-0103-4

Omarali, P. S. P. (2022). Considerations of game dynamics and mechanics in college students' use of mobile games for learning. *International Journal of Mobile and Blended Learning*, 14(2), 1–12. DOI: 10.4018/IJMBL.309739

Ouariachi, T., & Wim, E. J. L. (2020). Escape rooms as tools for climate change education: An exploration of initiatives. *Environmental Education Research*, 26(8), 1193–1206. DOI: 10.1080/13504622.2020.1753659

Palioura, M.-M., & Dimoulas, C. A. (2022). Digital storytelling in education: A transmedia integration approach for the non-developers. *Education Sciences*, 12(8), 559. Advance online publication. DOI: 10.3390/educsci12080559

Panchal, K., & Sharma, R. (2022). The effect of gamification on students' engagement and motivation in higher education: A meta-analysis. *Journal of Educational Technology & Society*, 25, 103–118. DOI: 10.1504/IJTEL.2021.10037644

Parra-Martínez. (2023). The limits of immersion: Case Study of Desolatium a VR Graphic Adventure. *Proceedings of DiGRA 2023*.

Paulauskas, L., Paulauskas, A., Blažauskas, T., Damaševičius, R., & Maskeliūnas, R. (2023). Reconstruction of industrial and historical heritage for cultural enrichment using virtual and augmented reality. *Technologies*, 11(2), 36. DOI: 10.3390/technologies11020036

Pecheranskiy, I. (2023). Technical peculiarities of the narrative action games of the 1990s–2010s as a factor of their development in the context of digital art. *Culture of Ukraine*, DOI: .DOI: 10.31516/2410-5325.081.02

Pirrone, M., & D'Ulizia, A. (2024). The localization of software and video games: Current state and future perspectives. *Information (Basel)*, 15(10), 648. DOI: 10.3390/info15100648

Polydorou, D. (2024). Immersive storytelling experiences: A design methodology. *Digital Creativity (Exeter)*, 35(4), 301–320. DOI: 10.1080/14626268.2024.2389886

Pratiwi, R., Andriati, R., & Indah, F. P. S. (2020). The positive effect of educative game tools (puzzle) on cognitive levels of pre-school children (4-5 years). *The Malaysian Journal of Nursing*, 11(3), 35–41. Advance online publication. DOI: 10.31674/mjn.2020.v11i03.006

Puga, R. U. (2022). Game-based learning: A tool that enhances the collaborative work. case study: undergraduate students. *European Conference on Games Based Learning*, DOI: DOI: 10.34190/ecgbl.16.1.419

Rahman, N. H. A., Mansor, S. N. A., Saad, S. M., Yahaya, S. H., & Omar, M. (2023). Stem-modules using gamification as induction teaching aids for non-academic tahfiz students. In *International Teaching Aid Competition 2023* (pp. 153–160). Universiti Teknologi MARA. [et al.]

Rawlinson, R. E., & Whitton, N. (2024). Escape rooms as tools for learning through failure. *Electronic Journal of e-Learning*, 00. Advance online publication. DOI: 10.34190/ejel.21.7.3182

Recke, M., & Perna, S. (2021). Emergent narratives in remote learning experiences for project based education. *Electronic Journal of e-Learning*, 19(2), 59–70. DOI: 10.34190/ejel.19.2.2142

Reed, J. and Ferdig, R. (2021). Gaming and anxiety in the nursing simulation lab: A pilot study of an escape room. *Journal of professional nursing: official journal of the American Association of Colleges of Nursing*, 37 2:298–305, DOI: .DOI: 10.1016/j.profnurs.2021.01.006

Reeves, L. E., Bolton, E., Bulpitt, M., Scott, A., Tomey, I., Gates, M., & Baldock, R. A. (2021). Use of augmented reality (ar) to aid bioscience education and enrich student experience. *Research in Learning Technology*, 29, 2572. DOI: 10.25304/rlt.v29.2572

Repetto, M., Bruschi, B., & Talarico, M. (2023). Key issues and pedagogical implications in the design of digital educational escape rooms. *Journal of E-Learning and Knowledge Society*, 19. Advance online publication. DOI: 10.20368/1971-8829/1135749

Ross, R., & Bennett, A. (2022). Increasing engagement with engineering escape rooms. *IEEE Transactions on Games*, 14(2), 161–169. DOI: 10.1109/TG.2020.3025003

Ross, S., Sabol, T., & Walker, A. (2022). Increasing engagement in stem education through narrative and game-based learning. *Journal of Science Education and Technology*, 31(5), 675–688.

Russell, J. (2020). Telling better stories: Toward critical, place-based, and multispecies narrative pedagogies in hunting and fishing cultures. *The Journal of Environmental Education*, 51(3), 232–245. DOI: 10.1080/00958964.2019.1641064

Sahin, G. (2023). Designing a digital escape room game: An experience of a digital learning tool in basic education. *Journal of Educational Technology and Online Learning*, DOI: .DOI: 10.31681/jetol.1334912

Salvador-Ullauri, L., Acosta-Vargas, P., González, M., & Luján-Mora, S. (2020). Combined method for evaluating accessibility in serious games. *Applied Sciences (Basel, Switzerland)*, 10(18), 6324. DOI: 10.3390/app10186324

Sánchez, R. M. (2023). *Transforming online education: the impact of gamification on teacher training in a university environment.* Metaverse Basic and Applied Research., DOI: 10.56294/mr202347

Sarage, D., O'Neill, B., & Eaton, C. (2020). There is no i in escape: Using an escape room simulation to enhance teamwork and medication safety behaviors in nursing students. *Simulation & Gaming*, 52(1), 40–53. DOI: 10.1177/1046878120976706

Saripudin, D., Komalasari, K., & Anggraini, D. (2021). Value-based digital storytelling learning media to foster student character. *International Journal of Instruction*, 14(2), 369–384. Advance online publication. DOI: 10.29333/iji.2021.14221a

Schimpf, C., Lachmann, J., Wetzel, M. H., Fischer, P. D. B., Leineweber, A., & Rafaja, D. (2023). Escape our lab: Creating an escape room game in the field of materials science and crystallography. *Journal of Applied Crystallography*, 56(5), 1544–1556. DOI: 10.1107/S1600576723006714 PMID: 37791356

Sereno, M., Wang, X., Besançon, L., McGuffin, M. J., & Isenberg, T. (2020). Collaborative work in augmented reality: A survey. *IEEE Transactions on Visualization and Computer Graphics*, 28, 2530–2549. DOI: 10.1109/TVCG.2020.3032761 PMID: 33085619

Shanta, S., & Wells, J. (2020). Tle design based learning: Assessing student critical thinking and problem solving abilities. *International Journal of Technology and Design Education*, 32(1), 267–285. DOI: 10.1007/s10798-020-09608-8

Shin, Y., Kim, J., Jin, K., & Kim, Y. B. (2020). Playtesting in match 3 game using strategic plays via reinforcement learning. *IEEE Access : Practical Innovations, Open Solutions*, 8, 51593–51600. DOI: 10.1109/ACCESS.2020.2980380

Shmelev, B. A. (2021). Linguodidactic potential of educational computer games. *Tambov University Review. Series. Humanities (Washington)*, (192), 58–69. Advance online publication. DOI: 10.20310/1810-0201-2021-26-192-58-69

Shoulson, A. (2021). Producing immersive interactive narratives in complex 3d open worlds. In *Proceedings of the AAAI Conference on Artificial Intelligence and Interactive Digital Entertainment*. DOI: DOI: 10.1609/aiide.v9i6.12604

Sidekerskiene, T., & Damaševičius, R. (2023). Out-of-the-box learning: Digital escape rooms as a metaphor for breaking down barriers in stem education. *Sustainability (Basel)*, 15(9), 7393. Advance online publication. DOI: 10.3390/su15097393

Skult, N. and Smed, J. (2020). Interactive storytelling in extended reality: Concepts for the design. pages 449–467, DOI: .DOI: 10.1007/978-3-030-37643-7_21

Stokes, J. and Pike, J. (2022). Future ready? engaging learners and building transferable skills through authentic assessment and digital literacy. *Pacific Journal of Technology Enhanced Learning*, DOI: .DOI: 10.24135/pjtel.v4i1.139

Stork, M. (2020). Supporting twenty-first century competencies using robots and digital storytelling. *Journal of Formative Design in Learning*, 4(1), 43–50. DOI: 10.1007/s41686-019-00039-w

Surapaneni, K. M. (2023). Creating an immersive learning experience through digital escape rooms in physiology education. DOI: .DOI: 10.1152/advan.00102.2023

Suter, B., Bauer, R., & Kocher, M. (2021). *Narrative mechanics*. Edition Medienwissenschaft., DOI: 10.14361/9783839453452

Talbot, T., Thiry, K. E., and Jenkins, M. P. (2020). Storyboarding the virtuality: Methods and best practices to depict scenes and interactive stories in virtual and mixed reality. pages 129–135, DOI: .DOI: 10.1007/978-3-030-51828-8_17

Tang, T., Vezzani, V., & Eriksson, V. (2020). Developing critical thinking, collective creativity skills and problem solving through playful design jams. *Thinking Skills and Creativity*, 37, 100696. DOI: 10.1016/j.tsc.2020.100696

Teixeira, I., Lima, D., & Vasconcelos, C. (2023). Storytelling as an interdisciplinary strategy in geoscience education. *International Conference on Education and New Developments 2023 – Volume 2*, DOI: DOI: 10.36315/2023v2end117

Terlouw, G., Kuipers, D., van 't Veer, J. T., Prins, J., & Pierie, J. (2021). The development of an escape room–based serious game to trigger social interaction and communication between high-functioning children with autism and their peers: Iterative design approach. *JMIR Serious Games*, 9(1), e19765. Advance online publication. DOI: 10.2196/19765 PMID: 33755023

Tian, R., Saito, D., Washizaki, H., Fukazawa, Y., Kobayashi, H., & Tsuji, A. (2024). Enhancing programming education through game-based learning: Design and implementation of a puyo puyo-inspired teaching tool. In *Proceedings of the 55th ACM Technical Symposium on Computer Science Education* V. 2, *SIGCSE 2024*, page 1838–1839. ACM, DOI: DOI: 10.1145/3626253.3635477

Ticala, R., Ciupe, A., Meza, S., & Orza, B. (2020). Augmenting learning through vr storytelling. In *2020 International Symposium on Electronics and Telecommunications (ISETC)*, pages 1–4. DOI: DOI: 10.1109/ISETC50328.2020.9301040

Torres, C., Liberato, H., Fernandes, A., Brito, L., & Covas, M. (2023). A microlearning approach by the# estudoemcasa apoia project. In INTED2023 Proceedings (pp. 5470-5474). IATED.

Treglia, E., Magnanini, A., & Caione, G. (2022). *Video games and accessibility*. Research Anthology on Developments in Gamification and Game-Based Learning., DOI: 10.4018/978-1-6684-3710-0.ch059

Turner, A., Tichter, A., & Pillow, T. (2021). Let's escape didactics: Virtual escape room as a didactic modality in residency. *Journal of Education & Teaching in Emergency Medicine*, 6(2). Advance online publication. DOI: 10.5070/M562052905 PMID: 37465713

Tzima, S., Styliaras, G., & Bassounas, A. (2020). Revealing hidden local cultural heritage through a serious escape game in outdoor settings. *Information (Basel)*, 12(1), 10. DOI: 10.3390/info12010010

Veldkamp, A., Daemen, J., Teekens, S., Koelewijn, S., Knippels, M. P. J., and Joolingen, W. (2020a). Escape boxes: bringing escape room experience into the classroom (extended abstract). *British Journal of Educational Technology*, pages 67–70, DOI: .DOI: 10.1111/bjet.12935

Veldkamp, A., Merx, S., & van Winden, J. (2020b). Educational escape rooms: Challenges in aligning game and education. DOI: DOI: 10.20944/preprints202010.0344.v1

Veldkamp, A., van de Grint, L., Knippels, M. P. J., & van Joolingen, W. V. (2020c). Escape education: A systematic review on escape rooms in education. *Educational Research Review*, DOI: DOI: 10.20944/preprints202003.0182.v1

Vemula, S. (2021). Leveraging vr/ar/mr and ai as innovative educational practices for "igeneration" students. pages 265–277. DOI: DOI: 10.4018/978-1-7998-4739-7.ch015

Vergne, M. J., Simmons, J. D., & Bowen, R. S. (2019). Escape the lab: An interactive escape-room game as a laboratory experiment. *Journal of Chemical Education*, 96(5), 985–991. DOI: 10.1021/acs.jchemed.8b01023

Verswijvelen, M., Sosa, R., & Martini, N. (2020). Designing game-inspired narratives for learning. In *DRS2020*. Synergy., DOI: 10.21606/drs.2020.294

Videnovik, M., Vold, T., Dimova, G., Kiønig, L., & Trajkovik, V. (2022). Migration of an escape room–style educational game to an online environment: Design thinking methodology. *JMIR Serious Games*, 10(3), e32095. Advance online publication. DOI: 10.2196/32095 PMID: 36155979

Viudes-Carbonell, S. J., Gallego-Durán, F. J., Llorens-Largo, F., & Molina-Carmona, R. (2021). Towards an iterative design for serious games. *Sustainability (Basel)*, 13(6), 3290. Advance online publication. DOI: 10.3390/su13063290

Vlachopoulos, D. (2021). Sustainable educational escape rooms for entrepreneurship students. In *INTED2021 Proceedings*, volume 1 of *INTED2021*, page 10240–10246. IATED, DOI: DOI: 10.21125/inted.2021.2136

Vodă, A. I., Căuti anu, C., Grădinaru, C., Tanasescu, C., & de Moraes, G. H. S. M. (2022). Exploring digital literacy skills in economics and social sciences and humanities students. *Sustainability (Basel)*, 14(5), 2483. DOI: 10.3390/su14052483

Voreopoulou, A., Mystakidis, S., & Tsinakos, A. (2024). Augmented reality escape classroom game for deep and meaningful english language learning. *Computers*, 13(1), 24. DOI: 10.3390/computers13010024

Vuillaume, L. A., Laudren, G., Bosio, A., Thévenot, P., Pelaccia, T., & Chauvin, A. (2021). A didactic escape game for emergency medicine aimed at learning to work as a team and making diagnoses: Methodology for game development. *JMIR Serious Games*, 9(3), e27291. Advance online publication. DOI: 10.2196/27291 PMID: 34463628

Wells, A. K., Mayer, R., Plass, J., & Homer, B. (2021). Playing a video game and learning to think: What's the connection? *Journal of Cognitive Enhancement: Towards the Integration of Theory and Practice*, 5(4), 459–467. DOI: 10.1007/s41465-021-00214-7

Xanthidis, D., Manolas, C., Paul, S., & Xanthidou, O. K. (2020). Virtual and augmented reality: Enhancing the learning experience in higher education in the u.a.e. current standing & research directions. In *2020 Seventh International Conference on Information Technology Trends (ITT)*, pages 206–211. DOI: DOI: 10.1109/ITT51279.2020.9320882

Yin, W. (2022). An artificial intelligent virtual reality interactive model for distance education. *Journal of Mathematics*, 2022(1), 7099963. Advance online publication. DOI: 10.1155/2022/7099963

Yousof, S. M. (2023). *Crossword Puzzle Games, Short Stories, and Mind Maps Assignments as Innovative Online Teaching Methods: Three Promising Applied Experiences During the COVID-19 Pandemic*. Emerald Publishing Limited., DOI: 10.1108/S2055-364120230000052005

Zahedi, L., Batten, J., Ross, M. S., Potvin, G., Damas, S., Clarke, P. J., & Davis, D. (2021). Gamification in education: A mixed-methods study of gender on computer science students' academic performance and identity development. *Journal of Computing in Higher Education*, 33(2), 441–474. DOI: 10.1007/s12528-021-09271-5

Zhang, J., Zou, G., & Zhang, G. (2021). "meet the deer king": "splash-ink" interaction in the innovative vr game based on dunhuang art and culture. In *SIGGRAPH Asia 2021 XR*. DOI: DOI: 10.1145/3478514.3487618

KEY TERMS AND DEFINITIONS

Augmented Reality (AR) and Virtual Reality (VR): Technologies used in DEERs to create immersive and interactive storytelling environments.

Branching Narratives: Interactive storylines in DEERs that allow player choices to influence the direction and outcome of the story.

Cognitive Skills Development: Enhancing abilities such as critical thinking, problem-solving, and decision-making through DEER puzzles and tasks.

Cohesive Narrative: A well-crafted storyline that aligns with educational objectives, providing context and enhancing the engagement of learning activities.

Collaborative Puzzles: Challenges in DEERs designed to foster teamwork, communication, and problem-solving among players.

Collaborative Story Development: The involvement of diverse teams, including educators, game designers, and students, in creating a DEER narrative.

Cultural Sensitivity: Adapting DEER content and narratives to respect and resonate with diverse cultural backgrounds.

Digital Educational Escape Rooms (DEERs): Interactive educational games that combine storytelling, puzzles, and gamification to provide immersive learning experiences.

Educational Alignment: The process of integrating learning objectives into game mechanics to ensure that every aspect of the game contributes to educational goals.

Educational Gamification: The integration of game design principles into educational activities to enhance learning outcomes and student engagement.

Feedback Loop: A mechanism in iterative design where user feedback is used to refine and improve the DEER.

Gamification Strategies: The use of game-like elements such as points, rewards, and challenges to increase engagement and motivation in learning.

Immersive Experience: An environment designed to fully engage learners by seamlessly integrating narrative, puzzles, and educational content.

Iterative Design Process: A cyclic approach to development involving repeated designing, testing, analyzing, and refining of the DEER based on feedback.

Localization: The process of adapting narratives, language, and content of DEERs to suit cultural and educational contexts for diverse audiences.

Playtesting: The process of testing a DEER with real users to gather feedback on gameplay, mechanics, and educational effectiveness.

Puzzles: Core components of escape rooms designed to challenge players and reinforce specific knowledge or skills through interactive tasks.

Storyboarding: A technique for visualizing the narrative, mapping out puzzles, and planning player interactions in a DEER

Time Constraints: A game mechanic that adds urgency and excitement by imposing a time limit on tasks, simulating real-world decision-making scenarios.

Chapter 5
Design Patterns and Integration Strategies for Implementing Digital Escape Rooms

ABSTRACT

This chapter explores design patterns and integration strategies for developing Digital Educational Escape Rooms (DEERs), focusing on enhancing learning experiences through interactive and immersive methods. Key elements, such as narrative integration, puzzle-based learning, adaptive difficulty, and collaborative challenges, are analyzed for their role in fostering engagement, cognitive development, and knowledge retention. The integration of multimedia is emphasized as a crucial component for creating effective DEERs, enhancing learner motivation and understanding. Puzzle-based learning is examined as a balance between entertainment and education, engaging learners through active problem-solving and critical thinking. The chapter also highlights the importance of aligning DEER content with clear educational objectives, ensuring the relevance and effectiveness of the activities. Techniques like adaptive difficulty, scaffolding, and feedback systems are discussed as methods to personalize learning experiences while maintaining an optimal level of challenge.

DEER DESIGN PATTERNS

Designing Digital Educational Escape Rooms (DEERs) involves incorporating various design patterns to create an engaging and effective learning experience.

DOI: 10.4018/979-8-3693-4219-0.ch005

Narrative Integration in DEER

In the design of DEERs, narrative integration is a crucial element that significantly enhances the educational experience. Crafting a compelling story that aligns with educational objectives involves creating a context or scenario that fully immerses learners in the activity, making the learning process both engaging and meaningful. This narrative integration is not just a backdrop for the puzzles and challenges but is interwoven with the educational content, thereby enhancing the overall impact of the learning experience. The power of narrative in learning environments is well-documented.

According to Bruner (Bruner, 1991), narratives are fundamental structures of human cognition and are particularly effective in educational settings because they are naturally engaging and easy to remember. In DEERs, the narrative serves as a framework within which the educational objectives are embedded. This could involve a storyline that complements the curriculum, such as a historical adventure for a history lesson or a science-fiction scenario for a STEM subject. The narrative creates a context that gives meaning to the tasks and challenges within the DEER, making the learning experience more coherent and relatable.

The integration of narrative in DEERs aids in cognitive engagement. Mayer's Cognitive Theory of Multimedia Learning (Mayer, 2001) suggests that learners engage more deeply with material when it is presented in a story format (see Table 1). A well-crafted narrative in a DEER captures the imagination of the learners, leading to increased motivation and engagement. This engagement is crucial in ensuring that the learners are not just passively absorbing information but actively interacting with the content. The integration of narrative in DEERs plays a crucial role in enhancing cognitive engagement, as outlined by Mayer's Cognitive Theory of Multimedia Learning. This theory emphasizes that effective learning occurs when information is presented in a way that aligns with how the human mind processes and stores information, utilizing both visual and auditory channels.

Firstly, according to the Multimedia Principle, people learn better from words and pictures than from words alone. In DEERs, narratives combine text, images, and audio within an immersive story context. This multimodal presentation engages multiple senses, allowing learners to form mental connections between the verbal and visual information. By embedding educational content within a story, learners are more likely to be interested and motivated, which enhances their ability to absorb and retain information.

The Personalization Principle suggests that learners benefit more from conversational language rather than formal language. Narratives in DEERs often use a conversational tone, making the learning experience more relatable and engaging. This approach reduces the psychological distance between the learner and the content, fostering a more intimate and effective learning environment. By feeling personally involved in the story, learners are more likely to invest cognitive resources in processing the material.

Mayer's Coherence Principle states that learning improves when extraneous material is excluded. In the context of DEERs, a well-crafted narrative focuses on relevant information and avoids unnecessary details that could overwhelm the learner. By streamlining the content within the narrative, cognitive overload is reduced, allowing learners to concentrate on essential concepts and tasks. This focused approach helps in better organizing information in the learner's mind, facilitating easier retrieval later on.

The Segmenting Principle emphasizes that people learn better when information is presented in manageable segments rather than a continuous stream. DEERs apply this principle by dividing the narrative into chapters or scenes, each containing specific learning objectives and challenges. This segmentation allows learners to process information at their own pace, ensuring they fully comprehend one part of the story before moving on to the next. It enhances comprehension and retention by preventing cognitive overload.

According to the Modality Principle, learners benefit more from graphics and narration than from graphics and on-screen text. DEERs often incorporate audio narration synchronized with visual elements within the narrative. This reduces the cognitive effort required to read text and allows learners to process auditory and visual information simultaneously. By freeing up cognitive resources, learners can focus more on understanding and integrating the information presented.

The Temporal Contiguity Principle asserts that learning is more effective when corresponding words and pictures are presented simultaneously rather than separately. In DEERs, narratives ensure that dialogue or descriptions are presented alongside relevant images or animations. This synchronization helps learners make immediate connections between different pieces of information, strengthening their understanding and aiding memory formation.

Table 1. Integration of narrative in DEERs aiding cognitive engagement according to Mayer's cognitive theory of multimedia learning

Mayer's Principle	Description	Integration of Narrative in DEERs Aiding Cognitive Engagement
Multimedia Principle	People learn better from words and pictures than from words alone.	Narratives in DEERs combine text, images, and audio within a story context, enhancing understanding through multiple representations and engaging both visual and auditory channels.
Personalization Principle	People learn better from conversational style rather than formal style.	Using conversational language in narratives makes content more relatable and engaging, increasing learner motivation and attention within DEERs.
Coherence Principle	Learning improves when extraneous material is excluded.	A focused narrative in DEERs eliminates irrelevant information, reducing cognitive overload and helping learners concentrate on essential content.
Segmenting Principle	People learn better when information is presented in manageable segments.	Narratives can be divided into chapters or scenes in DEERs, allowing learners to process information at their own pace, enhancing comprehension and retention.
Modality Principle	People learn better from graphics and narration than from graphics and on-screen text.	DEER narratives often use audio narration with visuals, reducing the need to read extensive text and enabling simultaneous processing of auditory and visual information.
Temporal Contiguity Principle	Better learning occurs when corresponding words and pictures are presented simultaneously.	In DEERs, synchronized narrative dialogue and visuals ensure that related information is processed together, improving understanding and memory recall.
Dual Coding Theory	Learning is enhanced when information is presented both visually and verbally.	Narratives engage both verbal and visual cognitive systems by presenting stories through spoken words and imagery, enhancing encoding and retrieval in DEERs.

Narrative integration in DEERs can foster empathy and emotional connection, which are important aspects of effective learning. A narrative that resonates with learners on an emotional level can make the educational experience more impactful. For instance, a DEER designed around social studies could include narratives that encourage learners to understand and empathize with different historical or cultural perspectives. Narrative integration aligns educational objectives with a compelling story, thereby enhancing learner immersion and engagement. This approach leverages the cognitive and emotional aspects of storytelling to create a rich and effective educational experience.

Puzzle-Based Learning in DEER

In the realm of DEERs, puzzle-based learning stands as a cornerstone, blending entertainment and education in a harmonious balance. The essence of this approach lies in designing puzzles and challenges that are intrinsically fun yet simultaneously serve as potent vehicles for reinforcing learning materials or skills. This amalgamation of enjoyment and education in DEERs engages learners in a manner that transcends conventional learning paradigms.

Puzzle-based learning in DEERs is predicated on the cognitive engagement theory, which posits that engaging tasks, such as solving puzzles, stimulate cognitive development and enhance learning (Shute and Zapata-Rivera, 2012). The design of puzzles in DEERs is orchestrated to ensure they are not only captivating but also inherently educational. Each puzzle is crafted to align with specific learning objectives, ensuring that the resolution of each challenge translates directly into a learning outcome. This design strategy requires a deep understanding of both the subject matter and effective puzzle design principles to create an experience where the entertainment value does not overshadow the educational intent.

Puzzle-based learning in DEERs leverages the principles of active learning. As postulated by Bonwell and Eison (Bonwell and Eison, 1991), active learning involves students in doing things and thinking about what they are doing. In DEERs, learners actively engage with the content through puzzle-solving, which encourages them to apply concepts, analyze problems, and think critically. This hands-on approach to learning fosters a deeper understanding of the material and enhances retention. Well-designed puzzles in DEERs can cater to different learning styles, accommodating a diverse range of learners. Gardner's theory of multiple intelligences (Gardner, 1983) suggests that individuals have unique learning preferences and strengths. By incorporating a variety of puzzle types – from logical and spatial challenges to linguistic and interpersonal tasks – DEERs can engage learners across this spectrum, making the learning experience more inclusive and effective.

Puzzle-based learning in DEERs represents an innovative educational approach that merges enjoyment with educational rigor. Through carefully designed puzzles that align with learning objectives and cater to diverse learning styles, DEERs offer an engaging and effective way to reinforce learning materials and skills.

Alignment with Learning Objectives in DEER

Ensuring alignment with learning objectives is a foundational principle in the design and implementation of DEERs. This alignment guarantees that the activities and challenges within DEER are purposeful, directly contributing to the intended educational outcomes. By carefully integrating curriculum standards, cognitive engagement strategies, and assessment techniques, DEER can be tailored to support meaningful and measurable learning experiences.

The first step in achieving alignment is to establish clear and specific learning objectives. These objectives should be well-defined, measurable, and directly linked to the desired outcomes of the educational activity. For instance, a DEER designed for a history class might include objectives such as "Students will be able to analyze primary source documents to deduce historical events." To translate these objectives into DEER activities, puzzles and challenges must directly reflect these goals, such as decoding a cipher containing historical dates or piecing together a timeline of events.

To further ensure alignment, the design of puzzles and tasks must integrate cognitive skills at appropriate levels of Bloom's Taxonomy. Lower-order objectives, such as knowledge recall, can be addressed through straightforward quizzes or matching games embedded in the DEER. Conversely, higher-order objectives, such as analysis and synthesis, require more complex tasks, such as solving multi-step problems or constructing arguments based on evidence. This hierarchical approach ensures that DEER challenges correspond to the depth and breadth of the intended learning outcomes.

Narrative integration also plays a critical role in aligning DEER with learning objectives. The storyline should serve as a contextual framework for embedding educational content. For example, a DEER designed to teach environmental science might place learners in a fictional scenario where they must solve real-world problems, such as analyzing soil samples or designing sustainable solutions to save an ecosystem. These narrative elements not only enhance engagement but also ensure that the learning objectives are deeply woven into the experience, making the educational purpose clear and relevant.

Adaptive difficulty is another technique that supports alignment with learning objectives by tailoring challenges to the learners' progress and abilities. By incorporating dynamic systems that adjust puzzle complexity in real time, DEER can maintain an appropriate level of challenge for diverse learners. For example, if a student struggles with a particular task, the system could provide additional hints or reduce the complexity, ensuring that the educational objective remains achievable without compromising cognitive engagement.

Assessment is integral to aligning DEER with learning objectives. Formative assessments, such as in-game quizzes and checkpoints, provide immediate feedback to learners, helping them understand their progress and areas for improvement. Summative assessments, such as a final challenge or reflective discussion, measure whether the learning objectives have been met. Embedding these assessments directly into the DEER experience ensures that the activities remain focused on educational outcomes.

Finally, collaboration and teamwork within DEER should be aligned with objectives that emphasize social and interpersonal skills. For instance, tasks requiring group problem-solving or communication can foster skills such as negotiation and critical discussion. These activities not only address content-specific goals but also support broader learning outcomes, such as teamwork and leadership.

By combining these techniques (summarized in Table 2) — clear objectives, cognitive skill integration, narrative-driven tasks, adaptive difficulty, embedded assessments, and collaborative challenges — DEER can ensure strong alignment with learning objectives. This alignment transforms DEER into powerful tools for education, creating immersive, engaging, and effective learning environments that achieve measurable academic and personal development goals.

Table 2. Techniques for ensuring alignment with learning objectives in DEER

Technique	Description and Implementation
Clear Learning Objectives	Define specific, measurable objectives that directly relate to the curriculum, ensuring that all tasks and puzzles reflect these goals. For example, decoding ciphers for historical event analysis.
Cognitive Skill Integration	Design tasks that align with levels of Bloom's Taxonomy, from simple recall (e.g., quizzes) to higher-order skills like analysis and synthesis (e.g., multi-step problem-solving).
Narrative Integration	Embed educational content into an immersive storyline to provide context and relevance, such as solving environmental problems in a science-based DEER.
Adaptive Difficulty	Use dynamic systems to adjust task complexity based on learner performance, ensuring accessibility and sustained engagement for diverse skill levels.
Embedded Assessment	Incorporate formative assessments (e.g., in-game quizzes) for real-time feedback and summative assessments (e.g., final challenges) to measure achievement of learning objectives.
Collaborative Challenges	Design group tasks that foster teamwork and interpersonal skills, aligning with objectives that emphasize communication, negotiation, and leadership.

Adaptive Difficulty in DEER

The implementation of adaptive difficulty in DEERs represents a sophisticated approach to personalized learning. This system dynamically adjusts the difficulty of puzzles based on the player's performance, ensuring that each learner is presented with challenges that are neither too easy nor too difficult. This customization is crucial in catering to the diverse skill levels and learning paces of different players, thereby optimizing the educational impact for each individual.

Adaptive difficulty in DEERs is grounded in the concept of differentiated instruction, a teaching philosophy that advocates for adjusting the teaching approach to accommodate the diverse needs of learners (Tomlinson, 1999). By incorporating this approach in DEERs, the puzzles and challenges can be calibrated in real-time to match the evolving competencies of the players. For example, if a player is swiftly solving puzzles, the system could incrementally increase the complexity of subsequent challenges to maintain engagement and ensure continued learning. Conversely, if a player is struggling, the system might offer simplified versions of puzzles or additional hints, thereby reducing frustration and promoting a continued sense of achievement.

This adaptive approach aligns with Vygotsky's (Vygotsky, 1978) Zone of Proximal Development (ZPD), which suggests that optimal learning occurs when tasks fall within a learner's ZPD - the range of tasks that a learner can perform with assistance but not yet independently. Adaptive difficulty in DEERs effectively keeps the learner within this zone, constantly adjusting the challenge level to ensure it is just beyond the learner's current ability, thereby promoting learning and development.

The implementation of adaptive difficulty in DEERs utilizes principles from educational psychology, particularly the concepts of formative assessment and feedback (Black and Wiliam, 1998). As learners progress through the DEER, their performance serves as a formative assessment that informs the adaptation of the puzzles. This continuous assessment and adjustment process ensures that each learner receives personalized feedback through the difficulty level of the tasks, facilitating a tailored educational experience.

Adaptive difficulty in DEERs is a pivotal design feature that enhances the educational effectiveness of these tools. By ensuring that each learner encounters challenges that are appropriately calibrated to their skill level and learning progress, DEERs can provide a more engaging, motivating, and effective learning experience for a diverse range of learners.

Collaborative Challenges in DEER

In the design of DEERs, the incorporation of collaborative challenges plays a pivotal role in enriching the learning experience. These challenges are crafted to necessitate teamwork and communication among participants, thereby fostering essential collaborative skills and group problem-solving abilities. This emphasis on collaboration transforms the DEER experience into a conduit for developing interpersonal skills and collective intelligence.

Collaborative challenges in DEERs are based on the educational principle that learning is a social process, as posited by Vygotsky (Vygotsky, 1978). According to his theory of social development, interaction with others plays a fundamental role in the development of cognition. In DEERs, tasks designed to require collaboration compel participants to engage in dialogue, exchange ideas, and collectively devise strategies to solve puzzles. This interaction not only facilitates the acquisition of knowledge but also enhances critical thinking and problem-solving skills.

The incorporation of collaborative challenges in DEERs aligns with Johnson and Johnson's (Jonassen, 1999) research on cooperative learning. Their studies have shown that cooperative learning experiences, where individuals work together to achieve common goals, lead to higher levels of achievement, improved interpersonal relationships, and greater psychological health. In DEERs, collaborative tasks create a cooperative environment where players must rely on each other's strengths, negotiate roles and responsibilities, and collectively work towards a shared objective.

Collaborative challenges in DEERs are instrumental in developing communication skills. Effective communication is crucial in these scenarios, as players must articulate their thoughts, listen to others, and synthesize diverse perspectives to progress through the escape room (Laal and Ghodsi, 2012). This aspect of DEERs is particularly valuable in preparing learners for real-world contexts where teamwork and communication are essential for success. The integration of collaborative challenges in DEERs serves as a vital component in fostering teamwork, communication, and group problem-solving skills. These challenges not only enhance the educational value of DEERs but also prepare learners for collaborative endeavors in academic and professional settings.

Multimedia Utilization in DEERs

The utilization of multimedia in DEERs represents a dynamic approach to engaging learners and catering to a diverse array of learning styles. By incorporating various forms of media, including audio, video, text, and interactive elements, DEERs can create a rich, multi-sensory learning environment that resonates with learners of all preferences. This incorporation of multimedia elements is not merely about adding aesthetic appeal but is strategically aligned with educational goals to enhance learner engagement and provide a more inclusive learning experience.

Multimedia utilization in DEERs is grounded in Mayer's Cognitive Theory of Multimedia Learning (Mayer, 2001), which posits that people learn more effectively from words and pictures than from words alone. In the context of DEERs, this theory suggests that incorporating visual and auditory elements along with textual information can significantly enhance the learning experience. For instance, video clips can be used to introduce a scenario or concept, audio elements can provide ambient sounds or narrative storytelling, and text-based clues can offer critical information or instructions. This multimodal approach aids in comprehension and retention by appealing to different sensory channels.

Interactive elements in DEERs, such as clickable areas, drag-and-drop tasks, and virtual simulations, contribute to active learning, a concept advocated by Bonwell and Eison (Bonwell and Eison, 1991). This active engagement with content through interactive media allows learners to experience hands-on learning, where they are not passive recipients of information but active participants in the educational process.

Feedback Systems in DEER

The incorporation of feedback systems in DEERs plays a crucial role in enhancing the learning experience. These systems, characterized by their immediacy and constructiveness, serve to guide learners through the escape room challenges while reinforcing their understanding and learning progress. Feedback in DEERs is not merely a tool for correction but a pivotal educational element that supports and motivates learners.

Immediate feedback is essential in maintaining learner engagement and momentum within the DEER. As posited by Hattie and Timperley (Hattie and Timperley, 2007), effective feedback is timely, allowing learners to quickly understand their mistakes and correct their course of action. In the context of DEERs, this immediacy is crucial as it keeps the learners engaged in the problem-solving process and prevents frustration. Immediate feedback can be integrated into the DEER design through automated responses to puzzle inputs, hints triggered by learner inactivity, or real-time feedback from facilitators.

Constructive feedback, on the other hand, goes beyond merely indicating right or wrong answers; it involves providing explanations and guidance that help deepen the learners' understanding. According to Narciss (Narciss, 2007), constructive feedback should be informative and help learners understand why an answer is correct or incorrect, thereby contributing to their learning process. In DEERs, constructive feedback can be designed to align with the learning objectives, providing learners with insights into the underlying concepts and encouraging reflection.

Feedback systems in DEERs can be utilized to foster a growth mindset among learners. Dweck (2006) emphasizes the importance of feedback in shaping learners' attitudes towards challenges and failure. Feedback in DEERs can be framed positively, emphasizing the effort and progress of learners, and encouraging them to view challenges as opportunities for learning. This approach not only enhances the learning experience but also builds resilience and perseverance in learners.

Feedback systems are integral to the educational effectiveness of DEERs. They provide immediate and constructive guidance, supporting and enhancing the learning journey. Well-designed feedback systems in DEERs ensure that learners are continuously engaged, understand the material, and are motivated to persevere through challenges, thereby maximizing the educational potential of these innovative learning environments.

Clue and Hint Mechanics in DEER

The integration of clue and hint mechanics in DEERs is a nuanced aspect of design that plays a pivotal role in balancing challenge and engagement. This system, essential for assisting players who might find themselves at an impasse, needs to be intricately woven into the DEER framework to maintain the educational integrity and immersive experience. In DEERs, clues and hints serve as scaffolding tools that guide learners through challenging segments without directly solving the puzzles for them. According to Vygotsky's (Vygotsky, 1978) concept of the Zone of Proximal Development (ZPD), learners can achieve higher understanding and skill acquisition through problem-solving with appropriate guidance or collaboration. Clues in DEERs act as this guidance, offering hints that are just enough to propel learners past a sticking point, thus aligning with the principles of ZPD.

Designing effective clue and hint systems requires a deep understanding of the educational content and the anticipated challenges learners may face. Clark and Mayer (2011) emphasize the importance of clues in instructional design that direct learners' attention to relevant information without overwhelming them. In DEERs, clues should be subtle yet meaningful, encouraging learners to connect the dots and apply their knowledge. This could involve thematic hints related to the educational content or prompts that require learners to recall and apply previously learned information.

The delivery of these clues should be innovative and contextually embedded within the DEER narrative. This could be achieved through various means such as hidden messages, interactive objects, or even through characters within the DEER storyline. Narrative-based hints not only add to the immersion but also contribute to the storytelling aspect of the DEER, making the learning experience more engaging (Barab et al., 2005).

The timing and accessibility of these hints are also crucial. Too frequent or easily accessible hints can diminish the challenge and sense of achievement. Therefore, a balance must be struck, possibly by implementing systems where hints become available only after certain conditions are met, or by allowing a limited number of hints to be used throughout the DEER (Kirschner et al., 2006).

Clue and hint mechanics in DEERs are essential for maintaining learner engagement and aiding in the educational process. These mechanics, when well-integrated, support learners through challenging aspects of the DEER, enhance the narrative experience, and align with effective educational practices.

Real-world Relevance in DEER

In the realm of DEERs, establishing real-world relevance is pivotal in enhancing the learning experience's perceived value and relevance. This focus on real-world applicability involves designing DEER content and tasks that not only engage students in problem-solving and critical thinking within the game environment but also connect these skills and knowledge to real-world scenarios and applications.

The emphasis on real-world relevance in DEERs is grounded in the educational philosophy of situational learning, which posits that learning is most effective when it is contextually embedded and relates to authentic tasks (Lave and Wenger, 1991). By linking DEER tasks to real-world situations, learners can see the practical applications of the knowledge and skills they are acquiring, thereby fostering a deeper understanding and appreciation of the subject matter.

Designing DEERs with real-world relevance requires careful consideration of the content and the learners' experiences. Tasks should be created not only to challenge the learners within the game environment but also to mirror the types of challenges they might encounter outside of the classroom. This could involve scenarios that require learners to apply scientific principles to solve a real-world environmental problem or to use mathematical reasoning in practical financial planning situations. The key is to bridge the gap between theoretical knowledge and practical application (Jonassen, 1999).

Incorporating real-world relevance in DEERs helps in developing transferable skills that are essential in various life and work contexts. According to Perkins and Salomon (Perkins and Salomon, 1992), the ability to transfer learning from one context to another is a crucial goal of education. DEERs that mimic real-life situations enable learners to practice and hone skills such as critical thinking, collaboration, communication, and problem-solving in contexts similar to those they will encounter in the real world.

The integration of real-world relevance in DEERs significantly enhances the learning experience. It not only makes the learning process more engaging and meaningful but also prepares learners to apply their knowledge and skills in practical, real-life situations. This approach aligns with the principles of situational learning and ensures that DEERs are not just an academic exercise but a valuable tool for life-long learning and application.

Time-Bound Challenges in DEERs

Incorporating time-bound challenges in DEERs is a strategic design choice that significantly enhances the learning experience. The implementation of time limits in DEERs serves to instill a sense of urgency and excitement, making the activity more engaging and dynamic. These time constraints provide an opportunity to teach and reinforce time management skills, a critical competency in both academic and real-world contexts. The concept of time-bound challenges aligns with the intrinsic motivational theories proposed by Csikszentmihalyi in his description of 'flow' (Csikszentmihalyi, 1990). According to this theory, when individuals are faced with tasks that are challenging yet achievable within a certain timeframe, they are more likely to enter a state of deep engagement and focus. In the context of DEERs, time limits create a scenario where learners are immersed in solving puzzles and completing tasks, driving them to engage more deeply with the content and work collaboratively under pressure.

Time-bound challenges in DEERs serve as a practical tool for teaching time management skills. As noted by Macan (1994), effective time management involves setting goals, prioritizing tasks, and allocating time efficiently. DEERs provide a controlled environment where learners can practice these skills. Learners must strategize on how to approach tasks, decide which puzzles to tackle first, and work collaboratively to solve problems within the allotted time, thereby honing their ability to manage time effectively.

The inclusion of time limits in DEERs mirrors real-world scenarios where individuals are often required to make decisions and solve problems within specific timeframes. This aspect of DEERs helps in preparing learners for real-life situations where effective time management is crucial. It teaches them to work under pressure, make quick decisions, and allocate their time and resources efficiently, skills that are valuable in academic settings and the workforce. However, it is important to balance the sense of urgency created by time constraints with the learners' ability to successfully engage with the educational content. Too stringent time limits might lead to undue stress and detract from the learning objectives. Therefore, the time frames should be carefully calibrated to be challenging yet achievable, allowing learners to experience a sense of accomplishment upon completion (Keller, 1987).

The integration of time-bound challenges in DEERs adds an important dimension to the learning experience. It not only makes the activity more engaging and exciting but also serves as an effective medium for teaching and practicing time management skills. These challenges prepare learners for real-life situations where they must operate under time constraints, enhancing their decision-making and prioritization skills. Careful consideration should be given to the duration of these time limits to ensure that they enhance, rather than detract from, the educational objectives of the DEERs.

Scaffolding Design in DEER

The concept of scaffolding design in DEERs is a critical pedagogical strategy, involving the gradual escalation of challenge complexity. This method allows learners to progressively build upon their knowledge and skills, facilitating deeper understanding and mastery of the subject matter. The essence of scaffolding in DEER lies in designing challenges that evolve in difficulty, aligning with the learners' growing competence and confidence.

Scaffolding in educational contexts is rooted in the work of Vygotsky (Vygotsky, 1978), who introduced the idea of the Zone of Proximal Development (ZPD). The ZPD represents the difference between what learners can do without help and what they can achieve with guidance and encouragement. In the DEER environment, scaffolding is implemented by initially presenting simpler tasks that cover the basic concepts, gradually introducing more complex challenges as the learners' understanding deepens. This progression ensures that the learners are continuously stretched beyond their current abilities but within their reach, fostering a productive learning experience (Wood et al., 1976).

To effectively implement scaffolding in DEER, the design should begin with an assessment of the learners' initial knowledge and skills. This assessment informs the starting difficulty level of the DEER challenges. As learners interact with the DEER, they apply and reinforce their existing knowledge, while the increasing complexity of subsequent challenges introduces new concepts and skills. This step-wise approach in difficulty not only promotes learning but also maintains engagement, as learners are less likely to feel overwhelmed or bored (HMELO-SILVER et al., 2007). The scaffolding design in DEERs can be complemented by providing timely feedback and support. Feedback plays a crucial role in guiding the learners through the challenges, offering hints or explanations when they struggle, and reinforcing their understanding as they progress (Sadler, 1989). This support is gradually withdrawn as learners become more proficient, encouraging independence and confidence in their problem-solving abilities.

The scaffolding design in DEERs is an effective educational strategy that aligns with fundamental learning theories. By gradually increasing the complexity of challenges, DEERs provide an environment where learners can build upon their existing knowledge and acquire new skills systematically. This approach not only enhances the learning experience but also fosters a sense of achievement and motivation among learners.

Summary of Core DEER Design Patterns

The design of DEERs encompasses a set of core patterns that are fundamental to their effectiveness as educational tools. These patterns include the complexities of puzzle design, the integration of narrative and storytelling elements, the complexities of gameplay mechanics, and the consideration of user interface and user experience. Each of these components plays a critical role in creating an immersive, engaging, and educationally valuable DEER experience.

Puzzle design in DEERs is pivotal in engaging learners and facilitating the learning process. Effective puzzles not only captivate the participants but also promote critical thinking, problem-solving, and collaborative skills. According to Nicholson (2014), successful puzzle design should balance challenge and solvability, ensuring that puzzles are neither too easy nor impossibly hard.

The puzzles should also be intrinsically linked to the educational objectives, enabling the reinforcement of specific skills or knowledge (Adams, 2010). By embedding educational content within the puzzles, DEERs make learning an active and engaging process. This approach aligns with Mayer's principles of multimedia learning, which suggest that people learn better from words and pictures when they are presented in a way that is actively engaging (Mayer, 2001).

Narrative and storytelling elements are crucial in DEER design as they provide context and motivation for the puzzles and activities. A compelling narrative can transform a simple puzzle into a meaningful and memorable learning experience (Barab et al., 2005). Storytelling in DEERs should be immersive, allowing learners to feel part of the story. This immersive quality can enhance engagement and motivation, making the learning experience more impactful (Egenfeldt-Nielsen, 2007). The narrative should be carefully crafted to align with the educational goals, ensuring that the story facilitates rather than distracts from the learning objectives.

The gameplay mechanics in DEERs refer to the rules and procedures that govern the player's interactions within the game. These mechanics are essential in creating an engaging and smooth gameplay experience. Garris et al. (Garris et al., 2002) emphasize the importance of integrating instructional content into the mechanics of the game to create a seamless blend of play and learning. The mechanics should be intuitive and align with the educational objectives, ensuring that players are learning through their interactions and decisions within the game. Well-designed gameplay mechanics can promote a range of skills, including strategic thinking, collaboration, and decision-making.

The user interface (UI) and user experience (UX) in DEERs are critical components that directly impact player engagement and learning effectiveness. The UI should be intuitive and user-friendly, allowing players to navigate the DEER with ease and focus on the learning tasks without unnecessary distractions (NIELSEN,

1993). The UX, on the other hand, encompasses the overall experience of the player, including the emotional and psychological aspects of interacting with the DEER (Norman, 2004). A positive UX in DEERs is achieved by creating an engaging and immersive environment, providing clear instructions and feedback, and ensuring that the DEER is accessible to a diverse range of learners. This approach aligns with the principles of universal design for learning, which advocate for creating educational experiences that are accessible and meaningful for all students (Meyer et al., 2014).

The core design patterns of DEERs—puzzle design, narrative and storytelling elements, gameplay mechanics, and user interface and user experience—are integral to creating an effective and engaging educational escape room. These patterns, when thoughtfully designed and integrated, can transform DEERs into powerful tools for learning, offering an immersive and interactive educational experience.

Technological Aspects of DEER Design

The technological facets of Digital Educational Escape Rooms (DEER) play a crucial role in their efficacy and appeal. This encompasses the judicious selection of software and platforms, the integration of multimedia elements, and the incorporation of adaptive learning technologies. These components collectively shape the DEER experience, ensuring that it is not only educationally sound but also technologically engaging.

Selecting the appropriate software and platform for DEER is fundamental to its success. The choice is influenced by various factors including user accessibility, ease of content creation, and the ability to support interactive elements. Platforms like Unity and Unreal Engine, known for their robustness in game development, offer extensive features for creating immersive DEER experiences (Rouse, 2005). Web-based platforms such as Moodle and WordPress provide versatility and ease of access, making DEER more widely available (Cole, 2009).

The integration of multimedia elements in DEER is essential for creating an engaging learning environment. This integration includes audio, video, text, and interactive tasks. Research in multimedia learning theory by Mayer (2001) emphasizes the importance of using both verbal and visual materials to enhance learning and retention. The use of multimedia in DEER caters to diverse learning styles and helps in maintaining high levels of engagement and motivation among learners (Moreno and Mayer, 2007).

Adaptive learning technologies are at the forefront of personalized education, and their integration into DEER is a significant step towards customized learning experiences. These technologies use algorithms to adjust the difficulty of tasks based on individual learner performance, thereby providing a tailored experience that meets each learner's unique needs (Shute and Zapata-Rivera, 2012). The incorporation of adaptive learning in DEER not only enhances the learning process but also ensures that learners remain challenged and engaged (Conati, 2009).

The technological aspects of DEER design are multifaceted, encompassing the careful selection of software and platforms, the integration of multimedia elements, and the adoption of adaptive learning technologies. These technological considerations are instrumental in creating DEER experiences that are not only pedagogically effective but also technologically advanced and engaging.

Pedagogical Considerations of DEER Design Patterns

The design patterns of DEER offer a compelling framework for engaging learners in an immersive and interactive environment. These patterns, when examined through the lens of pedagogical considerations, reveal deep-rooted connections with established learning theories, robust assessment and feedback mechanisms, and a commitment to inclusivity and accessibility.

The conceptualization of DEER inherently aligns with several educational theories, notably Constructivism, Social Learning Theory, and Flow Theory. Constructivism, as posited by Piaget and Vygotsky, suggests that knowledge is constructed through interaction with the environment and social contexts. DEER, by design, creates a learning environment where students actively engage with content, construct knowledge through problem-solving, and interact with peers, thus resonating deeply with Constructivist principles (Jonassen, 1999; Vygotsky, 1978) by providing interactive, immersive environments where learners engage directly with content, solve complex problems, and collaborate with peers (see Table 3).

One of the core principles of constructivism is active learning, which asserts that learners gain knowledge most effectively when they are actively involved in the learning process. DEERs embody this principle by requiring participants to engage with puzzles, challenges, and interactive elements within the escape room. Instead of passively consuming information, learners in a DEER must explore the environment, manipulate objects, and make decisions that affect the outcome of the game. This hands-on engagement promotes deeper understanding, as learners are directly involved in constructing their knowledge through experience.

Social interaction is another fundamental aspect of constructivist theory, particularly emphasized by Vygotsky. He introduced the concept of the Zone of Proximal Development (ZPD), which refers to the difference between what a learner can do independently and what they can achieve with guidance from more knowledgeable others. DEERs facilitate social interaction by incorporating collaborative tasks that require learners to work together to solve problems. This collaboration encourages communication, idea exchange, and peer learning, allowing participants to benefit from each other's strengths and knowledge. Through teamwork, learners can accomplish more than they could individually, effectively operating within their ZPD.

Constructivism also highlights the importance of problem-solving as a means of constructing knowledge. Learners make sense of information by engaging in authentic tasks that require critical thinking and application of concepts. DEERs present realistic challenges embedded within compelling narratives, requiring learners to analyze information, draw on prior knowledge, and devise solutions to progress. This problem-based learning approach ensures that knowledge is constructed in context, making it more meaningful and transferable to real-world situations.

The principle of scaffolding is integral to constructivist learning. Scaffolding involves providing support to learners as they develop new skills or understanding, gradually removing assistance as they become more proficient. In DEERs, scaffolding can be implemented through hints, prompts, or guidance systems that assist learners when they encounter difficulties. This support helps learners overcome obstacles without diminishing the challenge, allowing them to build confidence and competence. By tailoring the level of assistance to the learners' needs, DEERs ensure that participants remain engaged and are continuously challenged at an appropriate level.

Contextual learning is another key constructivist principle, emphasizing that knowledge is best acquired in meaningful contexts rather than in isolation. DEERs immerse learners in rich, contextual narratives that make the learning experience relevant and engaging. By situating tasks within a story or real-life scenario, learners can see the practical application of the concepts they are learning. This contextualization enhances motivation and aids in the retention of information, as learners are more likely to remember and understand material that is connected to a meaningful context.

A learner-centered approach is central to constructivism, focusing on the learner's experiences, interests, and prior knowledge. DEERs accommodate this by allowing learners to navigate through challenges at their own pace and make choices that affect their path. This autonomy enables learners to draw on their individual backgrounds and preferences, making the learning experience more personalized. By placing learners at the center of the educational process, DEERs promote ownership of learning and encourage intrinsic motivation.

Finally, reflection is a crucial component of constructing knowledge. Constructivism posits that learners need to reflect on their experiences to internalize new information and integrate it with existing knowledge structures. DEERs can facilitate reflection by incorporating debriefing sessions after activities or by prompting learners to consider the strategies they used to solve problems. This reflection helps learners solidify their understanding, recognize areas for improvement, and apply their learning to future situations.

Bandura's Social Learning Theory, which emphasizes learning through observation, imitation, and modeling, finds its application in DEER through collaborative challenges and peer learning (see Table 4). The social aspect of DEER, where learners work in groups and learn from each other's strategies and reasoning, embodies this theory (Bandura, 1977). Bandura's Social Learning Theory posits that people learn not only through direct experience but also by observing, imitating, and modeling the behaviors of others. This theory emphasizes the importance of social context and the interactions between individuals in the learning process. Digital Educational Escape Rooms (DEERs) effectively harness these principles by incorporating collaborative challenges and promoting peer learning, thereby enhancing the educational experience. One of the core components of Bandura's theory is observational learning, also known as modeling. In the context of DEERs, learners are placed in a virtual environment where they can observe their peers as they tackle puzzles and navigate challenges. This setting allows participants to witness different problem-solving strategies and approaches in real-time. By observing others, learners can pick up new techniques and ideas that they might not have considered independently. This shared experience enriches their understanding and equips them with diverse methods to apply in their own efforts.

For observational learning to be effective, attention is crucial. DEERs are designed to be engaging and immersive, capturing the learners' focus through intriguing storylines and interactive elements. The collaborative nature of these escape rooms requires participants to pay close attention to their peers' actions and contributions. This heightened attention ensures that learners are fully engaged with the material and with each other, setting the stage for effective observation and learning.

Table 3. Alignment of constructivist principles with DEER design

Constructivist Principle	Description	Alignment with DEER Design
Active Learning	Learners construct knowledge through active engagement with content and experiences.	DEERs require learners to actively participate in solving puzzles and challenges, fostering hands-on engagement and exploration.
Social Interaction	Learning is enhanced through collaboration and interaction with others.	DEERs often involve collaborative tasks where learners work together to solve problems, promoting communication and teamwork.
Problem-Solving	Knowledge is constructed through solving authentic, real-world problems.	DEERs present learners with realistic challenges that require critical thinking and application of knowledge to find solutions.
Scaffolding	Support is provided to learners to help them reach higher levels of understanding.	DEERs can include hints or guidance systems that provide assistance when learners are struggling, allowing them to progress while still challenging them.

continued on following page

Table 3. Continued

Constructivist Principle	Description	Alignment with DEER Design
Zone of Proximal Development (ZPD)	Learners can achieve more with guidance from more knowledgeable others.	DEERs facilitate cooperative learning, where peers can support each other, and instructors can provide assistance within the learners' ZPD.
Contextual Learning	Learning occurs in context and is connected to real-life situations.	DEERs immerse learners in contextual narratives that make learning relevant and meaningful, enhancing the applicability of knowledge.
Learner-Centered	Focus is on the learner's experiences and prior knowledge.	DEERs allow learners to navigate through the content at their own pace, drawing on their existing knowledge and experiences to solve challenges.
Reflection	Learners reflect on their experiences to construct understanding.	DEERs can include debriefing sessions or reflection prompts after activities, encouraging learners to think about what they have learned and how they solved problems.

Another key element is retention, the ability to remember the observed behaviors or strategies. DEERs facilitate retention by providing immediate opportunities to apply what has been observed. As learners progress through the escape room, they encounter new challenges where they can implement the techniques they have seen their peers use. This immediate application reinforces memory retention, as learners are actively practicing and embedding the new strategies into their cognitive repertoire.

Reproduction, or the capability to replicate the observed behavior, is also integral to the learning process. DEERs support this by offering interactive tasks that allow learners to practice and refine the strategies they have observed. The escape room environment encourages experimentation in a low-risk setting, where learners can try out new approaches without fear of significant negative consequences. This hands-on practice solidifies their ability to reproduce the behaviors effectively.

Motivation plays a significant role in whether learners choose to imitate observed behaviors. DEERs inherently motivate participants through gamification elements such as points, progress indicators, and the satisfaction of solving puzzles. The social aspect of working with peers adds an element of peer recognition and collective achievement. When learners see their peers succeed using certain strategies and receive positive feedback, they are more inclined to adopt those strategies themselves, driven by the desire to achieve similar success.

The concept of vicarious reinforcement is also present in DEERs. Learners observe the outcomes of their peers' actions—seeing which strategies lead to success and which do not. This observation helps them understand the consequences of different behaviors without having to experience them firsthand. By witnessing peers being rewarded for effective problem-solving, learners are encouraged to emulate those behaviors to attain similar rewards.

Engaging in collaborative challenges within DEERs can enhance learners' self-efficacy, or their belief in their own abilities to succeed. As participants observe their peers overcoming obstacles and contributing to the group's progress, they may feel more confident in their own capabilities. Successfully navigating challenges together reinforces the notion that they possess the skills necessary to achieve goals, which can translate into increased confidence in future learning endeavors.

The Csikszentmihalyi's concept of Flow, a state of complete immersion and focus, is integral to the success of DEER (see Table 5). The balance of challenge and skill, along with immediate feedback inherent in DEER, aligns with the conditions for achieving Flow, thereby making learning an engaging and enjoyable experience (Csikszentmihalyi, 1990). Csikszentmihalyi's concept of Flow is fundamental to the effectiveness of Digital Educational Escape Rooms (DEERs) because it describes a state of deep immersion and engagement that maximizes learning and enjoyment. Flow occurs when individuals are fully absorbed in an activity that provides a perfect balance between challenge and skill, offers clear goals, and delivers immediate feedback. DEERs are designed to align with these conditions, thereby facilitating the Flow state and enhancing the overall educational experience.

Firstly, DEERs establish clear goals by embedding explicit objectives within their narratives. Learners are informed about what they need to accomplish to "escape" or complete the game. This clarity ensures that participants understand the tasks at hand, which is essential for entering the Flow state. By knowing what is expected, learners can focus their attention and energies on achieving these goals without ambiguity or confusion.

Secondly, the balance of challenge and skill is carefully managed in DEERs. These games are crafted with adjustable difficulty levels and a variety of puzzles that cater to different skill sets. For learners to experience Flow, the challenges presented must match their abilities—neither too easy to cause boredom nor too difficult to induce anxiety. DEERs often allow educators to tailor the complexity of tasks to the proficiency levels of their students, ensuring that each learner is appropriately challenged and remains engaged.

Immediate feedback is another critical component of Flow that DEERs effectively incorporate.

Table 4. Application of Bandura's social learning theory in DEER through collaborative challenges and peer learning

Component of Theory	Description	Application in DEER through Collaborative Challenges and Peer Learning
Observational Learning (Modeling)	Learning by observing others' behaviors and the outcomes of those behaviors.	In DEERs, learners observe peers solving puzzles or employing strategies, which they can imitate or adapt in their own problem-solving efforts.
Attention	Focusing on the behavior being modeled to effectively learn from it.	DEERs capture learners' attention through engaging tasks that require them to pay attention to peers' actions and strategies during collaborative challenges.
Retention	The ability to remember observed behaviors to replicate them later.	Collaborative tasks in DEERs encourage learners to retain strategies observed from peers, which can be applied to future challenges within the game or in real-life contexts.
Reproduction	The capability to reproduce the observed behavior or strategy.	DEERs provide opportunities for learners to practice and replicate observed behaviors immediately in-game, reinforcing learning through action.
Motivation	The willingness to imitate a behavior based on anticipated outcomes.	DEERs motivate learners through game rewards, progress, and peer recognition, encouraging them to adopt effective strategies observed from others.
Vicarious Reinforcement	Learning about the consequences of behaviors by seeing others rewarded or punished.	Learners in DEERs see peers succeed or face setbacks based on their actions, informing their own decisions and encouraging adoption of successful behaviors.
Self-Efficacy	Belief in one's own ability to succeed in specific situations.	Collaborative success in DEERs enhances learners' self-efficacy as they observe peers' successes and realize their own potential to contribute effectively.

As learners interact with the escape room, they receive prompt responses to their actions, such as solving puzzles or making choices within the narrative. This instant feedback allows learners to adjust their strategies in real-time, reinforces learning through positive reinforcement or corrective guidance, and keeps them immersed in the activity. The continuous loop of action and response maintains the momentum of engagement essential for Flow. The immersive nature of DEERs fosters intense concentration on the task. The combination of compelling storylines, interactive challenges, and engaging multimedia elements captures learners' attention, helping them to focus deeply on the activity. By minimizing external distractions and providing a rich, immersive environment, DEERs encourage sustained attention, which is a hallmark of the Flow experience.

A sense of control is also integral to achieving Flow, and DEERs promote this by granting learners autonomy over their actions within the game. Participants can make decisions, solve problems in their own way, and see the direct consequences of their choices. This empowerment enhances their engagement and investment in the learning process, as they feel responsible for their progress and outcomes.

The engaging content and challenging tasks within DEERs often lead to a loss of self-consciousness, where learners become so absorbed in the activity that they lose awareness of themselves as separate from the action. This merging of action and awareness intensifies the immersive experience and contributes to the enjoyment and effectiveness of the learning process.

An altered sense of time is another indicator of the Flow state that DEERs can induce. When learners are deeply engaged, they may perceive time as passing more quickly than it actually does. This distortion occurs because their full attention is devoted to the task, leaving little cognitive capacity to monitor the passage of time. This phenomenon signifies a deep level of immersion and is associated with positive experiences in learning environments.

Finally, DEERs promote intrinsic motivation by making the learning experience inherently enjoyable. The game-based format and the thrill of solving puzzles and progressing through the narrative encourage learners to engage with the material for the sheer pleasure of the activity. This internal drive is crucial for achieving Flow, as it ensures that learners are fully committed and enthusiastic about the task, enhancing both engagement and retention of information.

Assessment in DEERs transcends traditional testing methods, encompassing formative assessment techniques that are embedded seamlessly into the gameplay. This approach aligns with contemporary educational research that emphasizes the importance of ongoing assessment and feedback for effective learning (Black and Wiliam, 1998). In DEER, feedback is immediate and often integrated into the game mechanics, allowing learners to reflect on their understanding and strategies in real-time. This constant feedback loop not only aids in reinforcing learning but also keeps the learners engaged and motivated (Hattie and Timperley, 2007).

Inclusivity and accessibility are paramount in the design of DEER. Universal Design for Learning (UDL) principles advocate for multiple means of engagement, representation, and expression to cater to diverse learners (Meyer et al., 2014). DEER, by employing multimedia elements and interactive challenges, can accommodate various learning styles and needs, adhering to UDL principles (see Table 6). The customization potential in DEER allows for adaptations to meet the specific requirements of learners with disabilities, thus ensuring that the learning experience is inclusive and accessible to all (Edyburn, 2010).

The pedagogical considerations in the design of DEERs are deeply interwoven with established learning theories, innovative assessment and feedback mechanisms, and a strong commitment to inclusivity and accessibility. These considerations are not only pivotal in ensuring the effectiveness of DEER as an educational tool but also in making learning an engaging, inclusive, and dynamic experience.

DEER CURRICULUM INTEGRATION STRATEGIES

The integration of DEERs into educational curricula demands a comprehensive understanding of the educational context, curriculum analysis, alignment of learning objectives and outcomes, and the identification of key learning areas suitable for DEER implementation. This holistic approach ensures that DEERs are not merely ancillary activities but are intrinsically woven into the educational fabric, enhancing the overall learning experience.

Table 5. Integration of Csikszentmihalyi's flow in DEERs for enhancing engagement and learning

Flow Condition	Description	DEER Alignment with Flow Condition
Clear Goals	Knowing what needs to be done.	DEERs provide explicit objectives within the game's narrative, ensuring learners understand the tasks required to "escape" or complete challenges.
Balance of Challenge and Skill	Tasks are neither too easy nor too difficult, matching the individual's skill level.	DEERs are designed with adjustable difficulty levels and varied puzzles to match the learners' abilities, maintaining engagement without causing frustration or boredom.
Immediate Feedback	Receiving prompt information about performance.	DEERs offer instant feedback on actions, such as solving puzzles or making decisions, allowing learners to adjust strategies in real-time and stay immersed in the experience.
Concentration on Task	Deep focus on the activity at hand.	The immersive environment and engaging storyline of DEERs capture learners' attention, promoting sustained concentration and minimizing external distractions.
Sense of Control	Feeling in command of one's actions and their outcomes.	DEERs allow learners to make choices, solve problems, and navigate the experience autonomously, fostering a sense of agency and control over the learning process.
Loss of Self-Consciousness	Lack of concern for self; merging of action and awareness.	Engaging narratives and challenging tasks in DEERs encourage learners to become fully absorbed, reducing self-awareness and enhancing immersion.
Altered Sense of Time	Perception of time is distorted; often time passes quickly.	The engaging nature of DEERs can lead learners to lose track of time, indicating deep immersion and enjoyment in the learning activity.
Intrinsic Motivation	Engaging in the activity for its own sake, not for external rewards.	DEERs are inherently enjoyable due to their game-based format, encouraging learners to participate willingly and fostering a love for learning through play.

Understanding the Educational Context of DEERs

Incorporating DEERs into an educational curriculum requires an in-depth analysis of the existing curriculum and its requirements. This analysis involves a careful examination of the curriculum's scope and sequence, ensuring that the introduction of DEERs aligns with the educational standards and goals of the institution (Wiggins and McTighe, 2005). This process also entails an assessment of the resources available, both in terms of physical materials and time, to effectively implement DEERs without disrupting the existing educational structure (Tomlinson and McTighe, 2006).

The alignment of learning objectives and outcomes is a critical aspect of integrating DEERs into the curriculum. Each DEER should be designed with clear educational objectives in mind, ensuring that the activities and challenges within the escape room directly contribute to the attainment of these objectives (Marzano, 2007). This alignment necessitates a pedagogically sound design where the skills and knowledge imparted through the DEER are directly relevant to the curriculum's learning outcomes, thereby reinforcing and augmenting the traditional learning experience (Bloom, 1956).

Table 6. How DEER accommodate learning styles and needs aligned with UDL principles

UDL Principle	Description	DEER Accommodation of Learning Styles and Needs
Multiple Means of Representation	Provide diverse ways of presenting information to address the varied learning preferences.	DEERs incorporate text, images, audio, and interactive media to present content, catering to visual, auditory, and kinesthetic learners. Puzzles and clues can be offered in multiple formats, ensuring accessibility for learners with different abilities, including those with disabilities.
Multiple Means of Action and Expression	Offer various methods for learners to interact with the material and demonstrate their knowledge.	DEERs allow learners to solve challenges through different strategies, promoting choice and autonomy. Interactive elements enable learners to express understanding via problem-solving, collaboration, and creative thinking, accommodating different expression styles and motor abilities.
Multiple Means of Engagement	Utilize multiple ways to engage learners and maintain motivation.	DEERs are designed to be engaging through gamification, storytelling, and interactive challenges. They can be adapted to include topics of interest to different age groups, adjustable difficulty levels, and rewards, thus sustaining motivation and catering to individual interests and cultural backgrounds.

Identifying the key learning areas where DEERs can be most effectively implemented is essential. This involves recognizing subjects or topics within the curriculum that lend themselves well to the DEER format, such as those requiring problem-solving, critical thinking, or collaborative learning (Jonassen, 1999). The integration of DEERs should focus on areas where they can provide the most educational value, either by addressing challenging concepts in an engaging manner or by enhancing the development of skills that are best learned through interactive and experiential methods (Kolb, 1984).

Designing DEER for Curriculum Fit

The effective design of DEERs for curriculum fit is a nuanced process that requires an alignment of DEER content with curriculum standards, a balance between educational goals and interactive gameplay, and the incorporation of diverse learning styles and needs. This approach ensures that DEERs are not only engaging and entertaining but also serve as a robust educational tool that complements and enhances the existing curriculum.

The alignment of DEER content with curriculum standards is fundamental to its educational efficacy. This involves a thorough understanding of the curriculum, including its core objectives, competencies, and expected outcomes (Wiggins and McTighe, 2005). The design of DEERs should be closely tied to these standards, ensuring that the puzzles, tasks, and challenges within the escape room directly contribute to the attainment of curriculum goals. This alignment requires a deliberate and thoughtful design process, where educational objectives are not peripheral but central to the DEER experience (Marzano, 2007).

Achieving a balance between educational goals and interactive gameplay is crucial in DEER design.

This balance involves creating a DEER experience that is not only pedagogically sound but also engaging and enjoyable for students (Gee, 2003). The gameplay elements should be designed to reinforce and augment the educational content, ensuring that the learning objectives are achieved through interactive and immersive experiences. The challenge lies in making the educational content compelling enough to maintain student engagement, while also ensuring that the gameplay does not overshadow the learning outcomes (Prensky, 2001).

Incorporating diverse learning styles and needs is essential in making DEERs accessible and effective for all students. According to Gardner's theory of multiple intelligences (Gardner, 1983), learners have different strengths and preferences in how they process information. DEERs should be designed to cater to these varied learning styles, providing a range of activities and challenges that appeal to different types of intelligences, such as visual-spatial, linguistic, logical-mathematical, and interpersonal intelligences. DEERs should be adaptable to accommodate students with varying needs, including those with physical, cognitive, or sensory disabilities, ensuring inclusivity and accessibility in line with universal design principles (CAST, 2018).

The design of DEERs for curriculum fit is a complex yet rewarding endeavor that necessitates an alignment with curriculum standards, a balance between educational content and gameplay, and the accommodation of diverse learning styles and needs. By adhering to these principles, DEERs can be effectively integrated into educational settings, providing a dynamic and enriching learning experience.

DEER in Different Educational Settings

The application of DEERs transcends traditional educational boundaries, finding its place in both formal education settings and informal learning environments. DEER's versatility allows for cross-curricular and interdisciplinary approaches, making it a multifaceted tool in the realm of education.

In formal education settings, DEERs serve as an innovative instructional strategy that complements traditional teaching methods. Within classrooms, DEERs are employed to reinforce subject-specific knowledge and to develop critical thinking, problem-solving, and collaborative skills among students (Van Eck, 2006). The interactive nature of DEERs aligns with constructivist learning theories, posited by Vygotsky (Vygotsky, 1978), which emphasize active learning through experience and social interaction. By integrating DEERs into the formal curriculum, educators can offer a more dynamic and engaging learning experience, thereby enhancing student motivation and participation (Prensky, 2001).

In informal learning environments, such as museums, community centers, and after-school programs, DEERs function as an engaging educational tool that fosters exploratory and self-directed learning. These settings provide a more relaxed atmosphere where DEERs can be used to stimulate curiosity, encourage exploration, and provide educational content in an entertaining format (Falk and Dierking, 2016). In such environments, DEERs are not bound by the constraints of traditional curricula, allowing for more creativity and flexibility in their design and implementation.

DEERs inherently support cross-curricular and interdisciplinary approaches to learning. By their nature, DEERs can integrate concepts from multiple disciplines, encouraging learners to make connections across different fields of study (Jacobs, 1989). This integration fosters a holistic understanding and allows students to apply knowledge in varied contexts, which is essential in developing critical thinking and problem-solving skills. For instance, a DEER themed around environmental science can incorporate elements of biology, geography, and even ethics, providing a rich and comprehensive learning experience.

Teacher Training and Professional Development

The implementation of DEERs in educational settings necessitates a focused approach to teacher training and professional development. This preparation is pivotal in equipping educators with the necessary skills and knowledge for effective DEER implementation, providing them with structured workshops and training programs, and ensuring ongoing support and resource sharing.

Preparing educators for DEER implementation is a multifaceted process that involves more than just an understanding of the technical aspects of DEERs. It requires educators to integrate new pedagogical approaches that align with the interactive and immersive nature of DEERs. According to Guskey (Guskey, 2000), effective professional development should result in changes in teaching practices that improve student learning outcomes. Therefore, training for DEER implementation should encompass strategies for incorporating DEERs into the curriculum, methods to facilitate student engagement and learning within a DEER, and assessment techniques suitable for this innovative learning tool. This preparation enables educators to seamlessly integrate DEERs into their teaching repertoire, enhancing the overall educational experience (Darling-Hammond et al., 2017).

Workshops and training programs play a crucial role in equipping educators with the skills needed for DEER implementation. These programs should be comprehensive, covering various aspects of DEER design and facilitation, including narrative development, puzzle creation, and the use of technology. Joyce and Showers (Joyce and Showers, 2002) emphasize that for professional development to be effective, it must include theory, demonstration, practice, and feedback. Training programs should therefore provide hands-on experiences where educators can collaborate, design, and trial DEER scenarios, receiving feedback from peers and trainers. This experiential learning approach ensures that educators not only understand the theoretical underpinnings of DEERs but also acquire practical skills in their implementation.

Continuous support and resource sharing are essential components of ongoing professional development for educators involved in DEER implementation. As Fullan (Fullan, 2007) notes, the sustainability of innovative educational initiatives is heavily dependent on continuous learning and support systems. This can be facilitated through online communities of practice, where educators can share experiences, resources, and best practices (Wenger, 1998). Providing access to a repository of DEER resources, such as templates, toolkits, and case studies, can significantly aid educators in the design and facilitation of DEERs. This ongoing support not only helps educators in refining their DEER strategies but also fosters a collaborative environment for continuous professional growth.

Teacher training and professional development for DEER implementation encompass comprehensive preparation, structured workshops and training programs, and ongoing support and resource sharing. By focusing on these areas, educators can be effectively equipped to harness the potential of DEERs, thereby enriching the educational landscape with this innovative teaching tool.

Assessment and Evaluation in DEER

The integration of DEER into educational practices necessitates a reevaluation of assessment and evaluation strategies. This reevaluation includes integrating both formative and summative assessments, employing observational and reflective assessment strategies, utilizing DEER for evaluating skills and competencies, and leveraging feedback mechanisms and learning analytics. These assessment approaches are crucial in gauging the effectiveness of DEERs in meeting educational objectives and in enhancing the learning experience.

Incorporating both formative and summative assessments in DEER allows for a comprehensive evaluation of student learning. Formative assessments, as described by Black and Wiliam (Black and Wiliam, 1998), are essential in providing ongoing feedback to students and educators about the learning process. These assessments can be integrated into DEERs through in-game quizzes, reflection logs, and peer discussions, offering insights into student understanding and progress. Summative assessments, on the other hand, evaluate student learning at the conclusion of the DEER experience. These can take the form of final puzzles or projects that require students to apply the knowledge and skills they have acquired (Popham, 2008). The combination of formative and summative assessments ensures a holistic understanding of student learning outcomes.

Observational and reflective assessments are particularly effective in the DEER context. Observation allows educators to assess students' problem-solving processes, teamwork, and engagement in real-time (Angelo and Cross, 1993). Educators can note how students interact with the DEER environment, collaborate with peers, and apply critical thinking skills. Reflective assessments, such as student journals or group debriefings, encourage learners to reflect on their experiences, challenges, and learning within the DEER (Schön, 1983). This reflective process not only deepens learning but also provides educators with valuable insights into the students' learning processes.

DEERs are particularly well-suited for evaluating a range of skills and competencies. These include critical thinking, collaboration, communication, and problem-solving skills. The immersive and interactive nature of DEERs provides a unique environment where students can demonstrate these skills in action (van Merrienboer et al., 2003). By designing DEERs with specific competencies in mind, educators can create targeted experiences that allow for the practical application and assessment of these skills.

The incorporation of feedback mechanisms and learning analytics in DEERs offers an advanced approach to assessment. Digital platforms used in DEERs can track student progress, providing real-time feedback and data on student performance (Siemens and Long, 2011). This data can be analyzed to identify learning trends, pinpoint areas of difficulty, and adapt the DEER experience to better meet student needs. Such analytics can also inform future DEER designs, ensuring they are more effectively tailored to educational objectives. Feedback from students, obtained through surveys or discussions post-DEER, is invaluable in assessing the effectiveness of the DEER and in making necessary adjustments to enhance the learning experience.

Assessment and evaluation in DEER require a multifaceted approach that encompasses formative and summative assessments, observational and reflective strategies, skills and competency evaluation, and the utilization of feedback mechanisms and learning analytics. By employing these diverse assessment methods, educators can gain a comprehensive understanding of student learning and the overall efficacy of DEER in educational settings.

DESIGNING EDUCATIONAL CONTENT WITHIN DEERS

Fundamentals of Educational Content Design in DEER Games

The design of educational content within DEER necessitates a nuanced approach, encompassing the principles of educational game design, the alignment of educational objectives with game mechanics, and the integration of narrative into educational content. These fundamentals are pivotal in creating DEERs that are not only engaging and immersive but also effective in achieving educational goals.

The principles of educational game design form the bedrock of DEER content development. According to Gee (Gee, 2003), good game design involves creating experiences that are deeply engaging and challenging, yet achievable. This involves constructing puzzles and challenges that are intrinsically linked to the learning objectives, ensuring that players are not only motivated to participate but also gain meaningful educational experiences. Schell (Schell, 2008) further elaborates that successful educational games should provide immediate feedback, allowing learners to understand the impact of their actions and decisions within the game. This immediate feedback loop is essential in DEERs for maintaining player engagement and reinforcing learning.

Aligning educational objectives with game mechanics is crucial in ensuring that DEERs serve their intended educational purpose. Squire (Squire, 2006) argues that the most effective educational games are those where learning objectives are seamlessly integrated into the game mechanics. In DEERs, this means that the puzzles and challenges players encounter should directly relate to the educational goals, whether they involve critical thinking, problem-solving, or subject-specific knowledge. This alignment ensures that the educational content is not peripheral but central to the game experience, making the learning process more meaningful and impactful.

Narrative plays a significant role in enhancing the educational content of DEERs. According to Barab et al. (Barab et al., 2005), narratives can provide context and relevance to educational content, making it more engaging and relatable for learners. In the context of DEERs, the narrative serves as a framework within which educational objectives are embedded. It helps in creating an immersive environment where learners can explore and interact with educational content in a story-driven context. This narrative-driven approach not only enhances learner engagement but also aids in the retention of information and concepts, as the storyline provides a memorable backdrop for the educational content (Egenfeldt-Nielsen, 2007).

The design of educational content in DEERs is a multifaceted process that requires adherence to the principles of educational game design, a strategic alignment of educational objectives with game mechanics, and the effective use of narrative. By integrating these elements, DEERs can provide an immersive, engaging, and educationally enriching experience for learners.

Developing Engaging and Effective Puzzles in DEERs

The development of engaging and effective puzzles in DEERs is a critical aspect of their design, significantly influencing their educational efficacy. This process involves a selection of puzzle types, balancing educational challenges with learning objectives, and ensuring that puzzle design is aligned with desired learning outcomes. Each of these elements plays a fundamental role in creating an immersive and intellectually stimulating environment conducive to learning.

The choice of puzzles in DEERs is not arbitrary; it is crucial to select types that reinforce educational goals. Research indicates that puzzles should be intrinsically linked to the learning material, thereby promoting deeper cognitive engagement (Adams, 2010). Logic puzzles, word-based challenges, and spatial reasoning tasks are commonly employed in educational settings. Logic puzzles, for instance, enhance deductive reasoning and problem-solving skills (Schell, 2008), while word-based puzzles can improve language and vocabulary skills (Anderson and Krathwohl, 2001). Similarly, spatial reasoning puzzles aid in developing visual-spatial understanding, which is fundamental in fields like geometry and architecture (Newcombe and Stieff, 2012).

The balance between challenge and learning in puzzle design is paramount. If a puzzle is too easy, it fails to engage learners effectively; if too difficult, it can lead to frustration and disengagement (Csikszentmihalyi, 1990). The concept of 'flow' - a state of deep focus where a person is fully immersed in an activity - is essential in this context. Achieving flow requires tasks that are neither too challenging nor too easy in relation to the individual's skill level (Csikszentmihalyi, 1990). In DEERs, this balance is critical as it ensures that learners are constantly engaged and motivated, while also ensuring that the learning objectives are being met. This balance can be achieved by adjusting the complexity of the puzzles in accordance with the learner's progress, ensuring a steady learning curve (Shute, 2008).

The alignment of puzzle design with learning outcomes is a key consideration in DEERs. The puzzles should be intricately designed to meet specific educational objectives. According to Bloom's Taxonomy, learning objectives can range from simple recall of facts to more complex analysis and evaluation skills (Bloom, 1956). Puzzles in DEERs should be designed to target these varying cognitive levels. For example, a puzzle that requires learners to recall specific information can address lower-order cognitive skills, whereas a puzzle that requires learners to analyze a problem or synthesize information from various sources can target higher-order cognitive skills (Anderson and Krathwohl, 2001). This alignment ensures that the puzzles are not only engaging but also effective in achieving the desired educational outcomes.

The development of puzzles in DEERs is a nuanced process that requires careful consideration of the types of puzzles used, the balance between challenge and learning, and the alignment of puzzle design with learning outcomes. By focusing on these aspects, DEERs can provide an engaging, challenging, and educationally effective experience.

In the realm of DEER, interactive storytelling techniques play a pivotal role in creating an enriching educational experience. These techniques, encompassing the creation of immersive storylines, the incorporation of story elements to enhance learning, and the use of narrative as a tool for engagement and motivation, are integral in crafting DEERs that captivate and educate simultaneously. This blend of narrative and education transforms the learning process into a dynamic and memorable journey.

The creation of immersive storylines is fundamental in DEER design. Jenkins (Jenkins, 2004) posits that effective storytelling in interactive media should create a world that users can enter and become part of, enhancing their engagement and learning experience. In DEERs, storylines are crafted not just to entertain but also to create a context for the educational content. These narratives often involve scenarios or challenges that mimic real-world situations or hypothetical scenarios, encouraging learners to immerse themselves fully in the experience. The storyline becomes a canvas upon which educational objectives are painted, making the learning process more engaging and relatable (Murray, 1997).

The incorporation of story elements into DEERs serves to enhance the learning experience. Bruner (Bruner, 1991) suggests that narrative structure aids in organizing and making sense of complex information. In DEERs, elements such as characters, settings, conflicts, and resolutions are not mere embellishments but tools that help embed educational content within a coherent and engaging narrative. These elements can be aligned with educational goals to reinforce learning. For instance, characters might represent different historical figures or scientific concepts, while conflicts and resolutions can mirror real-life challenges and solutions, thereby deepening the learner's understanding and retention of the subject matter (Egan, 1986).

Narrative in DEERs serves as a powerful tool for engagement and motivation. According to Ryan (Ryan, 2001), narrative can create a sense of agency and motivation in learners, as they become part of the story. In the context of DEERs, the narrative can be structured to provide learners with a sense of purpose and urgency, driving them to engage with the content and solve challenges. This sense of being part of a story not only motivates learners but also facilitates deeper emotional and cognitive engagement with the educational material, leading to a more effective and impactful learning experience (Dickey, 2005).

In conclusion, interactive storytelling techniques in DEERs, encompassing immersive storylines, the strategic incorporation of story elements, and the use of narrative as an engagement tool, are crucial in creating a learning environment that is both educational and captivating. These techniques not only enhance the learner's engagement and motivation but also facilitate a deeper connection with the educational content, making learning a dynamic and memorable adventure.

Multimedia Integration in DEER

The integration of multimedia in DEER represents a significant advancement in educational methodologies, leveraging visual and audio elements, interactive media, and adhering to best practices in multimedia content creation. This fusion of various media forms enhances learning experiences, making them more engaging, interactive, and effective.

The use of visual and audio elements in DEERs plays a crucial role in enhancing the learning experience. Mayer's Cognitive Theory of Multimedia Learning (Mayer, 2001) suggests that individuals learn better from words and pictures than from words alone. In the context of DEERs, this translates to using visual aids like diagrams, images, or animations, coupled with audio narrations or sound effects, to create a rich learning environment. These elements can help in simplifying complex concepts, maintaining learner engagement, and aiding memory retention (Clark and Mayer, 2011).

Interactive media in DEERs, encompassing elements like clickable areas, drag-and-drop activities, and interactive simulations, provide a hands-on learning experience. According to (Prensky, 2001), interactive media supports active learning, where learners engage directly with the material, leading to better understanding and retention of knowledge. Such interactivity also allows for instant feedback, which is essential in the learning process, enabling learners to understand their mistakes and learn from them immediately (Shute, 2008).

Creating effective multimedia content for DEERs involves adhering to best practices that ensure the content is educationally sound and engaging. This includes maintaining a balance between different media elements to avoid cognitive overload (Sweller, 1988), ensuring content accessibility for all learners (CAST, 2018), and aligning multimedia content with learning objectives to ensure educational efficacy (Gagné, 1985). These practices ensure that multimedia integration in DEERs is not only technologically innovative but also pedagogically effective.

The integration of multimedia in DEERs, through the use of visual and audio elements, interactive media, and adherence to best practices in multimedia content creation, significantly enhances the educational experience. This integration not only makes learning more engaging and interactive but also ensures that educational content is delivered effectively.

Adaptive Learning Strategies in DEER

Adaptive learning strategies in DEER signify a paradigm shift in educational methodologies, focusing on personalizing learning experiences, creating adaptive challenges based on learner performance, and implementing real-time feedback mechanisms and adjustments. These strategies are pivotal in crafting a learning environment that responds to the individual needs of each student, thereby enhancing the overall effectiveness of the educational experience.

The personalization of learning experiences in DEERs is central to catering to the diverse needs of learners. According to Brusilovsky and Millán (Brusilovsky and Millán, 2007), adaptive learning systems should be capable of adjusting the educational content based on the individual learner's knowledge, preferences, and learning style. In the context of DEERs, this means developing systems that can track a learner's progress and modify the difficulty level or type of puzzles and tasks accordingly. Such personalization not only makes the learning experience more relevant and engaging for each student but also ensures that the educational content is delivered at a pace and complexity that is most suitable for their learning journey. This approach is rooted in the principles of Constructivism, which advocate for education to be learner-centered and adaptive to individual learning processes (Piaget, 1970).

Incorporating adaptive challenges in DEERs that evolve based on learner performance is crucial for maintaining an optimal learning curve. This adaptation can be achieved through intelligent tutoring systems or adaptive algorithms that analyze student responses and performance in real-time, adjusting the difficulty of tasks to match the learner's current skill level (VanLehn, 2006). This ensures that the challenges remain within the zone of proximal development for each learner, a concept introduced by Vygotsky (Vygotsky, 1978), which describes the optimal space for learning just beyond the learner's current abilities.

Real-time feedback mechanisms and adjustments are essential components of adaptive learning in DEERs. Timely and specific feedback allows learners to understand their performance, recognize areas that need improvement, and adjust their strategies accordingly (Hattie and Timperley, 2007). In DEERs, these feedback mechanisms can be integrated into the game design, providing immediate responses to learner actions and decisions. Such systems not only enhance learning by providing continuous guidance but also contribute to a dynamic and responsive learning environment. This real-time adjustment and feedback align with the principles of formative assessment, emphasizing the importance of ongoing evaluation and feedback in the learning process (Black and Wiliam, 1998).

Ensuring Inclusivity and Accessibility in DEER Design

In the realm of DEER, designing for inclusivity and accessibility is paramount. This commitment extends to accommodating diverse learning needs and styles, integrating accessibility features based on universal design principles, and incorporating cultural sensitivity and global perspectives. These aspects are crucial in creating DEERs that are not only educational and engaging but also accessible and respectful to all learners, regardless of their background or abilities.

Addressing diverse learning needs and styles is essential in DEER design. Gardner's theory of multiple intelligences (Gardner, 1983) emphasizes the importance of recognizing and catering to different ways individuals learn, including linguistic, logical-mathematical, spatial, bodily-kinesthetic, musical, interpersonal, intrapersonal, and naturalistic intelligences. DEERs can be designed to engage these various intelligences by including a variety of puzzles and challenges that require different skills and ways of thinking. This approach not only ensures inclusivity but also enhances the educational value of DEERs by providing a rich, multifaceted learning experience (Tomlinson, 1999).

Incorporating accessibility features in DEER design is a critical aspect of ensuring all learners can participate fully. Universal Design for Learning (UDL) principles (Meyer et al., 2014) advocate for the creation of learning experiences that offer multiple means of engagement, representation, action, and expression. This includes designing DEERs with adjustable difficulty levels, alternative formats for presenting information (such as text-to-speech options for visually impaired learners), and multiple ways for students to express their understanding and solutions. By embedding these features, DEERs can become accessible to learners with a range of physical, cognitive, and sensory abilities, thereby creating an inclusive learning environment.

Cultural sensitivity and the inclusion of global perspectives are vital in DEER design. In an increasingly interconnected world, it is essential that educational content respects and reflects diverse cultures and viewpoints. Banks (Banks, 2008) emphasizes the importance of incorporating multicultural education principles to foster an understanding and appreciation of diverse cultural experiences. DEERs should avoid cultural stereotypes and instead present content that is culturally inclusive and sensitive. This not only helps in creating a respectful learning environment but also prepares learners to operate in a global context, enhancing their cultural awareness and competence (Gay and Banks, 2010).

Challenges and Considerations in DEER Game Design

Designing DEERs presents a unique amalgamation of challenges and considerations that blend the realms of education and game design. This landscape is marked by the need to address common design challenges, the development of strategies for effective implementation, and an anticipation of future trends and evolving design paradigms. These elements are crucial in crafting DEERs that are not only engaging and interactive but also pedagogically sound and forward-looking.

Common design challenges in DEER game design often revolve around balancing educational objectives with engaging gameplay. As noted by Prensky (Prensky, 2001), the key challenge lies in creating experiences that are both fun and educational. Designers must consider the diversity of learners, ensuring that the content is accessible and inclusive (Tomlinson, 1999). Technical constraints, such as limited resources and varying levels of technological infrastructure among the user base, also pose significant challenges (Egenfeldt-Nielsen, 2007). These issues require a thoughtful approach that integrates pedagogical goals seamlessly with engaging game mechanics.

Effective implementation of DEERs requires a multi-faceted strategy. This involves the application of sound instructional design principles, such as those outlined in Gagné's Nine Events of Instruction (Gagné, 1985), to ensure that educational outcomes are effectively met. Gagné's Nine Events of Instruction and

Their Relevance to Digital Educational Escape Room Design are summarized in Table 7. Incorporating elements of gamification, like points, badges, and leader boards, can significantly enhance engagement and motivation, as suggested by Kapp (Kapp, 2012). Iterative design and testing are critical, allowing for adjustments and refinements based on feedback from real-world implementations. Collaborative development processes, involving educators, game designers, and subject matter experts, can also ensure that the final product is both educationally effective and technically robust (Sheldon, 2020).

Looking ahead, DEER game design is poised to evolve with emerging technologies and changing educational paradigms. The increasing integration of virtual and augmented reality technologies offers new avenues for creating immersive and interactive learning experiences. The growing emphasis on personalized learning suggests that future DEERs will likely incorporate adaptive learning technologies, tailoring challenges and content to individual learner profiles (Baker and Siemens, 2014). As the field of educational gaming matures, there is a trend towards more nuanced and complex narratives, which can enhance both engagement and the depth of learning (Squire, 2011). These evolving trends point towards a future where DEERs are more immersive, personalized, and narratively rich, offering exciting possibilities for education.

DEER game design, while challenging, presents a rich opportunity to blend educational content with engaging gameplay. Addressing design challenges, implementing effective strategies, and staying abreast of future trends are key to creating DEERs that are both educationally impactful and technologically innovative.

Concluding Remarks

The exploration of DEER game design culminates in a synthesis of its key principles, a reflection on the interplay between education and gameplay, and a contemplation of future research and development in this innovative educational field. This conclusion draws together the core aspects that make DEER an impactful educational tool, while also looking ahead to how they might evolve and continue to transform learning experiences.

The design of DEER games is grounded in several fundamental principles that ensure their educational efficacy and engagement. Central to these is the alignment of educational objectives with immersive game mechanics, ensuring that each aspect of gameplay contributes meaningfully to the learning outcomes. The integration of narrative and multimedia elements further enhances this alignment, creating a rich and engaging context for learning. The principles of universal design and inclusivity are paramount, ensuring that DEER games are accessible and engaging for a diverse range of learners. These key principles, underscored by the theories of Vygotsky

(Vygotsky, 1978) on social learning and Gee (Gee, 2003) on game-based learning, form the cornerstone of effective DEER game design.

The integration of education and gameplay in DEER presents a unique and innovative approach to learning. This integration leverages the motivational elements of gameplay to enhance educational engagement, making learning an active and enjoyable experience. The balance between challenge and skill, as described by Csikszentmihalyi's concept of flow (Csikszentmihalyi, 1990), is crucial in this context, ensuring that learners are neither overwhelmed nor under-challenged. This interplay between education and gameplay not only supports the acquisition of knowledge and skills but also fosters critical thinking, problem-solving, and collaboration.

Table 7. Gagné's nine events of instruction and their relevance to DEER design

Event	Description	Relevance to Digital Educational Escape Room Design
Gain Attention	Capture learners' attention to prepare them for learning.	Achieved through engaging narratives, immersive graphics, or intriguing puzzles that draw learners into the experience.
Inform Learners of the Objectives	Inform learners about what they will learn and be able to do by the end.	Objectives are integrated into the storyline, making learners aware of the goals they need to achieve to "escape" or complete the mission.
Stimulate Recall of Prior Learning	Remind learners of relevant knowledge they already have.	Puzzles or challenges require application of prior knowledge, encouraging recall and application of previous learning.
Present the Content	Present the material to be learned.	New information is presented through clues, narratives, and interactive elements embedded within the game environment.
Provide Learning Guidance	Help learners understand and integrate new information.	Hints or guidance are provided within the game to assist learners in solving puzzles without giving away the answers.
Elicit Performance	Require learners to demonstrate what they've learned.	Learners solve puzzles that require them to apply new knowledge or skills within the escape room.
Provide Feedback	Offer immediate feedback on learners' performance.	The game provides immediate responses to actions, indicating correct or incorrect solutions to puzzles.
Assess Performance	Evaluate learners' understanding.	Escape room completion status and time serve as an assessment, or specific in-game assessments are embedded.
Enhance Retention and Transfer	Facilitate the application of knowledge to new situations.	Designed to promote reflection and encourage application of learned knowledge in future contexts.

Looking forward, the field of educational DEER games is ripe for further research and development. Future endeavors could explore the long-term impacts of DEER games on learning outcomes, the potential for personalized learning experiences through adaptive technologies, and the integration of emerging technologies like virtual and augmented reality. Research could focus on the scalability of DEER games in various educational settings, their role in interdisciplinary learning, and their impact on learners with different needs and backgrounds. The ongoing evolution of DEER game design will undoubtedly continue to push the boundaries of educational innovation.

CLASSROOM MANAGEMENT AND FACILITATION WHILE USING DEERS

Introduction to Classroom Management in the Context of DEERs

The integration of DEERs into educational settings represents a significant shift in teaching and learning dynamics, necessitating a reevaluation of classroom management strategies. DEERs, which combine the engaging elements of escape rooms with educational content, offer unique challenges and opportunities for classroom management. This section explores the overview of DEERs in educational settings, their role in classroom management, and the transition process from traditional learning environments to DEER-based learning.

DEERs have gained prominence as innovative educational tools that combine the thrill of escape room challenges with curriculum-aligned learning objectives (Nicholson, 2014). These interactive environments engage students in a narrative-driven, problem-solving experience, fostering collaboration, critical thinking, and creativity (Veldkamp et al., 2021). By integrating academic content into puzzles and tasks, DEERs offer an immersive learning experience that differs significantly from traditional classroom settings. This hands-on approach aligns with experiential learning theories, emphasizing active engagement and reflection in the learning process (Kolb, 1984).

Effective classroom management in DEERs is crucial for maximizing their educational potential. Unlike traditional settings, where the teacher often leads and controls the learning process, DEERs require a more facilitative approach. The teacher's role shifts to guiding, monitoring, and supporting students as they navigate through the tasks. Classroom management in this context involves creating an environment conducive to exploration and learning, ensuring that all students are engaged and that the activities align with learning objectives. Managing group dynamics, time constraints, and resource allocation, becomes essential for the smooth operation of DEERs.

The transition from traditional learning environments to DEER-based ones requires careful planning and adaptation. Educators must consider how to integrate DEERs into their existing curriculum while addressing the diverse needs and learning styles of students (Poon Teng Fatt, 2000). This transition involves not only the adoption of new technologies and methodologies but also a shift in pedagogical perspectives. Educators must embrace a learner-centered approach, where students have more autonomy and responsibility for their learning (Hannafin et al., 1997). Training and professional development are crucial for teachers to effectively facilitate DEER activities and manage the classroom dynamics that these activities entail. The transition requires a rethinking of assessment strategies to align them with the interactive and collaborative nature of DEERs (Ifenthaler et al., 2012).

Preparing the Classroom for DEER Activities

The preparation of a classroom for Digital Educational Escape Room (DEER) activities is a multifaceted process that encompasses the physical setup and resource allocation, the integration of necessary technological requirements, and the adherence to safety considerations and protocols. This preparation is crucial for the effective implementation of DEER activities, ensuring that they are not only educationally beneficial but also accessible, safe, and engaging for all students.

The physical arrangement of a classroom for DEER activities involves more than just the spatial organization; it requires careful consideration of the resources available and how they can be utilized to maximize student engagement and learning. According to Jonassen et al. (Jonassen, 1999), creating an environment conducive to active learning involves the strategic placement of furniture and resources to facilitate collaboration and interaction. In the context of DEERs, this might mean arranging desks in a way that promotes teamwork or setting up stations with different puzzles or challenges. The allocation of resources, such as physical props or printed materials, must be planned to ensure that each group of students has equal access to the tools necessary for solving the DEER challenges.

The integration of technology in DEERs is not just a supplementary aspect but a core component of their design and implementation. As outlined by Gee (Gee, 2003), effective educational games leverage technology to create immersive and interactive learning experiences. This requires a thorough assessment of the technological tools and software necessary for DEER activities, such as computers, tablets, or specific educational gaming platforms. The setup of these technologies must be tested prior to the activity to ensure functionality and ease of use, as technical difficulties can disrupt the flow of the game and impact the learning experience.

Safety in DEER activities extends beyond physical safety to include digital safety and the emotional well-being of students. Educators must establish clear protocols to ensure a safe learning environment. This involves not only the physical arrangement of the classroom to prevent accidents but also the implementation of guidelines for the responsible use of technology. As DEERs often involve problem-solving under time constraints, it is important to create an atmosphere where students feel comfortable taking risks and making mistakes, as these are integral parts of the learning process (Dweck, 2006).

The preparation of the classroom for DEER activities is a comprehensive process that requires careful planning and consideration of various factors. By effectively managing the physical setup, technological integration, and safety protocols, educators can create a conducive environment for DEERs that enhances the learning experience and ensures the safety and well-being of all students.

Facilitating DEER Activities

The facilitation of DEERs in a learning environment is an intricate process that requires educators to adopt multifaceted roles. This facilitation is not just about overseeing the activity but involves actively guiding students, managing group dynamics, and fostering collaboration. Each aspect of this facilitative role is crucial in ensuring that DEERs are not only engaging and enjoyable but also effective in achieving their educational objectives.

The role of the educator in DEER activities extends beyond traditional teaching; it embodies the role of a facilitator. This shift from a didactic approach to a facilitative one is essential for the success of DEERs. According to Vygotsky's theory of social constructivism, learning is a social process wherein the educator acts more as a collaborator than a director (Vygotsky, 1978). In DEERs, the educator must guide students through the process of discovery and problem-solving, encouraging exploration and inquiry. This facilitative approach empowers students, giving them the agency to take charge of their learning while the educator provides the necessary support and scaffolding.

Guiding students through the challenges posed in DEERs requires a delicate balance. The educator must ensure that the challenges are neither too easy nor too difficult, thus maintaining an optimal level of engagement and learning. Aachieving this balance is key to keeping students in a state of heightened focus and enjoyment (Csikszentmihalyi, 1990). The educator's role in this context is to provide subtle guidance, helping students navigate through challenges without overtly directing them. This involves providing hints, encouraging critical thinking, and prompting reflection, which aligns with constructivist principles of learning.

An integral part of facilitating DEER activities is managing group dynamics and fostering collaboration among students. Collaborative learning, as highlighted by Johnson and Johnson's cooperative learning theory, is predicated on the interdependence of group members and the development of social skills (Johnson and Johnson, 1994). In DEERs, the educator must ensure that group interactions are positive and productive, guiding students in effective communication, conflict resolution, and collaborative problem-solving. This not only enhances the learning experience but also teaches students valuable interpersonal skills.

Behavior Management Strategies in DEER Sessions

Implementing behavior management strategies in DEER sessions is a nuanced process that involves establishing clear expectations and norms, addressing behavioral challenges in interactive settings, and utilizing positive reinforcement and motivational techniques. These strategies are essential for maintaining a conducive learning environment and ensuring that all participants are engaged and focused on the educational goals of the DEER.

The foundation of effective behavior management in DEER sessions begins with the establishment of clear expectations and norms. As Jones and Jones (Jones and Jones, 2016) highlight, setting ground rules at the outset of any educational activity helps in creating a structured environment where students understand what is expected of them. In the context of DEER, this involves articulating the objectives of the session, the rules of engagement, and the desired conduct within the escape room scenario. These norms should be collaboratively developed with the students to foster a sense of ownership and responsibility (Marzano, 2007).

DEER sessions, by their nature, are highly interactive and dynamic, which can sometimes lead to behavioral challenges. Kagan et al. (Kagan et al., 2009) emphasize the importance of proactive strategies in managing group dynamics. This includes monitoring group interactions, identifying potential conflict areas, and intervening early to guide students back on track. In DEER activities, educators need to be vigilant and responsive, ensuring that the collaborative nature of the task does not devolve into competition or conflict, but rather fosters a cooperative learning environment (Slavin, 1995).

Positive reinforcement and motivational techniques are integral to maintaining student engagement and promoting desired behaviors in DEER sessions. The use of praise, rewards, and positive feedback, as suggested by Skinner (Skinner, 1957), can significantly influence student behavior and motivation. In DEER activities, educators can employ these techniques by acknowledging effective collaboration, problem-solving skills, and other positive behaviors exhibited during the session.

Such reinforcement not only encourages students but also creates a positive and supportive learning atmosphere (Deci et al., 1991).

Incorporating DEERs into Diverse Educational Contexts

The incorporation of DEERs into diverse educational contexts necessitates a thoughtful and adaptable approach. This involves tailoring DEERs to accommodate different age groups and abilities, integrating them across various subject areas, and customizing them to cater to both individual and group learning needs. Such adaptability not only enhances the inclusivity of DEERs but also maximizes their educational potential.

Adapting DEERs to suit various age groups and abilities is essential for ensuring that all learners can benefit from this innovative educational tool. Gardner's theory of multiple intelligences (Gardner, 1983) supports the idea that educational activities should cater to different learning styles and abilities (see Table 8).

Gardner's Theory of Multiple Intelligences suggests that individuals possess different kinds of intelligences through which they understand and interact with the world. In adapting DEERs to suit various age groups and abilities, leveraging these multiple intelligences can enhance engagement and learning outcomes.

Linguistic Intelligence involves sensitivity to spoken and written language and the ability to use language effectively. In DEERs, this can be supported by incorporating word puzzles, storytelling elements, and reading tasks. For younger participants or those with developing reading skills, the language used can be simplified, while more complex narratives and vocabulary can challenge older or more advanced learners.

Logical-Mathematical Intelligence is characterized by the capacity to analyze problems logically and perform mathematical operations. DEERs can cater to this intelligence by including logic puzzles, pattern recognition tasks, and mathematical challenges. Adjusting the difficulty level of these puzzles ensures that they are appropriate for the cognitive abilities of different age groups, making them accessible yet stimulating.

Spatial Intelligence refers to the ability to recognize and manipulate spatial patterns. To engage learners with strong spatial skills, DEERs can feature visual puzzles, mazes, and tasks that require spatial reasoning. The complexity of these tasks can be tailored to suit the developmental stages of the participants, ensuring that younger children are not overwhelmed while older participants remain challenged.

Bodily-Kinesthetic Intelligence involves using one's physical self to solve problems or create products. Although DEERs are digital, they can still engage this intelligence by incorporating interactive elements that simulate physical movement or manipulation of virtual objects. For example, drag-and-drop activities or virtual simulations can be designed with varying degrees of complexity to accommodate different age groups and physical abilities.

Musical Intelligence is the skill in performance, composition, and appreciation of musical patterns. DEERs can integrate this intelligence by including musical puzzles, rhythm-based challenges, and sound recognition tasks. Younger participants might enjoy simple melody matching games, while older learners could be challenged with composing sequences or identifying complex musical patterns.

Interpersonal Intelligence is the capacity to understand and interact effectively with others. DEERs can promote this intelligence by designing collaborative challenges that require teamwork and communication. These tasks encourage social interaction among participants, which can be particularly beneficial for learners across all age groups, fostering cooperation and enhancing problem-solving through collective effort.

Intrapersonal Intelligence involves self-awareness and the ability to understand one's own emotions and motivations. DEERs can support this intelligence by including reflective tasks and personal decision-making scenarios. These elements allow participants to engage in self-guided challenges, encouraging them to reflect on their choices and thought processes. The content can be adjusted to align with the maturity levels of different age groups, ensuring appropriateness and relevance.

Naturalistic Intelligence is the ability to recognize and categorize natural objects and phenomena. DEERs can engage this intelligence by integrating nature-themed puzzles, environmental challenges, and classification tasks. For younger learners, this might involve simple identification of animals or plants, while older participants could engage in more complex ecological problem-solving scenarios.

Existential Intelligence pertains to sensitivity and capacity to tackle deep questions about human existence. DEERs can address this intelligence by incorporating philosophical puzzles and ethical dilemmas. These are more suitable for older participants who are capable of abstract thinking and can appreciate discussions around meaning, purpose, and the human condition.

By thoughtfully integrating elements that cater to each of Gardner's multiple intelligences, DEERs can be adapted to meet the diverse needs of various age groups and abilities. This approach ensures that each participant can engage with the content in a way that aligns with their strengths, thereby enhancing motivation and facilitating deeper learning. It also promotes inclusivity by recognizing and valuing different types of intelligence, allowing all learners to contribute meaningfully and experience success within the escape room environment. For example, for younger students, DEERs can be simplified to focus on basic skills and concepts, using more tangible and less complex puzzles. For older or more advanced students, the challenges can be made more intricate, requiring higher-order thinking skills, as outlined in Bloom's taxonomy (Bloom, 1956). Considerations for students with special needs are crucial. DEERs should be designed with accessibility in mind, providing alternative ways to participate and engage with the content (Tomlinson and McTighe, 2006).

Table 8. Adapting DEERs using Gardner's multiple intelligences

Intelligence	Description	Adapting DEERs to Suit Various Age Groups and Abilities
Linguistic Intelligence	Sensitivity to spoken and written language; ability to learn languages.	Include word puzzles, storytelling elements, and reading tasks tailored to the participants' reading levels and age groups.
Logical-Mathematical Intelligence	Capacity to analyze problems logically and perform mathematical operations.	Incorporate logic puzzles and mathematical challenges appropriate for the cognitive abilities of different age groups.
Spatial Intelligence	Ability to recognize and manipulate spatial patterns.	Use visual puzzles, mazes, and spatial reasoning tasks suitable for the participants' developmental stages.
Bodily- Kinesthetic Intelligence	Using one's body to solve problems or create products.	Include interactive tasks requiring movement or manipulation of virtual objects, adapted to the physical abilities of various age groups.
Musical Intelligence	Skill in performance, composition, and appreciation of musical patterns.	Incorporate musical puzzles, rhythm challenges, and sound-based tasks that appeal to different age groups and abilities.
Interpersonal Intelligence	Understanding and interacting effectively with others.	Design collaborative challenges and social interaction tasks that encourage teamwork among participants of all ages.
Intrapersonal Intelligence	Understanding oneself and one's thoughts and feelings.	Include reflective tasks, personal decision-making scenarios, and self-guided challenges suitable for the participants' maturity levels.
Naturalistic Intelligence	Recognizing and categorizing natural objects.	Integrate nature-themed puzzles, classification tasks, and environmental challenges adjusted for the knowledge levels of different age groups.
Existential Intelligence	Sensitivity to deep questions about human existence.	Incorporate philosophical puzzles and ethical dilemmas appropriate for older participants capable of abstract thinking.

DEERs offer a versatile platform that can be integrated across various subject areas, enriching the curriculum and providing a multidisciplinary approach to learning. In science education, for instance, DEERs can simulate laboratory scenarios where students solve puzzles related to scientific concepts (Bybee, 2002). In humanities, they can be themed around historical events or literary works, encouraging students to engage deeply with the material (Bruner, 1996). The key is to align the DEERs' themes and challenges with the subject matter, ensuring that they complement and enhance the learning objectives of the curriculum.

Customizing DEERs to meet individual and group learning needs is pivotal in maximizing their educational impact. Vygotsky's social development theory, which emphasizes the importance of social interaction in learning (Vygotsky, 1978), suggests that DEERs should be designed to promote collaboration and communication among students. However, they should also allow for individualized learning paths, catering to students who may need more time or different types of challenges. This customization can be achieved through adaptive learning technologies that adjust the difficulty level based on the learner's performance (Shute and Zapata-Rivera, 2012).

In conclusion, incorporating DEERs into diverse educational contexts requires a multifaceted approach that considers the varied needs and abilities of learners, the integration with different subject areas, and the customization to individual and group learning needs. This adaptability ensures that DEERs are not only inclusive but also effective in providing enriching and engaging learning experiences.

Feedback Mechanisms in DEER Facilitation

In the realm of DEERs, the facilitation process is significantly enhanced by effective feedback mechanisms. These mechanisms involve providing constructive feedback in real-time, encouraging peer feedback and self-assessment, and utilizing feedback for the continuous improvement of DEERs. Each of these aspects plays a pivotal role in ensuring that DEERs are not only engaging and educational but also evolve in response to the needs and experiences of learners. The provision of real-time, constructive feedback in DEERs is essential for facilitating an effective learning experience. Hattie and Timperley (2007) emphasize the importance of feedback in enhancing learning, noting that it should be timely, relevant, and geared towards improvement. In the context of DEERs, this means offering immediate responses to students' actions and decisions within the game. Such feedback helps students understand where they are in the learning process, what they are doing well, and where they need improvement. It serves as a guide, helping them navigate the challenges of the escape room and linking their actions to learning outcomes.

Peer feedback and self-assessment are integral components of the learning process in DEERs. Black and Wiliam (1998) highlight the role of peer interaction in fostering learning, stating that students can benefit from evaluating each other's work and providing constructive feedback. In DEERs, students are often required to work in teams, making peer feedback a natural and valuable part of the experience. Self-assessment encourages learners to reflect on their own performance, fostering a deeper understanding of the material and a greater sense of responsibility for their learning (Sadler, 1989).

Feedback is not only beneficial for students but also for educators and DEER designers. Nicol and Macfarlane-Dick (2006) argue that feedback should be used as a tool for continuous improvement in educational settings. In the case of DEERs, feedback from students and educators can inform future design decisions, leading to enhancements in game mechanics, puzzle design, and educational content. This iterative process ensures that DEERs remain relevant, engaging, and educationally effective.

Professional Development for Educators in DEER Implementation

The implementation of DEERs in educational settings necessitates a concerted effort in professional development for educators. This development encompasses understanding the specific training needs for effective DEER facilitation, engaging in ongoing professional development opportunities, and fostering a culture of sharing best practices and collaborative learning among educators. Such comprehensive professional development is crucial to equip educators with the skills and knowledge necessary to effectively utilize DEERs as an innovative educational tool.

Effective facilitation of DEERs requires educators to possess a unique set of skills and knowledge. As noted by Guskey (2000), professional development should be tailored to meet the specific needs of educators in relation to the instructional strategies they are expected to implement. In the case of DEERs, this includes understanding the principles of game-based learning, developing skills in designing and managing DEER activities, and learning how to integrate these activities with curriculum objectives. Educators need training in technology tools used in DEERs, as well as in methods for assessing student learning within this context (Prensky, 2001).

Ongoing professional development is vital for educators to stay abreast of evolving practices in DEER implementation. Joyce and Showers (2002) argue that continuous learning opportunities are essential for the successful integration of new teaching strategies. For DEERs, this means regular workshops, webinars, and conferences where educators can learn about the latest developments in game-based learning, share experiences, and gain insights into effective DEER design and facilitation. Such opportunities not only enhance educators' skills but also keep them motivated and engaged in their professional growth (Fullan, 2007).

Collaborative learning and the sharing of best practices among educators play a pivotal role in the successful implementation of DEERs. Wenger's (Wenger, 1998) concept of communities of practice emphasizes the value of collaborative learning in professional development. By sharing experiences, challenges, and successes in DEER implementation, educators can learn from one another, develop new strategies, and collectively advance their practice. This collaborative approach fosters a supportive professional community that encourages innovation and continuous improvement in educational practices (Darling-Hammond et al., 2017).

Challenges and Solutions in DEER Classroom Management

The integration of DEERs into classroom management presents a unique set of challenges, necessitating strategic approaches for effective implementation. These challenges range from logistical constraints to ensuring educational efficacy, each requiring a thoughtful solution. Real-world examples and case studies further illuminate how these challenges can be effectively addressed, providing practical insights for educators.

Educators face several common challenges when implementing DEERs in classroom settings. Firstly, logistical issues such as time constraints, space limitations, and resource availability often pose significant hurdles (Eggen and Kauchak, 2010). Educators must grapple with ensuring that DEER activities align with curriculum standards and learning objectives, a challenge highlighted in the works of Marzano (Marzano, 2007). Another critical issue is the varying levels of student engagement and participation, which can impact the overall effectiveness of the DEER experience (Kagan et al., 2009).

Strategic approaches are essential to overcome these challenges. To address logistical issues, educators can adopt flexible DEER designs that can be adjusted based on available resources and time (Tomlinson and McTighe, 2006). For aligning DEERs with curriculum standards, backward design principles can be employed, where activities are created with the end goals in mind, ensuring they meet educational objectives (Wiggins and McTighe, 2005). To enhance student engagement, differentiated instruction strategies can be used, tailoring DEER activities to the diverse needs and interests of students (Tomlinson, 1999). This approach not only increases participation but also caters to different learning styles, making the DEER experience more inclusive and effective.

Effective classroom management and facilitation in the context of DEERs hinge on several strategic approaches. Primarily, it involves the careful planning and structuring of the classroom environment to accommodate DEER activities, as advocated by Weinstein and Mignano (Weinstein and Mignano, 2003) in their discussion of classroom management strategies. This includes the logistical aspects of space, resources, and technology integration. The role of the educator evolves from traditional instructional methods to a more facilitative role. This shift requires educators to develop skills in guiding and supporting students through DEER activities, managing group dynamics, and fostering a collaborative learning environment.

Looking ahead, the future of classroom management with DEERs is poised for continuous evolution and innovation. The growing interest in game-based learning and educational technology suggests that DEERs will become an increasingly integral part of classroom experiences. This future will likely involve further customization of DEERs to cater to a wider range of learners, incorporating adaptive learning technologies for personalized educational experiences (Shute and Zapata-Rivera, 2012). The integration of emerging technologies such as augmented reality (AR) and virtual reality (VR) could further enhance the immersive nature of DEERs, opening new avenues for interactive and experiential learning (Dunleavy et al., 2008). Educators will continue to play a crucial role in this evolution, necessitating ongoing professional development to equip them with the skills and knowledge required to effectively implement and facilitate DEERs (Guskey, 2000). The collaborative aspect of DEERs will also likely expand, with increased emphasis on global collaboration and cross-cultural exchanges, fostering a more interconnected and diverse learning environment.

DEERs represent a progressive and impactful approach in education, blending traditional pedagogical strategies with innovative game-based learning. As educators and institutions continue to embrace and refine this approach, DEERs are poised to significantly shape the future of classroom management and student engagement, offering enriching, inclusive, and dynamic learning experiences.

REFERENCES

Adams, E. (2010). *Fundamentals of Game Design*. New Riders.

Anderson, L. W., & Krathwohl, D. R. (2001). *A Taxonomy for Learning, Teaching, and Assessing: A Revision of Bloom's Taxonomy of Educational Objectives*. Longman.

Angelo, T., & Cross, P. (1993). *Classroom Assessment Techniques: A Handbook for College Teachers*. Jossey-Bass.

Baker, R., & Siemens, G. (2014). *Educational Data Mining and Learning Analytics*. Cambridge University Press., DOI: 10.1017/CBO9781139519526.016

Bandura, A. (1977). *Social Learning Theory*. Prentice Hall.

Banks, J. A. (2008). *An Introduction to Multicultural Education*. Pearson.

Barab, S. A., Thomas, M. K., Dodge, T., Carteaux, R., & Tuzun, H. (2005). Making learning fun: Quest atlantis, a game without guns. *Educational Technology Research and Development*, 53(1), 86–107. DOI: 10.1007/BF02504859

Black, P., & Wiliam, D. (1998). Assessment and classroom learning. *Assessment in Education: Principles, Policy & Practice*, 5(1), 7–74. DOI: 10.1080/0969595980050102

Bloom, B. S. (1956). *Taxonomy of Educational Objectives, Handbook I: The Cognitive Domain*. David McKay Co Inc.

Bonwell, C. C., & Eison, J. A. (1991). *Active Learning: Creating Excitement in the Classroom*. ASHE-ERIC Higher Education Report. School of Education and Human Development, George Washington University, Washington DC.

Bruner, J. (1991). The narrative construction of reality. *Critical Inquiry*, 18(1), 1–21. DOI: 10.1086/448619

Bruner, J. (1996). *The Culture of Education*. Harvard University Press. DOI: 10.4159/9780674251083

Brusilovsky, P., & Millán, E. (2007). *User Models for Adaptive Hypermedia and Adaptive Educational Systems*. Springer Berlin Heidelberg., DOI: 10.1007/978-3-540-72079-9_1

Bybee, R. W. (2002). *Learning Science and the Science of Learning*. NSTA press. DOI: 10.2505/9780873552080

CAST. (2018). Universal design for learning guidelines version 2.2. http://udlguidelines.cast.org

Clark, R. C., & Mayer, R. E. (2011). *E-Learning and the Science of Instruction: Proven Guidelines for Consumers and Designers of Multimedia Learning*. Pfeiffer. DOI: 10.1002/9781118255971

Cole, J. (2009). *Using Moodle: Teaching with the Popular Open-source Course Management System*. O'Reilly Media, Inc.

Conati, C. (2009). Intelligent tutoring systems: New challenges and directions. *Annual Review of Computer Science*, 3, 555–580.

Csikszentmihalyi, M. (1990). *Flow: The Psychology of Optimal Experience*. Harper & Row.

Darling-Hammond, L., Hyler, M. E., & Gardner, M. (2017). *Effective Teacher Professional Development*. Learning Policy Institute. DOI: 10.54300/122.311

Deci, E. L., Vallerand, R. J., Pelletier, L. G., & Ryan, R. M. (1991). Motivation and education: The self-determination perspective. *Educational Psychologist*, 26(3–4), 325–346. DOI: 10.1080/00461520.1991.9653137

Dickey, M. D. (2005). Engaging by design: How engagement strategies in popular computer and video games can inform instructional design. *Educational Technology Research and Development*, 53(2), 67–83. DOI: 10.1007/BF02504866

Dunleavy, M., Dede, C., & Mitchell, R. (2008). Affordances and limitations of immersive participatory augmented reality simulations for teaching and learning. *Journal of Science Education and Technology*, 18(1), 7–22. DOI: 10.1007/s10956-008-9119-1

Dweck, C. (2006). *Mindset: The New Psychology of Success*. Random House.

Edyburn, D. L. (2010). Would you recognize universal design for learning if you saw it? ten propositions for new directions for the second decade of udl. *Learning Disability Quarterly*, 33(1), 33–41. DOI: 10.1177/073194871003300103

Egan, K. (1986). *Teaching as Story Telling*. Althouse Press.

Egenfeldt-Nielsen, S. (2007). Educational Potential of Computer Games. *Continuum*.

Eggen, P., & Kauchak, D. (2010). *Educational Psychology: Windows on Classrooms*. Pearson.

Falk, J. H., & Dierking, L. D. (2016). *The Museum Experience Revisited*. Routledge., DOI: 10.4324/9781315417851

Fullan, M. (2007). *The New Meaning of Educational Change*. Teachers College Press.

Gagné, R. M. (1985). *The Conditions of Learning and Theory of Instruction*. Holt, Rinehart & Winston.

Gardner, H. (1983). *Frames of Mind: The Theory of Multiple Intelligences*. Basic Books.

Garris, R., Ahlers, R., & Driskell, J. E. (2002). Games, motivation, and learning: A research and practice model. *Simulation & Gaming*, 33(4), 441–467. DOI: 10.1177/1046878102238607

Gay, G., & Banks, J. A. (2010). *Culturally Responsive Teaching: Theory, Research, and Practice*. Teachers College Press.

Gee, J. P. (2003). *What Video Games Have to Teach Us About Learning and Literacy*. Palgrave Macmillan. DOI: 10.1145/950566.950595

Guskey, T. R. (2000). *Evaluating Professional Development*. Corwin Press.

Hannafin, M., Hill, J., & Land, S. (1997). Student-centered learning and interactive multimedia: Status, issues, and implications. *Contemporary Education*, 68, 94–99.

Hattie, J., & Timperley, H. (2007). The power of feedback. *Review of Educational Research*, 77(1), 81–112. DOI: 10.3102/003465430298487

Hmelo-Silver, C., Duncan, R. G., & Chinn, C. A.HMELO-SILVER. (2007). Scaffolding and achievement in problem-based and inquiry learning: A response to kirschner, sweller, and clark (2006). *Educational Psychologist*, 42(2), 99–107. DOI: 10.1080/00461520701263368

Ifenthaler, D., Eseryel, D., & Ge, X. (Eds.). (2012). *Assessment in Game-Based Learning: Foundations, Innovations, and Perspectives*. Springer. DOI: 10.1007/978-1-4614-3546-4

Jacobs, H. H. (1989). *Interdisciplinary Curriculum: Design and Implementation*. Association for Supervision and Curriculum Development.

Jenkins, H. (2004). Game design as narrative architecture. *Computer*, 44, 118–130.

Johnson, D. W., & Johnson, R. T. (1994). *Learning Together and Alone: Cooperative, Competitive, and Individualistic Learning*. Allyn and Bacon.

Jonassen, D. H. (1999). Designing constructivist learning environments. In Reigeluth, C. M. (Ed.), *Instructional-design Theories and Models: A New Paradigm of Instructional Theory* (Vol. II, pp. 215–239). Lawrence Erlbaum Associates.

Jones, V., & Jones, L. (2016). *Comprehensive Classroom Management: Creating Communities of Support and Solving Problems*. Pearson.

Joyce, B., & Showers, B. (2002). *Student Achievement through Staff Development*. ASCD.

Kagan, S., Kyle, P., & Scott, S. (2009). *Managing Diverse Classrooms: How to Build on Students' Cultural Strengths*. ASCD.

Kapp, K. (2012). *The Gamification of Learning and Instruction: Game-Based Methods and Strategies for Training and Education*. Pfeiffer.

Keller, J. M. (1987). Development and use of the arcs model of instructional design. *Journal of Instructional Development : JID*, 10(3), 2–10. DOI: 10.1007/BF02905780

Kirschner, P. A., Sweller, J., & Clark, R. E. (2006). Why minimal guidance during instruction does not work: An analysis of the failure of constructivist, discovery, problem-based, experiential, and inquiry-based teaching. *Educational Psychologist*, 41(2), 75–86. DOI: 10.1207/s15326985ep4102_1

Kolb, D. A. (1984). *Experiential Learning: Experience as the Source of Learning and Development*. Prentice Hall.

Laal, M., & Ghodsi, S. M. (2012). Benefits of collaborative learning. *Procedia: Social and Behavioral Sciences*, 31, 486–490. DOI: 10.1016/j.sbspro.2011.12.091

Lave, J., & Wenger, E. (1991). *Situated Learning: Legitimate Peripheral Participation*. Cambridge University Press., DOI: 10.1017/CBO9780511815355

Macan, T. H. (1994). Time management: Test of a process model. *The Journal of Applied Psychology*, 79(3), 381–391. DOI: 10.1037/0021-9010.79.3.381

Marzano, R. J. (2007). *The Art and Science of Teaching: A Comprehensive Framework for Effective Instruction*. ASCD.

Mayer, R. E. (2001). *Multimedia Learning*. Cambridge University Press. DOI: 10.1017/CBO9781139164603

Meyer, A., Rose, D. H., & Gordon, D. (2014). *Universal Design for Learning: Theory and Practice*. CAST Professional Publishing.

Moreno, R., & Mayer, R. E. (2007). Interactive multimodal learning environments. *Educational Psychology Review*, 19(3), 309–326. DOI: 10.1007/s10648-007-9047-2

Murray, J. H. (1997). *Hamlet on the Holodeck: The Future of Narrative in Cyberspace*. The MIT Press.

Narciss, S. (2007). Feedback strategies for interactive learning tasks. In Spector, J. M., Merrill, M. D., van Merrienboer, J. J. G., & Driscoll, M. P. (Eds.), *Handbook of Research on Educational Communications and Technology* (3rd ed.). Lawrence Erlbaum Assoc.

Newcombe, N. S., & Stieff, M. (2012). Six myths about spatial thinking. *International Journal of Science Education*, 34(6), 955–971. DOI: 10.1080/09500693.2011.588728

Nicholson, S. (2014). *A RECIPE for Meaningful Gamification*. Springer International Publishing., DOI: 10.1007/978-3-319-10208-5_1

Nicol, D. J., & Macfarlane-Dick, D. (2006). Formative assessment and self-regulated learning: A model and seven principles of good feedback practice. *Studies in Higher Education*, 31(2), 199–218. DOI: 10.1080/03075070600572090

NIELSEN, J. (1993). *The Usability Engineering Lifecycle*, page 71–114. Elsevier, DOI: .DOI: 10.1016/B978-0-08-052029-2.50007-3

Norman, D. (2004). *Emotional Design: Why We Love (or Hate) Everyday Things*. Basic Books.

Perkins, D. N., & Salomon, G. (1992). Transfer of learning. In Husén, T. and Postlethwaite, T. N., editors, *The International Encyclopedia of Education*, pages 425–441. Pergamon, Oxford, 2 edition.

Piaget, J. (1970). Piaget's theory. In Mussen, P. H. (Ed.), *Carmichael's Manual of Child Psychology* (3rd ed., Vol. 1). Wiley.

Poon Teng Fatt, J. (2000). Understanding the learning styles of students: Implications for educators. *The International Journal of Sociology and Social Policy*, 20(11/12), 31–45. DOI: 10.1108/01443330010789269

Popham, W. J. (2008). *Transformative Assessment*. ASCD.

Prensky, M. (2001). *Digital Game-Based Learning*. McGraw-Hill.

Rouse, R. (2005). *Game Design: Theory and Practice*. Wordware Publishing, Inc.

Ryan, M.-L. (2001). *Narrative as Virtual Reality: Immersion and Interactivity in Literature and Electronic Media*. Johns Hopkins University Press.

Sadler, D. R. (1989). Formative assessment and the design of instructional systems. *Instructional Science*, 18(2), 119–144. DOI: 10.1007/BF00117714

Schell, J. (2008). *The Art of Game Design: A Book of Lenses*. Morgan Kaufmann. DOI: 10.1201/9780080919171

Schön, D. (1983). *The Reflective Practitioner: How Professionals Think in Action.* Temple Smith.

Sheldon, L. (2020). *The Multiplayer Classroom: Designing Coursework as a Game.* CRC Press., DOI: 10.1201/9780429285035

Shute, V. J. (2008). Focus on formative feedback. *Review of Educational Research*, 78(1), 153–189. DOI: 10.3102/0034654307313795

Shute, V. J., & Zapata-Rivera, D. (2012). Adaptive educational systems. In Durlach, P. J., & Lesgold, A. M. (Eds.), *Adaptive Technologies for Training and Education* (pp. 7–27). Cambridge University Press. DOI: 10.1017/CBO9781139049580.004

Siemens, G., & Long, P. (2011). Penetrating the fog: Analytics in learning and education. *TD Tecnologie Didattiche*, 22, 132–137.

Skinner, B. F. (1957). *Verbal Behavior.* Appleton-Century-Crofts. DOI: 10.1037/11256-000

Slavin, R. E. (1995). *Cooperative Learning: Theory, Research, and Practice.* Allyn and Bacon.

Squire, K. (2006). From content to context: Videogames as designed experience. *Educational Researcher*, 35(8), 19–29. DOI: 10.3102/0013189X035008019

Squire, K. (2011). *Video Games and Learning: Teaching and Participatory Culture in the Digital Age.* Teachers College Press.

Sweller, J. (1988). Cognitive load during problem-solving: Effects on learning. *Cognitive Science*, 12(2), 257–285. DOI: 10.1207/s15516709cog1202_4

Tomlinson, C. A. (1999). *The Differentiated Classroom: Responding to the Needs of All Learners.* ASCD.

Tomlinson, C. A., & McTighe, J. (2006). *Integrating Differentiated Instruction & Understanding by Design.* ASCD.

Van Eck, R. (2006). Digital game-based learning: It's not just the digital natives who are restless. *EDUCAUSE Review*, 41, 1–16.

van Merrienboer, J. J. G., Kirschner, P. A., & Kester, L. (2003). Taking the load off a learner's mind: Instructional design for complex learning. *Educational Psychologist*, 38(1), 5–13. DOI: 10.1207/S15326985EP3801_2

VanLehn, K. (2006). The behavior of tutoring systems. *International Journal of Artificial Intelligence in Education*, •••, 227–265.

Veldkamp, A., Knippels, M.-C. P. J., & van Joolingen, W. R. (2021). Beyond the early adopters: Escape rooms in science education. *Frontiers in Education*, 6, 622860. Advance online publication. DOI: 10.3389/feduc.2021.622860

Vygotsky, L. S. (1978). *Mind in Society: The Development of Higher Psychological Processes*. Harvard University Press.

Weinstein, C. S., & Mignano, A. J. (2003). *Elementary Classroom Management: Lessons for Research and Practice* (3rd ed.). McGraw-Hill.

Wenger, E. (1998). *Communities of Practice: Learning, Meaning, and Identity*. Cambridge University Press. DOI: 10.1017/CBO9780511803932

Wiggins, G., & McTighe, J. (2005). *Understanding by Design*. ASCD.

Wood, D., Bruner, J. S., & Ross, G. (1976). The role of tutoring in problem solving*. *Journal of Child Psychology and Psychiatry, and Allied Disciplines*, 17(2), 89–100. DOI: 10.1111/j.1469-7610.1976.tb00381.x PMID: 932126

KEY TERMS AND DEFINITIONS

Adaptive Difficulty: A design strategy in DEERs that adjusts the complexity of tasks based on the learner's performance to maintain an optimal level of challenge.

Bloom's Taxonomy: A hierarchical classification of learning objectives that guides the design of educational activities, ranging from knowledge recall to higher-order thinking skills.

Clue and Hint Mechanics: Tools embedded in DEERs to assist learners in overcoming obstacles without compromising the challenge and engagement of the experience.

Cognitive Engagement: The mental effort and involvement of learners in understanding and interacting with educational content.

Collaborative Challenges: Tasks in DEERs requiring teamwork and communication among participants, fostering interpersonal skills and group problem-solving.

Constructivism: An educational theory emphasizing active learning, where learners construct knowledge through experiences, problem-solving, and social interaction.

Digital Educational Escape Rooms (DEERs): Interactive, game-based learning environments designed to enhance engagement and learning through puzzles, narratives, and collaborative challenges.

Feedback Systems: Mechanisms in DEERs that provide learners with immediate and constructive feedback to guide their learning and reinforce understanding.

Flow Theory: A psychological concept describing a state of deep focus and immersion in activities that balance challenge and skill level.

Mayer's Cognitive Theory of Multimedia Learning: A theory stating that people learn better when information is presented through multiple modalities (e.g., text, images, and audio) rather than through text alone.

Narrative Integration: The incorporation of storytelling elements into educational activities to provide context, engagement, and relevance to the learning experience.

Puzzle-Based Learning: An approach where solving puzzles is used as a method to reinforce educational concepts and develop critical thinking skills.

Scaffolding Design: A pedagogical strategy that gradually increases the complexity of challenges, supporting learners as they build competence and confidence.

Social Learning Theory: A theory that emphasizes learning through observation, imitation, and interaction with others in a social context.

Time-Bound Challenges: The use of time limits in DEERs to enhance urgency, engagement, and time management skills among learners.

Universal Design for Learning (UDL): An educational framework that ensures learning experiences are accessible and effective for all students, regardless of their individual abilities or preferences.

Chapter 6
Advanced Digital Escape Room Design Technologies in Educational Practice

ABSTRACT

The chapter discusses the integration of immersive technologies like Virtual Reality (VR) and Augmented Reality (AR) in educational contexts, focusing on building engaging escape room experiences for learning. It outlines a conceptual framework based on experiential and constructivist learning theories, which emphasize hands-on, interactive environments where learners actively construct knowledge. VR/AR escape rooms facilitate this by creating immersive scenarios that enhance learning engagement, collaboration, and cognitive development. The chapter also addresses design principles for Digital Educational Escape Rooms (DEERs), highlighting the need for narrative-driven, interactive tasks that align with educational objectives. Active learning, feedback mechanisms, and collaboration are central to creating engaging experiences.

CONCEPTUAL FRAMEWORK FOR USING VR/AR IN EDUCATION

The emergence of Virtual Reality (VR) and Augmented Reality (AR) technologies has revolutionized various sectors, including education. These immersive technologies offer a novel paradigm for interactive learning environments, such as escape room experiences, which are gaining traction in educational settings. The framework for using VR and AR in education is anchored in the principles of

DOI: 10.4018/979-8-3693-4219-0.ch006

experiential learning (Kolb et al., 2014) and constructivist theories (Taber, 2011). Experiential learning emphasizes the importance of creating real-world or simulated experiences where learners can engage interactively and reflectively. VR and AR technologies facilitate this by offering immersive environments that mimic real-life scenarios or create fantastical settings for exploration and problem-solving. These environments are particularly conducive to escape room experiences, where learners must navigate challenges, puzzles, and tasks, often collaboratively, to achieve specific learning objectives.

Constructivist learning theory, another pillar of this framework, posits that knowledge is constructed by learners as they interact with their environment. VR and AR escape rooms are ideal for this type of learning, as they allow students to build knowledge actively and collaboratively. These technologies enable the creation of complex, dynamic environments where learners can experiment, manipulate objects, and witness the consequences of their actions in real-time, thereby facilitating a deeper understanding of the subject matter.

The integration of VR and AR in educational escape rooms also aligns with the principles of cognitive load theory (Van Merriënboer and Sweller, 2005). This theory emphasizes the importance of managing the amount of information and stimuli presented to learners to avoid overwhelming their cognitive capacities. VR and AR can be designed to present information progressively and interactively, allowing learners to process and assimilate knowledge at a manageable pace. This is particularly important in escape room scenarios, where the complexity of tasks can be adjusted to suit different learning levels and objectives.

The use of VR and AR in education benefits from the principles of gamification (Dicheva et al., 2015; Seaborn and Fels, 2015). Gamification involves applying game-design elements in non-game contexts, like education, to make learning more engaging and motivating (Deterding et al., 2011). Escape rooms, by their very nature, are game-like, and when designed with educational objectives in mind, they can significantly enhance learner engagement and motivation. These environments can incorporate elements such as points, levels, and rewards, aligned with educational goals to promote a deeper engagement with the learning material.

The design and implementation of VR and AR escape rooms in educational settings require careful consideration of various factors, including curriculum alignment, technological accessibility, and the development of appropriate content. The experiences must be meaningfully integrated into the curriculum to ensure that they complement and enhance traditional learning methods. Consideration must be given to the accessibility of these technologies, ensuring that they are inclusive and can be effectively used by a diverse range of learners.

Using VR and AR in education, particularly within the context of escape room experiences, is grounded in a blend of educational theories and practical considerations. These immersive technologies provide an engaging and dynamic platform for experiential and constructivist learning, allowing students to interact with and manipulate their learning environment in meaningful ways. By leveraging the principles of cognitive load theory and gamification, educators can design VR and AR escape rooms that are not only educational but also engaging and accessible to a diverse range of learners. The implementation of such technologies, however, must be carefully planned and executed. It requires a deep understanding of the educational objectives, the technology's capabilities, and the learners' needs and preferences. This approach ensures that the technology serves as a tool to enhance learning rather than a distraction or a gimmick. The ultimate goal is to create immersive, interactive experiences that align with educational goals, stimulate engagement, and foster a deeper understanding of complex concepts. Therefore, building immersive VR and AR escape room experiences in educational practice is not just about the technology itself, but about how it is integrated into the learning process. It involves a thoughtful blend of theory and practice, aiming to create rich, interactive environments that stimulate learning in an engaging, effective, and inclusive manner. As these technologies continue to evolve, so too will their applications in education, offering ever more innovative ways to engage learners and enrich their educational experiences.

Figure 1 represents the conceptual framework for using VR (Virtual Reality) and AR (Augmented Reality) in education. It visually maps out the key components and their interconnections as explained in Table 1. This framework outlines how various theoretical and practical aspects come together in the use of VR and AR technologies in educational contexts, emphasizing the multi-faceted nature of this integration.

Figure 1. Conceptual framework of VR/AR escape rooms

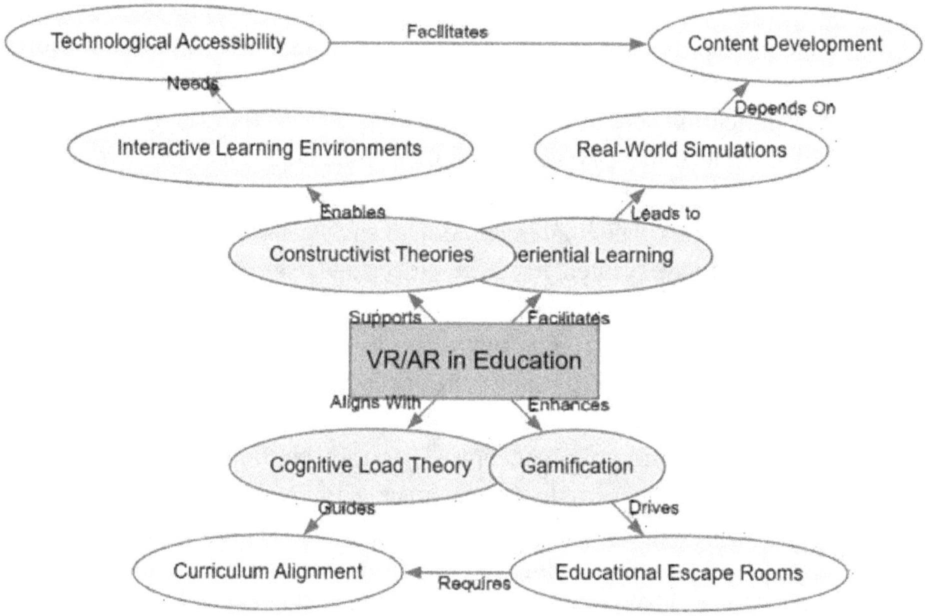

Design Principles for Digital Educational Escape Rooms

Central to the design of DEERs is the principle of active learning. Unlike traditional lecture-based pedagogies, escape rooms necessitate that learners engage directly with the material, often requiring problem-solving, critical thinking, and collaboration (see, for example, an escape room described by (Zou et al., 2024)). These interactive experiences place students in scenarios where they must apply their knowledge practically, thereby deepening their understanding of the subject matter. The design of these rooms should thus be inherently problem-based, with challenges and puzzles that are directly aligned with the educational objectives.

Narrative and storytelling play a pivotal role in the design of these digital spaces. A compelling story not only captivates the learners' interest but also provides a context within which the educational content can be explored. This narrative framework guides the design of puzzles and tasks, ensuring they are coherent and integrated seamlessly into the storyline. The narrative should be relevant to the subject matter and crafted in a way that it resonates with the target audience, making the learning experience both immersive and relatable.

Another essential principle is the integration of feedback mechanisms. In a digital educational escape room, immediate feedback is vital for guiding learners through the problem-solving process. This feedback can take various forms, from hints that help steer students back on track to detailed explanations following task completion. Such mechanisms not only aid in the learning process but also help maintain engagement and motivation, especially when learners face challenging tasks.

Collaboration is a cornerstone of escape room design. These digital environments should encourage teamwork and communication, fostering a collaborative learning atmosphere. Tasks might be designed in a way that requires input from multiple participants, promoting the sharing of knowledge and perspectives. This collaborative approach mirrors real-world scenarios, preparing students for professional environments where teamwork and communication are essential. By incorporating collaborative elements, DEERs not only enhance learning outcomes but also develop vital interpersonal skills.

Accessibility and inclusivity are also paramount in designing DEERs. The digital format offers unique opportunities to cater to a diverse range of learning styles and needs. However, it is crucial that these environments are designed with universal design principles in mind, ensuring that they are accessible to all learners, including those with disabilities. This might involve incorporating adjustable difficulty levels, providing alternative ways to access content, or including assistive technologies. By prioritizing accessibility, educational escape rooms can be a powerful tool for inclusive education.

Table 1. Components of the conceptual framework

Concept	Description
VR/AR in Education	This is the central node, representing the focus on integrating VR and AR technologies in educational settings.
Experiential Learning	VR and AR technologies are linked to experiential learning, emphasizing real-world or simulated experiences for interactive learning.
Constructivist Theories	These theories, which advocate for knowledge construction through interaction with the environment, are crucial in understanding how VR/AR can be used effectively in education.
Cognitive Load Theory	This theory is important in the design of VR/AR experiences, ensuring that the cognitive demands on learners are balanced and effective for learning.
Gamification	The application of game-design elements in VR/AR educational experiences to make learning more engaging.
Educational Escape Rooms	A specific application of VR/AR in education, providing immersive and interactive learning experiences.
Interactive Learning Environments	Highlighting the role of VR/AR in creating dynamic spaces for learning.

continued on following page

Table 1. Continued

Concept	Description
Real-World Simulations	The use of VR/AR to simulate real-life scenarios for educational purposes.
Curriculum Alignment	Ensuring that VR/AR experiences are aligned with educational objectives and standards.
Technological Accessibility	A critical aspect of implementing VR/AR in education, ensuring that these technologies are accessible to all learners.
Content Development	The creation of educational content for VR/AR platforms, tailored to meet specific learning objectives.

The use of technology in escape rooms must be purposeful and pedagogically sound. While the allure of using latest technological advancements is understandable, the technology chosen should serve the educational goals and enhance the learning experience. This might involve using VR or AR to create more immersive environments or integrating interactive elements that allow for hands-on engagement with the subject matter. The key is to ensure that the technology aligns with and supports the educational objectives, rather than overshadowing or complicating them.

The assessment of learning outcomes in DEERs requires careful consideration. While the primary focus of these environments is on the learning process, it is also important to evaluate the knowledge and skills acquired. This assessment should be integrated into the escape room experience in a way that feels natural and complements the learning objectives. For instance, the completion of certain tasks or puzzles can be used as a measure of understanding specific concepts.

Active learning

Active learning is a foundational design principle for DEERs, emphasizing learner engagement through hands-on, participatory experiences. Unlike passive learning, where learners are recipients of information, active learning requires them to interact with the content, collaborate with peers, and apply critical thinking to solve problems. In the context of DEERs, this principle is operationalized by creating immersive environments where learners take on roles, face challenges, and engage directly with tasks that promote deeper understanding and retention. The workflow of active learning is summarized in Figure 2.

Figure 2. Active learning process in DEERs

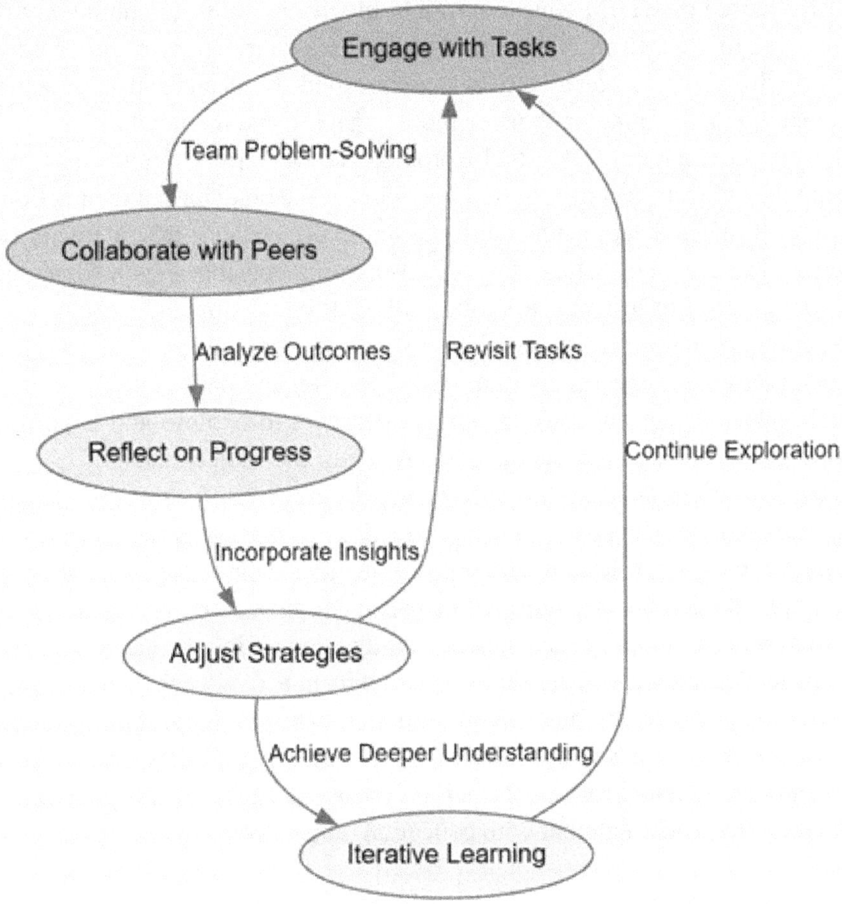

DEERs naturally lend themselves to active learning by placing learners in dynamic scenarios where they must solve puzzles, decode clues, and complete tasks to progress. These activities demand active participation, fostering cognitive engagement and problem-solving skills. For example, in a history-themed DEER, learners might be tasked with piecing together historical artifacts or deciphering ancient scripts to unlock the next stage. These tasks not only require learners to engage actively but also encourage them to apply knowledge, make connections, and synthesize information in meaningful ways.

Collaboration is another critical component of active learning in DEERs. By design, many escape room activities require teamwork, where participants must communicate effectively, delegate tasks, and build on each other's ideas to succeed. This collaborative element mirrors real-world problem-solving scenarios, helping learners develop essential interpersonal and communication skills. For instance, a DEER focused on STEM education might involve a group assembling a virtual machine, requiring each team member to contribute their expertise and coordinate efforts. Such activities foster a sense of shared responsibility and collective achievement, making the learning experience more engaging and memorable.

The interactive nature of DEERs also promotes active learning by allowing learners to manipulate objects, test hypotheses, and witness the outcomes of their actions in real time. Virtual and augmented reality technologies enhance this interactivity by immersing learners in realistic or fantastical environments that respond dynamically to their inputs. For instance, a VR-based DEER could simulate a chemistry lab where learners can mix virtual chemicals to observe reactions, providing a safe and controlled setting for experimentation. This hands-on approach not only reinforces theoretical knowledge but also encourages learners to explore and experiment, cultivating curiosity and a deeper understanding of the subject matter.

Active learning in DEERs is further supported by immediate feedback mechanisms. AI-powered systems can provide real-time feedback based on learner performance, guiding them toward solutions without diminishing their sense of achievement. For example, if a learner struggles with a particular puzzle, the system might offer contextual hints or simplify the challenge while maintaining the narrative's integrity. This feedback loop ensures that learners remain actively engaged, even when tasks become difficult, and helps them reflect on their problem-solving strategies.

Table 2 summarizes key considerations for designing DEERs that effectively implement the Active Learning principle, ensuring engagement, collaboration, critical thinking, and alignment with educational objectives.

Table 2. Checklist for active learning design principle in DEERs

Aspect	Checklist Item
Engagement and Participation	• Do the activities require learners to actively interact with the content? • Are tasks designed to encourage hands-on participation? • Do the scenarios promote sustained engagement and curiosity?
Collaboration and Teamwork	• Are activities designed to encourage group collaboration? • Do learners need to communicate and coordinate to solve challenges? • Are roles distributed to ensure all participants contribute meaningfully?
Critical Thinking and Problem-Solving	• Do tasks challenge learners to think critically and solve problems? • Are scenarios designed to simulate real-world problem-solving contexts? • Are learners encouraged to analyze, evaluate, and synthesize information?
Interactivity and Experimentation	• Do learners have opportunities to manipulate objects or test hypotheses? • Are tasks designed to provide immediate feedback on actions? • Does the environment support exploration and experimentation?
Alignment with Learning Objectives	• Are activities aligned with specific educational goals? • Do tasks reinforce theoretical knowledge through practical application? • Are diverse learning styles considered in the activity design?
Feedback and Reflection	• Is real-time feedback provided to guide learners without diminishing • engagement? • Are learners encouraged to reflect on their problem-solving strategies? • Does feedback align with learning outcomes?

Narrative and storytelling

Narrative and storytelling are central to the design of DEERs, serving as the foundation that binds the educational content with the interactive experience. A well-crafted narrative does more than entertain; it captivates learners' attention, creates emotional engagement, and provides a meaningful context within which learning occurs. This narrative framework (Figure 3) not only guides the progression of activities and puzzles but also ensures that the tasks are integrated seamlessly into the storyline, enhancing both coherence and immersion.

Figure 3. Narrative integration in DEERs

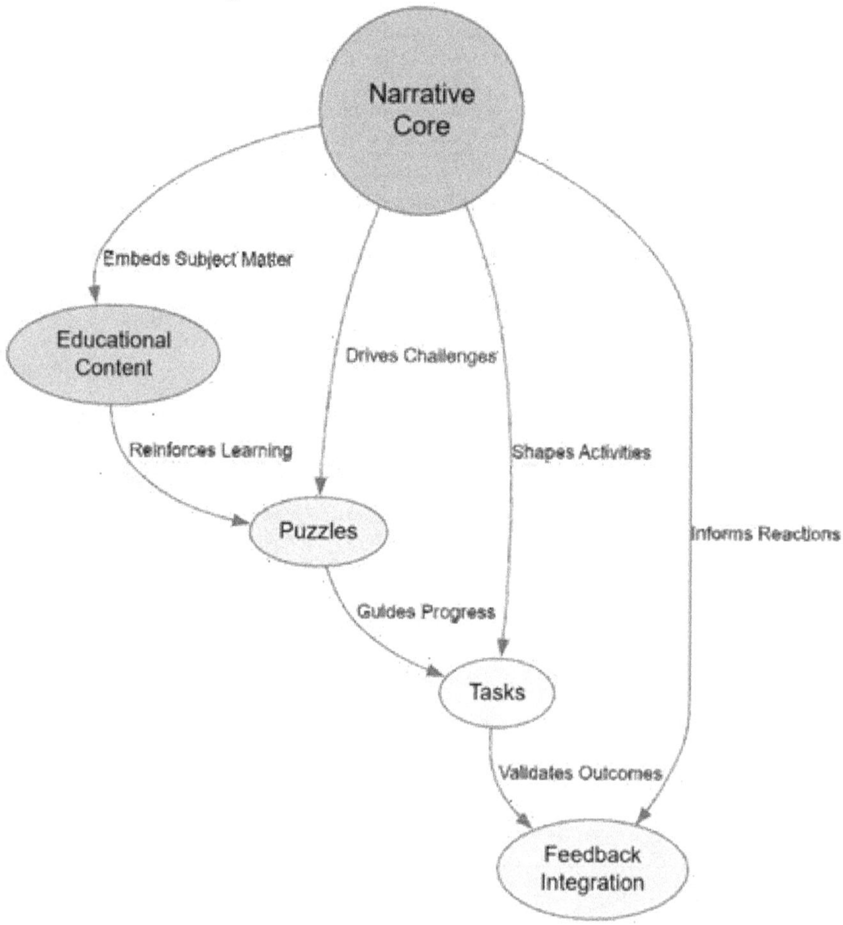

In DEERs, the narrative acts as a bridge between the subject matter and the learners. For instance, a science-based DEER might place learners in the role of researchers trying to save a virtual ecosystem, weaving environmental science concepts into the storyline. This approach not only introduces educational content in an engaging way but also helps learners understand its real-world relevance. By embedding puzzles and challenges into the narrative, learners are naturally drawn into the learning process, as they must apply their knowledge and skills to progress through the story.

The relevance of the narrative to the subject matter is crucial. A poorly aligned storyline can detract from the learning experience, making it feel disjointed or superficial. Instead, the narrative should be carefully crafted to resonate with the educational objectives and the interests of the target audience. For example, a history-focused DEER for high school students might revolve around unraveling a historical mystery, encouraging learners to explore key historical events and artifacts in the process.

Such an approach not only fosters curiosity but also deepens learners' engagement with the material.

The narrative must also consider the emotional and cognitive engagement of learners. A compelling story can evoke curiosity, empathy, and a sense of adventure, all of which contribute to a more immersive and memorable experience. Elements such as unexpected twists, time-sensitive missions, or the need to make critical decisions can heighten the stakes, drawing learners further into the experience. For instance, a medical-themed DEER might task learners with diagnosing and treating a virtual patient under a time constraint, fostering both emotional investment and critical thinking.

Another key aspect of narrative design in DEERs is its adaptability. As learners progress, the narrative can evolve based on their choices and performance, creating a personalized experience. This dynamic storytelling can be achieved using AI-driven narrative engines that adjust the storyline in real-time, ensuring it remains engaging and relevant to the learner's journey. Such adaptability not only enhances the learner's sense of agency but also aligns the narrative more closely with individual learning paths.

The storytelling elements in DEERs should leverage multimedia to create a rich and immersive environment. Visual and auditory cues, such as character dialogues, background music, and interactive visuals, can bring the narrative to life, making the learning experience more vivid and engaging. For example, a space exploration DEER might incorporate holographic projections of planets, audio messages from mission control, and visual representations of cosmic phenomena, immersing learners in the storyline while conveying complex scientific concepts.

Table 3 provides a comprehensive guide for designing compelling and educationally relevant narratives in DEERs, ensuring alignment with learning objectives, engagement, and seamless integration of content.

Table 3. Checklist for narrative and storytelling design principle in DEERs

Aspect	Checklist Item
Engaging Narrative	• Is the narrative captivating and able to sustain learners' interest? • Does the story include elements of curiosity, adventure, or mystery? • Are there emotional hooks or stakes that make the experience immersive?
Relevance to Subject Matter	• Does the narrative align with the educational objectives? • Is the storyline directly tied to the subject content being taught? • Does the narrative make the learning material relatable to the target audience?
Integration of Educational Content	• Are puzzles and tasks seamlessly embedded into the narrative? • Does each activity contribute to the progression of the story? • Is the educational content presented in a way that feels natural within the storyline?
Adaptability and Personalization	• Can the narrative evolve based on learners' choices or performance? • Are there mechanisms for dynamic storytelling to maintain engagement? • Does the narrative allow for personalized learning paths?
Immersive Multimedia Elements	• Are visual and auditory cues used to enhance the storytelling? • Do multimedia elements align with and support the narrative? • Are interactive visuals or character interactions included to deepen immersion?
Coherence and Flow	• Does the narrative progress logically and coherently? • Are transitions between tasks and story elements smooth and intuitive? • Does the narrative maintain a consistent tone and style throughout?

Feedback mechanisms

Feedback mechanisms are a cornerstone of effective DEERs, playing a crucial role in guiding learners through the problem-solving process and enhancing their educational experience (Figure 4). Immediate and well-structured feedback helps learners identify errors, refine their strategies, and stay engaged, even when faced with challenging tasks. By providing timely and relevant feedback, DEERs ensure that learners can progress through activities with confidence and a clear understanding of their performance.

Figure 4. Feedback loop in DEERs

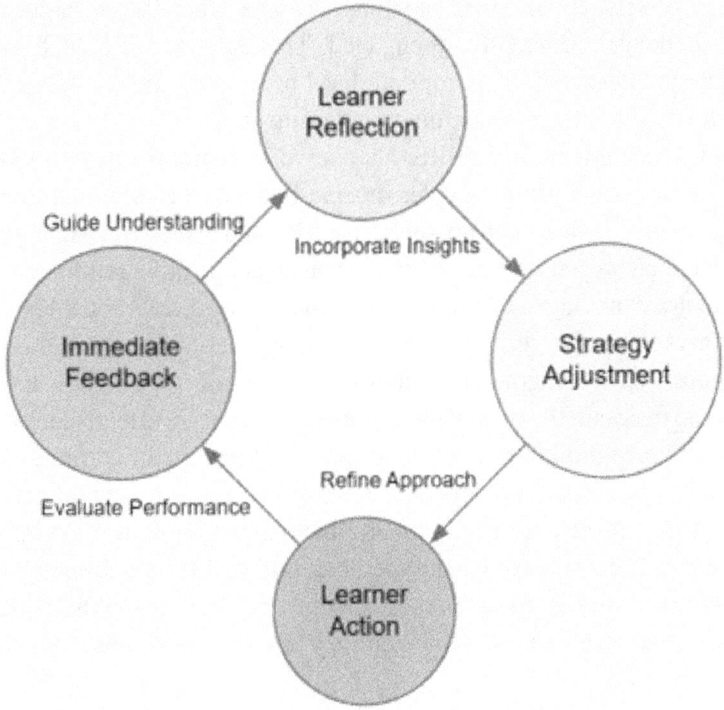

In DEERs, feedback often takes various forms, tailored to the specific needs of the learners and the nature of the tasks. Hints are one of the most common feedback mechanisms, designed to gently steer learners back on track without giving away the solution outright. For instance, in a math-themed DEER, if learners struggle with solving an equation, the system might provide a hint that recalls a key formula or highlights a relevant step in the solution process. Such targeted hints maintain the challenge of the activity while supporting learners in overcoming obstacles.

Another essential form of feedback is detailed explanations provided after task completion. These explanations can include a breakdown of the correct solution, the reasoning behind it, and its relevance to the overall learning objectives. For example, in a history-themed DEER, completing a puzzle about historical timelines might be followed by an explanation that contextualizes the events, offering learners deeper insights into their significance. This reflective feedback not only reinforces learning but also ensures that the educational value of each task is maximized.

Real-time adaptive feedback, powered by AI, can further enhance the learning experience in DEERs. Adaptive systems can monitor learners' actions and dynamically adjust feedback based on their performance. For example, if a learner consistently struggles with a particular type of puzzle, the system might offer progressively more detailed hints or adjust the difficulty level of subsequent tasks to match the learner's abilities. Conversely, for learners excelling at a task, the system might introduce additional challenges to keep them engaged. This personalized feedback ensures that each learner receives the appropriate level of support and challenge, fostering a more inclusive and effective learning environment.

Feedback mechanisms in DEERs also serve to maintain motivation and engagement, particularly during complex or time-intensive tasks. When learners face difficulties, timely feedback can prevent frustration and disengagement by providing encouragement and a sense of progress. For instance, positive reinforcement, such as congratulatory messages or unlocking a small reward, can boost learners' morale and motivate them to continue. Feedback integrated into the narrative, such as guidance from a virtual character or clues tied to the storyline, can further enhance immersion and maintain the flow of the experience. To ensure effectiveness, feedback mechanisms must be thoughtfully designed and aligned with the learning objectives. Feedback should be clear, actionable, and appropriately paced, providing just enough information to support learners without diminishing the sense of challenge. DEERs should incorporate diverse feedback types, catering to different learning styles and preferences. For visual learners, graphical feedback such as highlighted errors or progress bars may be effective, while auditory learners might benefit from spoken explanations or hints.

Table 4 provides a structured guide for designing effective feedback mechanisms in DEERs, ensuring they are timely, personalized, motivational, and aligned with educational objectives.

Table 4. Checklist for feedback mechanism design principle in DEERs

Aspect	Checklist Item
Timeliness of Feedback	• Is feedback provided immediately after a learner's action? • Does the feedback occur at appropriate moments to maintain the flow of the activity? • Are learners notified promptly when they make progress or errors?
Form of Feedback	• Are hints designed to guide learners without giving away the solution? • Are detailed explanations provided after task completion? • Does the feedback use various formats, such as visual, textual, or auditory cues, to • cater to different learning styles?
Personalization and Adaptability	• Does the feedback adapt to individual learner performance and needs? • Are progressively detailed hints provided for learners facing difficulties? • Is the difficulty level adjusted dynamically based on learner performance?

continued on following page

Table 4. Continued

Aspect	Checklist Item
Motivational Feedback	• Does the feedback include positive reinforcement to maintain learner motivation? • Are rewards or congratulatory messages incorporated into the feedback? • Does the feedback integrate into the narrative to enhance immersion?
Clarity and Actionability	• Is the feedback clear and easy to understand? • Does the feedback provide actionable steps for learners to improve or correct • mistakes? • Is the feedback concise and focused, avoiding unnecessary complexity?
Alignment with Learning Objectives	• Does the feedback reinforce the educational goals of the DEER? • Is the feedback designed to deepen learners' understanding of the subject matter? Are the feedback mechanisms aligned with the overall instructional strategy?

Collaboration

Collaboration is a foundational design principle of DEERs, offering a dynamic platform for learners to engage in teamwork and communication. By emphasizing collaborative learning, DEERs create an environment where participants work together to achieve common goals, fostering a sense of shared responsibility and mutual support. This approach (Figure 5) not only enhances the learning experience but also helps develop essential interpersonal skills such as teamwork, communication, and problem-solving, which are critical in both academic and professional contexts.

Figure 5. Collaboration workflow in DEERs

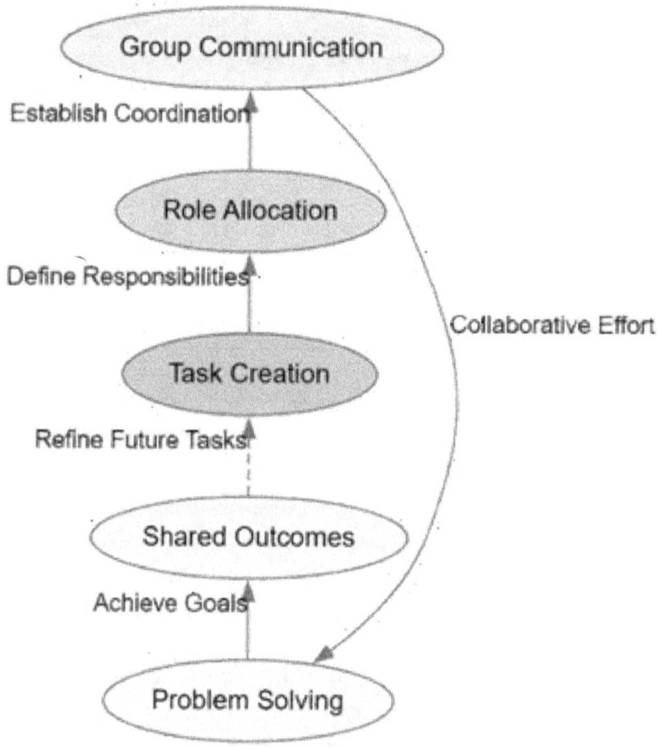

The design of DEERs should intentionally incorporate tasks and challenges that require input from multiple participants. Collaborative tasks often involve distributing roles or responsibilities among team members, ensuring that each learner contributes meaningfully to the group's progress. For example, a puzzle in a STEM-focused DEER might require one participant to analyze a graph, another to perform calculations, and a third to assemble the final solution. Such interdependence encourages the sharing of knowledge and perspectives, allowing learners to benefit from diverse approaches and insights. This mirrors real-world scenarios where teamwork is essential for tackling complex problems.

Collaboration in DEERs also fosters the development of communication skills. As learners work together to solve challenges, they must articulate their ideas clearly, listen actively, and build consensus. Effective communication is particularly crucial in scenarios where participants have different roles or access to unique information. For instance, in a history-themed DEER, one participant might possess a clue about a historical event, while another holds a piece of the timeline. To solve the puzzle, they must exchange information and piece together the solution collaboratively.

Such interactions simulate professional environments where clear and effective communication is key to success.

The collaborative nature of DEERs also promotes social learning, where participants learn from one another by observing, discussing, and sharing ideas. This social interaction can lead to deeper understanding and retention of knowledge, as learners are exposed to different perspectives and problem-solving strategies. For example, a DEER focusing on environmental science might involve a group debate about sustainable practices, with each participant contributing their research and ideas to formulate a collective solution. These collaborative discussions not only enhance cognitive engagement but also cultivate a sense of community and shared purpose among learners.

To optimize collaboration, DEERs should be designed to balance individual contributions with group dynamics. Tasks should be structured to ensure that all participants have opportunities to contribute and feel valued. For instance, time-sensitive challenges or multi-step puzzles can create a sense of urgency and encourage active participation, while tasks that require creative thinking or decision-making allow learners to showcase their unique strengths. Collaborative tools such as shared whiteboards, in-game chat systems, or virtual breakout rooms can facilitate communication and coordination in digital settings.

Incorporating collaboration as a core design principle in DEERs not only enhances learning outcomes but also prepares learners for real-world challenges. Professional environments often demand teamwork and the ability to navigate diverse perspectives and ideas. By simulating such scenarios, DEERs equip learners with the interpersonal and collaborative skills necessary for success in their future endeavors. The shared experience of solving puzzles and achieving goals together can strengthen relationships and foster a positive learning atmosphere. Table 5 provides a structured guide for designing effective feedback mechanisms in DEERs, ensuring they are timely, personalized, motivational, and aligned with educational objectives.

Table 5. Checklist for collaboration design principle in DEERs

Aspect	Checklist Item
Task Design for Teamwork	• Are tasks structured to require input from multiple participants? • Do tasks promote interdependence among team members? • Are roles or responsibilities distributed to ensure equal participation?
Communication and Coordination	• Do tasks encourage clear and effective communication among learners? • Are tools provided to facilitate collaboration, such as chatbots? • Do tasks require learners to exchange information or build consensus?
Development of Interpersonal Skills	• Do tasks support the development of teamwork and problem-solving skills? • Are opportunities provided for active listening and constructive feedback? • Do tasks promote mutual respect and understanding among participants?
Social Learning Opportunities	• Are learners encouraged to observe and learn from each other's approaches? • Do tasks involve group discussions or brainstorming activities? • Are diverse perspectives integrated into the problem-solving process?
Inclusivity in Collaboration	• Are tasks designed to ensure that all participants feel valued? • Do tasks consider different strengths and contributions of learners? • Are thre mechanisms to prevent dominant participants from overshadowing others?
Alignment with Real-World Scenarios	• Do tasks simulate professional teamwork scenarios? • Are collaborative challenges reflective of real-world problems? • Do tasks prepare learners for future collaborative environments?

Accessibility and Inclusivity

Accessibility and inclusivity are foundational principles in the design of DEERs, ensuring that these immersive learning environments cater to a diverse range of learners. DEERs leverage advanced technologies, such as Virtual Reality (VR) and Augmented Reality (AR), to provide engaging, interactive experiences (see Table 6). However, to maximize their educational impact, these environments must address the needs of learners with varying abilities, preferences, and constraints.

Table 6. Ensuring accessibility in VR/AR technologies

Accessibility Feature	Visual Needs	Auditory Needs	Cognitive Needs	Physical Needs
Text-to-Speech	✓			
Subtitles/Closed Captions				
High-Contrast Visuals				
Audio Descriptions				
Simplified Interfaces				
Adjustable Difficulty Levels				
Keyboard/Controller Support				
Motion-Free Modes (No VR Head Tracking)				
Alternative Input Methods (e.g., Voice Commands)				
Haptic Feedback				

Accessibility in DEERs begins with removing physical, sensory, and cognitive barriers that could hinder participation. For learners with visual impairments, features such as audio descriptions, high-contrast visuals, and text-to-speech functionality are critical. These adaptations ensure that visually impaired learners can navigate and interact with the environment effectively. Similarly, for individuals with auditory impairments, subtitles, closed captions, and visual cues replace or complement auditory information, enabling full engagement with narrative and instructional elements.

Cognitive accessibility is another vital consideration. DEERs should accommodate learners with varying levels of cognitive ability, including those with learning disabilities or attention challenges.

Simplified user interfaces, clear instructions, and adjustable difficulty levels allow learners to progress at their own pace, reducing cognitive overload. Real-time adaptive systems, powered by AI, can further enhance cognitive accessibility by dynamically adjusting task complexity based on individual performance and engagement levels. This ensures that all learners, regardless of their starting point, experience meaningful and achievable challenges.

Physical accessibility must also be a priority in DEER design. Motion-free modes, which remove the reliance on VR head tracking or extensive physical movement, are essential for individuals with mobility impairments. Alternative input methods, such as voice commands, keyboard controls, or adaptive devices, ensure inclusivity for learners who may not be able to use standard controllers or gestures.

Inclusivity (Table 7) extends beyond accommodating disabilities to creating culturally and socially responsive learning environments. DEERs should consider the diverse cultural backgrounds, languages, and experiences of learners. For example, multilingual support ensures that learners can interact with the content in their preferred language. Similarly, culturally relevant scenarios and narratives make the learning experience more relatable and engaging for a wide audience.

Table 7. Inclusivity features vs learner needs in DEERs

Inclusivity Feature	Cultural Needs	Linguistic Needs	Cognitive Needs	Sensory and Physical Needs
Multilingual Support	✓	✓		
Culturally Relevant Scenarios				
Customizable Narratives				
Simplified Interfaces				
Adjustable Difficulty Levels				
Alternative Input Methods				
Audio Descriptions				
Haptic Feedback				
Subtitles/Closed Captions				
Universal Design Principles				

The design process for accessible and inclusive DEERs requires a proactive approach, integrating universal design principles from the outset. Universal design emphasizes creating environments that are usable by all people, to the greatest extent possible, without the need for adaptation. By incorporating accessibility features as standard rather than optional add-ons, DEERs can avoid marginalizing certain groups of learners. Iterative user testing with diverse participant groups ensures that accessibility and inclusivity goals are met effectively.

Purposeful Use of Technology

The purposeful use of technology is a cornerstone design principle for DEERs) which is summarized in Table 8 and illustrated in Figure 6. In educational contexts, technology should not merely serve as an aesthetic enhancement or a novelty but must be deliberately chosen and implemented to support pedagogical objectives. While the appeal of incorporating the latest technological innovations, such as virtual reality

(VR), augmented reality (AR), or artificial intelligence (AI), is undeniable, the emphasis must remain on their capacity to enhance learning experiences meaningfully.

Table 8. Checklist for purposeful use of technology in DEERs

Aspect	Checklist Item
Alignment with Educational Goals	• Does the technology directly support the learning objectives? • Is the technology used to enhance understanding of the subject matter? • Does the technology facilitate experiential or active learning?
Learner Engagement	• Does the technology create an engaging and immersive learning environment? • Are interactive elements included to encourage hands-on participation? • Is the content presented in a way that stimulates curiosity and critical thinking?
Ease of Use	• Is the user interface intuitive and learner-friendly? • Are instructions for using the technology clear and accessible? • Does the technology avoid unnecessary complexity or distraction?
Adaptability and Feedback	• Does the technology provide real-time feedback to learners? • Can the technology adapt to different skill levels or learning paces? • Are features included to address diverse learning styles?
Inclusivity and Accessibility	• Are alternative input methods provided for learners with physical disabilities? • Does the technology include features like text-to-speech, subtitles, or translations? • Is the design inclusive for a diverse range of cultural and linguistic backgrounds?
Technological Feasibility	• Is the chosen technology suitable for the available resources? • Does the implementation avoid over-reliance on high-end or inaccessible tools? • Are there fallback options for low-tech or offline scenarios?

Figure 6. Purposeful use of technology in DEERs

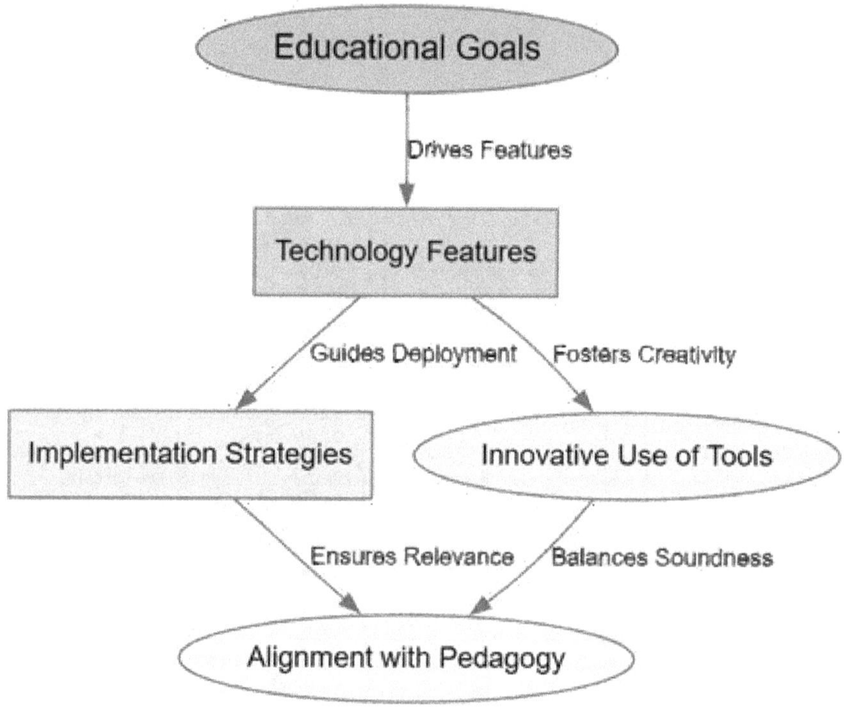

The integration of technology into DEERs should be guided by its alignment with specific learning goals. For instance, VR and AR technologies can be used to create immersive environments that enable learners to explore and interact with complex scenarios that would be impossible or impractical in a traditional classroom. A history-based DEER, for example, might employ VR to transport learners to ancient civilizations, where they solve puzzles embedded within historically accurate reconstructions. This immersive approach not only engages learners but also provides a hands-on, experiential learning opportunity that deepens understanding and retention of the subject matter.

Interactive elements powered by technology are equally critical in fostering hands-on engagement. Features such as drag-and-drop interfaces, digital simulations, or AR overlays can allow learners to manipulate objects, test hypotheses, or visualize abstract concepts in real time. For instance, in a physics-themed DEER, learners might use an AR-enabled device to assemble a virtual circuit and observe its functionality, providing an intuitive grasp of electrical concepts that might otherwise seem abstract. Such interactive technologies not only make learning more engaging but also support constructivist principles by enabling learners to build knowledge through active participation.

The technology used in DEERs should enhance the learning experience without overshadowing the educational objectives or introducing unnecessary complexity. Poorly implemented, technology can frustrate learners, distract from the learning goals, and ultimately detract from the educational value of the experience. For instance, while a highly sophisticated VR environment may seem impressive, if the user interface is unintuitive or the tasks within the environment are poorly aligned with the learning objectives, the technology becomes a barrier rather than a facilitator of learning.

To ensure technology serves a purposeful role, it is essential to adopt a learner-centered design approach. This involves carefully evaluating the educational objectives of the DEER, the needs and abilities of the target audience, and the affordances of the chosen technologies. The goal is to create an environment where the technology seamlessly integrates with the content, supporting learners in achieving the desired outcomes. For example, in a mathematics DEER, AI-powered adaptive learning systems might provide real-time feedback and tailored challenges based on individual performance, ensuring that the technology dynamically supports the learner's progress.

Summary and Discussion

The design principles for DEERs (summarized in Figure 7 and Table 9) revolve around creating immersive, interactive, and inclusive learning experiences. By focusing on active learning, narrative integration, feedback mechanisms, collaboration, accessibility, purposeful use of technology, and assessment of learning outcomes, educators can create engaging and effective educational environments. These principles not only guide the creation of DEERs but also contribute to a broader understanding of how digital technologies can be harnessed to enhance education.

Figure 7. Design principles of digital educational escape rooms

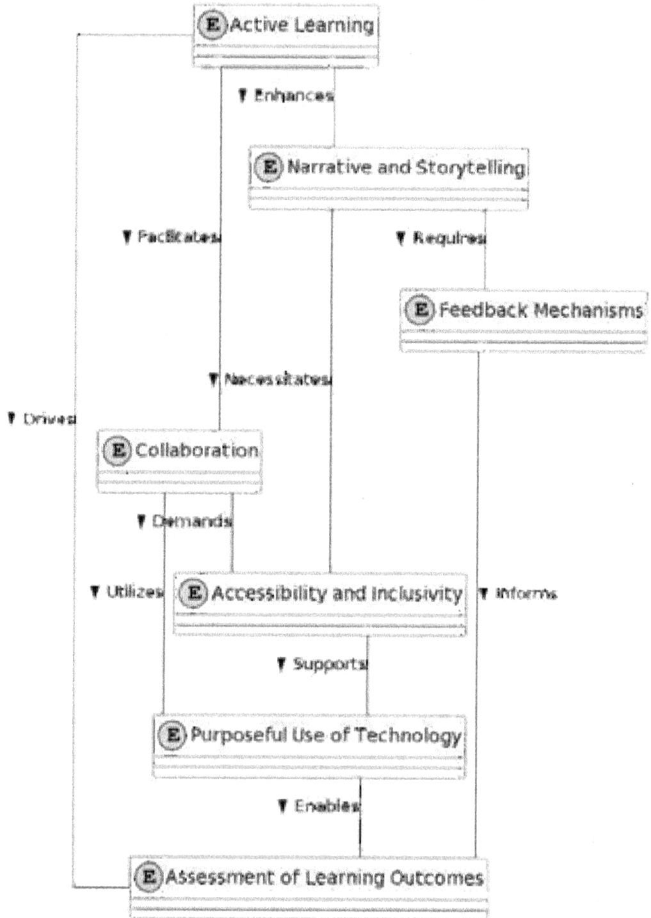

Table 9. Design principles

Design Principle	Relationships and Descriptions
Active Learning	• Enhances Narrative and Storytelling: Active learning strategies are enriched through immersive storytelling. • Facilitates Collaboration: Active learning promotes collaborative activities and group problem-solving. • Drives Assessment of Learning Outcomes: Active learning techniques are essential for effective assessment in educational escape rooms.
Narrative and Storytelling	• Requires Feedback Mechanisms: Engaging narratives need responsive feedback to guide learners. • Necessitates Accessibility and Inclusivity: Storytelling should be accessible and inclusive to all learners.
Feedback Mechanisms	• Informs Assessment of Learning Outcomes: Immediate feedback helps in assessing learner understanding and progress.
Collaboration	• Demands Accessibility and Inclusivity: Collaborative tasks must be designed to be accessible to all learners. • Utilizes Purposeful Use of Technology: Collaboration is enhanced through the use of appropriate technology.
Accessibility and Inclusivity	• Supports Purposeful Use of Technology: Accessible design principles guide the technology choices for educational escape rooms.
Purposeful Use of Technology	• Enables Assessment of Learning Outcomes: Technology aids in the efficient and effective assessment of learning outcomes.

USING AI FOR PERSONALIZED LEARNING PATHS IN DEERS

The Use of AI in DEERs: An Overview

The foundational concept underpinning the use of AI in DEERs is the notion of personalized learning, an approach that seeks to customize learning activities and content according to the unique characteristics of each learner. This concept, deeply rooted in the constructivist theories of Piaget (Piaget, 1954) and Vygotsky (Vygotsky, 1978), posits that learning is most effective when it aligns with the learner's existing knowledge base and cognitive abilities. AI technologies in DEERs enhance this personalized approach by leveraging data-driven insights to adapt learning scenarios and challenges in real-time, thereby catering to the evolving educational needs of each student.

AI in DEERs typically involves the use of machine learning algorithms and data analytics to analyze learners' interactions within the escape room environment. This analysis can include tracking decision-making patterns, problem-solving approaches, and the time taken to complete various tasks. These data points, as illustrated in the work of Xie et al. (Xie et al., 2019), provide valuable insights into individual learning styles and areas where a student may require additional support or challenge. By harnessing these insights, AI can dynamically adjust the difficulty level of tasks, introduce new challenges tailored to individual learning progress, or provide customized feedback and hints to guide the learner effectively.

Another critical aspect of AI in DEERs is its ability to foster collaborative learning environments. As noted by (Johnson and Johnson, 1994), cooperative learning strategies are instrumental in developing key interpersonal skills and enhancing knowledge retention. AI systems can facilitate this by grouping learners with complementary skills and learning styles, thereby promoting effective teamwork and collaborative problem-solving within the escape room.

The implementation of AI in DEERs also raises important considerations regarding ethical issues and data privacy. As highlighted by (Shneiderman, 2020), the ethical use of AI in educational settings necessitates transparency, accountability, and respect for learner privacy. This involves ensuring that data collected from learners are used responsibly and with explicit consent, and that AI-driven interventions are transparent in their operation and objectives.

The role of educators in AI-integrated DEERs cannot be understated. While AI can provide personalized learning paths and insights, the teacher's role in interpreting these insights and integrating them into broader educational strategies is vital. As argued by (Selwyn, 2019), technology should augment, not replace, the teacher's expertise, particularly in understanding the nuanced educational needs of students.

Using Large Language Models and chatbots in DEERs

Large language models, like GPT (Generative Pre-trained Transformer) series developed by OpenAI, represent the forefront of natural language processing (NLP) technology. These models have demonstrated remarkable capabilities in understanding, generating, and interpreting human language (Vaswani et al., 2017). In the context of DEERs, such models can be employed to create intelligent chatbots that interact with learners. For example, a generative pretrained transformer (GPT), developed by OpenAI, was employed as an assistant in low-cost virtual escape games (Rychert et al., 2024). These chatbots can serve various functions, from guiding learners through the escape room, providing hints, engaging in educational dialogues, to even assessing learner responses and offering feedback. For example, Escapeling is a Telegram bot aimed at improving English skills outside of a classroom setting (Johnson et al., 2022).

The application of large language models in DEERs aligns with the principles of adaptive learning, where educational experiences are tailored to the needs and abilities of individual learners. As argued by (Xie et al., 2019), adaptive learning systems, powered by AI, have the potential to significantly enhance personalized learning experiences. In DEERs, this personalization can be achieved through chatbots that adapt their interactions based on the learner's responses, learning pace, and preferences, thus providing a customized learning journey.

The use of chatbots in DEERs can also promote active learning and engagement. According to (Chi, 2009) active learning involves activities that compel students to think deeply about the content, thereby enhancing understanding and retention. Chatbots, through interactive dialogues and problem-solving exercises, can engage learners in such active learning processes, making the DEER experience more immersive and educationally fruitful. However, the integration of large language models and chatbots in educational settings is not devoid of challenges. One significant concern is the accuracy and appropriateness of the content generated by these models. As (Bender et al., 2021) highlight, ensuring the reliability and educational suitability of AI-generated content is crucial. This necessitates continuous monitoring and fine-tuning of these models to align with educational standards and objectives.

The use of large language models and chatbots in DEERs represents a promising frontier in educational technology. By harnessing the power of advanced NLP, these tools can offer personalized, adaptive, and engaging learning experiences. However, this integration must be approached with a keen eye on content reliability, educational relevance, and ethical considerations.

AI-based Personalization Algorithms

The incorporation of AI-based personalization algorithms in educational contexts heralds a transformative shift in how learning content is tailored to meet the diverse needs of individual learners. This approach leverages the advanced capabilities of AI to analyze vast datasets about learners' behaviors, preferences, and performance, thereby enabling the creation of customized learning experiences.

The theoretical underpinning of AI-based personalization in education is rooted in the concept of adaptive learning, which posits that educational experiences should adjust to the unique needs of each learner. This approach is grounded in the constructivist theories of learning, which emphasized the importance of individual experiences and social contexts in the learning process. AI algorithms, through their ability to process and learn from large volumes of data, can identify patterns and insights that inform the creation of personalized learning paths. This capability aligns with the principles of adaptive learning, where instructional content and activities are dynamically adjusted in response to the learner's evolving needs.

The practical application of AI-based personalization algorithms extends across various educational settings, from K-12 to higher education and corporate training. In these environments, AI algorithms analyze data points such as learners' interaction with educational content, assessment results, and learning pace. This analysis enables the algorithm to predict learners' needs and adapt the instructional content accordingly. For instance, if a learner is struggling with a particular concept, the algorithm can provide additional resources or adjust the difficulty level of the tasks to better suit the learner's understanding.

The use of AI in personalizing learning also encompasses the development of intelligent tutoring systems (ITS). As elucidated by (VanLehn, 2006), ITS are computer systems that provide immediate and customized instruction or feedback to learners, without human intervention. These systems utilize AI algorithms to mimic the pedagogical strategies of human tutors, offering a scalable approach to personalized education. However, the deployment of AI-based personalization algorithms in education raises important considerations regarding ethical issues, such as data privacy and algorithmic bias. As highlighted by (Shneiderman, 2020), ensuring the ethical use of AI in education involves addressing concerns related to transparency, accountability, and fairness. It is crucial to ensure that these algorithms do not perpetuate biases or inequalities, and that they uphold the highest standards of data privacy and security.

AI Real-Time Adaptation, Player-Centric Design, and Adaptive Learning

Artificial intelligence (AI) significantly enhances engagement in DEERs by enabling real-time adaptation, player-centric design, and personalized learning experiences. These capabilities are driven by advanced computational techniques, including machine learning algorithms, data analytics, and real-time decision-making systems. By continuously responding to learner behavior, AI ensures that the gameplay remains dynamic, interactive, and tailored to individual needs.

Real-time adaptation relies on reinforcement learning (RL) and predictive modeling to dynamically adjust gameplay based on learner interactions. For instance, RL agents, such as Deep Q-Learning or Proximal Policy Optimization (PPO), analyze player data, including response times, success rates, and engagement levels, to make decisions about the next task or narrative element. If a player struggles with a specific challenge, the system can adapt by reducing task complexity or providing context-sensitive hints. For example, in a physics-focused DEER, if a learner repeatedly fails to calculate the force required to lift an object, the AI may substitute the problem with a simpler variant, such as calculating the force for a lighter object. This ensures that learners remain engaged without feeling overwhelmed. NLP models, such as GPT or T5, can dynamically adjust the storyline. If players take too long on a task, the system might introduce a narrative extension, such as a "power surge" that grants additional time, thus blending real-time gameplay adjustments with immersive storytelling.

AI also enhances DEERs through player-centric design, creating a responsive environment tailored to diverse learner profiles. Player modeling techniques, such as Bayesian Networks or Hidden Markov Models, classify players into categories (e.g., novice, intermediate, or advanced) based on their performance. This classification allows the AI to adapt tasks and roles dynamically. For instance, in team-based DEERs, clustering algorithms like K-Means can group players with complementary skills, assigning roles such as "Calculator" for numerically skilled participants and "Story Weaver" for creative contributors. Interactive dialogues powered by NLP further enhance the player experience by offering adaptive feedback. If the AI detects frustration from repeated errors, it can respond with encouraging messages or simplified instructions, maintaining player motivation and engagement.

Personalization and adaptive learning form the cornerstone of AI-driven DEERs, ensuring that challenges are tailored to individual performance data. AI systems monitor key metrics, such as response time, accuracy, and error patterns, to dynamically adjust the difficulty of tasks. Adaptive Difficulty Scaling (ADS) techniques, powered by Gradient Boosting Machines (GBMs) or XGBoost, predict the optimal task difficulty for each learner. For instance, if a learner solves problems with high accuracy and speed, the AI increases the difficulty by introducing multi-step problems, reducing available hints, or imposing stricter time limits. Conversely, if a learner struggles, the AI simplifies tasks or provides scaffolding, such as breaking a complex integral into smaller, more manageable components.

An example of this in practice is a DEER designed for teaching integration in mathematics. A learner begins with a basic problem, such as solving $\int x\, dx$. If completed successfully within a short time, the AI escalates the challenge to a more complex integral, such as $\int x^2 e^x\, dx$. However, if the learner struggles, the system simplifies the problem or offers guided hints, such as emphasizing key steps in the integration process. Similarly, in a language learning DEER, NLP models analyze incorrect word choices and introduce simpler vocabulary tasks while embedding definitions within the storyline to support learning.

AI systems also employ reinforcement learning to continuously adapt the learning environment. By observing engagement metrics, such as time spent on tasks or the frequency of hint requests, the AI dynamically adjusts game elements to sustain player interest. For example, a system detecting disengagement might introduce a time-sensitive puzzle or a narrative twist, such as a simulated "crisis" that requires immediate problem-solving. This approach ensures that learners remain immersed and motivated throughout the experience.

Real-world Implementations of AI in DEERs and Results

The real-world application of AI in DEERs encapsulates a range of techniques, from data analytics and machine learning to natural language processing, all aimed at augmenting the learning experience. The AI's role in DEERs primarily revolves around enhancing interactivity, personalization, and adaptivity. By leveraging AI algorithms, DEERs can adapt to individual learning styles and needs, providing a more engaging and effective educational experience. This adaptive learning approach, rooted in the theories of Piaget (Piaget, 1954) and Vygotsky (Vygotsky, 1978), emphasizes the importance of tailoring educational experiences to the learner's cognitive abilities and social context. AI enables this tailoring by analyzing learner data, such as performance metrics (Table 10), engagement levels, and interaction patterns, to dynamically adjust the difficulty levels of tasks, provide personalized feedback, and suggest learning paths that align with individual learner profiles.

Table 10. Performance metrics for adaptive AI in DEERs

Metric	Description and Use Case
Response Time	Measures how quickly learners solve puzzles, informing the system whether to increase or reduce task complexity.
Success Rate	Tracks the percentage of correct answers to gauge proficiency. High success rates trigger more challenging tasks.
Error Patterns	Identifies common mistakes to tailor feedback or adapt future puzzles.
Engagement Metrics	Monitors time spent, hint requests, and abandonment rates to adjust gameplay for better immersion.
Learning Progression	Assesses improvement over time, ensuring tasks align with learner growth.

In terms of practical implementations, AI has been used in DEERs to create dynamic puzzles and challenges that evolve based on the learner's performance. For example, an AI system might analyze a student's problem-solving approach in a DEER and adjust subsequent challenges to better suit the student's learning style. This approach not only makes the learning process more engaging but also more effective, as students are more likely to understand and retain information presented in a manner tailored to their learning preferences.

AI-driven analytics in DEERs offer educators valuable insights into student learning processes. These insights can inform instructional strategies and curriculum development, allowing educators to identify common difficulties faced by students and adjust the instructional content accordingly. This application of AI in DEERs aligns with the findings of Xie et al. (Xie et al., 2019), who emphasized the potential of AI in providing educators with data-driven insights to enhance teaching and learning processes. However, the implementation of AI in DEERs is not without challenges. One of the primary concerns is ensuring the reliability and accuracy of AI algorithms. As Bender et al. (Bender et al., 2021) highlight, AI systems, particularly those involving natural language processing, must be carefully designed to avoid biases and inaccuracies that could negatively impact the learning experience. This concern necessitates rigorous testing and continuous refinement of AI algorithms to ensure their effectiveness and reliability in educational settings. Another significant challenge is the integration of AI into existing educational infrastructures. Successful integration of technology in education requires not only technological resources but also a shift in educators' attitudes and instructional practices. Educators must be adequately trained to use AI-powered DEERs effectively and to interpret the data and insights generated by these systems.

CASE STUDIES: DESIGN OF DIGITAL ESCAPE ROOM IN PRACTICE

Case Study: Digital Escape Room for Mathematics

The designed digital escape room, which has been first described in (Sidekerskiene and Damaševicius, 2023) is a carefully structured educational tool aimed at enhancing the learning experience for university students studying mathematics, particularly in the context of the "Mathematics 2" course. This course focuses on several key mathematical concepts, including indefinite integrals, definite integrals, improper integrals, and double integrals, which are fundamental topics in higher education mathematics. The escape room serves as an innovative platform to reinforce these concepts in an engaging and interactive manner, encouraging students to apply their knowledge in practical, problem-solving scenarios.

The digital escape room was designed using Google Forms, a platform chosen for its accessibility and ease of use. The primary objective of this escape room is to provide students with an opportunity to review and apply their understanding of the course material in a game-based environment. The entire experience is structured around a series of rooms, each representing a different challenge or mathematical puzzle that students must solve to progress. The ultimate goal is to "escape" from the room by successfully completing all the tasks within a set time limit of 100 minutes.

Each room within the escape room is designed to focus on a specific aspect of the mathematics curriculum. For instance, one room may require students to solve an indefinite integral using integration by parts, while another might challenge them to compute the area under a curve through definite integration. The problems are not only mathematical in nature but also incorporate elements that require logical reasoning and sometimes even external research, such as converting numerical answers into different formats like binary numbers.

One of the key features of the escape room is the use of the "Lock and Key" design pattern. This pattern is central to the structure of the escape room and involves a series of puzzles that students must solve to find "keys" that unlock subsequent rooms. This design not only makes the experience more engaging but also ensures that students must thoroughly understand the mathematical concepts to progress. For example, in one of the digital corridors, students must first solve a mathematical problem and then convert the result into a binary number to unlock the next door. This integration of different skills makes the learning experience more holistic and comprehensive.

The escape room is divided into several themed rooms, each with a distinct mathematical focus. For example, the "Yellow Room" (Figure 8) challenges students to apply their knowledge of symmetric bounds and the properties of odd and even functions within definite integrals. Here, students must use their understanding to solve integrals and then convert their answers using double factorials to find the key that allows them to proceed. This room, like others, combines mathematical rigor with an element of puzzle-solving that goes beyond straightforward calculation, adding depth to the learning experience.

Figure 8. An example of a room screen

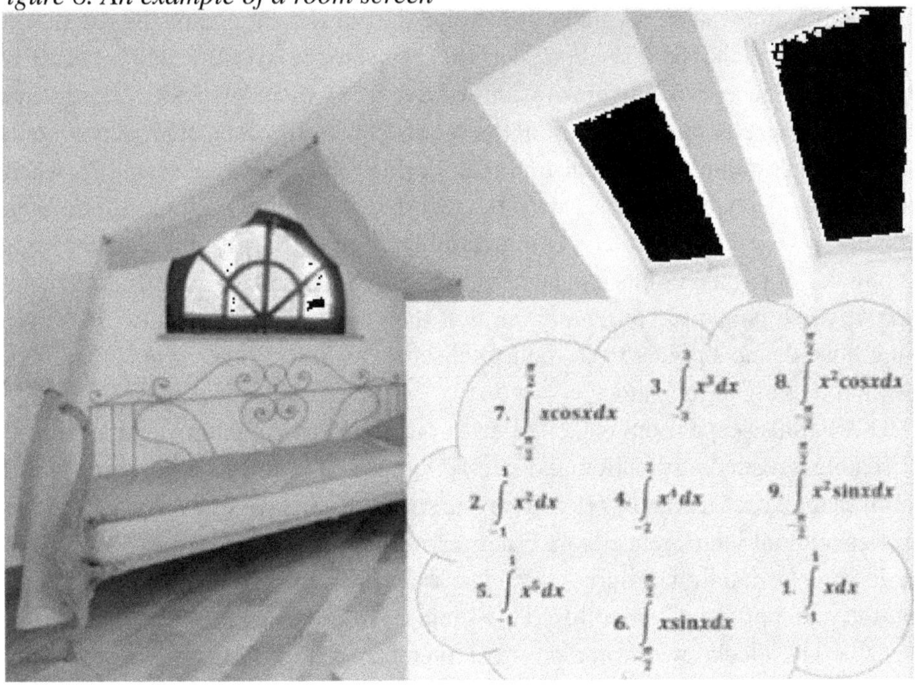

The layout of the escape room is designed to be both intuitive and challenging. Most rooms are "locked," meaning that students must solve the puzzles within them to advance. However, there are also "unlocked" rooms where students can enter without solving a preceding puzzle, though they risk losing points if they fail to complete the tasks correctly. This design choice adds an element of strategy to the game, as students must decide whether to focus on solving all puzzles for maximum points or to skip certain challenges to save time.

To illustrate, one of the unlocked rooms, the "Grey Room," presents students with graphs of even, odd, and neither even nor odd functions. The challenge here is for students to correctly match these graphs with the appropriate mathematical formulas for calculating definite integrals with symmetric bounds. While students can leave this room without completing the task, doing so without selecting the correct answers results in a penalty, thereby encouraging students to engage fully with the material.

The escape room also includes obstacles that require students to use their mathematical knowledge creatively. For example, in the "Pirate Room," students are presented with Wallis' integrals and must multiply the answers, excluding zero, to find a mathematical constant that unlocks the next door. In another room, students must calculate the mass of a plate using double integrals, applying their knowledge of the applications of these integrals to solve the puzzle and escape the room.

In addition to the individual puzzles, the escape room is designed to be a collaborative experience. Students form teams of five to six members and work together to solve the puzzles and escape from the room. This team-based approach not only fosters collaboration but also mirrors real-world problem-solving scenarios where multiple perspectives and skills are often required to arrive at a solution. The game is competitive, with teams striving to not only escape but also accumulate the highest number of points. Points can be earned by solving puzzles correctly, and teams can use these points to "purchase" hints if they find themselves stuck. However, using hints deducts points from their total score, adding another layer of strategic decision-making to the game.

The digital escape room is facilitated through the Zoom platform, which allows for seamless team coordination and interaction. The use of breakout rooms within Zoom enables each team to work in a private space, discussing strategies and solving puzzles without interference from other teams. The platform's features also allow the lecturer to monitor progress, provide assistance when requested, and ensure that the game proceeds smoothly. This setup proved particularly effective during the COVID-19 lockdown, when physical interactions were limited, and provided a valuable means for students to engage in collaborative learning remotely.

Educational Goals

The implementation of digital escape rooms in mathematics education is underpinned by a set of carefully crafted design principles and educational goals. The primary aim is to enhance student engagement and facilitate deeper learning by leveraging the immersive and interactive nature of digital escape rooms. These educational tools are designed to encourage students to think critically, solve complex problems, and collaborate effectively with peers. The overarching goal is to transform the often abstract and challenging concepts of mathematics into a more tangible and accessible learning experience.

To achieve these goals, the design of digital escape rooms for mathematics is guided by several key principles. First, the escape rooms are built around core mathematical concepts, such as integration and calculus, ensuring that all tasks and puzzles directly relate to the curriculum. This alignment ensures that students not only enjoy the learning experience but also reinforce their understanding of critical mathematical theories.

Second, the escape rooms are designed to be challenging yet achievable, with tasks that progressively increase in difficulty. This scaffolding approach helps maintain student motivation while ensuring that they build confidence as they progress through the game. The design also incorporates a variety of puzzle types, catering to different learning styles and ensuring that all students can engage with the content meaningfully.

Finally, the educational goals also emphasize the development of soft skills such as teamwork, communication, and time management. By working in teams, students are encouraged to discuss strategies, share insights, and collectively solve the problems presented. This collaborative aspect of the escape room experience mirrors real-world problem-solving scenarios, preparing students for future academic and professional challenges.

Design Lifecycle

The design lifecycle of a Digital Educational Escape Room (DEER) (Figure 9) for teaching mathematics at the university level is a carefully planned process that ensures the educational tool effectively meets its learning objectives. The lifecycle begins with the Planning phase, which is foundational to the entire process. During this stage, educators focus on identifying the specific learning goals that the escape room is intended to achieve. This involves selecting the key mathematical concepts that will be taught, such as integration or calculus, and designing puzzles and challenges that align with these objectives. The planning phase is crucial as it sets the direction for the development of the escape room, ensuring that the activities are not only educational but also engaging and challenging for the students.

Figure 9. Design lifecycle of a digital escape room

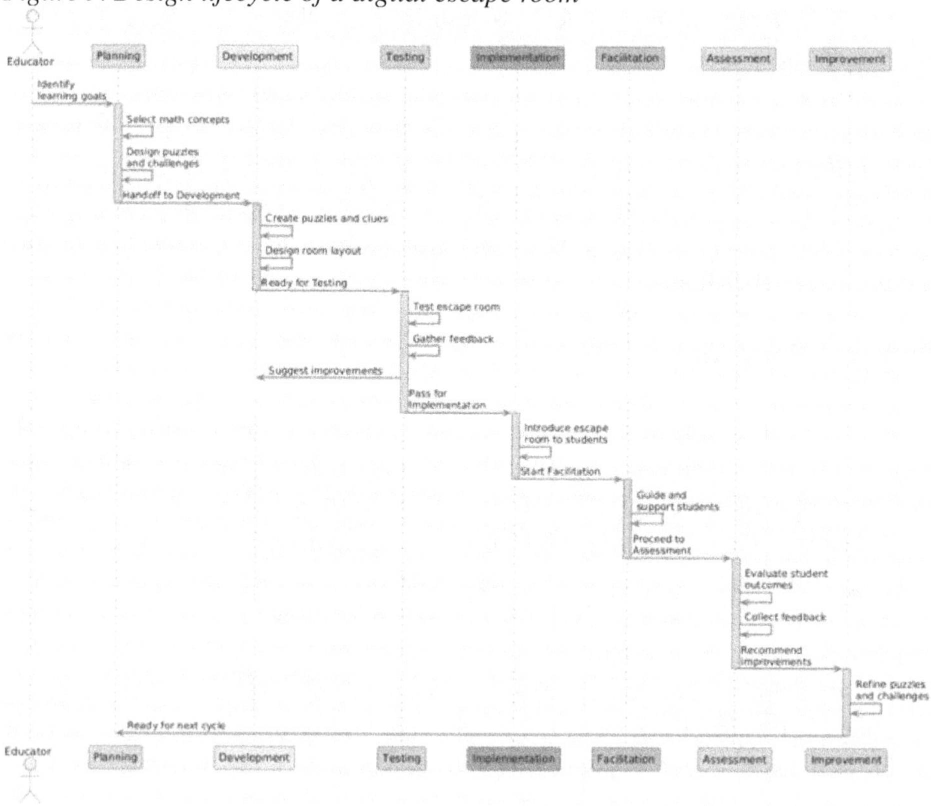

Following the planning phase, the lifecycle moves into the Development stage. Here, the theoretical ideas conceived during planning are transformed into tangible elements. This stage involves the creation of the actual puzzles, clues, and the overall design of the escape room. The room layout is carefully crafted to support the learning goals identified in the previous phase, with a particular emphasis on ensuring that the puzzles are both engaging and challenging. The development stage is highly creative, requiring both educational insight and design skills to create an environment that is both educationally effective and enjoyable for students.

Once the escape room has been developed, it enters the Testing phase. This critical stage involves rigorously testing the escape room to ensure that it meets the intended learning objectives and provides the right level of challenge for students. During testing, the escape room is typically piloted with a group of participants who provide feedback on their experience. This feedback is essential as it helps identify any issues with the puzzles or the overall flow of the room. Based on this feedback, improvements are made to refine the experience, ensuring that the escape room is both effective in teaching the mathematical concepts and engaging for the students.

After successful testing and any necessary adjustments, the escape room is ready for the Implementation phase. In this stage, the finalized escape room is introduced to the students as part of their mathematics course. The implementation phase is where the escape room comes to life, and students are provided with an introduction to both the escape room and the learning goals it is designed to achieve. This stage is crucial as it marks the point where the educational tool is integrated into the curriculum, and its effectiveness in enhancing student learning is observed in a real classroom setting.

During the Facilitation phase, which occurs concurrently with implementation, the educator plays a key role in guiding and supporting the students as they navigate the escape room. The facilitator ensures that students remain on track, providing assistance and answering questions as needed. This role is essential in helping students overcome challenges and ensuring that the escape room experience is both productive and educational. The facilitator also monitors the progress of the students, making real-time adjustments to support learning as required.

Once the escape room session is completed, the process moves into the Assessment phase. During this stage, students' learning outcomes are evaluated to determine how effectively the escape room has met its educational objectives. Assessment can take various forms, including traditional exams, quizzes, self-reflection, or group reflection. The results of these assessments provide valuable data on the students' understanding of the mathematical concepts and highlight areas where the escape room could be improved. The assessment phase is also an opportunity to gather further feedback from students about their experience, which can be used to enhance future iterations of the escape room.

The final stage of the lifecycle is Improvement, which is driven by the data and feedback collected during the assessment phase. In this stage, the escape room is refined and adjusted to address any identified weaknesses or to enhance its educational effectiveness. This could involve modifying puzzles that were too easy or difficult, redesigning parts of the room to better align with learning objectives, or incorporating new ideas based on student feedback. The improvement phase ensures that the escape room evolves and remains an effective educational tool for future cohorts of students. After these refinements, the cycle can begin anew with the next iteration of planning, development, and implementation, continually improving the quality and impact of the digital educational escape room.

Create Escape Room Storyline

The process of creating the storyline of a digital escape room, as illustrated by Figure 10, begins with a foundational step: Defining the Learning Objectives. This is a critical phase where educators or designers establish the specific educational goals that the escape room is intended to achieve. By clearly defining these objectives, the entire development process is guided toward ensuring that the experience is not only engaging but also educationally valuable. The learning objectives serve as the compass for the storyline, ensuring that every element of the escape room is aligned with the desired outcomes. For example, if the goal is to teach students about integrals in mathematics, the puzzles and challenges within the escape room must be designed to reinforce and deepen understanding of this concept.

Figure 10. Creating storyline of escape room

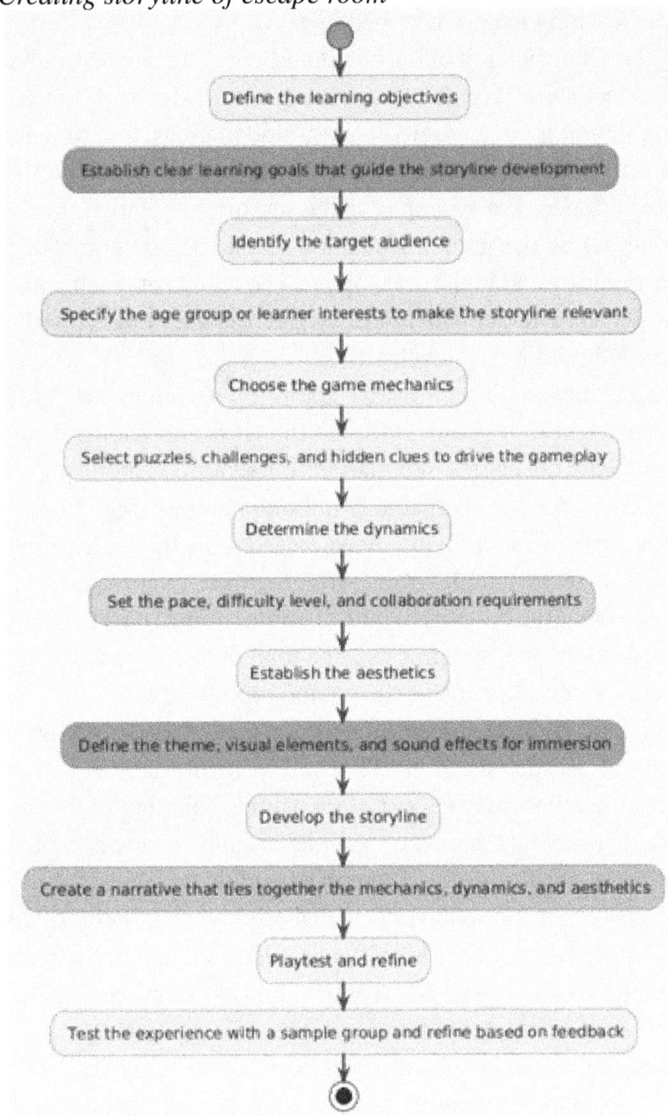

Once the learning objectives are set, the next step is to Identify the Target Audience. This involves determining the specific group of learners who will participate in the escape room, such as university students, high school students, or professionals in training. Understanding the target audience is crucial for tailoring the storyline to be engaging and relevant. For instance, a storyline designed for younger students might include more narrative-driven elements and simpler challenges, while one aimed at older students or professionals could involve more complex problems and scenarios. The storyline must resonate with the participants, capturing their interests and making the learning experience meaningful.

Following the identification of the target audience, the process moves to Choosing the Game Mechanics. Game mechanics are the rules and systems that drive the gameplay, including puzzles, challenges, hidden clues, and timed tasks. These mechanics must be carefully selected to align with the learning objectives and create an engaging experience. For example, if the learning objective involves teaching problem-solving skills, the game mechanics might include complex puzzles that require critical thinking and logical reasoning. The choice of mechanics will heavily influence how players interact with the game and how they engage with the learning material embedded within the storyline.

With the mechanics in place, the next step is to Determine the Dynamics of the escape room experience. Dynamics refer to the flow and pacing of the game, the level of difficulty, and the degree of collaboration required among players. This step is essential for ensuring that the game is not only challenging but also enjoyable. For instance, the dynamics might involve increasing the difficulty of puzzles as the game progresses or requiring players to work together to solve certain challenges. This stage ensures that the game is balanced, maintaining player engagement without causing frustration. The dynamics set the tone of the experience, dictating how intense, cooperative, or competitive the escape room will be.

Once the dynamics are determined, the process involves Establishing the Aesthetics of the escape room. Aesthetics cover the visual elements, theme, and sound effects that create the immersive environment of the game. This step is about bringing the storyline to life in a way that captivates the players and enhances the learning experience. For example, a mathematics escape room could be set in a futuristic lab, with a sleek, high-tech design and sound effects that reinforce the theme of discovery and exploration. The aesthetics play a significant role in engaging the players' senses, making the escape room not just a learning exercise but an immersive experience that players will remember.

With the mechanics, dynamics, and aesthetics defined, the next phase is to Develop the Storyline. This involves creating a narrative that ties together all the elements and provides context for the escape room experience. The storyline should align with the learning objectives and be engaging for the target audience. For instance, the narrative could involve a scenario where players must solve a series of mathematical puzzles to unlock a secret formula that will save the world. The storyline provides a framework that makes the puzzles and challenges meaningful, motivating players to engage deeply with the content as they progress through the game.

The final step in the process is to Playtest and Refine the escape room. This involves testing the game with a sample group of learners to evaluate its effectiveness and identify any areas that need improvement. Playtesting is crucial for gathering feedback on how well the storyline, mechanics, dynamics, and aesthetics work together to create a cohesive and educational experience. Based on the feedback received, the game is refined, making necessary adjustments to improve the overall flow, challenge level, and engagement. This iterative process ensures that the final version of the escape room is both educationally effective and enjoyable for the target audience.

Puzzle Design

The puzzle design using the Search and Discovery pattern, as illustrated in Figure 11, involves a structured yet dynamic process where players must engage deeply with their environment to uncover hidden elements and solve complex challenges. This pattern is particularly effective in creating an immersive and interactive experience that tests players' problem-solving abilities, observational skills, and teamwork.

Figure 11. Puzzle design using search and discover pattern

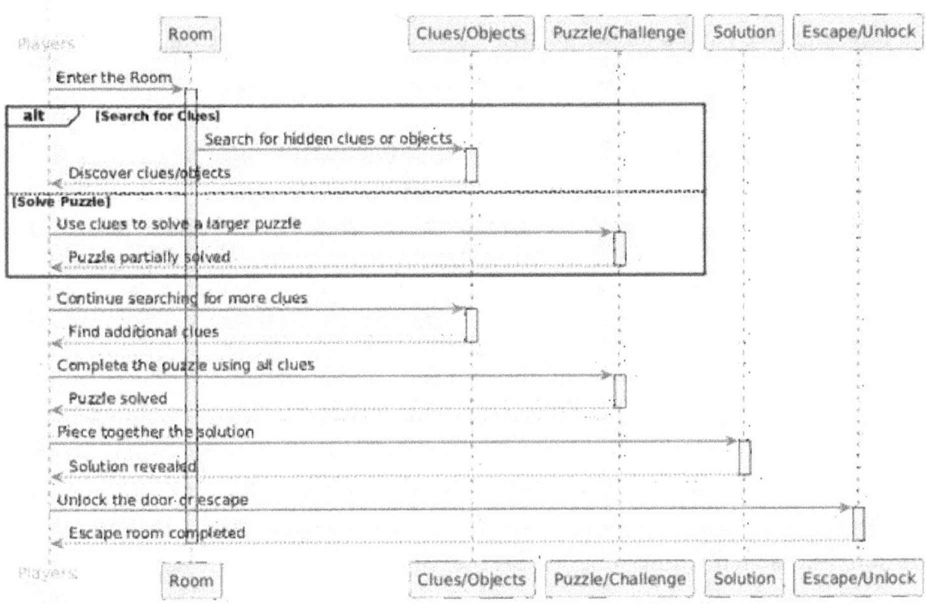

The process begins when players Enter the Room, marking the start of their journey. Upon entry, the room presents an environment filled with potential clues and objects, some of which are deliberately hidden. These elements are integral to the design as they form the basis of the challenges that players will face. The initial task for the players is to thoroughly explore their surroundings, actively searching for these hidden clues or objects. The room itself may contain a variety of hiding spots—some clues might be in plain sight, requiring only keen observation, while others may be more obscure, demanding creative thinking or the use of specific tools.

Once players begin their Search for Clues, they interact with various elements within the room. This interaction could involve examining bookshelves, inspecting wall decorations, or even manipulating furniture. Each interaction is designed to either reveal a clue directly or provide a hint towards the location of a clue. For instance, a painting on the wall might tilt to reveal a hidden compartment, or a seemingly random set of numbers found on an object might correspond to a lock combination. The discovery of these clues is a crucial step in the puzzle design, as it keeps players engaged and continuously drives the narrative forward.

As players collect these clues, the puzzle design moves into the Solve Puzzle phase. The clues gathered are not just random; they are pieces of a larger puzzle or challenge that players need to solve. For example, the clues might be different parts of a code that, when combined, unlock a safe, or they could be symbols that form a sequence needed to activate a mechanism. The challenge here is to correctly interpret the clues and understand how they fit together to solve the larger puzzle. This phase tests the players' ability to think critically and make connections between seemingly unrelated pieces of information.

If the puzzle isn't fully solved after the initial search, players must Continue Searching for more clues. This aspect of the design ensures that the game remains challenging and that players stay engaged by requiring them to revisit parts of the room they may have overlooked. It also adds depth to the puzzle, as some solutions may only become apparent after additional clues are found, or once a broader understanding of the room's layout and hidden features is gained.

When players have gathered all the necessary clues, they move on to Complete the Puzzle. At this point, all the pieces should fit together, allowing the players to solve the larger challenge. This stage is particularly satisfying, as it brings together the various threads of the narrative and rewards the players for their thoroughness and persistence in exploring the room. The puzzle design at this stage ensures that the solution is challenging but achievable, providing a sense of accomplishment upon completion.

After solving the puzzle, players must Piece Together the Solution. This involves synthesizing all the information they've uncovered and applying it to reveal the final key or code needed to escape the room. For instance, the clues might form a sequence that opens a lock, or they could reveal a hidden mechanism that triggers the door's release. The design here is focused on ensuring that the final solution is logical and that it ties together all the elements players have interacted with throughout the game.

Finally, the players use the solution to Unlock the Door or Escape. This marks the culmination of the Search and Discovery pattern, where the successful completion of the puzzle allows the players to exit the room, bringing the game to a close. The escape or unlocking sequence is designed to be the ultimate payoff for the players' efforts, providing a satisfying conclusion to the narrative and a rewarding end to their problem-solving journey.

Throughout this entire process, the puzzle design emphasizes a balance between challenge and accessibility. Clues are hidden in a way that requires careful observation and thought but are not so obscure as to be frustrating. The pattern also encourages collaboration among players, as discussing observations and pooling knowledge is often the key to uncovering all the necessary clues and solving the puzzle. This cooperative element not only enhances the gameplay experience but also reinforces the educational or team-building objectives of the escape room.

Puzzle Sequencing

The puzzle sequencing in a digital escape room designed using the Sequence and Order pattern challenges participants to identify, analyze, and apply a specific order of clues to solve the overarching puzzle and ultimately escape the room. This process is represented in Figure 12, which outlines the various stages that participants go through, from entering the room to solving the final puzzle.

Figure 12. Puzzle sequencing process

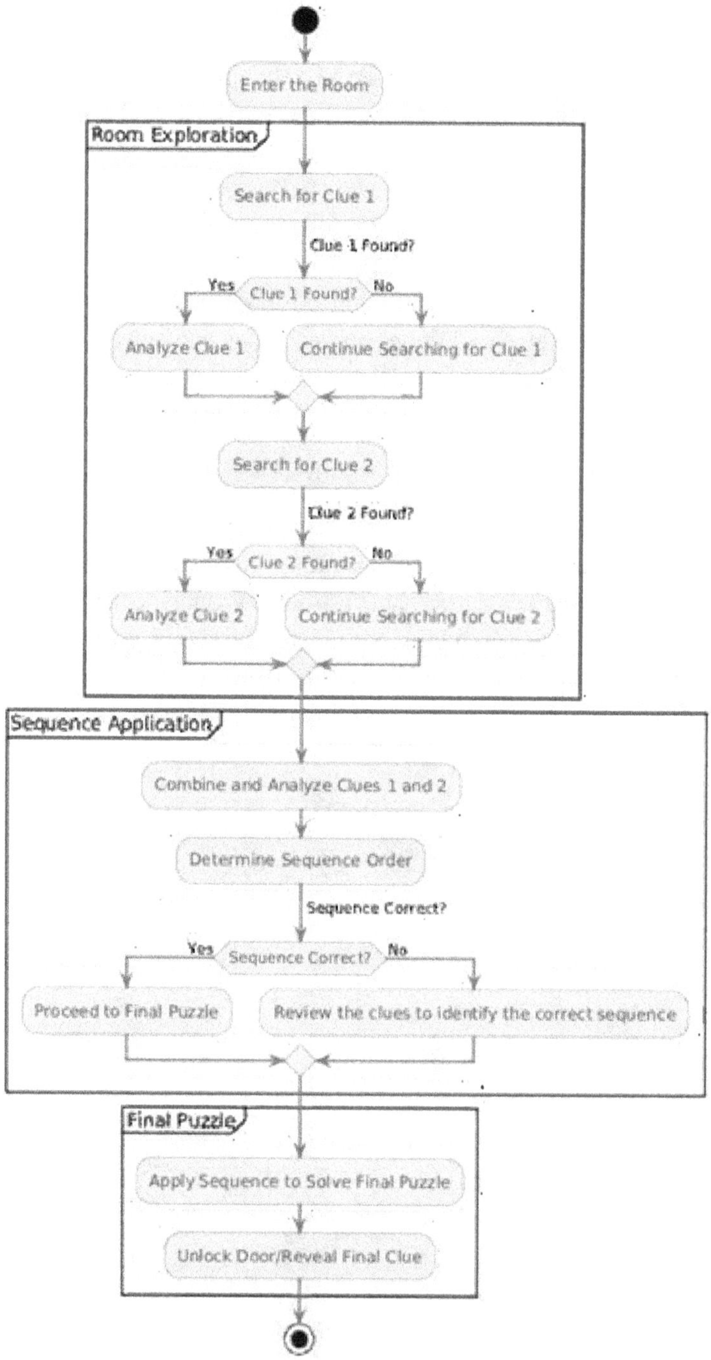

The sequence begins as participants Enter the Room, setting the stage for their journey through a series of interconnected puzzles. The first major phase in the diagram is Room Exploration, where participants must actively search the room for hidden clues. The search is not a straightforward task; it requires careful observation and critical thinking, as each clue is essential for progressing to the next stage. The diagram illustrates this process through a sequence of actions where participants first search for Clue 1.

Upon finding Clue 1, participants must analyze its relevance and consider how it might fit into a larger sequence. This phase emphasizes the importance of both discovery and initial analysis, as understanding the clue's role in the sequence is crucial for later stages.

The exploration phase continues with the search for Clue 2 and any subsequent clues. Each clue follows a similar pattern: participants search, determine if the clue has been found, and, once discovered, analyze its significance. The design of this phase ensures that participants remain engaged, as they must piece together the clues they find and begin to form hypotheses about how these clues might relate to one another. The repetition of searching and analyzing clues not only builds suspense but also reinforces the learning objective of identifying patterns and sequences.

After all three clues have been found and analyzed, the process transitions into the Sequence Application phase. Here, participants must Combine and Analyze Clues to determine the correct order or sequence in which these clues should be applied. This is a critical juncture in the puzzle sequencing process, as the correct identification of the sequence is key to progressing. The activity diagram highlights a decision point where participants must assess whether they have correctly determined the sequence. If they believe the sequence is correct, they proceed to the next phase. However, if they are unsure or if their sequence is incorrect, they must Reanalyze the Clues, revisiting their previous findings to adjust their approach. This iterative process ensures that participants are deeply engaged in critical thinking and problem-solving, as they must continuously evaluate and refine their understanding of the clues.

Once participants have confidently determined the correct sequence, they move on to the Final Puzzle phase. In this stage, the correct sequence is applied to solve the final challenge. The solution could involve arranging items in a specific order, entering a sequence of numbers or symbols, or using the correct sequence to unlock a mechanism within the room. The application of the sequence is the culmination of all the previous efforts, bringing together the clues and the analysis into a coherent solution. Successfully applying the sequence results in the Unlocking of the Door or the revelation of a Final Clue that allows the participants to escape the room.

The careful sequencing of puzzles in this pattern is designed to progressively build the challenge while maintaining engagement and excitement. Each step is interdependent, meaning that the success in later stages hinges on the thoroughness and accuracy of the earlier stages. The participants' journey through the escape room is a reflection of their ability to think critically, work collaboratively, and apply logic to complex problems. The activity diagram captures this complexity by showing the flow of actions and decisions that guide participants from the initial discovery of clues to the final, satisfying solution.

In summary, the Sequence and Order pattern in puzzle sequencing is a complex process that requires participants to methodically search for, analyze, and apply clues in a specific order. The activity diagram illustrates the thoughtful design behind this process, emphasizing the importance of each phase in building toward the final solution. By structuring the escape room in this way, designers create a challenging yet rewarding experience that teaches participants valuable skills in pattern recognition, logical sequencing, and problem-solving.

Develop Interactive Elements and Problem-Solving Tasks

The digital escape rooms designed for mathematics education are rich in interactive elements that are carefully curated to engage students and promote active learning. Each escape room typically consists of a series of rooms or levels, each of which presents a unique challenge related to a specific mathematical concept. For instance, students might be required to solve an indefinite integral by parts or calculate the area under a curve using definite integration.

One of the key interactive elements is the "Lock and Key" pattern, where students must solve a mathematical problem to unlock the next stage of the game (Figure 13). This element not only reinforces the learning material but also adds a gamified aspect to the experience, making the learning process more engaging. In addition to this, puzzles often require students to apply their mathematical knowledge in innovative ways, such as converting a calculated answer into a binary number to unlock a digital door.

Figure 13. Playing process in the locked escape rooms

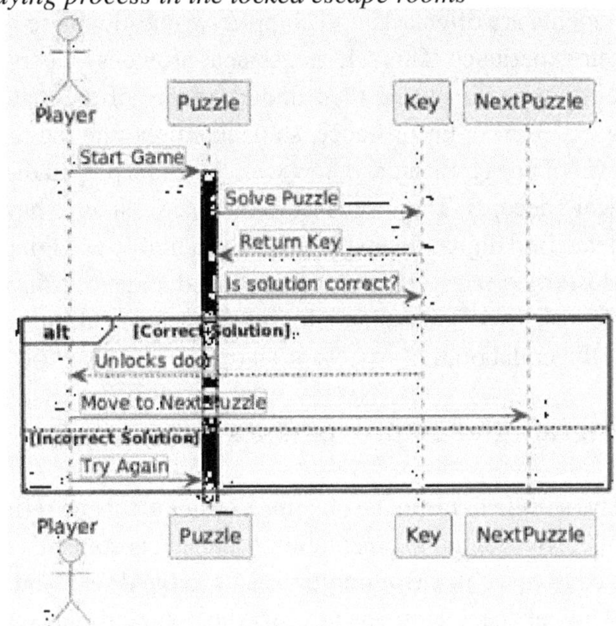

The problem-solving tasks are designed to be multi-faceted, requiring students to use a combination of mathematical skills, logical reasoning, and sometimes external resources like the internet. For example, in one scenario, students might need to integrate a function with symmetric bounds and then apply their understanding of even and odd functions to determine the correct solution. These tasks are not isolated; they build upon one another, requiring students to apply knowledge gained from previous puzzles to succeed in subsequent challenges. This interconnectedness ensures that students are continuously engaged and that their learning is cumulative, reinforcing the retention of mathematical concepts.

Assessment of Skill Development

The assessment of skill development in the context of digital escape rooms is multi-dimensional, focusing not only on the mastery of mathematical concepts but also on the enhancement of critical thinking, problem-solving, and teamwork skills. The effectiveness of these escape rooms in developing these skills is assessed through both direct observation during gameplay and through post-game evaluations. During the game, lecturers monitor student progress, observing how teams approach and solve each puzzle. This real-time assessment allows educators to gauge how well students are applying their mathematical knowledge, as well as how effectively they are communicating and collaborating with their peers. In addition to these observations, students are often asked to complete a questionnaire after the game, reflecting on their experience. This self-assessment provides valuable insights into how the escape room has impacted their understanding of the material and their overall learning experience. For instance, students might rate the difficulty of the puzzles, their level of engagement, and how well the game helped them grasp complex mathematical concepts. The results from these assessments have consistently shown that students find digital escape rooms to be a highly engaging and effective learning tool. Most importantly, they report improved understanding of mathematical concepts, increased confidence in their problem-solving abilities, and a greater appreciation for the collaborative aspects of learning.

Feedback and Iterative Improvements

Feedback plays a critical role in the ongoing development and refinement of digital escape rooms. After each implementation, feedback is collected from students, which is then used to make iterative improvements to the design and content of the escape rooms. This feedback loop ensures that the escape rooms remain relevant, challenging, and effective in achieving their educational goals.

Student feedback typically focuses on the difficulty of the puzzles, the clarity of the instructions, and the overall enjoyment of the experience. Based on this feedback, educators might adjust the difficulty level of certain tasks, clarify ambiguous instructions, or introduce new interactive elements to enhance engagement.

Data analytics are often employed to track how students interact with the escape room. Metrics such as the time taken to solve puzzles, the number of hints used, and the success rates for different challenges provide quantitative data that can inform further refinements. These iterative improvements ensure that the escape rooms continue to be an effective tool for teaching mathematics, adapting to the needs and preferences of students over time.

DEER DESIGN GUIDELINES AND IMPLEMENTATION PITFALLS

DEER Design Guidelines for Teacher Implementation

The implementation of DEERs in educational contexts demands a careful and theoretically grounded approach to ensure alignment with pedagogical objectives and the cognitive needs of learners. Teachers, as the principal facilitators of these learning environments, must adhere to specific design guidelines that balance educational effectiveness, technological integration, and learner engagement. These guidelines, derived from foundational theories in education and cognitive science, aim to enable educators to design DEERs that are both immersive and pedagogically sound.

To begin with, the design of DEERs must be anchored in clear and measurable learning objectives. Educators are tasked with identifying the key competencies and knowledge areas that the escape room experience should reinforce. This involves a precise articulation of educational goals, ensuring that every task, puzzle, and interaction aligns with the intended outcomes. For instance, in a mathematics-focused DEER, challenges must directly address specific mathematical concepts, such as integration techniques or geometric problem-solving, to ensure relevance and coherence within the curriculum. The objectives should cater to a range of cognitive levels, allowing both foundational reinforcement and the development of higher-order thinking skills.

The narrative structure of DEERs constitutes a pivotal aspect of their design, as it provides context and motivation for learner engagement. A well-crafted narrative serves to immerse participants, guiding them through the educational experience while maintaining their focus on the underlying pedagogical goals. Teachers must ensure that the storyline is coherent, engaging, and directly tied to the learning material. For example, in a history-oriented DEER, the narrative could involve unraveling a mystery set in a specific historical period, with each puzzle designed to deepen learners' understanding of key events, figures, or societal dynamics.

Collaboration and teamwork are integral to the DEER experience, reflecting the social constructivist principles that underpin modern educational theories. Teachers should design tasks that necessitate cooperative problem-solving, encouraging learners to pool their collective knowledge and perspectives. This can be achieved through puzzles that require diverse skill sets, fostering an environment where learners actively contribute and benefit from the strengths of their peers. Such collaborative elements not only enhance the educational value of DEERs but also promote the development of critical interpersonal skills, including communication, negotiation, and conflict resolution.

Accessibility and inclusivity must remain at the forefront of DEER design. Teachers must consider the diverse needs of learners, including those with physical, cognitive, or linguistic challenges. This entails incorporating principles of universal design for learning (UDL), such as providing multiple means of representation, engagement, and expression. For example, visual puzzles might be supplemented with auditory clues, and instructions should be available in multiple languages or formats to accommodate a broad spectrum of learners. By ensuring that DEERs are accessible, teachers can create inclusive environments that provide equitable learning opportunities for all participants.

The integration of technology in DEERs requires thoughtful consideration to enhance rather than overshadow the educational objectives. Teachers should select tools and platforms that align with the pedagogical goals and are feasible within the technological constraints of their institutions. For instance, the use of augmented reality (AR) or virtual reality (VR) should be justified by its ability to offer unique educational benefits, such as enabling learners to interact with three-dimensional models or simulated environments. However, the adoption of such technologies must be accompanied by adequate training for educators and learners to ensure their effective use.

Assessment strategies within DEERs must be embedded seamlessly into the gameplay to evaluate learners' progress without disrupting their immersion. Teachers should design puzzles and tasks that naturally elicit evidence of learning, such as demonstrating problem-solving strategies or applying theoretical knowledge in practical scenarios. Mechanisms for immediate feedback should be incorporated to guide learners and reinforce key concepts during the experience. For example, a correctly solved puzzle might trigger a brief explanation of the underlying principle, solidifying the educational takeaways.

Finally, iterative refinement is crucial. Teachers must collect and analyze data on learner interactions, engagement levels, and outcomes to identify areas for improvement. This iterative process ensures that DEERs remain dynamic and responsive to the evolving needs of learners. Feedback from participants, both qualitative and quantitative, should inform adjustments to the narrative, task complexity, and technological elements. By adopting a reflective and adaptive approach, educators can continually enhance the effectiveness and appeal of DEERs as a pedagogical tool.

The design and implementation of DEERs demand a rigorous and multifaceted approach that integrates educational theories, technological considerations, and a commitment to inclusivity. Teachers, as facilitators of these environments, must ensure that DEERs are not only engaging and innovative but also aligned with the overarching goals of education. By adhering to these design guidelines, summarized in Table 11, teachers can leverage the potential of DEERs to create engaging learning experiences that foster deep understanding, collaboration, and critical thinking.

Table 11. DEER design guidelines for teacher implementation

Guideline	Description
Clear Learning Objectives	Define measurable educational goals that align with curriculum requirements and target specific competencies.
Engaging Narrative	Develop a coherent and motivating storyline connected to the learning material to enhance immersion.

continued on following page

Table 11. Continued

Collaboration and Teamwork	Design tasks that require cooperative problem-solving to promote interpersonal skills and knowledge sharing.
Accessibility and Inclusivity	Apply universal design principles to ensure equitable participation for all learners, including those with diverse needs.
Thoughtful Technology Integration	Choose tools that support pedagogical goals and provide unique educational benefits, such as AR/VR, with appropriate training.
Embedded Assessment	Integrate natural assessment mechanisms into gameplay, with immediate feedback to guide and reinforce learning.
Iterative Refinement	Collect feedback and analyze data to continuously improve DEER design, adapting to learner needs.

Common DEER Implementation Pitfalls and How to Avoid Them

The implementation of DEERs presents educators with an opportunity to create engaging and interactive learning environments. However, poorly executed designs can undermine their educational potential and lead to ineffective or even counterproductive outcomes. Recognizing common pitfalls and adopting strategies to mitigate them is essential for successful DEER implementation.

One frequent pitfall is the lack of alignment between the DEER tasks and the intended learning objectives. In poorly designed DEERs, puzzles and challenges often emphasize entertainment over education, resulting in a disconnection between gameplay and academic content. For example, a DEER intended to teach mathematical concepts might include generic riddles or irrelevant trivia that fail to reinforce the targeted skills. This misalignment can leave students feeling frustrated or confused about the purpose of the activity. To avoid this, educators must ensure that each puzzle or task directly supports the learning objectives. For instance, a puzzle in a mathematics DEER might require students to calculate the volume of a solid using integration, seamlessly combining problem-solving with curriculum-aligned content.

Another common issue is the overloading of cognitive demands on participants. Educators sometimes create escape rooms with excessively complex or poorly structured puzzles, overwhelming students and detracting from the learning experience. For instance, presenting multiple complex tasks simultaneously, without clear instructions or progression, can lead to confusion and disengagement. This violates the principles of cognitive load theory, which emphasizes the importance of balancing task complexity with learners' capabilities. To address this, educators should scaffold tasks, starting with simpler puzzles and gradually increasing difficulty. Clear instructions and visual cues can further help students navigate the escape room effectively.

A lack of inclusivity and accessibility is another significant challenge in DEER implementation. Many escape rooms are designed without considering the diverse needs of participants, such as those with physical disabilities, cognitive impairments, or limited language proficiency. For instance, a visually impaired student might struggle with puzzles that rely solely on visual clues, or a non-native speaker might find text-heavy instructions difficult to follow. To mitigate this, educators should apply principles of universal design for learning (UDL), incorporating multiple modes of representation and interaction. For example, providing audio descriptions for visual elements or offering text translations can make the experience accessible to a broader range of learners.

The over-reliance on technology is another pitfall that can detract from the educational value of DEERs. While technologies like augmented reality (AR) and virtual reality (VR) can enhance immersion, their inappropriate use can result in technical difficulties and distract from learning objectives. For example, relying on complex VR setups without proper training or reliable hardware may lead to frequent disruptions, frustrating both students and educators. To prevent this, teachers should prioritize technological tools that are reliable, user-friendly, and aligned with the learning goals. When using advanced technologies, sufficient training and contingency plans should be in place to address potential issues.

Poorly designed feedback mechanisms also contribute to ineffective DEER experiences. Some escape rooms fail to provide timely or constructive feedback, leaving students unaware of their progress or the reasons behind their mistakes. For instance, a puzzle that allows multiple incorrect attempts without offering hints or explanations may lead to frustration and disengagement. To counter this, educators should integrate immediate and meaningful feedback into the escape room design. For example, solving a puzzle could trigger a brief explanation of the underlying concept, reinforcing learning while maintaining engagement.

Neglecting collaboration and team dynamics is another common issue. Some DEERs are structured in a way that allows dominant team members to take over, sidelining quieter participants and reducing the overall educational impact. For instance, a puzzle that only requires one person to interact with a central interface can exclude other team members from active participation. To foster collaboration, educators should design tasks that require input from multiple participants, such as splitting clues across different team members or creating puzzles that require collective problem-solving. This ensures that all learners are actively engaged and contributing to the experience.

Finally, insufficient post-activity reflection and assessment can diminish the long-term educational value of DEERs. Without a structured debriefing session, students may fail to connect the activities with broader learning outcomes or reflect on their performance. For example, completing a DEER on physics principles without discussing how the puzzles illustrated concepts like inertia or energy transfer misses an opportunity for deeper understanding. To avoid this, educators should include a debriefing phase where students reflect on their experience, discuss the puzzles, and connect them to the curriculum. This phase can also provide valuable feedback for refining future iterations of the escape room.

In conclusion, while DEERs have the potential to transform educational experiences, their effectiveness hinges on thoughtful design and implementation. By aligning tasks with learning objectives, managing cognitive load, ensuring inclusivity, integrating appropriate technology, providing timely feedback, fostering collaboration, and incorporating reflective practices, educators can avoid common pitfalls (Summarized in Table 12) and maximize the impact of DEERs as a pedagogical tool. Addressing these challenges proactively ensures that DEERs remain both engaging and educational, fostering meaningful learning experiences for all learners.

Table 12. Common DEER implementation pitfalls and remedies

Pitfall	**Real-World Example**	**Potential Remedies**
Misalignment with Learning Objectives	Puzzles focus on generic riddles or trivia instead of curriculum-aligned content, e.g., a math DEER including irrelevant word puzzles.	Design puzzles directly tied to learning goals, such as using integration problems to teach calculus.
Overloading Cognitive Demands	Excessively complex puzzles presented simultaneously without clear instructions or progression.	Scaffold tasks by starting with simpler puzzles and increasing difficulty gradually; provide clear instructions and visual cues.
Lack of Inclusivity and Accessibility	Visually impaired students struggle with puzzles relying solely on visual clues, or text-heavy instructions hinder non-native speakers.	Apply universal design principles, e.g., provide audio descriptions for visuals and offer multilingual or alternative formats.
Over-reliance on Technology	Complex VR setups result in technical issues, disrupting the flow and frustrating participants.	Use reliable, user-friendly technology aligned with learning goals; provide training and have contingency plans.
Poor Feedback Mechanisms	Puzzles allow multiple incorrect attempts without providing hints or explanations.	Incorporate immediate, constructive feedback, e.g., solving a puzzle triggers a brief explanation of the underlying concept.
Neglecting Collaboration and Team Dynamics	One team member dominates interactions while others are sidelined due to single-user task designs.	Design tasks requiring input from multiple participants, such as splitting clues or fostering collective problem-solving.
Insufficient Post-Activity Reflection	No debriefing session to connect puzzles to broader learning outcomes, reducing long-term educational impact.	Include a debriefing phase for reflecting on activities, discussing solutions, and linking them to the curriculum.

SYNTHESIS AND FUTURE DIRECTIONS

Synthesizing Key Learnings from Case Studies

The exploration of digital escape rooms in mathematics education, as evidenced by the case studies, offers several valuable insights into how immersive and gamified learning environments can enhance student engagement and learning outcomes. One of the key learnings is the effectiveness of digital escape rooms in transforming the abstract and often challenging concepts of mathematics into more accessible and engaging experiences. By embedding mathematical problems within the context of a game, students are not only motivated to engage with the material but also encouraged to apply their knowledge in practical, problem-solving scenarios. This method fosters deeper understanding and retention of mathematical concepts, as students are required to think critically and creatively to solve the puzzles presented.

The case studies highlight the importance of collaboration in digital escape rooms. The team-based nature of these experiences mirrors real-world scenarios where collaboration and communication are essential for problem-solving. Students working together in teams are able to pool their collective knowledge and skills, leading to more effective and innovative solutions to the challenges presented. This collaborative approach not only enhances the learning experience but also helps in the development of essential soft skills such as teamwork, communication, and leadership.

Another significant learning from these case studies is the role of immediate feedback and iterative problem-solving in enhancing student learning. The structure of the escape rooms, which often includes immediate consequences for incorrect answers (such as point deductions or the inability to progress), helps students quickly identify and correct their mistakes. This iterative process of trial and error is critical in reinforcing learning, as it allows students to refine their understanding and approach to problem-solving in real time. The flexibility and adaptability of DEERs are notable strengths. These tools can be tailored to suit different educational contexts, subjects, and student needs. Whether focusing on fundamental concepts or more advanced topics, digital escape rooms can be customized to provide an appropriate level of challenge for students at various levels of understanding. This adaptability makes them a versatile tool in the educator's repertoire, capable of being integrated into a wide range of teaching strategies.

Emerging Trends in VR and AR Education

As digital technologies continue to evolve, the fields of Virtual Reality (VR) and Augmented Reality (AR) are emerging as significant trends in educational innovation. These technologies offer new dimensions to the learning experience by providing immersive and interactive environments that go beyond traditional digital platforms. For example, "Escape VR: The Guilt" serves as a successful example of how immersive VR gaming can evoke emotions and captivate players (Oliveira et al., 2023). In the context of STEM education, VR and AR have the potential to further enhance the learning experiences provided by digital escape rooms and similar educational tools.

One of the most promising trends is the use of VR to create fully immersive learning environments. In these virtual spaces, students can interact with 3D models, conduct experiments, and explore complex concepts in a hands-on manner. For example, in a VR-enhanced mathematics escape room, students might be able to manipulate 3D geometric shapes, visualize complex functions, or walk through a virtual environment that represents a mathematical landscape. This level of immersion can significantly enhance understanding by providing students with a tangible experience of abstract concepts. AR, on the other hand, offers the ability to overlay digital information onto the real world, creating opportunities for blended learning experiences that combine physical and digital elements. For instance, AR could be used to enhance physical escape rooms with digital clues, puzzles, and interactive elements that appear when students use their devices to scan certain objects or areas. For example, a mobile-driven augmented reality (AR) escape room described by (Antoniou et al., 2024) aimed to acclimate the students with the workflows of basic life support.

This integration of physical and digital worlds can make learning more dynamic and engaging, catering to diverse learning styles and preferences. The integration of AI with VR and AR such as using digital twin (as in Galaxy Escape game (Nikkarikoski et al., 2024)) is another emerging trend, enabling more personalized and adaptive learning experiences. AI-driven systems can analyze student performance in real-time, offering customized feedback and adjusting the difficulty of tasks based on the student's progress. This capability enhances the effectiveness of VR and AR learning environments and ensures that each student receives a tailored learning experience that meets their individual needs.

Challenges for Widespread Adoption

Despite the potential benefits, several challenges must be addressed for the widespread adoption of digital escape rooms, VR, and AR in education. One of the primary challenges is the issue of accessibility. High-quality VR and AR systems, as well as the infrastructure needed to support them, can be expensive, making it difficult for schools and institutions with limited budgets to implement these technologies. Not all students may have access to the necessary hardware, such as VR headsets or AR-capable devices, which can exacerbate existing inequalities in educational access.

Another significant challenge is the need for educator training and support. Integrating digital escape rooms and VR/AR technologies into the curriculum requires educators to have a certain level of technical expertise and familiarity with these tools. However, many educators may not have the necessary training or confidence to effectively use these technologies in their teaching. This highlights the need for comprehensive professional development programs that can equip educators with the skills and knowledge required to leverage these tools effectively.

The development and implementation of digital escape rooms and VR/AR content can be time-consuming and resource-intensive. Designing these experiences requires not only subject matter expertise but also technical skills in software development, instructional design, and game design. For institutions without access to these resources, creating high-quality digital learning experiences can be a significant barrier.

There are also pedagogical challenges to consider. DEERs must be carefully integrated into the broader curriculum to ensure that they complement, rather than distract from, the learning objectives. Educators must also be mindful of the potential for these technologies to overwhelm or confuse students, particularly if the experiences are not well-aligned with their existing knowledge and skills.

Prospects for Future Research and Development

Looking forward, there is significant potential for future research and development in the use of digital escape rooms and immersive technologies in education. One promising area of research is the exploration of how these tools can be used to support differentiated instruction, where educational experiences are tailored to meet the diverse needs of students. Studies could investigate how digital escape rooms and VR/AR environments can be customized to provide varying levels of difficulty, pacing, and content based on individual student profiles.

Another important area for future research is the long-term impact of these technologies on learning outcomes. While there is growing evidence that digital escape rooms and immersive environments can enhance engagement and immediate understanding, further studies are needed to determine their effectiveness in promoting long-term retention of knowledge and the development of higher-order thinking skills. Research could also explore how these tools impact different student populations, such as students with learning disabilities, to ensure that they are inclusive and accessible to all learners.

The development of new frameworks and best practices for integrating these technologies into the curriculum is another critical area for research. As more educators begin to experiment with digital escape rooms and VR/AR, there is a need for clear guidelines on how to design and implement these experiences effectively. This includes understanding how to align these tools with curriculum standards, how to assess student learning in these environments, and how to ensure that they are used in a way that supports, rather than replaces, traditional teaching methods.

Finally, there is potential for further development of the technologies themselves. Advances in AI, machine learning, and natural language processing could lead to more sophisticated and adaptive digital learning environments. For example, future digital escape rooms could feature AI-driven characters that interact with students, providing hints, feedback, and guidance based on their progress. Similarly, improvements in VR and AR hardware could make these technologies more affordable and accessible, paving the way for broader adoption in educational settings.

REFERENCES

Antoniou, P. E., Papamalis, F., Dafli, E., Poultourtzidis, I., Schwarz, D., Woodham, L., Dimitriadis, S., Tagaras, K., Kyriakidis, N., David, P., Nikolaidou, M., Skríšovská, T., Poulton, T., & Bamidis, P. D. (2024). Medical education escape room aligned with flipped classroom and powered by mobile augmented reality. *Electronics (Basel)*, 13(12), 2367. Advance online publication. DOI: 10.3390/electronics13122367

Bender, E. M., Gebru, T., McMillan-Major, A., & Shmitchell, S. (2021). On the dangers of stochastic parrots: Can language models be too big? *Proceedings of the 2021 ACM Conference on Fairness, Accountability, and Transparency*. DOI: 10.1145/3442188.3445922

Chi, M. T. H. (2009). Active-constructive-interactive: A conceptual framework for differentiating learning activities. *Topics in Cognitive Science*, 1(1), 73–105. DOI: 10.1111/j.1756-8765.2008.01005.x PMID: 25164801

Deterding, S., O'Hara, K., Sicart, M., Dixon, D., and Nacke, L. (2011). Gamification: Using game design elements in non-gaming contexts. page 2425 – 2428. DOI: .DOI: 10.1145/1979742.1979575

Dicheva, D., Dichev, C., Agre, G., & Angelova, G. (2015). Gamification in education: A systematic mapping study. *Journal of Educational Technology & Society*, 18(3), 75–88.

Johnson, C., Urazov, M., and Zanoli, E. (2022). Escapeling: A gamified, ai-supported chatbot for collaborative language practice. *Lecture Notes in Networks and Systems*, 349 LNNS:141 – 148, DOI: .DOI: 10.1007/978-3-030-90677-1_14

Johnson, D. W., & Johnson, R. T. (1994). *Learning together and alone: Cooperative, competitive, and individualistic learning*. Allyn and Bacon.

Kolb, D. A., Boyatzis, R. E., & Mainemelis, C. (2014). *Experiential learning theory: Previous research and new directions*. Perspectives on Thinking, Learning, and Cognitive Styles., DOI: 10.4324/9781410605986-9

Nikkarikoski, I., Pehkonen, J., Kivelä, I., Alavesa, P., & Korkiakoski, M. (2024). Galaxy escape: A game as a tabletop holographic digital twin interface. In Companion Proceedings of the 2024 Annual Symposium on Computer-Human Interaction in Play (pp. 215–220). CHI PLAY '24: The Annual Symposium on Computer-Human Interaction in Play. ACM. DOI: DOI: 10.1145/3665463.3678810

Oliveira, I., Carvalho, V., Soares, F., Novais, P., Oliveira, E., & Gomes, L. (2023). Development of a virtual reality escape room game for emotion elicitation. *Information (Basel)*, 14(9), 514. Advance online publication. DOI: 10.3390/info14090514

Piaget, J. (1954). *The construction of reality in the child*. Routledge., DOI: 10.1037/11168-000

Rychert, A., Ganuza, M. L., & Selzer, M. N. (2024). Integrating gpt as an assistant for low-cost virtual reality escape-room games. *IEEE Computer Graphics and Applications*, 44(4), 14–25. DOI: 10.1109/MCG.2024.3426314 PMID: 38990754

Seaborn, K., & Fels, D. I. (2015). Gamification in theory and action: A survey. *International Journal of Human-Computer Studies*, 74, 14–31. DOI: 10.1016/j.ijhcs.2014.09.006

Selwyn, N. (2019). *Should robots replace teachers?: Ai and the future of education*. Polity Press.

Shneiderman, B. (2020). Human-centered artificial intelligence: Reliable, safe and trustworthy. *International Journal of Human-Computer Interaction*, 36(6), 495–504. DOI: 10.1080/10447318.2020.1741118

Sidekerskiene, T., & Damaševicius, R. (2023). Out-of-the-box learning: Digital escape rooms as a metaphor for breaking down barriers in stem education. *Sustainability (Basel)*, 15(9), 7393. Advance online publication. DOI: 10.3390/su15097393

Taber, K. S. (2011). Constructivism as educational theory: Contingency in learning, and optimally guided instruction. *Educational Theory*, •••, 39–61.

Van Merriënboer, J. J. G., & Sweller, J. (2005). Cognitive load theory and complex learning: Recent developments and future directions. *Educational Psychology Review*, 17(2), 147–177. DOI: 10.1007/s10648-005-3951-0

VanLehn, K. (2006). The behavior of tutoring systems. *International Journal of Artificial Intelligence in Education*, 16(3), 227–265.

Vaswani, A., Shazeer, N., Parmar, N., Uszkoreit, J., Jones, L., Gomez, A. N., Kaiser, L., & Polosukhin, I. (2017). Attention is all you need. *Advances in Neural Information Processing Systems*, •••, 30.

Vygotsky, L. S. (1978). *Mind in society: The development of higher psychological processes*. Harvard University Press., DOI: 10.2307/j.ctvjf9vz4

Xie, H., Zhu, W., Vong, L., Wong, T.-L., & Liao, S. S. (2019). Ai in education: Current insights and future perspectives. *Journal of Computer Science and Technology*, 34(4), 869–889.

Zou, R., Yin, S., Song, T., Qin, P., and Lee, Y.-C. (2024). Mitigating ageism through virtual reality: Intergenerational collaborative escape room design. DOI: .DOI: 10.1145/3613905.3650914

KEY TERMS AND DEFINITIONS

Active Learning: A learner-centered approach emphasizing engagement through hands-on, participatory experiences, fostering critical thinking, collaboration, and problem-solving.

Adaptive Learning: Customizing educational content and tasks dynamically to suit the unique needs and abilities of individual learners.

Assessment of Learning Outcomes: The process of evaluating students' knowledge and skills gained through DEERs to ensure alignment with educational goals.

Augmented Reality (AR): A technology that overlays digital elements like images, sounds, or data onto the real-world environment to enhance user interaction and understanding.

Cognitive Load Theory: A framework for designing instructional materials that minimize extraneous cognitive processing, ensuring effective learning by managing information complexity.

Collaboration: The integration of teamwork and communication as foundational elements of DEER design to promote social learning and collective problem-solving.

Constructivist Learning Theory: An educational theory positing that learners construct knowledge actively through interaction with their environment and experiences.

Digital Educational Escape Rooms (DEERs): Interactive, narrative-driven environments designed to engage learners through problem-solving and tasks that align with specific educational goals.

Experiential Learning: Learning through reflection on doing, which involves creating real-world or simulated experiences to deepen understanding.

Feedback Mechanisms: Structured systems within DEERs that provide immediate, targeted guidance or hints to learners, enhancing motivation and learning effectiveness.

Gamification: The application of game design elements, such as points, levels, and rewards, in non-game contexts to increase engagement and motivation.

Immersive Technologies: Advanced technologies like VR and AR that create highly interactive and engaging environments, simulating real-world or fantastical experiences.

Interactive Learning Environments: Dynamic spaces where learners engage with content, peers, and technology interactively to achieve educational objectives.

Narrative and Storytelling: The use of structured, context-rich stories within DEERs to create immersive and relatable learning experiences.

Player-Centric Design: A design approach in DEERs that adapts gameplay to individual learning profiles and preferences to enhance engagement.

Purposeful Use of Technology: Integrating technology in DEERs to directly support pedagogical objectives and enhance learning, avoiding unnecessary complexity.

Real-Time Adaptation: The use of AI to adjust tasks, challenges, and narratives on-the-fly based on learners' interactions and performance.

Technological Accessibility: Ensuring that technological solutions are inclusive and usable by learners of diverse backgrounds and abilities.

Virtual Reality (VR): A fully immersive digital environment where users can interact with a computer-generated 3D world, often using specialized equipment.

Chapter 7
Example of Digital Educational Escape Room Design Using Genially Platform

ABSTRACT

This chapter explores the design and implementation of digital educational escape rooms using the Genially platform. Focusing on two case studies—one on differential equations and the other on integral calculus—it demonstrates how escape rooms can be structured to enhance learning in mathematical disciplines. The chapter outlines the design process, from template selection and customization to the creation of interactive puzzles, narrative elements, and feedback mechanisms. By using Genially's flexible tools, educators can create immersive, gamified learning environments that promote problem-solving, critical thinking, and persistence. The chapter highlights the educational outcomes of these escape rooms, emphasizing their role in reinforcing theoretical knowledge through active engagement, fostering student collaboration, and motivating learners to apply mathematical concepts in innovative ways. Additionally, advanced customization techniques using Genially, such as embedding external code, are discussed to further personalize the escape room experience.

DOI: 10.4018/979-8-3693-4219-0.ch007

INTRODUCTION

Motivation for Using Genially in Escape Room Design

The Genially platform offers a versatile and interactive approach to designing educational escape rooms, which has been shown to enhance student engagement, foster critical thinking, and improve learning outcomes across diverse disciplines. Recent studies highlight the efficacy of gamified learning using Genially, making it a valuable tool for educators aiming to create immersive learning experiences. For example, Martinez et al. (2023) demonstrated that incorporating escape rooms into an introductory physics course significantly increased students' motivation and understanding of fundamental concepts. By integrating elements such as interactive puzzles and narrative-driven challenges within the Genially platform, the study found that students actively engaged with the material and demonstrated improved learning outcomes. Similarly, (Vera-Morales et al., 2023) applied Genially to gamify ecology education, achieving a 62.5% improvement in student grades compared to traditional teaching methods. This study underscores Genially's ability to facilitate active learning and professional skill development, including teamwork and innovation. In the context of analytical chemistry, (Santiago López, 2024) employed a Genially-based escape room titled "The Time Machine," which activated students' metacognition and reinforced theoretical and methodological knowledge in a fun and interactive manner. Students reported that this approach not only made learning enjoyable but also solidified their understanding of complex topics. This finding is supported by (Abdul Rahim, 2023), who created blended escape rooms in pharmaceutical chemistry courses. These rooms, hosted on Genially, introduced critical concepts while fostering camaraderie and teamwork among students in problem-based learning environments. (Aldalur, 2025) highlighted Genially's role in transforming software development education through gamified board games, while (Mauri-Medrano et al., 2024) emphasized that escape rooms outperform other gamified methods in fostering creative thinking and mastery, despite potential stress-related drawbacks. These studies collectively illustrate Genially's adaptability in meeting diverse educational goals. From a technical perspective, the Genially platform's user-friendly interface, customizable templates, and multimedia integration capabilities allow educators to design highly interactive and visually dynamic escape rooms (Jiménez et al., 2020). These features make it possible to tailor content to specific educational objectives, embed external simulations, and track student engagement through data analytics, ensuring that the learning experience is both effective and engaging.

Before Starting

Creating a digital escape room with the Genially app offers a dynamic and engaging way to facilitate learning experiences. In this chapter, we will explore the process of designing and implementing a successful digital escape room, from using prepared templates to customizing your own unique experience. To begin, let's delve into the various options available for creating your digital escape room. You can start by either using prepared templates or crafting your escape room from scratch. We will provide two examples to illustrate these methods: one using a predefined template and another demonstrating how to customize it to suit your specific needs. In addition to the flexibility of template usage, the Genially app offers extensive customization options for your digital escape room. You can personalize the theme, background, and layout according to your preferences and learning objectives. The app provides a variety of interactive elements, including images, videos, audio files, and interactive hotspots, to enhance the immersive experience of your escape room.

When designing your digital escape room, it's crucial to consider the educational goals and target audience. Tailoring the content and difficulty level of the puzzles and challenges can significantly enhance engagement and effectiveness for learners. Incorporating narrative elements or a storyline can further deepen the immersive experience and encourage participants to fully immerse themselves in the adventure.

Before diving into the creation process, ensure that you have an account on the Genially platform or sign in using your Google account. Once logged in, you will find yourself in the Genially environment, where a world of creative possibilities awaits. This environment is depicted in Figure 1, providing a visual reference for navigating the platform.

Figure 1. Overview of the Genially platform interface, showcasing the environment for creating digital escape rooms

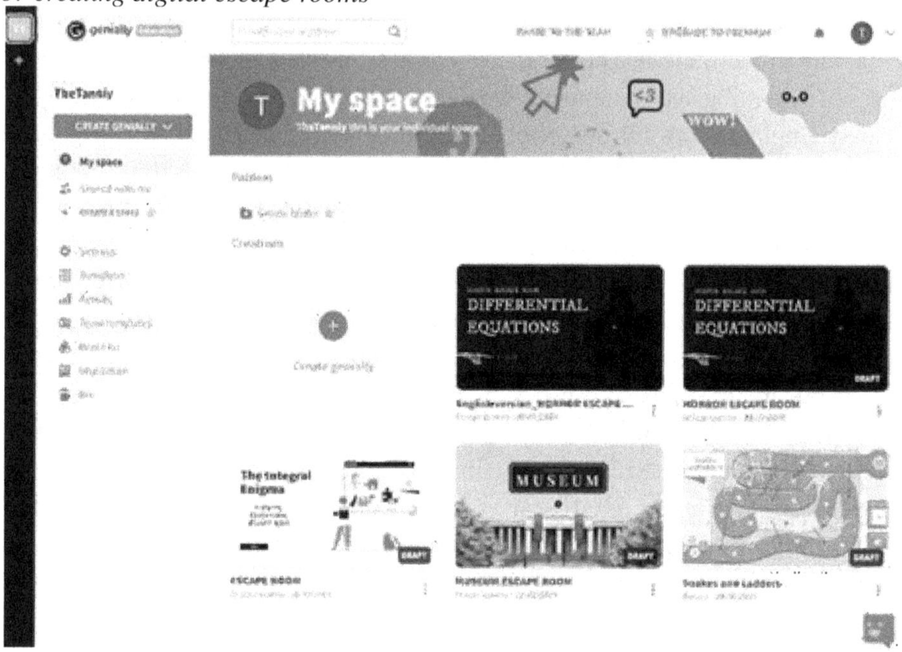

Within the Genially app, you will discover a plethora of tools and features that extend far beyond just digital escape rooms. While our focus in this chapter is on creating escape rooms, it's worth noting that you can also design engaging presentations, interactive quizzes, and much more. However, for the purposes of this discussion, we will concentrate solely on crafting a captivating digital escape room. To initiate the creation process, click on the "Create genially" button and navigate to the "Games and challenge" category, followed by "Escape games" (Figure 2). Here, you will find a variety of prepared templates tailored for crafting immersive escape room experiences. Simply select your desired template and fill it with your questions, puzzles, and clues to bring your digital escape room to life.

In this chapter, we explore the process of designing and implementing a digital escape room using the Genially app, a tool that enables educators to create dynamic, interactive learning experiences. The chapter begins by introducing two primary methods for creating a digital escape room: using predefined templates and customizing them or starting from scratch. Both approaches are detailed, with practical examples provided to illustrate each method. For those using templates, we guide readers through selecting and modifying the content to align with their specific educational goals. For those creating escape rooms from scratch, we walk through the design process, from crafting a compelling theme to embedding interactive elements. Next, we delve into the flexibility Genially offers, including the ability to personalize themes, backgrounds, and layouts. Interactive components such as images, videos, and audio files can be easily incorporated to create an immersive learning experience. This section emphasizes the importance of aligning the content of the escape room with the educational objectives and the target audience, ensuring the complexity of puzzles matches the learners' level.

The Genially Platform

The Genially platform https://genially.com/ is a powerful tool for creating interactive and visually dynamic content, designed to engage audiences across various educational, corporate, and creative sectors. Known for its flexibility and ease of use, Genially allows users to craft visually rich presentations, infographics, interactive lessons, digital games, quizzes, and much more, without requiring advanced programming or design skills. Its user-friendly interface and broad range of customizable templates make it particularly suitable for educators looking to incorporate multimedia elements into their teaching, offering a modern approach to knowledge dissemination and student engagement.

Genially was developed with the idea that education and communication should be both engaging and interactive. It supports a shift from traditional passive learning environments, where students are expected to absorb information, to more dynamic spaces that foster active participation. With tools that integrate images, text, video, and audio seamlessly, Genially allows educators to create content that speaks to multiple learning styles. Whether students prefer visual aids, interactive games, or video-based instruction, the platform has tools to cater to a variety of preferences and needs.

One of the key strengths of Genially lies in its vast collection of templates, which provide a starting point for users to design their content. These templates are pre-structured and come with built-in functionalities, offering a range of interactive options such as clickable buttons, animation effects, quizzes, and gamified elements. For example, in the context of education, Genially offers templates for digital escape rooms, interactive quizzes, and gamification strategies that can be used to promote student engagement and create an immersive learning experience. These templates are not rigid; they can be customized and adapted to meet specific educational goals, making the platform highly flexible and versatile.

Figure 2. Steps to initiate the creation of a digital escape room in Genially by selecting the "escape games" template under the "games and challenges" category

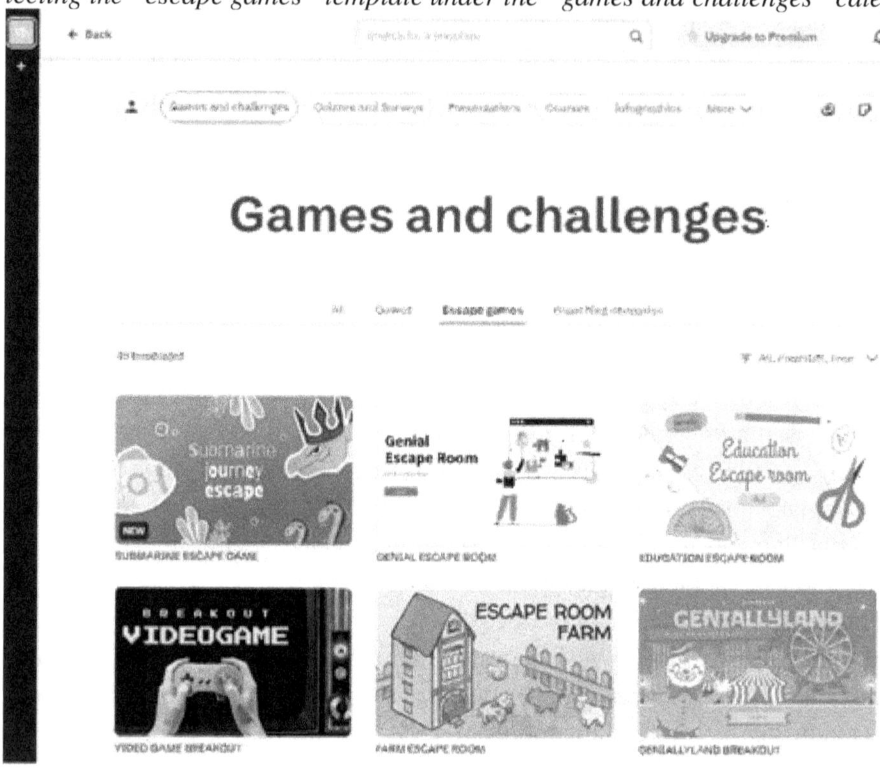

For educators, one of the standout features of Genially is its ability to incorporate multiple layers of interactivity within a single project. Users can embed hotspots, clickable elements, and pop-ups, making it possible to create deeply interactive experiences. This level of interactivity promotes exploration and discovery, encouraging students to engage more fully with the material. For instance, in a digital escape room, students can click on different areas of a slide to reveal hidden clues, unlock puzzles, or trigger videos that further explain a concept. These types of interactions allow students to learn through doing, fostering critical thinking and problem-solving skills that are often difficult to achieve through traditional lecture-based learning.

Genially's animation tools provide an added layer of engagement. With options to animate text, images, and other elements, users can create presentations or lessons that feel dynamic and alive. The platform offers a variety of animation styles, from simple fades and slides to more complex motion effects. These animations can be timed to occur automatically or triggered by user actions, making it possible to guide students through a learning experience at their own pace. For example, in a lesson on history or science, animations could be used to show the progression of a timeline or illustrate complex processes, allowing students to visualize information in a more meaningful way.

In addition to interactivity and animations, the Genially platform excels in multimedia integration. It allows for the seamless embedding of videos from platforms such as YouTube or Vimeo, as well as the inclusion of audio files, either for background music or narration. This multimedia approach enables educators to appeal to auditory and visual learners simultaneously. A science teacher, for instance, could enhance a lesson on the solar system by embedding a video tour of the planets, while adding a voice-over that explains each celestial body in greater detail. Such features make learning more dynamic and engaging, particularly in subjects where traditional textbooks or lectures may struggle to capture students' attention.

The platform's collaborative features also stand out, particularly for educational environments. Genially allows multiple users to work on the same project simultaneously, making it a valuable tool for group work and collaborative learning. Teachers can assign group projects where students collaborate on creating interactive presentations or escape rooms, fostering teamwork and collective problem-solving. The platform also offers tracking features, allowing educators to monitor how students interact with the content. This data can provide valuable insights into how well students are grasping the material, enabling teachers to make real-time adjustments to their teaching strategies.

Genially is also optimized for a range of devices, from desktop computers to tablets and smartphones. This cross-device compatibility means that students can access interactive content from anywhere, whether in the classroom or at home. For educators looking to implement flipped classroom models or blended learning environments, this is particularly advantageous. Lessons created on Genially can be shared via links, embedded in websites, or integrated with learning management systems (LMS) like Moodle or Google Classroom, making it easy to distribute educational content and track student progress.

The Genially platform's customization potential plays a crucial role in adapting the escape room to meet specific educational goals. Educators can personalize the content to align with the learning objectives of their course, such as focusing on first-order or second-order differential equations, specific solution methods, or real-world applications. The platform allows for the integration of multimedia elements like videos, audio, and animations, which can enhance the learning experience by providing additional context or explanations where needed. For example, educators can insert brief instructional videos or animated sequences that explain key concepts before students tackle a challenging puzzle.

The platform offers interactive elements such as hotspots, hidden clues, and branching scenarios, which can be customized to guide students through the escape room in a way that aligns with their learning progress. For instance, students who solve a particular puzzle correctly can be directed to a more advanced challenge, while those who struggle may be provided with additional hints or simpler tasks. This flexibility ensures that the escape room remains challenging and educationally appropriate for students at different levels of understanding.

Genially also integrates well with external content, allowing users to embed code from other platforms or applications. This feature opens up possibilities for incorporating live data, interactive simulations, or external quizzes, which can enhance the depth and interactivity of the content. For example, a business educator could embed a real-time stock market feed within a presentation on economics, allowing students to see live updates and apply theoretical knowledge to real-world scenarios.

The capabilities of Genially platform are summarized in Table 1.

Table 1. Capabilities of Genially platform

Capability	Description
Interactive Content Creation	Allows users to create highly interactive presentations, quizzes, infographics, and escape rooms with clickable elements, hotspots, and interactive navigation.
Template Library	Provides a wide range of customizable templates for gamification, presentations, quizzes, and digital games, making it easy to start creating professional-looking content.
Multimedia Integration	Supports embedding videos from platforms like YouTube, adding audio files for narration or background music, and incorporating images and animations for a dynamic experience.
Capability	**Description**
Animation Effects	Offers a variety of animation options such as fade, slide, and motion effects, which can be triggered automatically or based on user interactions.
Customization and Personalization	Provides extensive customization options for fonts, colors, layouts, and interactivity, enabling users to create tailored experiences that align with branding or educational objectives.
Collaborative Work	Enables multiple users to work on the same project simultaneously, making it ideal for group work, collaborative presentations, and student projects.
Cross-Device Compatibility	Optimized for use across different devices, including desktop computers, tablets, and smartphones, allowing content to be accessed from anywhere.
External Embedding and Integration	Supports embedding external content such as code, live data feeds, and external quizzes, and integrates with learning management systems (LMS) like Google Classroom and Moodle.
Data Tracking and Analytics	Provides tools to track how users interact with the content, offering insights into engagement and understanding, allowing educators to make data-driven adjustments.

Puzzles and narrative techniques

Different types of puzzles in digital escape rooms can be strategically designed to target and support the development of various cognitive skills in students, offering a dynamic approach to learning. By incorporating a variety of puzzle formats, educators can engage students in higher-order thinking processes, enhancing their problem-solving abilities, critical thinking, and conceptual understanding.

Multiple-choice puzzles are commonly used to assess students' understanding and recall of key concepts. In the context of digital escape rooms, these puzzles can be used to reinforce foundational knowledge by requiring students to identify correct answers from a set of options. For example, a multiple-choice question might ask students to determine the correct classification of a differential equation. This type of puzzle supports cognitive skills related to recall and comprehension, as students must draw on previously learned material and apply it in a specific context. The immediate feedback offered in escape rooms helps students recognize gaps in their knowledge, facilitating reinforcement and deeper understanding.

Step-by-step calculation puzzles present a sequence of problems, where each solution builds upon the previous step. These puzzles engage students in more complex cognitive processes such as problem-solving, critical thinking, and analysis. For instance, in a puzzle where students solve a differential equation in stages, they must carefully analyze each step, ensure accuracy, and synthesize the results to progress. This type of puzzle encourages students to think critically about each part of the process, fostering perseverance and careful attention to detail. It also helps them develop sequential reasoning, as they must understand how different parts of a problem are interconnected.

Matching puzzles, which require students to pair concepts with definitions or properties, are effective for reinforcing comprehension and conceptual understanding. In digital escape rooms, matching tasks might involve linking mathematical functions to their respective characteristics. This type of puzzle requires students to go beyond rote memorization, as they must demonstrate an understanding of relationships between different concepts. By working through matching puzzles, students develop the cognitive skill of classification, which is essential for organizing knowledge and applying it in more complex situations.

Interactive tasks that involve searching for hidden objects or solving physical riddles within the escape room add an element of critical thinking and attention to detail. For example, students may need to find a hidden key or solve a visual puzzle to unlock the next stage of the escape room. These types of puzzles promote exploratory learning, encouraging students to actively investigate their environment and use observation skills to uncover clues. This process requires students to think critically about how the pieces of the puzzle fit together and fosters spatial reasoning, as they navigate the virtual space to solve problems.

Musical puzzles, where students must play a sequence of notes correctly to advance, target application and sequential reasoning skills. After solving related questions, students are required to apply their knowledge by performing a task, such as playing specific notes on a virtual piano. This type of puzzle reinforces fine motor skills and pattern recognition, as students must identify the correct sequence and execute it accurately. The integration of music adds a multisensory dimension to the learning experience, making the escape room more engaging and promoting a deeper connection between abstract concepts and their real-world applications.

The final puzzles in a digital escape room, which often involve solving complex problems like integration by parts, are designed to challenge students' problem-solving, analysis, and synthesis skills. These puzzles require students to integrate all the knowledge and skills they have acquired throughout the escape room, culminating in a high-stakes challenge. By engaging with these complex tasks, students develop the ability to synthesize information from multiple sources and apply it to solve novel problems. The iterative nature of escape rooms, where students can retry tasks after failure, also encourages resilience and strategic thinking, as they refine their approach based on feedback. These types of puzzles are summarized in Table 2.

Table 2. Puzzle types and targeted cognitive skills

Puzzle Type	Description	Targeted Cognitive Skills
Multiple-Choice Puzzle	Students are asked to identify correct differential equations from a set of options.	Understanding, Recall, Application
Step-by-Step Calculation Puzzle	Students solve a sequence of differential equations where each step builds on the previous result.	Problem-solving, Critical Thinking, Analysis
Matching Puzzle	Students must match mathematical terms with their correct definitions or properties.	Recall, Comprehension, Conceptual Understanding
Interactive Task (Key Search)	Students search for hidden objects, which advances the narrative and opens the next set of puzzles.	Attention Critical Thinking
Musical Puzzle (Play Notes)	After solving questions, students are required to play the correct musical notes in sequence.	Application, Fine Motor Skills, Sequential Reasoning
Final Puzzle (Integration by Parts)	Students apply advanced integration techniques to solve hidden integrals and escape the final room.	Analysis, Synthesis, Problem-solving

Genially is a versatile and powerful platform that offers educators an opportunity to create interactive, multimedia-rich content that engages students in new and innovative ways. By integrating interactivity, animation, multimedia, and customization into a single platform, Genially helps transform traditional learning environments into dynamic, student-centered experiences. Whether used for digital escape rooms, interactive lessons, or collaborative group projects, the platform empowers educators to enhance engagement, encourage active learning, and ultimately improve educational outcomes.

Design Process of Educational Escape Room

In this chapter, you will learn the step-by-step process of designing a digital escape room using the Genially app. Below is a clear outline of the key stages in the design process, which will guide you through creating an engaging and educational digital escape room:

- **Choosing a Template or Starting from Scratch:** Begin by deciding whether to use one of Genially's predefined templates or design your escape room from scratch. Templates provide a quick start, while creating from scratch allows for complete customization.
- **Setting Up the Theme and Layout:** Personalize the theme, background, and layout of your escape room. This includes selecting a design that aligns with the educational goals and engaging your target audience.
- **Customizing Content:** Add questions, puzzles, and challenges that align with your learning objectives. You can edit or replace the default content in templates or create your own from scratch, ensuring that the difficulty level suits the learners' abilities.
- **Incorporating Interactive Elements:** Integrate interactive components such as images, videos, audio files, and hotspots. These elements add layers of engagement and make the experience immersive for learners.
- **Building a Narrative:** Develop a storyline or narrative to guide participants through the escape room. A compelling narrative increases engagement and provides structure to the sequence of challenges.
- **Adjusting Feedback and Penalties:** Customize how learners receive feedback on their answers. For wrong answers, consider adding a penalty time feature to encourage careful thinking and prevent guessing.
- **Creating and Managing Stages:** Design different stages or levels within the escape room. Each stage should increase in complexity and challenge, helping students gradually build their understanding of the topic.
- **Testing and Refining the Escape Room:** Test the escape room with a sample group of learners or colleagues to identify areas for improvement. Adjust the difficulty, flow, or technical issues based on feedback to optimize the learning experience.

- **Launching the Escape Room:** Once finalized, deploy the digital escape room to your students or target audience. Ensure that all elements function as intended, and be available for any troubleshooting during its use.

By following these steps, you will create a fully functional and engaging digital escape room that meets your educational objectives while keeping learners motivated and active throughout the experience. Each stage will be discussed in detail in the chapter, providing tips and examples along the way.

In the next section, we will provide an example of a Digital Educational Escape Room called the horror escape room "Differential Equations". In the last sections, we will delve deeper into designing and customizing your digital escape room, exploring various techniques for crafting captivating puzzles, optimizing user experience, and seamlessly integrating educational content into the gameplay. By the end of this chapter, you will be equipped with the knowledge and skills to create your own immersive escape room experiences that inspire learning and creativity. This process is summarized as workflow in Figure 3.

Figure 3. Workflow of escape room design process

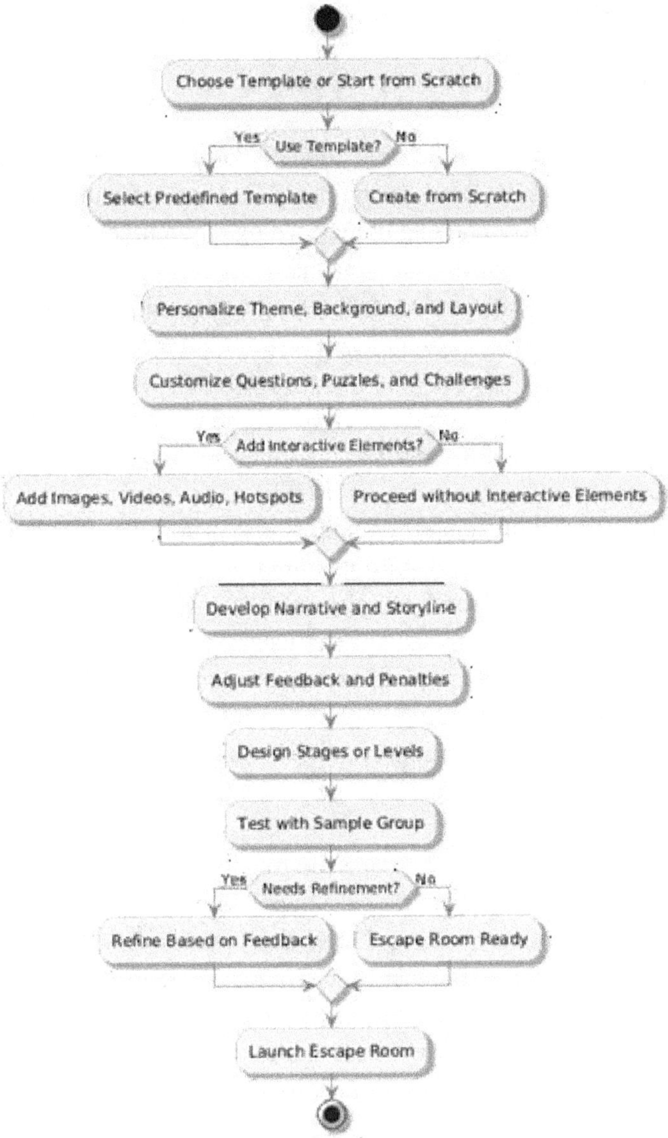

DIGITAL EDUCATIONAL ESCAPE ROOM "DIFFERENTIAL EQUATIONS"

Overview of the Escape Room Design

The "Differential Equations" digital educational escape room was designed to enhance university students' understanding of key concepts in differential equations through an interactive and engaging learning environment. The primary goal of this escape room is to provide an innovative approach to reinforcing students' knowledge of differential equations, encouraging them to apply theoretical concepts in a practical and immersive setting. The escape room targets university students studying mathematics or related fields, particularly those who are exploring differential equations as part of their coursework. By integrating challenging puzzles, interactive elements, and narrative-driven tasks, the escape room allows students to engage with the subject matter in a meaningful and stimulating way.

The escape room is built using the Genially platform, a versatile tool that allows educators to design interactive, multimedia-rich experiences. For this particular project, the "horror escape room" template was selected as the foundation, which offers a pre-structured game format that includes a series of questions, challenges, and clues. The horror-themed template adds an element of excitement and suspense to the learning process, motivating students to immerse themselves fully in the escape room experience. The template is highly customizable, allowing educators to modify the content, design, and structure to suit their educational objectives and the specific needs of their students.

The purpose of using a digital escape room in this context is to shift the learning process from a traditional lecture-based approach to one that promotes active problem-solving and critical thinking. Differential equations, while fundamental to many areas of mathematics and science, can often seem abstract and difficult for students to grasp when taught in a conventional classroom setting. By embedding differential equation problems into an interactive escape room, students are required to apply their knowledge in real-time as they solve puzzles and overcome obstacles. This not only reinforces their understanding of differential equations but also helps them see the relevance of these concepts in a dynamic, problem-solving context.

This Digital Educational Escape Room was developed by a lecturer at Kaunas University of Technology for the Mathematics 2 course. The primary goal of this escape room was to reinforce university students' understanding of differential equations.

Template Selection and Customization

Creating and implementing a digital escape room with Genially was incredibly easy. For this particular digital escape room, we chose the "horror escape room" template. You can see this template in the development stage in Figure 4.

Figure 4. The "horror escape room" template in development, illustrating the interface for creating interactive puzzles and questions

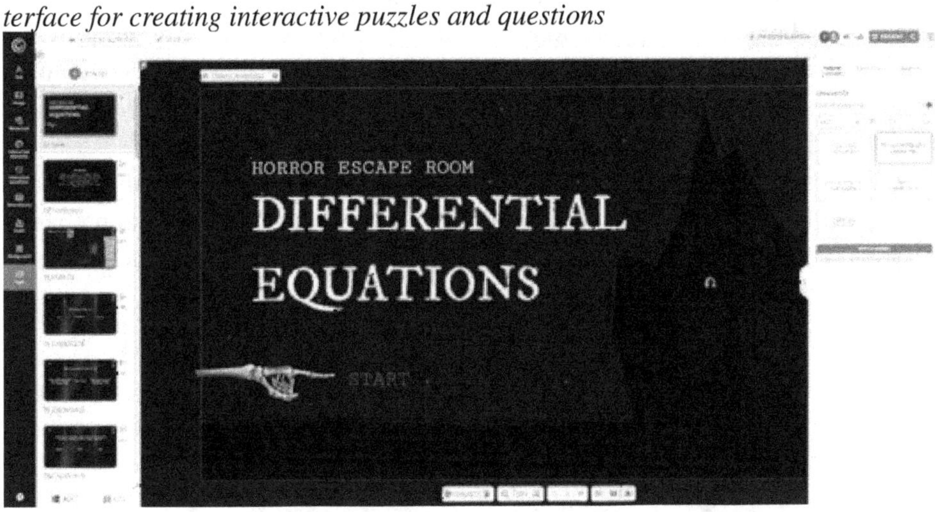

The decision to use the "horror escape room" template for the "Differential Equations" digital educational escape room was based on its potential to engage students in an immersive and motivating learning environment. The "horror" theme brings a sense of urgency and excitement to the experience, elements that are often missing from traditional learning approaches. This template not only captures students' attention through its narrative and atmospheric design but also encourages them to approach the subject matter with a heightened sense of focus and involvement. When learning complex topics such as differential equations, students often benefit from environments that stimulate engagement and reduce the monotony often associated with abstract mathematical concepts.

The "horror escape room" template was chosen for its pre-built structure that facilitates the design of a gamified learning experience while allowing for significant customization to align with educational goals. The template's interactive elements, such as hidden clues, time-based challenges, and locked doors that require solving puzzles to progress, are inherently suited to the study of differential equations, where problem-solving and critical thinking are essential. The narrative-driven structure of the horror template helps guide students through progressively challenging stages, requiring them to solve increasingly complex problems to advance in the game. This aligns well with the step-by-step learning required for mastering differential equations, where foundational knowledge must be applied before moving on to more advanced concepts.

The customization potential of the Genially platform ensures that the horror template can be easily adapted to suit the specific needs of the course content. Educators can modify the puzzles embedded in the template to focus on key differential equation concepts, such as first-order and second-order equations, classification methods, and real-world applications. The ability to replace default questions with tailored content allows educators to align the escape room with their curriculum, ensuring that students are not only engaged but also consistently working towards mastering the intended learning outcomes.

All clues games and scenarios have been implemented already in this template (Figure 5). The only thing you need is to replace the existing text with your prepared questions and answers. But be cautious, you should replace the "Right" and "Wrong" answers carefully, as shown in the slide (Figure 6).

Figure 5. (a) An example of the pre-implemented clues and games in the escape room template (b) The customization options available for replacing default questions and answers in the escape room

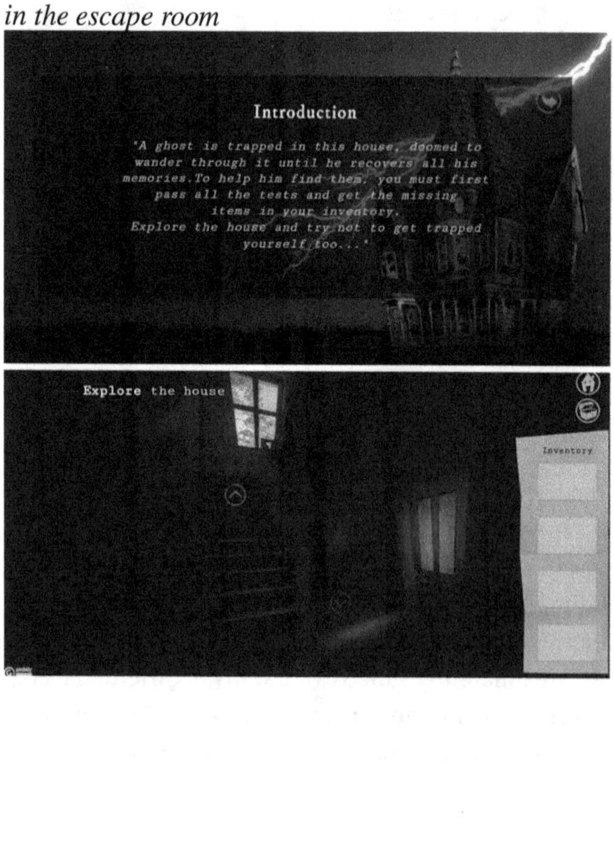

(a)

(b)

Figure 6. Illustration of replacing "right" and "wrong" answers in the template, showing the interaction elements and their customization

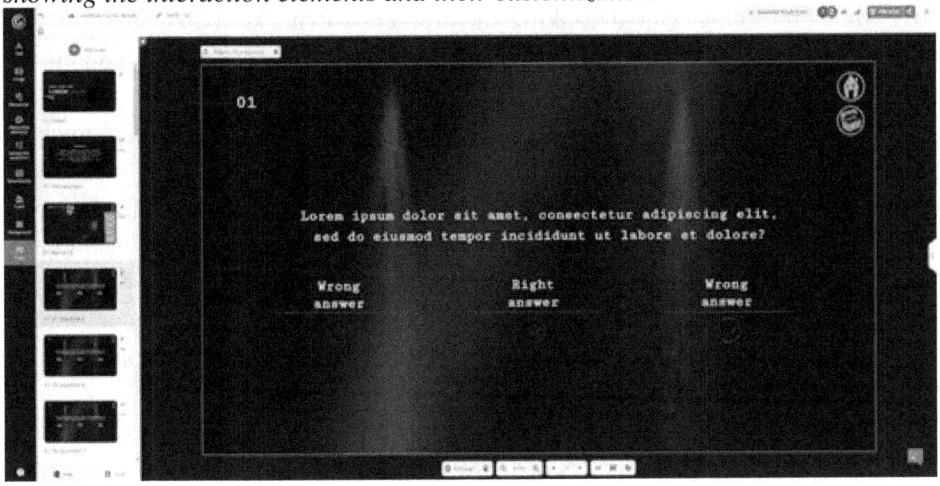

Customizing the visual elements of the template also adds to the educational experience. For example, images and background settings that evoke a sense of mystery or suspense can be used to emphasize the problem-solving nature of mathematics. Animations and transitions can be adjusted to create smooth progressions between different stages of the escape room, ensuring that students remain immersed in the experience without being distracted by technical aspects. The horror theme also introduces the opportunity to integrate creative storytelling, where mathematical challenges become part of the plot, thereby making abstract concepts like differential equations more relatable and easier to grasp.

The "horror escape room" template also allows for flexible adaptation to different levels of difficulty, making it a versatile tool for a range of student abilities. For more advanced students, the puzzles can be customized to include complex, multi-step differential equation problems, while for students who may need more support, hints or simplified tasks can be embedded to aid their progression. This adaptability ensures that the escape room remains inclusive and challenging for all students, promoting a sense of accomplishment as they work through the puzzles at their own pace.

Stages of the Escape Room

The "Differential Equations" digital escape room is structured in a series of stages, each designed to progressively challenge students and deepen their understanding of differential equations. Each stage builds upon the previous one, ensuring that students apply the concepts they've learned while encountering new problems that increase in complexity. This gradual progression helps solidify foundational knowledge while promoting critical thinking and problem-solving skills. The three stages of the escape room are designed to cover different aspects of differential equations, each with specific learning outcomes.

Each stage of the escape room is designed with specific learning outcomes in mind. Stage 1 focuses on fundamental skills such as identifying and classifying differential equations, ensuring that students can distinguish between different types of equations. Stage 2 builds on this by introducing problem-solving techniques for first-order equations, reinforcing students' ability to apply methods such as separation of variables. Finally, Stage 3 challenges students to tackle second-order differential equations, testing their ability to handle more complex problems and apply advanced solving techniques.

Together, these stages provide a comprehensive learning experience that promotes critical thinking and problem-solving at every level. The progression from basic identification to advanced problem-solving ensures that students build a strong conceptual foundation while developing the practical skills necessary to solve differential equations in real-world contexts. By the end of the escape room, students should feel confident in their ability to approach differential equations systematically, applying the appropriate methods to solve a variety of problems.

In the prepared template, there are 3 sets of questions: 4, 4, and 3 in each batch. But of course, you can duplicate or delete slides, so you can manage as many questions as you want. The questions that were in the "Differential Equations" digital escape room are presented below.

Stage 1: Identifying Differential Equations and Determining the Order

In the first stage of the escape room, students are introduced to fundamental concepts, focusing on identifying differential equations and determining their order. The puzzles in this stage are primarily multiple-choice questions that present students with different equations, asking them to classify whether they are differential equations and, if so, to identify their order.

1. Which of the provided options is a differential equation?
   a. $1 + y = x$
   b. $(1 + y)dx = xdy$ ✓
   c. $y'' - 6y' + 9y$
2. According to what do we determine the order of a differential equation?

   a. According to the highest-order derivative present in that equation, the order ✓
   b. According to the moon:)
   c. According to the lowest-order derivative present in that equation, the order
3. A problem that requires finding a solution to a differential equation satisfying initial conditions is called
   a. Newton
   b. Gauss
   c. Cauchy ✓
4. The function $f(x,y) = \sqrt{x^2 + y^2}$ is homogeneous of the m-th degree. What is the value of m?
   a. 0
   b. 1 ✓
   c. 2

Here, the checkmark symbol (✓) indicates the correct answer.

This stage is crucial for establishing a foundational understanding of differential equations. By correctly identifying and classifying equations, students demonstrate their grasp of the basic principles of differential calculus. The immediate feedback provided for correct and incorrect answers helps reinforce learning, as students can see where they may have misunderstood a concept and correct it before moving on to more complex problems. This stage ensures that all students are starting with the same fundamental knowledge, preparing them for the more advanced puzzles to come.

Stage 2: First-Order Differential Equations and Substitution Methods

The second stage of the escape room delves into first-order differential equations and introduces substitution methods as a tool for solving them. In this stage, students encounter a variety of puzzles that require them to classify first-order equations and apply appropriate substitution methods to solve them.

1. What kind of first-order differential equation $\frac{\sin^2 x}{x} + y'\sin x = y\cos x$ is it?
   a. with separate variables
   b. homogeneous
   c. Bernoulli
   d. linear with respect to y and y' ✓
2. Choose a substitution that should be used to solve this differential equation: $y' - \frac{2y}{x} = x^3 e^x - 1$
   a. $y = uv$ ✓
   b. $y = ux$

c. $y' = uv$
d. $y' = ux$

3. What kind of first-order differential equation $y' + \frac{y}{x} = -xe^{x}y^{2}$ is it?
   a. with separate variables
   b. homogeneous
   c. Bernoulli ✓
   d. linear with respect to y and y'
4. Solve Bernoulli equation $y' - \frac{y}{x-1} = \frac{y^2}{x-1}$. Solution is:
   a. $y = \frac{x-1}{C-x}$ ✓
   b. $y = \frac{1}{(x+C)\cos x}$
   c. $y = x$
   d. $y = \frac{x+1}{C-x}$

This stage not only tests students' ability to recognize different forms of first-order equations but also challenges them to apply mathematical techniques to find solutions. The puzzles in this stage might involve filling in missing steps in a solution process or choosing the correct substitution from a list of options. The aim is to reinforce the practical application of solving techniques, helping students internalize these methods through active problem-solving. By successfully completing this stage, students will have demonstrated their ability to tackle first-order differential equations using standard methods and will be ready to progress to more advanced problems.

Stage 3: Second-Order Linear Differential Equations

The final stage of the escape room presents students with second-order linear differential equations, testing their knowledge of solving higher-order equations and applying advanced techniques. In this stage, students might encounter a problem such as solving the homogeneous equation, where they are required to find the general solution. To solve these problems, students need to apply techniques such as finding the characteristic equation and using the roots to construct the solution.

1. Which equation is a second-order linear homogeneous differential equation with constant coefficients?
   a. $3y'' - y' = 5yy'$
   b. $3y'' + 12y' = 4y$ ✓
   c. $7y'' - 3y' - y - \cos x = 0$
2. Which equation is a second-order linear nonhomogeneous differential equation with constant coefficients?
   a. $3y' + 9y' + 20y - \ln x = 0$ ✓

b. $2y'' - 10y' + 25y = 0$
c. $y'' + 9y'y + y = x$

3. The roots of the characteristic equation of the second-order linear homogeneous differential equation with constant coefficients are real and distinct: $k_1 = 2$, $k_2 = 3$. The solution will be?
   a. $y = e^{2x}(C_1 + C_2x)$
   b. $y = C_1 e^{2x} + C_2 e^{3x}$ ✓
   c. $y = e^{2x}(C_1 \cos 3x + C_2 \sin 3x)$
3. The roots of the characteristic equation of the second-order linear homogeneous differential equation with constant coefficients are complex: $k_1 = 2 + 3i$, $k_1 = 2 - 3i$. The solution will be?
   a. $y = e^{2x}(C_1 + C_2x)$
   b. $y = C_1 e^{2x} + C_2 e^{3x}$
   c. $y = e^{2x}(C_1 \cos 3x + C_2 \sin 3x)$ ✓

This stage represents the culmination of the escape room, as it tests students' ability to synthesize all the knowledge they've gained from the previous stages. Solving second-order equations requires a deeper understanding of both the theory and the methods, including how to handle initial conditions, particular solutions, and general solutions. The puzzles in this stage might also include non-homogeneous equations, where students must apply methods such as undetermined coefficients or variation of parameters. The difficulty level of this stage is higher, reflecting the need for students to demonstrate not only mastery of the basics but also the ability to tackle more complex and abstract problems. Successfully completing this stage indicates that students have a strong understanding of differential equations and can apply advanced techniques to solve them.

Feedback and Penalty Mechanisms

In the "Differential Equations" digital escape room, feedback and penalty mechanisms play a crucial role in guiding students through the learning process while encouraging them to approach problems with more care and critical thinking. These mechanisms are designed to not only provide immediate responses to students' answers but also to challenge them to reflect on their mistakes and attempt to correct them. The escape room environment creates a balance between maintaining engagement and promoting meaningful learning by incorporating timed penalties for incorrect answers and animations that reinforce the learning process.

The feedback mechanism is immediate and dynamic, providing students with a clear indication of whether their answers are correct or incorrect. For correct answers, positive feedback is given in the form of visual cues—such as green checkmarks or congratulatory messages—and progression to the next stage of the escape room. This positive reinforcement motivates students to continue solving puzzles with confidence. On the other hand, incorrect answers trigger negative feedback, such as red crosses or short explanatory messages that hint at the mistake without giving away the correct answer. This immediate feedback helps students understand where they went wrong and encourages them to rethink their approach to the problem.

A key feature of the penalty mechanism is the use of time penalties for incorrect answers. When students make a mistake, a timer is triggered that imposes a delay before they can attempt the puzzle again. This penalty time can be adjusted depending on the complexity of the puzzle or the overall difficulty level of the escape room. For example, a minor mistake in a first-order differential equation puzzle might result in a short time penalty of 15 to 30 seconds, while more significant errors in later stages involving second-order differential equations could lead to longer penalties, such as 60 seconds or more. These penalties serve to discourage random guessing and encourage students to take a more thoughtful, deliberate approach to problem-solving.

If students choose the wrong answer, they will automatically receive a notification stating that it is incorrect and will be given the option to go back and attempt again. To prevent students from simply guessing the correct answer, it is essential to include a penalty time feature (Figure 7).

Figure 7. Setting up the penalty time feature for incorrect answers in Genially, showcasing the animation parameters and timing options

Setting up these penalties within Genially is straightforward, thanks to the platform's built-in timer widget. To implement a time penalty, the educator can assign specific triggers to incorrect answers. Once an incorrect answer is selected, the timer is activated, freezing the interface for the predetermined penalty duration. This delay prevents students from immediately attempting another answer, giving them time to reflect on their mistake. Educators can also use this period to display additional hints or animations that guide students toward the correct solution. For instance, during the penalty time, an animation might display a simplified version of the problem-solving process, offering a subtle hint without directly providing the answer.

The use of animations in feedback and penalty mechanisms is another powerful tool for reinforcing learning. Instead of merely telling students that their answer is wrong, the escape room can use animations to visually illustrate the problem-solving steps they may have missed. For example, if a student struggles with a first-order differential equation, an animation might show the separation of variables and integration steps required to solve the equation. This dynamic form of feedback encourages deeper thinking and helps students visualize abstract concepts more clearly. Animations can also be used as a form of reward for correct answers, where students see the result of their success in unlocking the next part of the escape room or advancing in the storyline.

To set this up, begin by selecting the "back" picture and adjusting the animation parameters as shown in Figure 8. For instance, set the timer to 30 seconds instead of 1 in the "start" section.

Figure 8. Adjusting animation settings for the penalty time effect, with a focus on the "fade in" feature and time adjustments

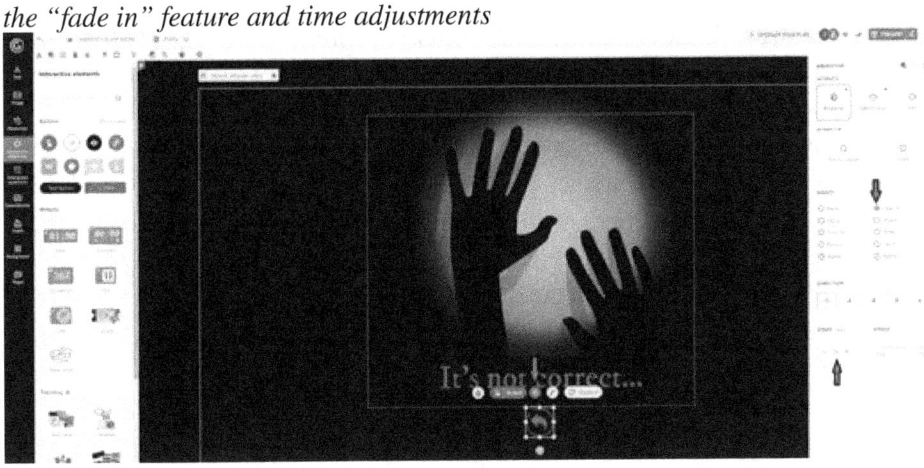

Ensure that the "Fade in" parameter is selected in the "Effect" section. Then, you can effortlessly include a timer by clicking on the "Timer" widget located in the "Interactive elements" section, which can be accessed from the left menu and also set the same period.

These feedback and penalty mechanisms work together to promote a learning environment where mistakes are seen as part of the learning process rather than simply failures. The penalty times, in particular, are effective at discouraging random guessing, as they impose a meaningful consequence for incorrect answers. By introducing a slight delay between attempts, students are encouraged to take their time, think critically about their approach, and use the feedback they receive to improve their understanding of the material. This helps to shift the focus from merely completing the escape room to mastering the mathematical concepts involved.

Interactive Elements and Challenges

The interactive elements and challenges in the "Differential Equations" digital escape room are central to creating an engaging and immersive learning experience. By integrating puzzles, hidden clues, and dynamic feedback systems, these components encourage active participation and deepen students' understanding of mathematical concepts. One of the key interactive elements is the search for hidden keys, which adds an exploratory dimension to the learning process. As students solve puzzles or complete tasks, they are required to search the virtual environment for hidden objects, such as keys or other clues, that unlock the next stage of the escape room. This gamified approach keeps students engaged and motivates them to persist through increasingly difficult mathematical problems, all while maintaining a sense of excitement and curiosity.

The search for the hidden key is an example of how interactive elements can transform a traditional math lesson into a more immersive experience. In a typical classroom setting, students may solve differential equation problems on paper or in a static online environment. However, the escape room requires students to not only solve the problem but also interact with the environment by locating hidden clues or completing physical challenges (in the virtual sense) to advance. For instance, solving a differential equation might lead to a clue about where the key is hidden on the screen, prompting students to actively explore the interface. This shifts the focus from passive learning to active engagement, making the process of solving equations feel more like a real-world problem-solving challenge.

When university students complete the first set of questions, they encounter an interactive task where they must search for a key (Figure 9). This interactive task adds an element of excitement and engagement to the escape room experience, as students actively participate in solving the puzzle. It provides a seamless transition between question sets, maintaining the flow of the escape room and keeping students immersed in the challenge. The digital escape room developer doesn't need to take any action; everything is already implemented by default in this template. Once students find the key and click on it, it is added to their inventory, allowing them to progress to the next stage of the digital escape room.

Figure 9. (a) An interactive puzzle where students search for a hidden key to advance to the next stage (b) Example of how the key is added to the player's inventory once found

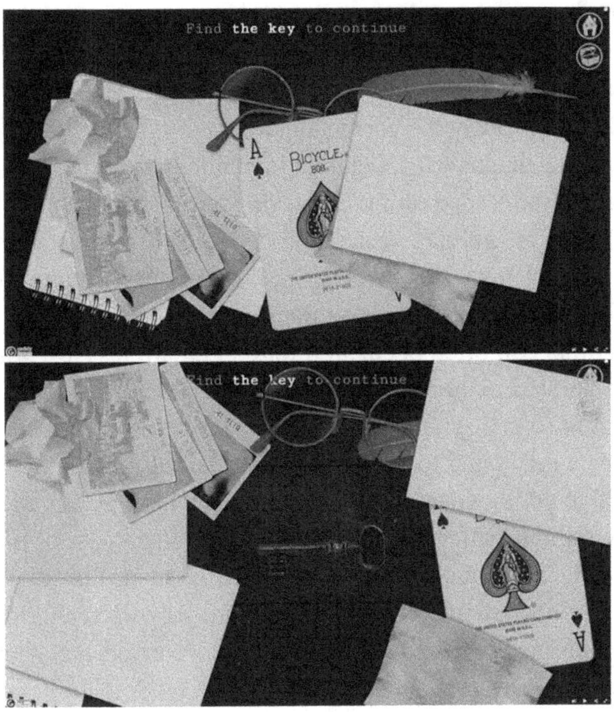

(a)

(b)

In addition to the search elements, multimedia components such as animations and interactive feedback play a crucial role in maintaining immersion and reinforcing learning. Animations are used to visualize mathematical processes or illustrate the consequences of students' choices. For example, when students answer a problem correctly, a congratulatory animation might appear, or a door may open to reveal the next stage. On the other hand, when a student answers incorrectly, feedback in the form of animations might show the mathematical steps they missed or present visual hints to guide them in the right direction. These dynamic elements not only keep students engaged but also provide immediate, visually appealing feedback that enhances the learning experience.

The integration of multimedia also helps accommodate different learning styles. Visual learners, for example, benefit from animations that demonstrate how to approach a problem, while auditory feedback can be integrated for students who respond better to sound cues. By combining these various forms of media, the escape room ensures that a diverse range of learners can remain engaged and supported throughout the experience. Additionally, interactive feedback for wrong answers is particularly important in reinforcing the learning process. Instead of simply telling students they are wrong, the escape room offers constructive feedback, helping students understand their mistake and encouraging them to try again with a better understanding of the material. For instance, if a student selects an incorrect solution for a differential equation, a small animation could play that breaks down the solution process step by step, allowing the student to see where they went wrong without explicitly giving away the correct answer.

These interactive elements work together to maintain a high level of student engagement throughout the escape room. By turning the abstract nature of differential equations into a hands-on, interactive experience, the challenges become more than just mathematical problems—they become part of an ongoing narrative in which students are active participants. The continual immersion helps prevent the kind of disengagement that can often occur in traditional math lessons, where students might become frustrated or bored when they encounter difficulties. Instead, the dynamic nature of the escape room, combined with its built-in feedback and visual cues, keeps students motivated to solve problems and learn from their mistakes.

These interactive components add layers of complexity and challenge that encourage students to think critically and strategically. The process of finding hidden keys or unlocking the next puzzle based on their performance in solving equations teaches students how to navigate complex problem-solving scenarios, where the path to the solution is not always straightforward. This mirrors real-world problem-solving, where multiple factors must be considered, and solutions often require careful analysis and reflection.

Musical Challenge Integration

The second stage offers a musical challenge, adding an engaging and immersive element to the escape room experience.

The musical challenge in the "Differential Equations" digital escape room introduces a creative and multisensory element to the learning process, where students must play correct piano notes based on their answers to mathematical problems. This puzzle not only engages students in an unexpected and enjoyable way, but it also adds an auditory and tactile dimension to the escape room experience, helping to reinforce the concepts of differential equations through a non-traditional medium. By blending music with mathematics, this challenge creates a unique learning opportunity that encourages students to think critically while using their auditory and motor skills.

To create this challenge, the puzzle is structured so that students first solve a series of differential equation-related problems. Each correct solution provides them with a part of the information needed to play a corresponding musical note or a sequence of notes on a virtual piano. For example, solving a first-order differential equation may yield a specific note, and the successful completion of a second-order problem could reveal a sequence of notes that must be played in the correct order to unlock the next stage of the escape room.

The technical setup of this challenge in Genially involves integrating an interactive piano keyboard that responds to user input. Using Genially's interactive element tools, hotspots can be placed on each piano key, making them clickable and able to produce corresponding sounds. The clues obtained from solving the mathematical puzzles guide students in selecting the correct keys. If students play the correct sequence of notes, they successfully move on to the next challenge. However, if they play the incorrect notes, they receive feedback indicating that their attempt was unsuccessful, prompting them to revisit the clues and reconsider their approach.

University students are presented with a series of four questions that they must successfully answer to obtain the coveted secret note (Figure 10, 11). This task not only tests their knowledge but also introduces a sense of anticipation and excitement as they progress through the room.

Figure 10. (a) Musical challenge interface in the escape room, featuring a series of questions students must answer to obtain a secret note (b) Example of the musical puzzle, where students use the acquired secret note to play correct piano notes

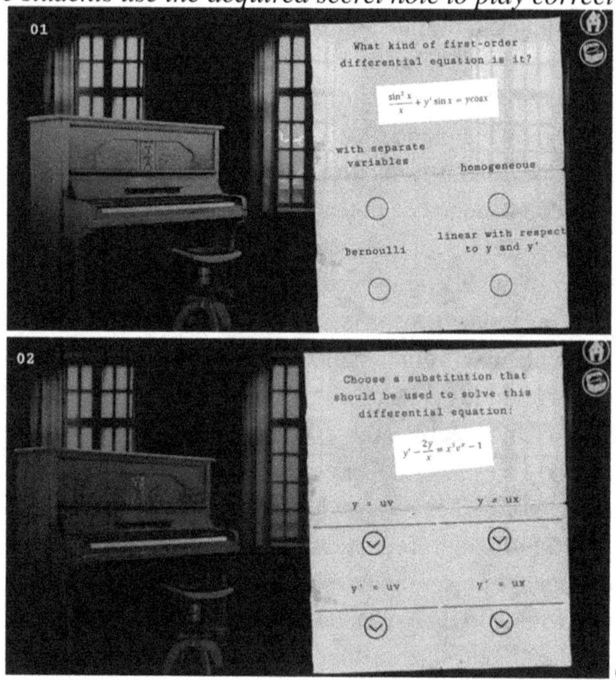

(a)

(b)

Figure 11. Interface for students playing musical notes on a virtual piano as part of an interactive challenge in the escape room

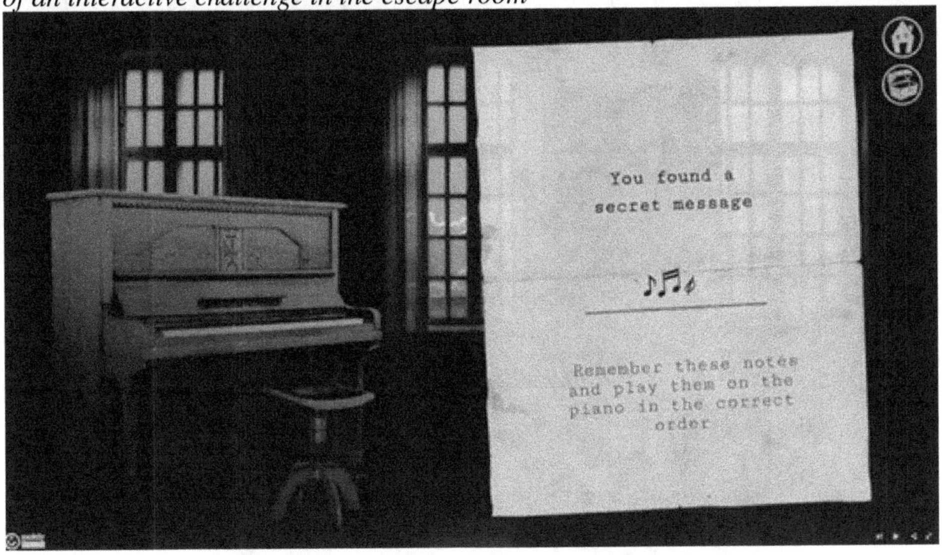

In the event of a misstep, students are granted the opportunity to try again, fostering a sense of perseverance and determination. This iterative process encourages students to learn from their mistakes and persist until they achieve success, instilling valuable problem-solving skills and resilience.

Upon acquiring the secret note, the challenge intensifies as students must translate their knowledge into action by playing the correct notes on the piano (Figure 12). This hands-on activity not only reinforces their understanding of the material but also provides a tactile and auditory dimension to the escape room, further enhancing its immersive nature.

Figure 12. (a) Introduction to the magnifying glass tool, used to find hidden clues and unlock the next phase of the escape room (b) Using the magnifying glass to reveal hidden elements, adding a detective-like dimension to the experience. (c) Successful discovery of a hidden clue using the magnifying glass

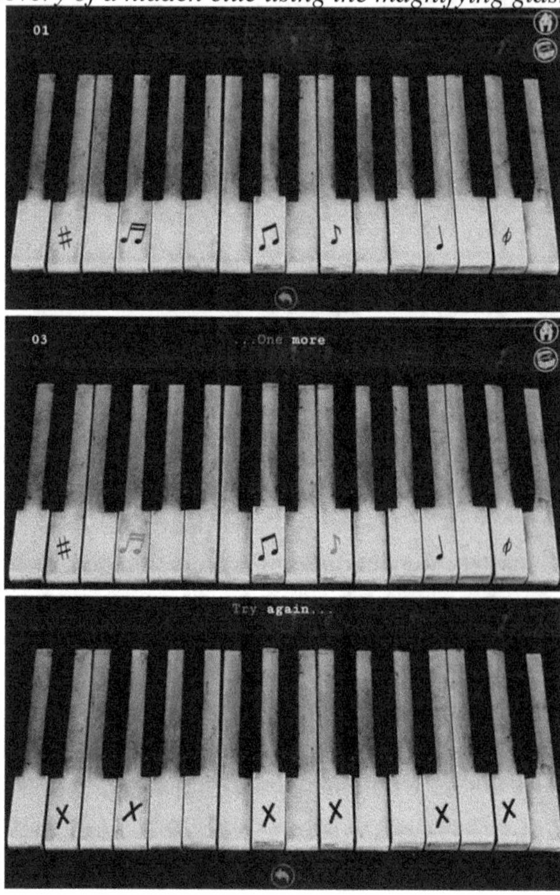

(a)
(b)
(c)

The musical stage adds depth and excitement to the escape room, captivating students' attention and motivating them to actively participate in the learning experience. By combining interactive challenges with educational content, the escape room creates a dynamic and engaging environment that encourages exploration, discovery, and learning.

This musical puzzle serves multiple purposes within the educational framework of the escape room. First, it breaks up the traditional learning format, offering a different type of problem-solving that relies on auditory feedback rather than purely visual or logical thinking. This auditory and tactile engagement can be particularly beneficial for students who are auditory learners or those who thrive on interactive, hands-on tasks. It offers a break from the more abstract and conceptual nature of differential equations, making the learning experience more well-rounded and engaging for a diverse range of learners.

The integration of music also plays a significant role in reinforcing the differential equation concepts themselves. By linking mathematical solutions to specific musical notes, the puzzle encourages students to think carefully about each step in their problem-solving process. They cannot simply rush through the equation without understanding it; each correct mathematical solution is necessary for the successful completion of the musical puzzle. This connection between the abstract (mathematical problem) and the concrete (musical notes) strengthens students' grasp of the material, as they must apply the knowledge gained from the equations in a different, creative context.

In addition, the musical challenge promotes critical thinking and persistence. As students solve each equation, they must process the clues and apply them to the musical puzzle in an orderly manner. If they make mistakes, the escape room provides feedback that encourages them to revisit their mathematical solutions and think more deeply about where they might have gone wrong. This iterative process, where students continually test their understanding and refine their approach, fosters resilience and perseverance, important traits in both mathematical problem-solving and in broader learning contexts.

After mastering the piano tasks, students uncover a pivotal tool: the "magnifying glass." This item isn't just a trinket; it's the key to unlocking the next phase of the digital escape room. Without it, the subsequent stage remains tantalizingly out of reach, beckoning with its mysteries and challenges. With the magnifying glass in hand, students embark on a new quest, scouring their surroundings for fine details, hidden clues, and cryptic messages. This pivotal moment marks a transition from mere observation to active investigation as students become detectives in their own thrilling adventures. Each discovery brings them one step closer to unraveling the enigma that lies at the heart of the escape room, fueling their curiosity and determination to overcome every obstacle in their path (Figure 13).

Figure 13. (a) Inputting the secret code into the typewriter to unlock the next stage of the escape room (b) The typewriter interface, where students enter the code to proceed

(a)

(b)

Once students uncover the secret code, they must input it into the typewriter, unlocking the gateway to the next stage of their journey (Figure 14). This pivotal moment heralds a new chapter in their quest, granting them passage to uncharted territories and untold mysteries. With bated breath and nimble fingers, they punch in the code, each keystroke echoing with anticipation and excitement. As the typewriter accepts their input, a sense of accomplishment washes over them, spurring them onward in their adventure. Now, armed with determination and newfound knowledge, they press forward into the unknown, eager to unravel the secrets that lie ahead.

Figure 14. The final puzzle in the "differential equations" escape room, testing students' knowledge of second-order linear differential equations

Final Puzzle and Escape Completion

The final puzzle in the "Differential Equations" escape room serves as the ultimate test of students' understanding of second-order linear differential equations with constant coefficients. At this stage, students are presented with a complex, multi-step problem that requires them to apply all of the concepts and methods they have learned throughout the escape room. The puzzle is designed to assess their ability to solve higher-order differential equations, including recognizing the appropriate form of the solution, determining characteristic equations, and working through the mathematical steps needed to find general and particular solutions. Students are tasked with demonstrating their understanding of second-order differential equations with constant coefficients, a fundamental concept in mathematics.

As they grapple with these advanced questions, they draw upon the knowledge and skills they've acquired throughout their adventure. With each correct answer, they inch closer to the ultimate goal of completing the escape room challenge.

The questions presented in Figure 15 serve as both a test of their mathematical prowess and a reflection of their journey thus far. Each problem represents a unique puzzle, requiring careful analysis and strategic thinking to solve. As students work through these challenges, they not only reinforce their understanding of differential equations but also experience a sense of accomplishment with every correct response.

Figure 15. (a) Sample question from the final puzzle set in the escape room, focusing on second-order linear homogeneous differential equations (b) Students' interaction with the puzzle-solving process, emphasizing engagement and active learning

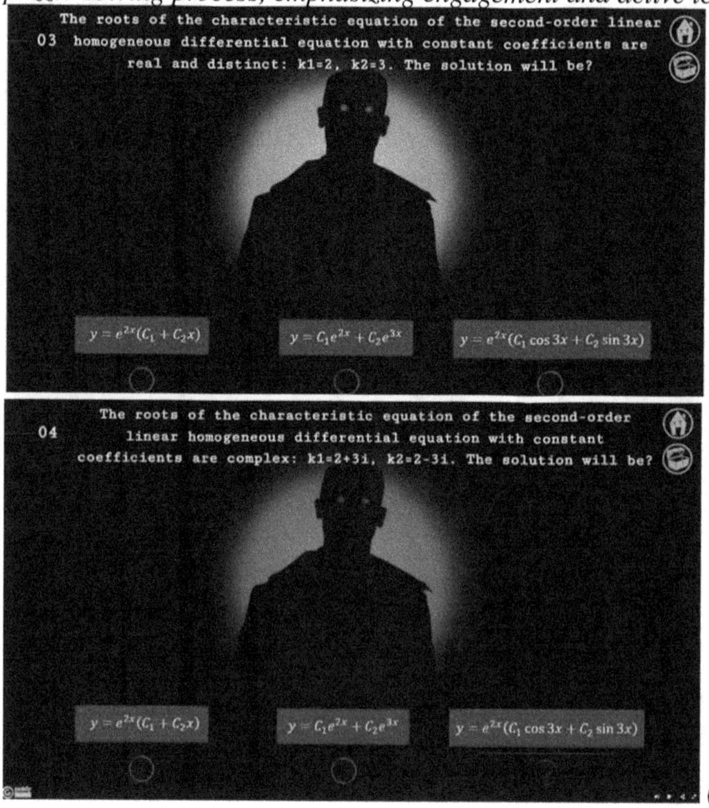

(a)

(b)

This final puzzle challenges students to synthesize their knowledge and apply the full range of mathematical tools they have learned, from solving characteristic equations to understanding the structure of solutions for second-order equations. The complexity of the puzzle, combined with the multi-step solution process, ensures that students must approach the problem methodically and demonstrate a deep understanding of the material.

Upon successfully answering all four questions, students reach the culmination of their digital escape room adventure, symbolized by Figure 16. Here, they reflect on their accomplishments and celebrate their triumphs, knowing that they have conquered each obstacle with perseverance and determination. It's a moment of pride and satisfaction, marking the conclusion of their journey filled with discovery, challenge, and growth.

Figure 16. The completion screen of the digital escape room, celebrating students' successful completion of all tasks and puzzles

Successfully completing the escape room by solving the final puzzle symbolizes mastery of the topic. By this point, students will have demonstrated their ability to tackle increasingly complex differential equations, from first-order problems to advanced second-order equations with constant coefficients. The structure of the escape room, with its escalating challenges and interactive elements, ensures that students must engage with the material in a hands-on, active way. The sense of achievement they feel upon completing the final puzzle not only reflects their mastery of the topic but also reinforces their confidence in applying mathematical concepts in a problem-solving context.

The completion of the escape room serves as a powerful reinforcement of the learning process. The final stage is designed to bring all the learning outcomes together, ensuring that students have not only retained the material but also applied it in a meaningful, integrated way. The interactive nature of the escape room allows students to reflect on their procession through the various stages, recognizing how each puzzle built on the previous one and prepared them for the final challenge. By solving the last puzzle and inputting the secret code, students are rewarded for their persistence, critical thinking, and mathematical prowess.

Conclusion and Educational Outcomes

The digital educational escape room on "Differential Equations" offers a compelling and engaging way to reinforce students' understanding of complex mathematical concepts. By integrating puzzles, interactive challenges, and narrative-driven tasks, this escape room provides a unique environment where learning is immersive and active, rather than passive.

The carefully designed stages of the escape room progressively challenge students, ensuring that they apply critical thinking and problem-solving skills while tackling differential equations. Through the various puzzle types, students are not merely recalling definitions or formulas; they are required to deeply engage with the material, analyze problems, and synthesize solutions in real-time.

One of the major educational benefits of this escape room is its capacity to enhance critical thinking. The puzzles encourage students to approach problems from different angles, considering the logic and principles underlying each question. For instance, when students are asked to classify differential equations or solve step-by-step calculations, they must not only recall mathematical rules but also think critically about how to apply them in novel situations. This aspect of the escape room fosters analytical thinking, as students must evaluate each option, consider potential solutions, and make informed decisions based on their knowledge.

The escape room strengthens problem-solving abilities by immersing students in a series of interactive tasks that require both logical reasoning and perseverance. Each stage of the escape room presents progressively difficult challenges that encourage students to approach problems methodically, breaking them down into manageable parts. For example, solving a second-order differential equation with constant coefficients requires not just recalling formulas, but also synthesizing multiple steps to arrive at the correct solution. The interactive elements, such as time penalties for wrong answers and the need to find hidden clues, add additional layers of complexity that demand a strategic approach. This kind of problem-solving experience mirrors real-world mathematical tasks, where persistence and adaptability are key to success.

In addition to promoting critical thinking and problem-solving, the escape room solidifies students' understanding of differential equations. By embedding mathematical problems in an interactive, game-like setting, the escape room allows students to apply their knowledge in a hands-on manner. The variety of puzzles—ranging from multiple-choice questions to matching and musical challenges—ensures that students engage with the content in multiple ways, reinforcing their comprehension of differential equation principles. The instant feedback provided by the platform, combined with the opportunity to retry incorrect answers, creates a supportive learning environment where mistakes become opportunities for further learning and growth.

DIGITAL EDUCATIONAL ESCAPE ROOM "THE INTEGRAL ENIGMA"

Overview of the Escape Room Design

The "Integral ENIGMA" digital educational escape room was designed with the goal of providing an engaging, interactive way for university students to deepen their understanding of indefinite integrals. This escape room focuses specifically on the concepts of integration, challenging students to apply their knowledge of calculus in a series of puzzles that range from basic antiderivatives to more advanced integration techniques. By transforming the study of integrals into an immersive learning experience, the escape room seeks to make abstract mathematical concepts more accessible, encouraging students to approach integration with confidence and creativity.

The primary educational objective of the "Integral ENIGMA" escape room is to reinforce students' ability to solve indefinite integrals while enhancing their problem-solving and critical thinking skills. Throughout the escape room, students encounter a variety of problems that require them to recognize and apply different integration techniques, including basic rules of integration, u-substitution, and integration by parts. These challenges are designed to simulate real-world applications of calculus, prompting students to think critically about how to approach each problem and apply the appropriate method. In doing so, the escape room helps students consolidate their learning by requiring them to actively engage with the material rather than passively memorizing formulas.

The target audience for this escape room is primarily university-level students, particularly those studying calculus in mathematics, engineering, or the physical sciences. The design of the puzzles assumes that students already possess a foundational understanding of integration, making it well-suited for learners at an intermediate level who are ready to apply their knowledge in more complex and interactive ways. By presenting the challenges within a gamified context, the "Integral ENIGMA" escape room is able to maintain student engagement and motivation, helping them to persist through difficult problems in a way that traditional classroom exercises might not.

This specific Digital Educational Escape Room was developed for the same course as the previous one. However, the primary goal of this escape room differed, as it aimed to reinforce university students' understanding of indefinite integrals.

The choice to use the "Genial Escape Room" template was strategic, as it offers a pre-built structure that is highly adaptable for educational purposes. This template provides an interactive, narrative-driven format that guides students through a series of challenges while keeping them immersed in the storyline of the escape room.

The narrative framework of the escape room, combined with Genially's flexible tools for creating multimedia content, allows for the integration of animations, feedback mechanisms, and interactive elements that enhance the learning experience. The template was customized to focus specifically on integrals, with each puzzle carefully designed to align with key calculus concepts, ensuring that students are continuously applying what they have learned in class.

Although the "Genial Escape Room" template was initially chosen, this time, it underwent significant modifications (Figure 17). Therefore, further we will focus on how to create an escape room using an existing template but updating it with your ideas.

Figure 17. Example of the "integral enigma" digital escape room template with modifications to enhance student learning about indefinite integrals

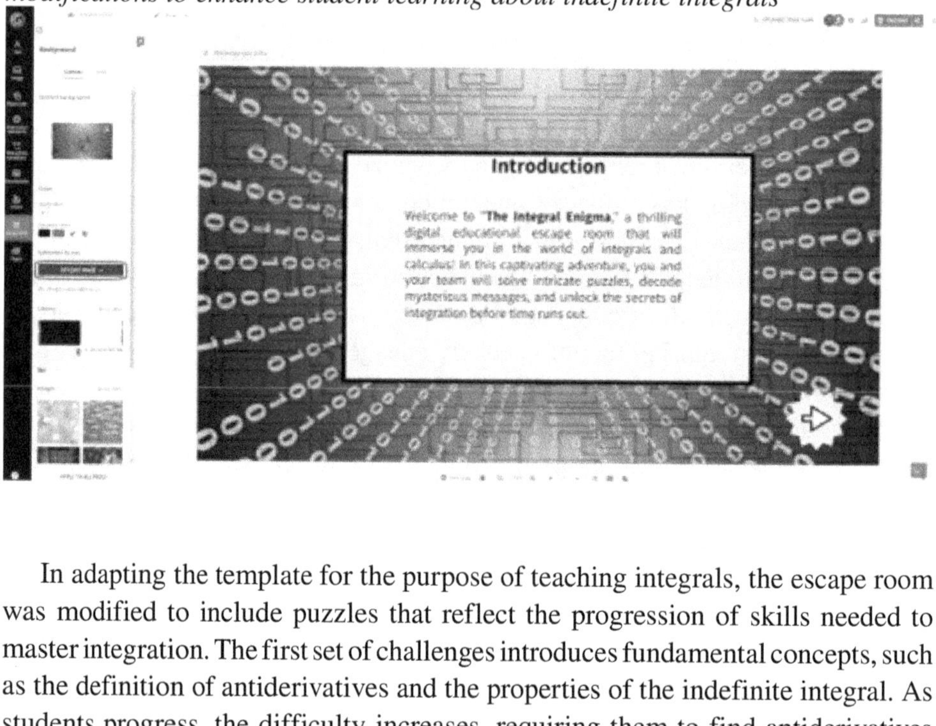

In adapting the template for the purpose of teaching integrals, the escape room was modified to include puzzles that reflect the progression of skills needed to master integration. The first set of challenges introduces fundamental concepts, such as the definition of antiderivatives and the properties of the indefinite integral. As students progress, the difficulty increases, requiring them to find antiderivatives of basic functions, apply substitution methods, and tackle more complex integrals using techniques like integration by parts. By gradually increasing the complexity of the puzzles, the escape room mimics the structure of a well-planned curriculum, ensuring that students build their skills incrementally while remaining engaged with the content.

If you want to change the background, you can easily do so by selecting "Background" instead of "Pages" and then clicking on "Upload image," as illustrated in the accompanying Figure 18.

Figure 18. Steps for customizing the escape room background by selecting the "background" option in Genially and uploading a new image

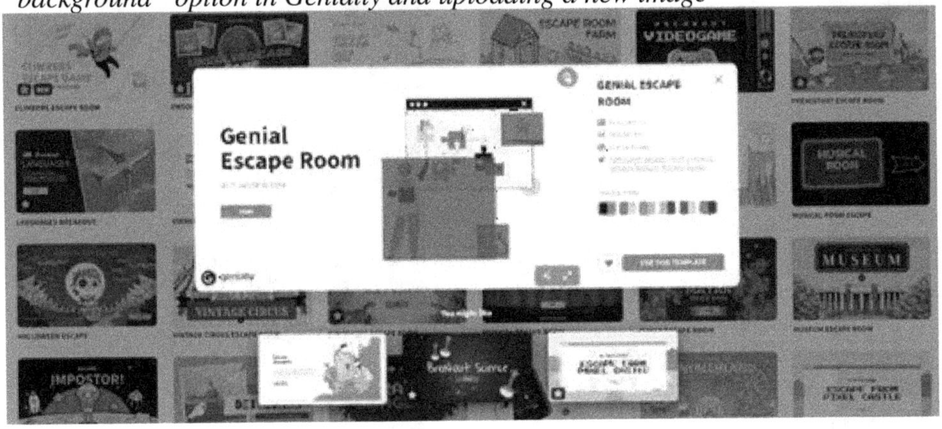

In this escape room, a narrative plot and characters were incorporated to enhance the immersive experience for the participants. This process included duplicating the introduction slide to effectively introduce the narrative elements. Duplicating slides in Genially is a straightforward process that facilitates the seamless integration of new content.

To duplicate a slide in Genially, you can easily accomplish this by moving the mouse cursor over the desired slide, as shown in Figure 19 a), and clicking on the designated icon marked in red. This action creates an identical copy of the slide, preserving the layout and design elements.

Once the page is duplicated, you have the flexibility to customize it according to your preferences and requirements. This includes changing the background, adjusting text, and adding any relevant images or graphics. By selecting the "Image" option from the left menu bar, you can effortlessly incorporate visual elements that contribute to the overall theme and narrative plot of the escape room (Figure 19, b)).

Figure 19. (a) Duplicating slides in Genially to introduce new narrative elements or puzzles in the escape room (b) Incorporating new images and text to develop the narrative and theme of the escape room

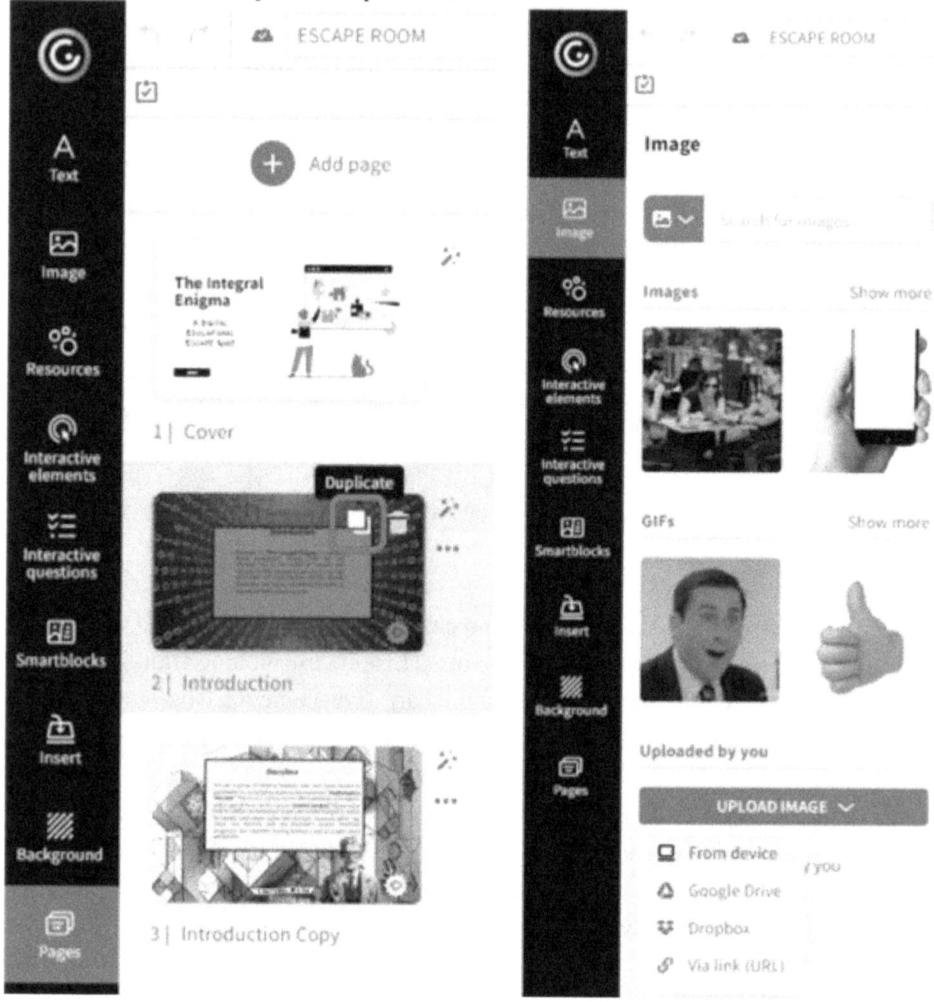

Narrative Plot

The narrative plot of the "Integral ENIGMA" escape room is designed to immerse students in a compelling storyline that intertwines seamlessly with the mathematical challenges they must solve. Adding a narrative plot and characters not only enhances the visual appeal of the escape room but also provides a narrative structure that guides participants through the challenges and puzzles. This storytelling element adds depth and engagement to the experience, immersing players in a captivating adventure that unfolds with each successive slide.

The plot revolves around the discovery of a long-lost manuscript, "The Integral Enigma," hidden deep within a mysterious, ancient museum. The manuscript is said to hold the secrets of mastering integrals, a fundamental mathematical tool, but to unlock its power, students must solve a series of puzzles that protect the manuscript's content. Each puzzle represents a challenge or riddle that has stumped mathematicians for centuries, and only those who can correctly solve the integrals can progress through the museum's rooms and ultimately access the manuscript.

The escape room begins with students finding themselves in the first chamber of the museum, where they are introduced to the lore of the Integral Enigma. They are tasked with retrieving the manuscript to gain ultimate mastery over integral calculus, but the catch is that the museum is rigged with puzzles designed to prevent anyone without sufficient mathematical knowledge from obtaining it. As the students move from room to room within the museum, they encounter increasingly difficult integration challenges that test their problem-solving abilities and understanding of calculus. The narrative's tension builds as the puzzles become more complex, mirroring the deeper mathematical concepts students are required to grasp.

Each room in the museum is dedicated to a specific type of integration problem, and the narrative is woven into these mathematical challenges. In the early rooms, students encounter ancient scrolls and artifacts that pose simple integration problems, like finding the antiderivative of basic functions. As they solve these, the storyline reveals that these initial puzzles are only the beginning of the greater challenge. The deeper students go into the museum, the more they uncover about the history of the manuscript and the mathematicians who tried and failed to solve the enigma before them. Each step forward brings new revelations, fueling their motivation to continue solving the puzzles and unlock the next stage.

The narrative heightens as students enter the second part of the museum, where more challenging integrals await. Here, the plot introduces historical figures—fictionalized versions of famous mathematicians—who each left behind clues or puzzles that must be solved to move forward. The narrative hints that these mathematicians, despite their genius, could not solve the final enigma, leaving the students with the task of surpassing even the great minds of the past. The stakes are raised with every puzzle, as students solve integrals involving trigonometric functions, logarithms, and advanced substitutions, drawing them further into the mystery of the manuscript.

In the final room, the tension of the plot reaches its peak. The students discover the final piece of the manuscript, but it is protected by the most complex puzzle of all: a multi-step integral that incorporates several advanced techniques, including integration by parts and trigonometric identities. The narrative at this point suggests that only those who have truly mastered the art of integration will be able to unlock the final secret of the manuscript. The puzzle is framed as the ultimate test, the final obstacle that stands between the students and the knowledge contained within "The Integral Enigma."

The plot concludes with the students' success or failure in solving the final puzzle. Upon solving the last integral, the manuscript is fully unlocked, revealing its secrets in a dramatic finale. The narrative positions this moment as a triumph of mathematical mastery, celebrating the students' perseverance and problem-solving skills. If students fail to solve the final puzzle within the time limit, the narrative takes a different turn, suggesting that the mystery remains unsolved for now, and invites the students to try again, motivating them to sharpen their skills and return for another attempt.

By incorporating characters into the escape room, participants can develop a connection to the narrative plot and feel personally invested in solving the puzzles. Whether it's guiding them through the narrative or presenting challenges along the way, characters play a crucial role in driving the plot forward and keeping players actively engaged.

The decision to add a narrative plot and characters to the escape room demonstrates the creator's commitment to creating an immersive and memorable experience for the participants. Through strategic use of storytelling elements and visual design, the escape room becomes more than just a series of puzzles—it becomes an interactive game that captivates and challenges players at every turn. If you lack the imagination to create your narrative plot, you can generate one using Generative AI. This cutting-edge technology uses advanced algorithms to craft an engaging narrative plot, providing you with endless possibilities to fuel your creativity.

Structural organization and learning path

The structural organization of "The Integral ENIGMA" digital educational escape room is carefully designed to provide a clear and logical learning path that progressively guides students through the complexities of integral calculus. By dividing the escape room into three themed rooms, each focusing on increasingly difficult integration techniques, and culminating with the Final Conundrum, the escape room ensures that students build their knowledge step by step. This progression allows learners to first master the foundational skills needed for solving integrals before tackling more advanced and challenging problems. The structured approach reinforces the learning process, ensuring that students not only grasp the mathematical concepts but also apply them in progressively more complex contexts.

The Digital Educational Escape Room "The Integral ENIGMA" was divided into 3 themed rooms and the Final Conundrum. Each themed room has 3 main puzzles and additional clues if students fail to choose the right answer. The decision to divide the escape room into three distinct themed rooms was driven by pedagogical considerations.

The first room serves as an introduction to integration, focusing on fundamental concepts such as the definition of antiderivatives and the properties of indefinite integrals. This initial stage is critical in building confidence among students, as it allows them to engage with integrals in a manageable and non-intimidating way. This room provides the necessary foundation for tackling more complex puzzles in subsequent stages and ensures that students of varying skill levels can engage with the material from the outset.

The second room introduces integration problems, such as finding the antiderivatives of simple functions and applying substitution methods. This stage is designed to deepen students' understanding of integrals by requiring them to apply substitution techniques and recognize more complex patterns in the functions they are integrating. The decision to isolate these challenges into their own room allows students to focus on mastering these techniques without being overwhelmed by the more advanced methods introduced later. By compartmentalizing the learning process, the escape room provides a scaffolded approach to problem-solving that encourages a gradual build-up of skills.

The third room shifts the focus to advanced integration techniques, such as integration by parts and other specialized methods. At this stage, students are expected to synthesize the skills they have developed in the previous rooms to solve more complex, multi-step integrals. This room challenges students to apply higher-order thinking and problem-solving skills, requiring them to deconstruct difficult problems and use advanced techniques. By isolating these more difficult problems in a separate room, the escape room ensures that students have the necessary skills and confidence to tackle these challenges, having already reinforced their foundational and intermediate knowledge. The organization of the rooms allows for a smooth and logical progression of difficulty, preventing cognitive overload and fostering a deeper understanding of integral calculus.

The Final Conundrum serves as the culminating challenge of the escape room, where students must apply everything, they have learned to solve a highly complex integral. This final puzzle is designed to test not only their ability to recall and use specific techniques but also their capacity for critical thinking and problem-solving. By integrating multiple methods into a single, multi-step problem, the Final Conundrum requires students to demonstrate mastery over the entire range of integral techniques covered in the escape room. This structure reinforces the idea that mastering integral calculus is not about memorizing isolated rules but about understanding how to apply a variety of strategies to different types of problems.

First Themed Room: Introduction to Integration Concepts

The first themed room of 'The Integral ENIGMA' escape room serves as an introduction to the basic concepts of integration, focusing primarily on fundamental skills such as the properties of the indefinite integral and the relationship between integrals and derivatives. The puzzles in this room are designed to build a strong foundation for students by emphasizing the core principles of indefinite integration, which are essential for more advanced calculus problems. By solving these initial puzzles, students reinforce their understanding of basic integration techniques, allowing them to approach more complex problems with confidence later in the escape room.

The first puzzle in this room introduces the concept of integrals by posing questions such as

1. The action of finding the anti-derivative function of a given function is called?
   a. Integration ✓
   b. Differentiation

The learning objective of this puzzle is to ensure that students are comfortable with fundamental concepts, a skill that is critical for solving more complex integrals in later stages. Through immediate feedback and hints, students are guided to recall that integration is the reverse process of differentiation, helping them solidify this foundational concept.

Another key puzzle in the first room focuses on recognizing the connection between derivatives and integrals, particularly how integration can be seen as the reverse operation of differentiation. Students are presented with a function that they must differentiate, and are then tasked with finding its integral, thereby returning to the original function. This puzzle emphasizes the inverse relationship between the two operations, guiding students to understand how each step in the differentiation process corresponds to an inverse step in integration. The learning objective here is to deepen students' understanding of the fundamental theorem of calculus, which establishes this connection between integrals and derivatives.

2. The derivative of the indefinite integral is equal to the integrand function
   a. $d\left(\int f(x)dx\right) = f(x)dx$
   b. $\left(\int f(x)dx\right)' = f(x)$✓

c. $\int d(F(x)) = F(x) + C$

The learning objective of this puzzle is to ensure that students are comfortable with the properties of the indefinite integral, a skill that is critical for solving more complex integrals in later stages.

The use of simple, straightforward problems allows students to build their confidence and understanding in a supportive, low-pressure environment, where they can receive immediate feedback and hints if needed. Focusing on antiderivatives, the properties of the indefinite integral, and the relationship between integrals and derivatives provides students with a comprehensive overview of the foundational skills necessary for tackling indefinite integrals.

By the end of this room, students should have reinforced their knowledge of basic integration techniques and feel prepared to take on the more complex puzzles in the next stages of the escape room. This introductory section serves as a critical stepping stone in the learning path, ensuring that all students — regardless of their prior experience — are equipped with the essential tools needed to succeed in the later, more advanced themed rooms.

Second Themed Room: Integration Problems

The second themed room of "The Integral ENIGMA" escape room introduces students to more complex integration problems, significantly increasing the difficulty level compared to the first room. This section is designed to challenge students to find the antiderivatives of simple functions and to apply substitution techniques. As students progress through these puzzles, they are required to apply their foundational knowledge of integration while engaging with more advanced mathematical concepts. This room is pivotal in helping students transition from basic antiderivatives to a deeper understanding of how to handle more complex integrals, thus preparing them for the final stages of the escape room.

One of the initial puzzles in this room focuses on the integration of trigonometric functions, such as

1. Find and Select the right answer $\int \sin(3x + 5)dx$ (Figure 20)
   a. $\sin(3x + 5) + C$
   b. $\cos(3x + 5) + C$
   c. $-\frac{1}{3}\cos(3x + 5) + C$✓
   d. $\frac{1}{3}\cos(3x + 5) + C$
   e. $\frac{1}{3}\cos(3x + 5)$

This puzzle introduces students to the concept of substitution as a technique to simplify complex integrals, helping them understand how to manipulate functions involving trigonometric terms. The learning objective here is to ensure that students can identify when substitution is necessary and apply it correctly to solve basic integrals, a skill that will be further tested in subsequent puzzles.

Figure 20. Example of an integration problem presented in the second themed room of the "integral enigma" escape room

2. Find and Select the right answer $\int \frac{dx}{cos^2(3x + 1)} dx$
   a. $\tan(3x + 5)$
   b. $\frac{1}{3}\tan(3x + 1) + C$✓
   c. $-\frac{1}{3}\cos^{-1}(3x + 1)$
   d. $\frac{1}{3}\cos^{-1}(3x + 1) + C$
   e. $-\frac{1}{3}\tan(3x + 1) + C$

The next set of puzzles focuses on logarithmic expressions, challenging students to integrate functions involving natural logarithms.

3. Find and Select the right answer $\int \frac{\ln x}{x} dx$ (Figure 21)
   a. $\frac{\ln^2 x}{2}$
   b. $\ln(\ln x) + C$
   c. $\frac{\ln^2 x}{2} + C$✓

By solving these problems, students solidify their knowledge of basic integrals, which are essential for tackling more advanced calculus problems that involve combinations of logarithmic and algebraic terms.

Figure 21. Puzzle challenge from the "integral enigma" escape room, where students must solve integrals concealed in the room

In summary, the second themed room of "The Integral ENIGMA" challenges students with intermediate integration problems that require a deeper level of mathematical reasoning and technique. By focusing substitution methods, this room provides students with the tools they need to solve more complex integrals while reinforcing their understanding of the foundational principles introduced in the first room. Through these puzzles, students gain confidence in handling more complex calculus problems, preparing them for the advanced challenges that await in the third room.

Third Themed Room: Advanced Integration Techniques

The third themed room in "The Integral ENIGMA" escape room is designed to challenge students with the most complex and advanced integration techniques, primarily focusing on methods like integration by parts, as well as more complex applications of substitution. This room represents a culmination of the skills students have developed in the previous stages, demanding higher-order thinking and problem-solving abilities. Each puzzle in this room is carefully crafted to reinforce the application of advanced integration strategies, pushing students to synthesize their knowledge and apply it in more complex, multi-step problems.

The first set of puzzles in this room introduces integration by parts, a method derived from the product rule of differentiation.

1. Integration by parts formula
   a. $\int udv = vu - \int ktu$
   b. $\int vdu = vu - \int udv$
   c. $\int udv = vu - \int vdu$✓

Another key puzzle in this room may involve integrals that require choosing the correct function u in integration by parts.

2. In integral $\int x \ln^2 x dx$, $u = \ln x$?
   c. True ✓
   d. False

The idea behind this puzzle is that the student should determine which function to choose as u in the integral $\int udv$ when logarithmic and algebraic terms are involved. The learning objective is to build students' perseverance and problem-solving resilience, as these puzzles challenge them to navigate multi-layered problems methodically.

The next set of puzzles in the third room introduces students to advanced trigonometric integrals, such as

3. In integral $\int x \tan^2 x dx$, $u = \tan x$?
   a. True
   b. False ✓

This puzzle challenges students to determine which function to choose as u in the integral $\int udv$when working with inverse trigonometric and algebraic terms.

Overall, the third themed room in "The Integral ENIGMA" escape room is designed to reinforce advanced integration techniques while promoting higher-order problem-solving skills. Each puzzle requires students to apply sophisticated strategies like integration by parts. By the end of this room, students will have developed a deep understanding of advanced integral calculus, equipped with the tools and problem-solving abilities needed to tackle even the most challenging integration problems. This room serves as a bridge to the final stage of the escape room, where all the skills developed here will be put to the ultimate test in the Final Conundrum.

Final Conundrum: Integration Mastery Test

Final Conundrum (see Figure 22) marks the ultimate challenge for students attempting to escape the digital escape room themed around the topic of integration by parts. Within this immersive digital environment, students encounter a series of puzzles and tasks designed to test their comprehension of integration techniques.

Figure 22. Final conundrum in the "integral enigma" escape room, requiring students to solve complex integrals to escape

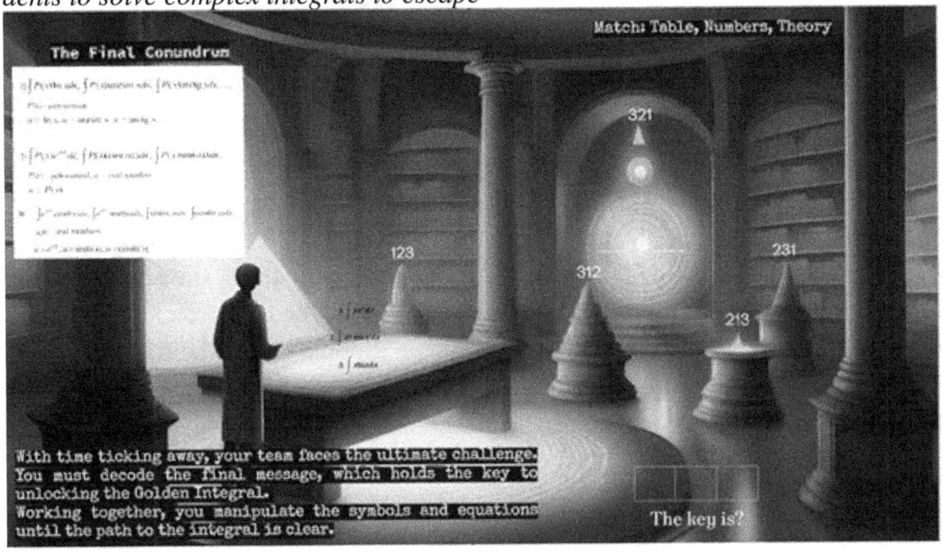

In this stage, students face the task of unraveling the final mystery: a set of integrals crucial for their escape. However, these integrals are cunningly concealed within the room and remain hidden from view until students actively engage with their surroundings by moving the cursor over them. Before confronting this challenge, students are equipped with essential theoretical knowledge outlining the fundamental principles and rules governing integration by parts. This theoretical foundation is embedded somewhere within the digital landscape, requiring students to explore diligently to uncover it. Armed with this theoretical understanding, students navigate to the final stage, where they encounter a table. Suspended above the table are three enigmatic integrals, shrouded from view until students direct their cursor over them. Each integral corresponds to a distinct theory group, as elucidated by the theoretical principles discovered earlier. To successfully progress, students must discern the relationships between the integrals and the corresponding theory groups. For instance, if the first integral aligns with the principles outlined in the second theory group, the second integral relates to the third theory group, and the third integral corresponds to the first theory group, the resulting code would be '231.' Students then input this code by clicking on the numbers in the correct sequence. However, an erroneous selection incurs a time penalty, compelling students to retry until they successfully input the correct code.

Successfully navigating this final challenge not only signifies mastery of integration by parts but also unlocks the path to freedom from the digital escape room, showcasing students' adeptness in tackling complex mathematical concepts within immersive environments. Further, we will focus on demonstrating how to create a question from scratch, providing step-by-step guidance. To begin, navigate to the "Pages" menu within your presentation software. Once there, locate the slide after which you wish to add a new slide, then click on the "Add page" option situated at the top left corner of the interface. Upon clicking, a "Add page" window will appear, offering a variety of page templates to choose from. For our purposes, select the "Blank" template, as depicted in Figure 23, and proceed by clicking the "ADD" button. Now that the blank slide has been created, let's move on to setting the background. Access the "Background" menu and apply the desired background, following the instructions provided in Figure 18. Once the background is in place, it's time to introduce the escape room character "Sophia" - an AI-driven museum guide. This can be accomplished by simply inserting an appropriate picture onto the slide, as demonstrated in Figure 24.

Figure 23. Steps to add a new blank page in Genially to introduce a new character or task to the escape room

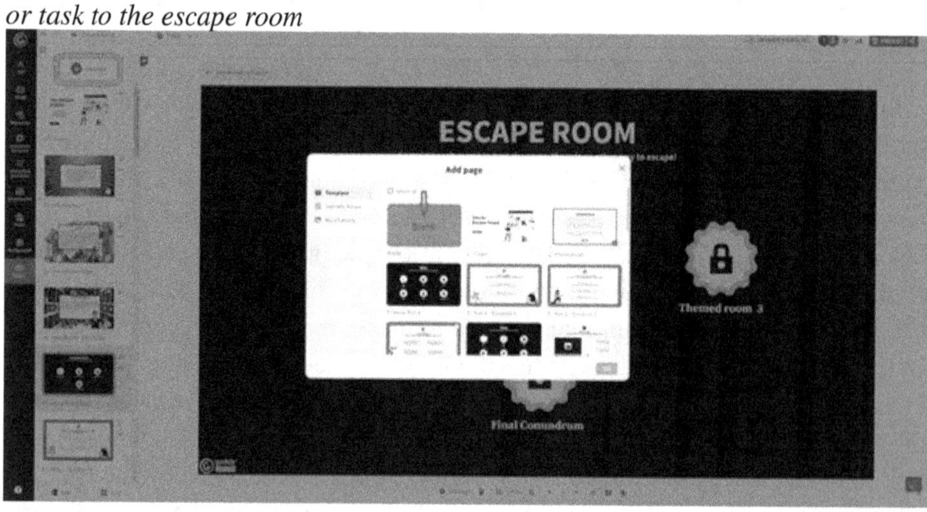

Figure 24. Adding the AI character "Sophia" to the escape room, guiding participants through challenges and clues

With Sophia now present, we can add clues or tasks to engage the audience. Click on "Text" and input the relevant clues or tasks, as shown in Figure 25. Remember to personalize the text to suit your specific question or scenario. Adjusting the size, font, and other text parameters is straightforward, as illustrated in Figure 26.

Figure 25. Inputting text for clues and tasks in Genially, personalizing the escape room challenges for participants

Figure 26. Adjusting text parameters such as size, font, and alignment to enhance the readability and design of the escape room

Moving forward, we need to include the correct answer while ensuring it remains hidden until prompted. To achieve this, select the "Edit animation" icon and adjust the parameters accordingly, as detailed in Figure 27.

Figure 27. Setting up animations for revealing correct answers using the "fade in" effect and mouse hover triggers

Choose "Hover mouse" as the trigger and apply the "Fade in" effect. By doing so, students will only reveal the answers when they hover their mouse over the designated area. And also, we need to consider what actions to take if the correct answer is chosen. It's essential to incorporate interactivity into our presentation to enhance engagement and provide immediate feedback to the audience. As depicted in Figures 28, 29 we can implement various interactive features to respond to both correct and incorrect answers.

Figure 28. Implementing interactive feedback features for correct and incorrect answers in the escape room

Figure 29. Directing participants to the next stage of the escape room after they select the correct answer

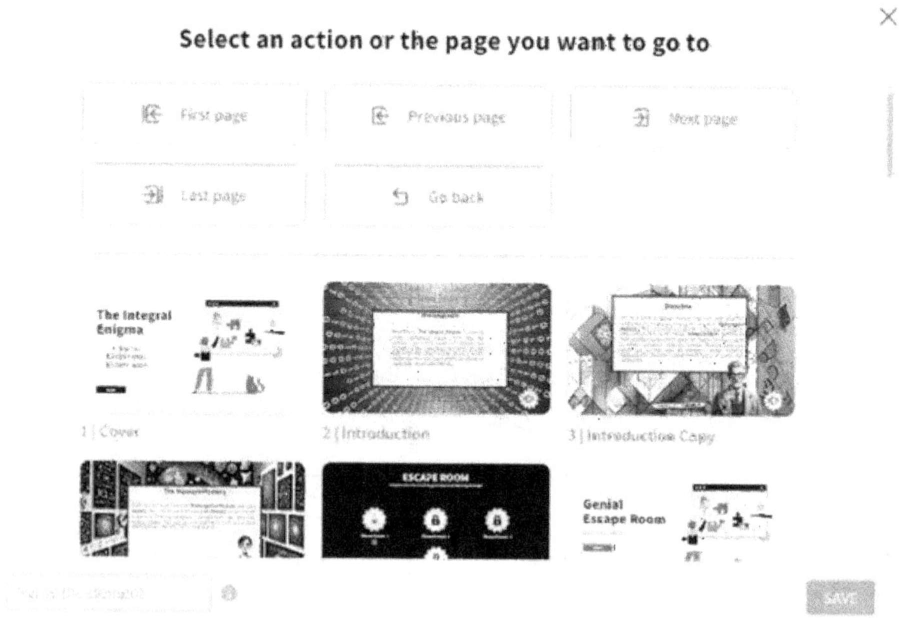

When a correct answer is selected, we have the opportunity to guide the audience further through the presentation. One option is to allow them to proceed to the next page seamlessly, continuing their learning journey. Alternatively, we can provide the flexibility for them to navigate to a specific page of interest, tailoring their experience to their preferences and needs. Conversely, if an incorrect answer is chosen, we must address it appropriately to reinforce learning objectives. In this scenario, the navigation should be directed towards a penalty page or a remedial section designed to provide additional context or guidance. By doing so, we create a supportive learning environment where mistakes are opportunities for growth and reflection.

Customizing and Personalizing the Experience

Beyond the traditional presentation tools, Genially provides the means to incorporate dynamic elements that can transform your escape room into an immersive experience like no other. Among these features is the capability to integrate sound into your escape room. By adding background music, ambient noises, or even character dialogue, you can heighten the atmosphere and draw players deeper into the adventure (Figure 30). Whether it's the ominous creaking of a door or the suspenseful ticking of a clock, sound can play a pivotal role in setting the mood and enhancing the overall immersion.

Figure 30. Adding background music and sound effects to enhance the immersive atmosphere of the digital escape room

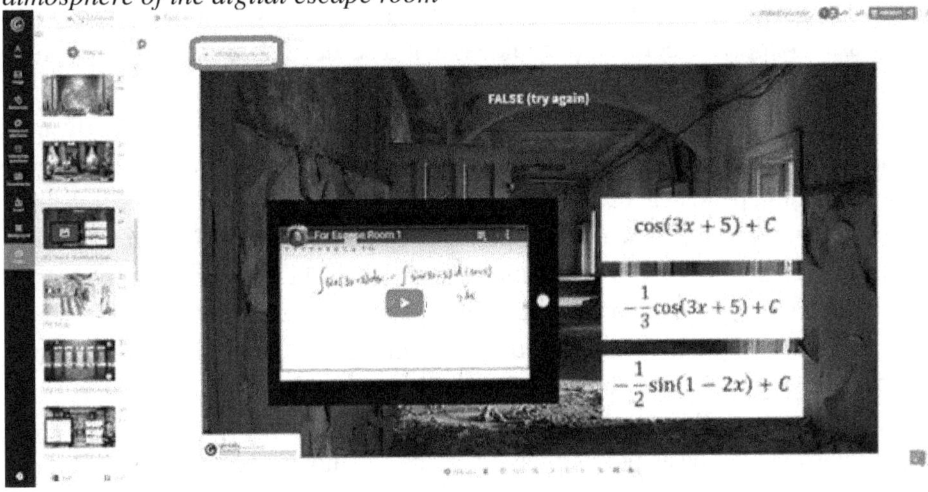

Genially allows you to seamlessly embed video content into your escape room puzzles. Whether it's a cryptic video message from a mysterious character or a clue hidden within a surveillance footage, videos can add layers of storytelling and intrigue to your escape room experience. With Genially, you have the flexibility to integrate videos seamlessly into your puzzles, enhancing both the narrative and the game-play.

But the features of Genially don't end there. From interactive elements like puzzles, riddles, and hidden clues to customized animations and transitions, there's a wealth of tools at your disposal to create a truly memorable escape room. Don't hesitate to experiment with these features and push the boundaries of traditional escape room design.

So, as you embark on your path to create your own escape room masterpiece, remember to leverage the full potential of Genially features. With creativity as your guide and Genially as your canvas, the possibilities are endless. Good luck and may your escape room captivate and delight adventurers far and wide!

Educational Outcomes and Reflection

The "Integral ENIGMA" escape room is designed to provide a comprehensive learning experience that enhances students' understanding of integral calculus while fostering key cognitive and problem-solving skills. By guiding students through a series of progressively challenging puzzles, the escape room ensures that they not only learn to apply integration techniques but also develop a deeper conceptual understanding of how these techniques fit into the broader framework of calculus.

The educational outcomes of this escape room extend beyond mastering the mechanics of solving integrals, as the interactive, game-like environment encourages critical thinking, persistence, and collaboration.

One of the primary educational outcomes of the "Integral ENIGMA" escape room is the reinforcement of foundational and advanced integration skills. From the first themed room, where students focus on simple antiderivatives, to the final conundrum that requires them to synthesize multiple techniques, students are continually applying their knowledge in different contexts. The structured progression of the escape room allows them to build confidence and competence in solving integrals, starting with basic polynomial functions and moving toward more complex applications involving integration by parts. This incremental learning path ensures that students fully grasp each concept before moving on to more difficult problems, ultimately fostering a solid understanding of integral calculus.

The escape room encourages students to think critically about the strategies they use to solve problems. Unlike traditional classroom exercises where problems are often presented in isolation, the escape room presents puzzles within a narrative framework that requires students to decide how to approach each problem strategically. For example, they must determine when to apply substitution or when integration by parts is the most effective method. This decision-making process enhances their ability to analyze problems and select the appropriate tools, a skill that is essential not only in mathematics but in any field that requires logical reasoning and problem-solving.

Another significant educational outcome of the escape room is the development of persistence and resilience in learning. The puzzle-based structure of the escape room ensures that students face challenges that require multiple attempts to solve. When students encounter difficult problems, they are encouraged to review their approach, learn from their mistakes, and try again, fostering a growth mindset. This iterative process of trial and error helps students build resilience, as they learn that failure is a part of the learning process. The interactive nature of the escape room, combined with immediate feedback and hints, supports this learning process by providing a safe space for students to make mistakes and correct them without the fear of formal assessment.

Collaboration is another key aspect of the educational outcomes achieved through the "Integral ENIGMA" escape room. While each puzzle can be solved individually, the design of the escape room encourages group work and peer-to-peer learning. Students often work together to solve complex problems, discussing different approaches and sharing strategies. This collaborative environment not only enhances their understanding of the material but also helps them develop communication and teamwork skills, which are critical for success in both academic and professional settings. The escape room's engaging format fosters a sense of camaraderie and shared accomplishment, making the learning experience both enjoyable and productive.

Reflection is an important component of the educational outcomes of the escape room as well. Upon completing the escape room, students are encouraged to reflect on the strategies they used, the challenges they faced, and the knowledge they gained throughout the experience. This reflection helps solidify their understanding of the material and promotes metacognitive skills, as students assess their own learning process and identify areas where they can improve. By reflecting on their performance, students become more aware of their strengths and weaknesses in solving integrals, which can guide their future studies and enhance their overall mathematical ability.

These educational outcomes have been summarized in Table 3. In summary, the educational outcomes of "The Integral ENIGMA" escape room go far beyond the mastery of integration techniques. The escape room fosters critical thinking, problem-solving, persistence, collaboration, and reflection, all of which contribute to a deeper and more holistic understanding of integral calculus. By engaging students in a narrative-driven, interactive learning environment, the escape room creates an immersive experience that motivates students to actively participate in their learning process, ensuring that they not only acquire the necessary mathematical skills but also develop the cognitive and interpersonal abilities that will serve them well in future academic and professional endeavors.

Table 3. Educational outcomes of "the integral ENIGMA" and related design elements and Genially features

Educational Outcome	Escape Room Design Element	Genially Feature Used
Mastery of Integration Techniques	Progressive puzzles from basic to advanced integrals (e.g., polynomials, trigonometric, logarithmic, integration by parts)	Multi-stage structure with increasingly complex puzzles, customizable quiz elements
Critical Thinking and Problem-Solving	Students must decide which integration method to use (e.g., substitution, integration by parts, trigonometric identities)	Interactive elements allowing for branching based on user choices, hotspots for clues
Persistence and Resilience	Puzzles designed to require multiple attempts, encouraging trial and error	Feedback mechanisms, hints, and penalty timers for incorrect answers, animated feedback
Collaboration and Teamwork	Escape room can be solved in groups, encouraging discussion and shared strategies	Collaboration features, multiple participants simultaneously
Metacognition and Reflection	Reflection encouraged after puzzle completion, assessing strategies and understanding	Integration of reflection prompts and quizzes for assessment, embedded interactive forms
Engagement and Motivation	Narrative-driven plot with a story that unfolds as students progress through puzzles	Genially's customizable narrative-driven templates, animations, and multimedia integration

TIPS FOR DIGITAL ESCAPE ROOMS IN GENIALLY

General Tips

Start by choosing a predefined escape room template under the "Escape Games" category within Genially's template library. These templates come with interactive features like clickable hotspots, animations, and transition effects. To maintain consistency across levels, duplicate slides by right-clicking on a slide in the left panel and selecting the "Duplicate" option. This ensures that the layout and design remain uniform across all stages, saving time and effort. Additionally, use pre-built templates to reduce the technical workload, customizing only the content relevant to your learning objectives.

When modifying the content, replace the default placeholders for "Right" and "Wrong" answers carefully. Each interactive element in Genially comes with an "Interactivity" menu that allows you to define the outcome of an action. For instance, a correct answer can trigger a slide transition to the next stage, while an incorrect answer might display a hint. To include complex content, such as mathematical equations, embed LaTeX-rendered images or upload PNG files of equations created using external tools. Ensure interactive hotspots are resized and aligned precisely by using the drag-and-drop functionality in the properties menu, providing intuitive navigation for players.

Visual elements set the tone for your escape room. Use the "Background" feature to upload custom images that align with the room's theme, such as a haunted mansion or a futuristic laboratory. To enhance engagement, apply animations like "Fade In" or "Zoom" to clues or interactive objects, available in the "Effects" tab of each element. Adjust timing and triggers to ensure smooth transitions. To avoid visual clutter or overlapping functionality, organize elements in layers using the "Layers" panel in the bottom-right corner of the editor.

Incorporating timers and hidden clues adds excitement and interactivity. Use the "Timer" widget from the "Interactive Elements" menu to set countdowns for challenges. Pair this with penalty-based feedback by linking the timer to a "Retry" slide for incorrect answers. Hidden clues can be created by adjusting the "Opacity" slider in the properties menu, making elements invisible until triggered by a specific action, such as clicking on a magnifying glass tool. For branching scenarios, link slides to different outcomes depending on the player's choices, allowing for non-linear gameplay.

A compelling storyline transforms your escape room into an immersive experience. Add characters using transparent PNG images and interactive text boxes that appear or disappear based on user actions. To enhance the auditory experience, embed background music or narrative voice-overs in MP3 format. Use conditional visibility triggers for invisible buttons to unveil narrative twists or new challenges when specific conditions are met.

Complex puzzles can elevate the challenge. Create dynamic locks requiring players to input specific sequences into text fields. For example, a combination lock puzzle can trigger a transition to the next slide if the correct code is entered. Use Genially's "Drag & Drop" feature to design puzzles where users must drag a key object to a specific slot to unlock the next stage. For auditory puzzles, use clickable hotspots on a virtual keyboard and assign each key a note. Players must click the correct sequence based on clues to progress.

Genially allows the integration of external content for enhanced functionality. Use the "Insert" option to embed HTML or iframe widgets, such as external simulations or quizzes. Interactive PDFs can be uploaded, with specific pages linked to hotspots for document-based challenges. Additionally, you can incorporate external puzzle generators (e.g., crossword or word search tools) by embedding them as iframes, providing diverse puzzle types within your escape room.

Feedback mechanisms are crucial for maintaining engagement and guiding players. Use pop-up layers to provide immediate feedback. For instance, clicking the correct answer might display an animation or congratulatory message, while an incorrect answer could trigger a hint or explanation. To simulate scoring, track collected items like keys or coins by visually representing them in an inventory area on the screen. For incorrect answers, create a retry mechanism by linking to a slide that resets the challenge, ensuring players can learn from their mistakes without feeling stuck.

Frequent testing is essential to identify and resolve issues. Use Genially's "Preview" mode to test interactions, animations, and transitions. Check for broken links or non-functional elements by reviewing the "Interactivity" panel for each slide. To ensure compatibility across devices, test the escape room on desktops, tablets, and mobile devices. Adjust element sizes and alignments for responsiveness, ensuring a seamless experience for all users.

When the escape room is ready, deploy it thoughtfully. Use Genially's "Share" settings to generate private links, ensuring controlled access. For educational purposes, embed the escape room into learning management systems (LMS) like Moodle or Google Classroom using the embed code provided by Genially. To improve future iterations, enable tracking features to analyze how users interact with the content, such as which puzzles they struggle with or complete quickly.

These tips are summarized in Table 4.

Table 4. Summary of tips for designing digital escape rooms in Genially

Category	Tips and Tricks
Template Selection	Use "Escape Games" templates for structure; duplicate slides for consistent levels.
Content Customization	Replace default answers carefully; embed equations using LaTeX or images; adjust hotspot sizes for intuitive navigation.
Visual Elements	Customize backgrounds with relevant images; use animations like "Fade In" for clues; manage element layers to avoid overlap.
Interactive Features	Add timers for time-limited challenges; hide clues with opacity and reveal using interactive tools; create branching paths with linked slides.
Narrative Enhancements	Add characters with transparent PNGs; embed audio for narratives or ambiance; trigger conditional content with invisible buttons.
Puzzle Mechanics	Simulate locks with text inputs; use draggable keys for progression; create audio challenges with clickable notes on a virtual keyboard.
External Embeds	Embed HTML or iframe widgets; link interactive PDFs to specific hotspots; integrate external puzzle generators via embeds.
Feedback and Scoring	Use pop-ups for instant feedback; track scores with visual objects (keys, coins); provide retry buttons linked to reset slides.
Testing and Debugging	Use preview mode frequently; test across devices for compatibility; debug by checking the interactivity panel for errors.
Deployment	Share via private links; embed escape rooms into LMS like Moodle or Google Classroom; enable tracking to analyze user interactions.

Genially App Customization Tips

The Genially app offers a variety of tools and features that enable educators and designers to create highly interactive and engaging digital experiences. In this section, we provide both novice and advanced customization tips, focusing on key features such as timers, animations, and conditional logic. These tips will help you personalize your digital escape room or any other interactive content to suit your educational goals.

Timers and Time Penalties

Timers are essential for adding urgency to challenges in digital escape rooms. Genially provides several ways to incorporate timers into your project:

Novice Tip: Basic Timer Setup For novice users, the easiest way to add a timer is by using the pre-built Timer widget available under the Interactive Elements menu. You can configure the following settings:

- **Duration**: Set the total time for the challenge (e.g., 60 seconds).
- **Countdown style**: Choose between different countdown styles, such as circular or numeric.
- **End action**: Define what happens when the time runs out (e.g., advance to a penalty screen, lock the user out).

Advanced Tip: Customizing Time Penalties For advanced users, you can incorporate custom time penalties when users provide incorrect answers. This can be achieved by linking a timer to specific interactive elements:

- Add a hidden element (e.g., an image or button) that appears only when a wrong answer is selected. Set an animation delay on the "show" action, specifying the penalty time (e.g., add a 30-second delay before the next action can be performed).
- To enhance control, integrate JavaScript code within Genially's embedded HTML features to create more complex timing scenarios.

Animations and Transitions

Animations play a critical role in creating an engaging and immersive experience. Genially provides a wide array of animation options for every element on your screen.

Novice Tip: Basic Animation Effects Novice users can apply simple animations to objects (e.g., text, images) by following these steps:

- Select an object and click on the Animations tab in the toolbar.
- Choose from basic effects such as Fade In, Slide, or Zoom.
- Configure the animation's start trigger (on page load, on click, or on hover).
- Adjust the duration and delay to ensure smooth transitions.

Advanced Tip: Conditional Animations Advanced users can create conditional animations where the visibility or movement of objects depends on user actions. Here's how:

- Combine multiple animations and set them to trigger based on specific events (e.g., an object only fades in when a correct answer is selected).
- Use the "Show/Hide" effect to control which elements are visible at any point, allowing for dynamic content that changes based on user interactions.
- Apply the "Loop" feature to continuously animate certain elements, adding subtle visual cues or hints to guide users.

For even more control, advanced users can integrate external animation libraries through embedded HTML code.

Conditional Logic and Branching Scenarios

Conditional logic allows you to create branching scenarios, where the user's choices determine the subsequent path in the escape room.

Novice Tip: Simple Branching Using Links A basic way to create conditional logic is by using links:

- Create multiple pathways by duplicating slides.
- Add interactive buttons or hotspots that link to different slides based on the user's choice (e.g., correct answers take them to the next puzzle, while wrong answers lead to a retry page).
- You can add feedback messages using simple "Show/Hide" animations triggered by these buttons.

Advanced Tip: Complex Branching with Variables Advanced users can employ variables and more complex logic to create fully dynamic branching paths:

- Define variables within Genially to track user progress (e.g., correct answers, completed tasks).
- Set conditions for slide navigation based on these variables. For instance, allow users to proceed only after collecting a certain number of points or clues.

- Use hidden elements or overlays that become visible when specific conditions are met, providing a more advanced user experience.
- For further customization, external scripts (JavaScript) can be embedded to handle more advanced logical scenarios and branching outcomes.

Personalizing the User Interface

In addition to interactive elements, customizing the user interface (UI) improves the overall experience.

Novice Tip: UI Customization Basics For beginners, personalizing the UI involves selecting a cohesive theme and applying it consistently:

- Choose from pre-made themes or create a custom color palette and font scheme that aligns with your escape room's narrative.
- Make sure navigation buttons are intuitive, using clear labeling (e.g., "Next", "Retry").

Advanced Tip: Custom User Interfaces Advanced users can design custom UIs by layering multiple elements and using overlays:

- Use transparent hotspots over images or interactive buttons to create a hidden navigation menu that is revealed when clicked.
- Add dynamic menus that update based on user progression (e.g., locked icons that change to unlocked after completing a task).
- To create a polished, game-like experience, consider embedding external CSS or JavaScript to build custom-designed interfaces that interact seamlessly with the escape room mechanics.

Audio and Video Integration

Adding multimedia elements enhances the immersive experience of your digital escape room.

Novice Tip: Embedding Audio and Video Novice users can easily embed audio or video by using the built-in tools:

- Add background music by embedding an audio file in the Interactive Elements section.
- Use video to provide visual clues or narrative cutscenes by inserting YouTube or locally uploaded video files into specific slides.

Advanced Tip: Dynamic Multimedia Advanced users can integrate multimedia more dynamically:

- Set audio or video to play only after certain triggers (e.g., after answering a question or finding a hidden clue).
- Synchronize audio with animations for more cohesive storytelling, adjusting start and stop points to align with the escape room's narrative.
- Use external audio or video hosting platforms with API integration for more control over playback conditions.

Using External Code (HTML, CSS, and JavaScript)

While Genially provides a wide array of built-in features for interactive content creation, users who are comfortable with coding can take customization to the next level by embedding HTML, CSS, and JavaScript. This allows for highly complex interactions that go beyond the standard functionality of the platform.

Embedding Custom HTML and CSS To further customize the layout, structure, and appearance of your escape room, you can embed custom HTML and CSS:

- Access the HTML widget from the "Insert" menu.
- Insert the required HTML structure to add new elements, such as customized buttons, forms, or complex tables.
- Use CSS to style these elements, allowing for finer control over fonts, colors, spacing, and overall visual design.
- Example: You can create a custom navigation bar that dynamically highlights the current stage of the escape room based on student progress.

Incorporating JavaScript for Complex Interactions For users interested in adding dynamic interactivity, JavaScript can be used to manage user inputs, calculate scores, and track progress:

- Use JavaScript to create conditional logic that is more complex than Genially's built-in branching capabilities. For example, you can set up a scoring system that tracks the number of correct answers across multiple puzzles and provides real-time feedback.
- Embed JavaScript directly in the HTML widget by wrapping it in <script> tags.
- Example: A timer that counts down and penalizes students for incorrect answers can be customized using JavaScript, where you can dynamically adjust the penalty based on performance.

Customizing Visual Transitions Between Puzzles

Smooth visual transitions help create a more immersive and seamless user experience. Genially offers a variety of animations, but advanced users can push this further by customizing transitions with external code or creatively using Genially's native tools.

Advanced Transition Techniques

- Use the **Animation Timeline** in Genially to choreograph complex animations between puzzles. This allows you to sequence multiple elements to appear, move, or fade in synchrony.
- Combine **fade-in and slide** animations for a gradual reveal of puzzles, creating a smooth flow between stages. Adjust the timing and delays for a natural transition that maintains user engagement.
- To take it a step further, use CSS to apply custom animations. Embedding keyframe animations in the HTML widget allows for highly specific transitions. For instance, using CSS's @keyframes, you can create unique entry effects like bouncing or rotating transitions between puzzles.
- Example: A puzzle might fade out slowly while the next one slides into place, giving the illusion of continuous progression through the escape room.

Embedding Live Data or External Quizzes for Adaptive Learning

Another advanced customization feature in Genially is the ability to embed live data or external quizzes that adapt based on user input. This allows for real-time updates and more personalized learning experiences.

Embedding Live Data

- Genially allows for the integration of live data from external sources. Using iFrames, you can embed live feeds, interactive charts, or forms that update in real time.
- Example: Embed a Google Sheets document that dynamically displays progress or scores from multiple participants as they complete the escape room.
- This can also be applied to integrate real-time dashboards, where students can view their results in comparison with their peers, motivating them to improve their performance.

Integrating External Quizzes

- You can use external quiz tools like Google Forms, Quizlet, or Kahoot! to create adaptive quizzes. Embed these tools using an iFrame within Genially, allowing for more sophisticated branching scenarios.
- Example: An embedded quiz can adjust difficulty levels based on the user's performance. If a student answers several questions correctly, the next set of puzzles can automatically become more challenging, while wrong answers trigger simpler review questions.
- For advanced users, APIs from external quiz platforms can be integrated into Genially, allowing you to pull real-time data on performance and adjust the escape room content accordingly.

Adaptive Learning with Conditional Logic and Data

For an even more personalized experience, Genially's internal logic can be combined with live data sources to create adaptive learning paths:

- Use variables in Genially to track student performance and adjust the path they follow through the escape room. For instance, students who excel in early puzzles can be directed to advanced stages, while others receive additional hints or simpler tasks.
- Integrate external data feeds to dynamically alter content. For example, quiz results from a learning management system (LMS) can be used to adjust the difficulty or provide tailored content in real-time.
- This approach enhances both engagement and learning efficacy by creating an escape room that adapts to each student's unique progress and learning style.

The Genially app provides powerful customization tools for both novice and advanced users. By following these tips, you can create highly personalized and interactive digital escape rooms that cater to a wide range of educational objectives. From simple timers and animations to complex branching logic and custom UIs, the possibilities for customization are nearly endless, allowing you to design immersive experiences that captivate and engage learners. By employing advanced customization techniques such as embedding external code, customizing transitions, and integrating live data or external quizzes, educators and designers can create highly personalized and interactive digital escape rooms. These methods enable more complex interactions and smoother transitions, offering a tailored, adaptive learning experience that engages students at multiple levels. As users become more proficient with these tools, the potential for creating immersive and sophisticated educational experiences in Genially becomes almost limitless.

CONCLUSION

In this chapter, we have explored how the thoughtful design of digital escape rooms can serve as a powerful educational tool, engaging students while targeting specific learning outcomes. By closely examining the design decisions involved in this case study, such as puzzle creation and narrative structure, we have demonstrated how each element contributes to the achievement of pedagogical goals.

The puzzles within the escape room were designed to progressively challenge students, reinforcing their understanding of key mathematical concepts. For instance, multiple-choice questions and logic-based puzzles targeted both understanding and application skills by requiring students to classify differential equations or solve step-by-step problems. These puzzles serve as formative assessments, enabling students to actively apply their knowledge while receiving immediate feedback. The cognitive demands of these puzzles align with Bloom's Taxonomy, moving from basic recall of information to higher-order thinking tasks like analysis and problem-solving. The creation of interactive and progressively difficult puzzles also encourages the development of critical thinking and problem-solving skills. By confronting increasingly complex challenges, students are required to employ deeper levels of analysis and reflection, promoting cognitive engagement and enhancing their ability to transfer learned concepts to new contexts.

The narrative structure of the escape room plays a critical role in maintaining student motivation and fostering engagement. The use of a storyline, in which students take on the role of historical mathematicians or embark on a mystery-solving adventure, helps contextualize abstract mathematical problems. This connection between content and narrative supports contextual learning, enabling students to see the relevance of mathematical concepts in real-world or historical contexts. By integrating the puzzles into a broader narrative, the escape room creates a sense of progression and achievement, keeping students invested in their learning process. This form of engagement through storytelling is particularly effective in fostering interdisciplinary thinking, as the narrative often links mathematical problems to broader themes from history, science, or other fields. By embedding educational content in a compelling storyline, students are more likely to engage deeply with the material and retain the knowledge they acquire.

The incorporation of interactive elements, such as time-bound challenges and peer-supported feedback mechanisms, further enhances the learning experience by promoting collaboration and active learning. Timed puzzles encourage students to develop their time management and collaborative problem-solving skills, as they must communicate and divide tasks to complete challenges within the allotted time. The use of interactive feedback systems also fosters peer learning, encouraging students to evaluate each other's contributions and provide constructive feedback. These interactive components support the development of social and emotional learning, as students practice teamwork, communication, and persistence in the face of challenges. By creating a cooperative learning environment, the escape room format not only promotes the acquisition of content knowledge but also builds essential soft skills that are transferable to real-world scenarios.

The design elements of this digital escape room are explicitly aligned with the desired educational outcomes, ensuring that each decision contributes to meaningful learning experiences. By integrating progressive puzzles, engaging narratives, and interactive challenges, the escape room facilitates the achievement of a variety of learning objectives, from cognitive skills such as understanding, application, and critical thinking to social competencies such as collaboration and communication. For educators, this clear alignment between design elements and educational outcomes provides a framework for developing their own escape rooms. Whether the goal is to reinforce subject-specific knowledge, promote interdisciplinary thinking, or cultivate collaboration and problem-solving skills, the techniques demonstrated in this chapter offer a versatile and effective approach to instructional design.

Digital escape rooms represent a dynamic and engaging approach to education, offering students the opportunity to learn through exploration, collaboration, and problem-solving. The design decisions detailed in this chapter—puzzle creation, narrative structure, and interactive elements—are not only central to creating a captivating experience but also crucial to achieving the intended learning outcomes. By carefully considering how each design choice contributes to educational objectives, educators can harness the full potential of digital escape rooms to foster meaningful and lasting learning experiences in their classrooms. The case study presented here serves as a guide for educators seeking to apply these techniques in their own teaching contexts. By aligning design elements with clear educational goals, educators can create personalized escape room experiences that both engage students and enhance their mastery of academic content. Ultimately, the flexibility of digital escape rooms allows for endless possibilities in educational innovation, limited only by one's creativity and imagination. The creation of digital escape rooms using Genially offers educators a powerful tool for fostering engagement, critical thinking, and active learning among students. By leveraging interactive elements, customizable templates, and immersive storytelling, educators can design captivating escape room experiences that not only reinforce academic concepts but also cultivate problem-solving skills and collaboration. As we've explored throughout this section, the process of creating a digital escape room is both accessible and versatile, allowing educators to tailor experiences to their specific learning objectives and student needs. Whether used as a standalone activity or integrated into broader instructional units, digital escape rooms have the potential to transform traditional learning environments into dynamic and immersive spaces where students are empowered to explore, discover, and succeed. With Genially, the possibilities are endless, and the result of digital escape room creation is limited only by one's imagination.

REFERENCES

Abdul Rahim, A. S. (2023). Escape the desert island: Blended escape rooms in the first-semester problem-based learning. *Journal of Chemical Education*, 100(6), 2459–2465. DOI: 10.1021/acs.jchemed.3c00119

Aldalur, I. (2025). Enhancing software development education through gamification and experiential learning with genially. *Software Quality Journal*, 33(1), 1. Advance online publication. DOI: 10.1007/s11219-024-09699-9

Jiménez, C., Arís, N., Magreñán Ruiz, Á., & Orcos, L. (2020). Digital escape room, using genial.ly and a breakout to learn algebra at secondary education level in spain. *Education Sciences*, 10(10), 271. DOI: 10.3390/educsci10100271

Martinez, E., Alvarez, J., Carbonell, V., Florez, M., & Montoya, E. (2023). Escaping the lab: Increasing motivation in the physics classroom through educational escape rooms. *Romanian Reports in Physics*, 75(1).

Mauri-Medrano, M., González-Yubero, S., Falcón-Linares, C., & Cardoso-Moreno, M. J. (2024). Gamifying the university classroom: A comparative analysis of game dimensions through educational escape room and a digital board game. *Frontiers in Education*, 9, 1354674. Advance online publication. DOI: 10.3389/feduc.2024.1354674

Santiago López, N. R. (2024). Gamification for analytical chemistry: A digital escape room. *Educación en la Química*, 35(2), 187–195. DOI: 10.22201/fq.18708404e.2024.2.86355

Vera-Morales, M., García, G. C., Lucas, V. S., & León-Castro, L. (2023). Gamification in the teaching of ecology for university students. In 2023 IEEE 3rd International Conference on Advanced Learning Technologies on Education & Research (ICALTER), DOI: DOI: 10.1109/ICALTER61411.2023.10372921

KEY TERMS AND DEFINITIONS

Digital Educational Escape Room: An interactive, game-like learning environment where students solve puzzles and challenges to progress, reinforcing subject knowledge in an engaging way.

Feedback Mechanism: A feature that provides immediate responses to user actions, such as indicating whether an answer is correct or incorrect, often with suggestions for improvement.

Final Conundrum: The ultimate challenge in a digital escape room, designed to test participants' mastery of all learned concepts in a comprehensive, complex puzzle.

Gamified Learning: An educational approach that uses game design elements, such as challenges and rewards, to increase student engagement and motivation.

Hotspot: An interactive element in digital content that users can click to reveal additional information, unlock puzzles, or navigate to other sections.

Narrative Elements: Story-driven components that enhance the immersive experience of a digital escape room by providing context and guiding participants through the challenges.

Penalty Time: A time delay imposed on users after incorrect answers, designed to discourage guessing and encourage deliberate problem-solving.

Template: A pre-designed structure or layout within the Genially platform that provides a starting point for creating interactive content.ww

Chapter 8
Future Directions and Innovations in Digital Educational Escape Room Games

ABSTRACT

This chapter explores the future directions and innovations in Digital Educational Escape Room (DEER) games, focusing on the integration of emerging technologies such as Virtual Reality (VR), Augmented Reality (AR), Large Language Models (LLMs), and Generative AI. These advancements are revolutionizing educational approaches, offering immersive, interactive, and adaptive learning experiences. DEERs that incorporate VR and AR provide enhanced engagement through fully immersive environments and real-world simulations, fostering deeper understanding and practical application of complex concepts, particularly in STEM education. The introduction of AI-driven tools like LLMs allows for personalized learning pathways and real-time feedback, further improving student outcomes. By focusing on soft skill development—such as problem-solving, critical thinking, and collaboration—DEERs offer a transformative learning approach aligned with the demands of the 21st-century workforce.

DOI: 10.4018/979-8-3693-4219-0.ch008

EMERGING TECHNOLOGICAL ADVANCEMENTS AND INNOVATIONS IN DIGITAL ESCAPE ROOMS

Virtual Reality (VR) and Augmented Reality (AR)

Virtual Reality (VR) and Augmented Reality (AR) are revolutionizing the educational landscape, offering immersive and interactive learning experiences (Algerafi et al., 2023). DEERs that leverage VR and AR technologies can transform educational content into engaging, interactive experiences, thereby enhancing student motivation and learning outcomes (Mystakidis et al., 2021a). VR, in particular, provides a fully immersive environment, allowing learners to explore and interact with digital content in a three-dimensional space, which is especially effective in STEM education (Zataraín-Cabada et al., 2022). An example of such fully immersive environment is Social VR-based 360° spaces (Kalvakolu et al., 2025). AR, on the other hand, blends digital elements with the physical world, creating an enriched learning environment that enhances the understanding of complex concepts (Stolzenberger et al., 2021).

The integration of VR and AR in DEERs not only enhances the engagement and interactivity of learning experiences but also offers new avenues for personalized and adaptive learning (Jantjies et al., 2018). These technologies support the creation of dynamic, contextually rich scenarios that can be tailored to individual learner needs, thus providing a more effective educational tool (Hevko et al., 2023). The use of VR and AR in DEERs aligns with contemporary educational theories that emphasize active learning, problem-solving, and the application of knowledge in real-world contexts (Papanastasiou et al., 2018). The immersive nature of VR allows learners to experience simulations that closely replicate real-world environments, making abstract concepts more tangible and easier to grasp (Matveev et al., 2021). In contrast, AR's ability to overlay digital information onto the physical world enables learners to interact with their surroundings in novel ways, reinforcing their understanding of the subject matter (Kazandis and Pellas, 2019). The educational potential of VR and AR in DEERs extends beyond traditional classroom settings. For instance, these technologies are being utilized in vocational training, where students can engage in simulated environments that closely mimic real-world tasks, thereby enhancing their practical skills and readiness for professional challenges (Mekacher, 2019). Similarly, in environmental education, AR can be used to visualize and interact with complex ecological systems, providing students with a deeper understanding of environmental processes and the impact of human activities on the planet (Patil et al., 2022). Such applications demonstrate the versatility of VR and AR in supporting diverse educational goals and catering to a wide range of learning preferences (Purwaningtyas et al., 2022).

The integration of VR and AR into DEERs is poised to play a significant role in the future of educational assessment (Cristea et al., 2020). These technologies can facilitate the development of innovative assessment tools that measure not only content knowledge but also critical thinking, problem-solving, and decision-making skills (Maas and Hughes, 2020). For example, VR-based DEERs can be designed to assess how well students apply theoretical knowledge to practical situations, providing educators with valuable insights into student learning outcomes (Pantelidis et al., 2018). Similarly, AR-enhanced DEERs can offer real-time feedback and adaptive challenges, allowing for continuous assessment and personalized learning paths (Rajaram and Nebeling, 2022).

Large Language Models and Generative AI

The integration of Large Language Models (LLMs) and Generative AI into DEERs represents a significant evolution in educational technology, promising to revolutionize the way educational content is delivered and experienced (Joseph, 2023; Damaševicius, 2023). These AI-driven technologies have the capacity to generate dynamic, contextually relevant content that can be tailored to meet the specific needs of individual learners (Yan et al., 2023). By leveraging the vast processing power of LLMs, DEERs can offer a more personalized learning journey, where content and challenges are adjusted in real-time based on a learner's progress, understanding, and engagement levels (Tzirides et al., 2023). This level of personalization is crucial in addressing the diverse learning styles and paces of students, thereby enhancing the overall efficacy of the educational process (Abd-alrazaq et al., 2023). The ability of Generative AI to create a virtually limitless variety of scenarios and problems within DEERs ensures that each learner can experience a unique and customized path through the educational material, promoting deeper understanding and retention (Chen et al., 2023).

The potential of LLMs and Generative AI to provide real-time feedback is a game-changer in the realm of digital education (Su et al., 2023). Traditional educational environments often suffer from delays in feedback, which can hinder learning and demotivate students. In contrast, AI-driven DEERs can offer instant, context-sensitive feedback that not only corrects mistakes but also explains the reasoning behind the correct answers, thereby reinforcing learning (Azaiz et al., 2023). This immediate feedback loop is essential in helping students recognize and address their misconceptions promptly, leading to a more effective learning experience (Laato et al., 2023). The adaptive capabilities of AI enable DEERs to modify the difficulty level of tasks in response to a learner's performance, ensuring that students remain challenged but not overwhelmed (Karabacak et al., 2023). This adaptability is key to maintaining student engagement and motivation, which are critical factors in successful learning outcomes (Phung et al., 2023).

Potential implications of the application of LLMs and Generative AI in DEERs for educational methodologies are profound (Lindsay et al., 2023). These technologies are poised to facilitate a shift from one-size-fits-all educational approaches to more individualized learning experiences that cater to the specific needs of each student (Su et al., 2023). In this context, DEERs can serve as an ideal platform for implementing and testing these advanced AI technologies, given their inherently interactive and immersive nature (Gan et al., 2023). The combination of AI-driven content generation and real-time adaptability in DEERs could lead to more effective educational practices that are not only engaging but also deeply personalized, making learning both more accessible and more impactful (Owan et al., 2023).

Expanding Interdisciplinary Applications of Digital Educational Escape Rooms

The expanding interdisciplinary applications of DEERs highlight their adaptability and effectiveness in various educational domains, moving beyond traditional learning environments. DEERs have proven to be highly adaptable tools in a wide range of educational fields, reflecting their inherent versatility and capacity to enhance learning across diverse disciplines (Cheng et al., 2019). The adoption of DEERs in STEM (Science, Technology, Engineering, and Mathematics) education has been particularly impactful, as these environments allow students to engage deeply with complex and abstract concepts through hands-on, interactive experiences (Lestari, 2021). By transforming theoretical knowledge into practical challenges, DEERs in STEM education foster an immersive learning environment where students can explore scientific principles, test hypotheses, and witness the immediate consequences of their actions in a controlled, virtual setting (Cheng et al., 2019). This approach not only strengthens conceptual understanding but also enhances student engagement by making learning more active and participatory (Istiana et al., 2023). The gamified nature of DEERs encourages students to approach problem-solving with creativity and curiosity, driving motivation and promoting a deeper, more sustained interest in STEM subjects.

In the realm of environmental education, DEERs have similarly demonstrated their effectiveness by providing an innovative platform for teaching complex global issues such as climate change (Ouariachi and Wim, 2020). Traditional methods of environmental education often struggle to convey the immediacy and scale of environmental challenges, leading to disengagement or a lack of understanding among students (Shutaleva, 2023). DEERs address this challenge by immersing students in scenarios that simulate real-world environmental dilemmas, requiring them to apply critical thinking and problem-solving skills to navigate these challenges successfully (Ouariachi and Wim, 2020). Through experiential learning, students gain a more profound understanding of environmental issues, as they are not merely passive recipients of information but active participants in the learning process (Markowitz et al., 2018). This method of teaching is particularly effective in fostering a sense of responsibility and urgency about global environmental issues, making the learning experience more impactful and relevant (Ouariachi and Wim, 2020). By engaging students in this way, DEERs not only educate them about the facts of climate change but also empower them to think critically about solutions and the role they can play in addressing these global challenges (Ouariachi and Wim, 2020).

The application of DEERs in media literacy education highlights their broad applicability and effectiveness in promoting essential 21st-century skills such as reasoning (Debchaudhury, 2024) and computational thinking (van Rensburg and Matthee, 2024). In an age where digital media is pervasive, the ability to critically analyze and navigate media content is crucial. DEERs provide an interactive platform where students can engage with scenarios that challenge them to distinguish between credible and misleading information, understand the impact of media on public opinion, and recognize the ethical implications of media production and consumption. This active engagement in the learning process helps students develop critical media literacy skills that are essential for responsible and informed citizenship in the digital age. The use of DEERs in media literacy education exemplifies how these digital tools can be tailored to meet the specific needs of various disciplines while maintaining their core strengths of interactivity, engagement, and experiential learning.

The exploration of DEERs in non-traditional educational settings further highlights their remarkable adaptability and potential to engage a diverse range of learners (Pamungkas et al., 2020). In vocational education, where practical, hands-on experience is crucial, DEERs have emerged as a particularly effective pedagogical tool. For example, in the field of computer network mechanics, DEERs have been utilized to create immersive learning environments that not only impart technical knowledge but also foster essential cognitive, behavioral, and affective skills. Through carefully designed scenarios, students are required to troubleshoot, diagnose, and solve complex network issues within a simulated environment that mirrors real-world challenges (Pamungkas et al., 2020). This approach ensures that students are not merely passive recipients of information, but active problem-solvers who apply theoretical knowledge in practical contexts, thereby enhancing their understanding and retention of the material (Huraj et al., 2022).

The interactive nature of DEERs in vocational training addresses some of the limitations of traditional educational methods, which often rely heavily on lectures and rote memorization (Mayombe, 2023b). By incorporating elements of gamification and experiential learning, DEERs transform the educational experience into one that is more engaging and enjoyable for students (Kladchuen and Srisomphan, 2021). The use of DEERs allows for a more dynamic learning process, where students can experiment, make mistakes, and learn from them in a risk-free environment (Pamungkas et al., 2020). This interactive approach not only enhances the learning experience but also increases student motivation and participation, as learners are more likely to engage deeply with content that is presented in an interesting and challenging manner. The ability to simulate real-world tasks and environments within DEERs provides students with a valuable opportunity to develop practical skills that are directly applicable to their future careers, making these tools well-suited for vocational education (Mayombe, 2023a).

DEERs in non-traditional settings like vocational training demonstrate their capacity to cater to diverse learning styles and needs (Collins et al., 2019). Vocational education often attracts students with varied educational backgrounds and learning preferences, and DEERs offer a flexible platform that can accommodate these differences (Latheesh, 2019). For instance, students who benefit from visual and kinesthetic learning styles can engage with interactive, visually rich environments, while those who excel in problem-solving can tackle complex scenarios that require critical thinking and strategic planning (Pamungkas et al., 2020). This flexibility makes DEERs an inclusive educational tool that can support a broad spectrum of learners, ensuring that each student has the opportunity to succeed according to their individual strengths and preferences (Pamungkas et al., 2020). The use of DEERs in vocational education also underscores the potential of these tools to bridge the gap between theoretical knowledge and practical application (Amolloh et al., 2018). In traditional educational settings, there is often a disconnect between what is taught in the classroom and the skills required in the workplace (Wart et al., 2020). DEERs address this issue by providing a context in which students can directly apply what they have learned in a simulated environment that closely resembles their future professional settings (Pamungkas et al., 2020). This not only reinforces the relevance of their education but also better prepares them for the challenges they will face in their careers (Mayombe, 2023b). By integrating DEERs into vocational training programs, educators can create a more holistic learning experience that prepares students not only for academic success but also for practical, real-world challenges (Mayombe, 2023b).

Future Pedagogical Shifts in Digital Educational Escape Rooms

Shift from Content Delivery to Skill Development

The pedagogical shift within DEERs from a traditional content delivery model based on Bloom's taxonomy to a more skill-focused approach signifies a significant evolution in contemporary educational methodologies (Wirnata et al., 2021). This transition reflects the growing recognition that the development of soft skills—such as critical thinking, problem-solving, and decision-making—is crucial in preparing students for the complexities of the 21st-century workforce and society (Putri and Dwikoranto, 2022). Traditional educational models, which often prioritize the hierarchical acquisition of knowledge from basic recall to complex evaluation, as outlined by Bloom's taxonomy, are increasingly seen as insufficient for equipping learners with the skills necessary to navigate today's dynamic and often unpredictable global environment (Damaševicius, 2023). In contrast, DEERs offer a more holistic and integrated approach, where knowledge is not just accumulated but actively applied in scenarios that mimic real-world challenges (Pujawan et al., 2022).

This pedagogical shift is further reinforced by the interactive and immersive nature of DEERs, which provide an ideal platform for cultivating these essential soft skills (Huilan et al., 2020). In a DEER, learners are not passive recipients of information; instead, they are active participants in a narrative-driven experience that requires them to engage deeply with the material, make decisions, and solve problems in a context that closely mirrors real-life situations (Foo and Foo, 2022). This experiential learning approach is highly effective in promoting the transfer of skills and knowledge to real-world applications, thereby enhancing both the relevance and retention of the content being taught (Sun, 2023). By placing students in scenarios where they must apply their knowledge to overcome challenges, DEERs foster a deeper understanding of the material and encourage learners to think critically about the information they are using (Mystakidis et al., 2021a). This not only aids in the development of problem-solving and decision-making skills but also reinforces the importance of these skills in a practical context (Paralikar et al., 2022).

The shift towards a skill-focused approach in DEERs is in line with contemporary educational theories that advocate for active learning and the application of knowledge in real-world contexts. These theories emphasize that learning is most effective when it involves active engagement with content, where students are encouraged to explore, experiment, and reflect on their experiences. DEERs are well-suited to this approach, as they naturally incorporate elements of exploration and experimentation within their design (Prasad, 2021). For instance, learners may be required to navigate through complex puzzles, work collaboratively to solve problems, or make decisions under pressure—all of which are activities that promote active learning and the practical application of knowledge (Qasrawi and BeniAbdelrahman, 2020). This method of learning is not only more engaging for students but also more effective in helping them develop the skills they will need to succeed in both their academic and professional lives (Foo and Foo, 2022).

The emphasis on developing critical thinking, problem-solving, and decision-making skills within DEERs also aligns with the broader educational objective of preparing students for the challenges of the 21st century. As the demands of the global workforce continue to evolve, there is an increasing need for individuals who can think critically, adapt to new situations, and make informed decisions in complex environments. DEERs, by their very nature, provide a training ground for these skills, offering learners the opportunity to practice and refine their abilities in a controlled, yet challenging, setting. This not only enhances the immediate learning outcomes but also contributes to the long-term development of competencies that are essential for success in a rapidly changing world (Wogu et al., 2019).

Enhancing 21st Century Soft Skills

DEERs have increasingly been recognized as potent tools for the enhancement of essential soft skills, which are critical for success in both professional and personal contexts (Mancheno-Smoak, 2019). These skills, including collaboration, communication, and creative problem-solving, are often underemphasized in traditional educational settings, where the focus tends to be on the acquisition of hard, technical knowledge. As the modern workforce demands individuals who can work effectively in teams, communicate across diverse platforms, and approach problems with innovative solutions, the cultivation of these competencies has become indispensable (Majid, 2021). DEERs, with their immersive and interactive design, provide an ideal environment for fostering these skills. In these scenarios, learners are placed in situations where they must collaborate with others, communicate their ideas clearly and persuasively, and think creatively to overcome challenges and achieve shared goals. This process mirrors real-world professional settings, making DEERs not only an engaging educational tool but also a preparatory ground for future workplace dynamics. A study conducted (Vekua et al., 2020) underscores the significant impact that DEERs can have on the development of these high-demanded competencies. The study highlighted how the application of DEERs in educational settings led to marked improvements in students' ability to collaborate, communicate, and solve problems creatively (Vekua et al., 2020). These findings are particularly noteworthy as they suggest that DEERs do more than just engage students in content learning—they actively contribute to the development of soft skills that are increasingly valued in the global job market. The study's results are a testament to the effectiveness of experiential learning approaches in cultivating these skills. By requiring students to engage in tasks that demand teamwork, negotiation, and innovative thinking, DEERs help to bridge the gap between theoretical knowledge and practical application, preparing students for the complex and collaborative nature of modern professional environments.

The interactive and collaborative nature of DEERs inherently fosters communication skills. In the process of navigating escape rooms, participants must articulate their thoughts, listen to the ideas of others, and build on these ideas to solve problems. This dynamic not only improves their ability to communicate effectively but also enhances their capacity to work as part of a team, understanding and integrating diverse perspectives to achieve a common objective. In today's increasingly interconnected and multicultural work environments, the ability to communicate clearly and collaborate across differences is a critical skill. DEERs provide a safe yet challenging space for students to practice and refine these abilities, contributing to their overall personal and professional development.

The creative problem-solving aspect of DEERs offers a unique advantage over more traditional educational methods. In a DEER, the challenges are often complex and multi-faceted, requiring participants to think outside the box and devise innovative solutions under time constraints. This mirrors real-world scenarios where problems are rarely straightforward and often require creative thinking and adaptability (Tan and Chong, 2023). By engaging in these kinds of activities, students not only enhance their problem-solving skills but also build confidence in their ability to tackle challenges in creative and resourceful ways. This skill is increasingly important in the modern workforce, where the ability to innovate and adapt is highly valued.

The role of DEERs in enhancing soft skills such as collaboration, communication, and creative problem-solving cannot be overstated. As the demands of the professional world continue to evolve, the importance of these skills has become more pronounced, and DEERs offer a compelling way to develop them in an educational context. The findings from studies like that of Vekua et al. (Vekua et al., 2020) provide strong evidence of the efficacy of DEERs in this regard, highlighting their potential to not only engage students in the learning process but also to equip them with the essential skills they need to succeed in their future careers and personal lives. As educational institutions seek to prepare students for the challenges of the 21st century, DEERs represent a powerful tool for bridging the gap between academic learning and practical, real-world skill development.

Innovative DEER Design for Accessibility

Innovations in DEER design are increasingly focusing on enhancing the educational impact while ensuring the experiences are engaging and accessible. A key aspect of this innovation is the integration of digital technologies that cater to a wider range of learning styles and abilities. This includes the use of VR/AR and interactive software platforms that can adapt to the needs of diverse learners. For example, the GDevelop environment has been used to create digital educational escape rooms that offer a variety of tasks and learning activities, suitable for different learning preferences and styles (Huraj and Hudáková, 2022). Such approach to DEER design is ensuring that these educational tools are not only innovative but also inclusive and effective for all learners.

Accessibility in DEERs is a fundamental consideration that ensures these innovative learning environments are inclusive and usable by all learners, including those with disabilities (Cressey, 2020). The commitment to accessibility in DEERs reflects a broader educational imperative to provide equitable learning opportunities for students of varying abilities, recognizing that diversity in the classroom includes a wide range of physical, sensory, and cognitive differences (O'shaughnessy, 2021; Wiesemüller et al., 2024). Designing DEERs that are adaptable and can be tailored to individual needs is essential for creating an inclusive educational experience (Dalton et al., 2019). This requires a thoughtful integration of accessibility features such as audio descriptions for visually impaired students, subtitles for those with hearing impairments, and adjustable difficulty levels to accommodate different learning paces and abilities (Seo and Richard, 2021). These modifications are not merely add-ons but integral components of the DEERs' design, ensuring that all students can fully engage with and benefit from the educational content (Sujathamalini et al., 2022).

Incorporating universal design principles into the creation of DEERs is a crucial strategy for achieving accessibility (Arendale, 2018). Universal design is an approach that aims to make environments usable by all people, to the greatest extent possible, without the need for adaptation or specialized design (Oswald et al., 2018). When applied to DEERs, this means creating learning experiences that are inherently accessible, considering the diverse needs of all potential users from the outset (Nieminen and Pesonen, 2019). For example, a DEER might include flexible input methods that allow students to interact with the content using a mouse, keyboard, voice commands, or touchscreens, depending on their physical abilities (Smith-Carrier et al., 2021). Similarly, providing multiple forms of content delivery—such as visual, auditory, and text-based formats—ensures that students with different sensory preferences or disabilities can access the material in the way that best suits them (Burgstahler, 2022). By adhering to universal design principles, DEERs can accommodate a broad spectrum of learning styles and abilities, thereby reaching a wider audience and fostering a more inclusive learning environment (Craddock et al., 2022).

The prioritization of accessibility in DEER design is not only a matter of compliance with educational standards and legal requirements but also an ethical commitment to educational equity (Dalton et al., 2019). When educators and designers make accessibility a central focus, they contribute to the creation of learning environments that do not marginalize students based on their abilities (O'shaughnessy, 2021). Instead, they promote a culture of inclusivity where all students are given the tools and opportunities to succeed (Sujathamalini et al., 2022). This approach aligns with contemporary educational philosophies that emphasize the importance of meeting students where they are and providing differentiated instruction to support diverse learning needs (Cressey, 2020). In the context of DEERs, this might involve designing challenges that can be completed in multiple ways, allowing students to choose the method that aligns with their strengths while still achieving the same educational outcomes (Oswald et al., 2018).

Accessible DEERs have the potential to enhance the learning experience for all students, not just those with disabilities (Seo and Richard, 2021). Features such as adjustable difficulty levels, customizable interfaces, and multimodal content delivery can benefit a wide range of learners by allowing them to engage with the material in ways that match their individual preferences and learning styles (Smith-Carrier et al., 2021). For example, a student who struggles with reading might benefit from audio descriptions, while another who learns best through hands-on interaction might appreciate the option to manipulate virtual objects using a touchscreen (Burgstahler, 2022).

By designing DEERs that are flexible and responsive to the diverse needs of all students, educators can create more engaging and effective learning experiences that support deeper understanding and retention of the material (Craddock et al., 2022).

Ensuring accessibility in DEERs is a critical aspect of creating inclusive and equitable educational experiences (Arendale, 2018). By designing DEERs that are adaptable and aligned with universal design principles, educators and designers can make these learning environments accessible to the widest possible audience, regardless of physical or cognitive abilities (Oswald et al., 2018). This commitment to accessibility not only enhances the educational outcomes for students with disabilities but also enriches the learning experience for all students, fostering a more inclusive and supportive educational environment (Dalton et al., 2019). As DEERs continue to evolve and expand their reach, prioritizing accessibility will be essential to ensuring that all learners have the opportunity to engage with and benefit from these innovative educational tools (Sujathamalini et al., 2022).

CROSS-INDUSTRY COLLABORATIONS IN DEER-BASED EDUCATION

Cross-industry collaborations in DEERs have potential, especially in STEM education and climate change awareness. Studies indicate that DEERs offer immersive and experiential learning, which is valuable in addressing global issues like climate change (Ouariachi and Wim, 2020). The concept of DEERs marks a significant evolution in educational methodologies, blending the experience of gaming with structured learning. DEERs are immersive, problem-solving experiences where participants navigate challenges in a virtual environment to achieve specific educational objectives. This innovative approach extends beyond traditional education, offering potential applications across various industries. The integration of gamification in learning through DEERs provides an engaging platform that can capture the essence of real-world scenarios, thereby enhancing the learning experience (Queirós et al., 2022).

In the realm of cross-industry collaborations, DEERs stand out for their versatility and adaptability. They are not confined to any single field, making them an excellent tool for interdisciplinary learning. Industries ranging from computer network mechanics to climate change education have adopted DEERs to convey complex concepts in an interactive and engaging manner. For instance, in the field of computer network mechanics, DEERs have been used to enhance cognitive, behavioral, and affective skills of students (Huraj and Hudáková, 2022). Similarly, in the context of climate change education, DEERs provide experiential and immersive learning experiences that foster problem-solving and critical thinking skills (Ouariachi and Wim, 2020).

The interdisciplinary nature of DEERs also plays a crucial role in enhancing educational outcomes. By merging elements from different disciplines, DEERs encourage a holistic learning approach. This integration of various fields into a single platform allows learners to develop a comprehensive understanding of complex subjects (Almeida and Simões, 2019). It also fosters collaboration and teamwork, as participants from diverse backgrounds come together to solve problems and achieve common goals (Parody et al., 2022). The adaptability of DEERs to different educational settings and objectives makes them a potent tool for learning and development across a wide spectrum of industries (Saleem et al., 2021). DEERs open up multiple possibilities for cross-industry collaboration and interdisciplinary learning. Their ability to simulate real-world scenarios in an engaging virtual environment makes them an invaluable tool in modern education.

Technological Integration and Innovation

The integration of VR and AR technologies in DEERs offers immersive and engaging learning experiences. VR allows users to immerse themselves in a completely digital environment, enhancing the depth and realism of educational content. This immersive experience is particularly effective in fields like STEM education, where complex concepts can be visualized and interacted with in a virtual three-dimensional space. Studies have shown positive perceptions among teachers regarding the use of VR escape rooms for STEM education, highlighting their potential to support creative, student-centered teaching methods (Mystakidis et al., 2021b). AR, on the other hand, blends digital elements with the real world, offering a unique way of enhancing the learning experience by overlaying digital information onto physical objects or environments.

The creation and implementation of DEERs are significantly enhanced by the availability of various digital tools and platforms, which play a crucial role in making these innovative learning environments accessible and effective across a range of educational contexts. These technologies offer educators the necessary resources to design, customize, and implement DEERs tailored to specific educational goals and the diverse needs of learners. The use of platforms such as GDevelop and Escape This! Canvas, a visual inquiry tool that supports the process of designing educational escape games (Eulenstein et al., 2025), exemplifies the effectiveness of these tools in vocational education, where they have been employed to create immersive and interactive learning experiences. GDevelop, in particular, stands out as a robust tool for developing DEERs, allowing educators to incorporate a wide variety of tasks and activities that align with vocational training objectives. This platform's user-friendly interface and extensive features enable educators to design complex, scenario-based challenges that not only engage students but also enhance their practical skills and theoretical understanding (Huraj and Hudáková, 2022).

The adaptability of these digital platforms is instrumental in overcoming some of the key barriers to the widespread adoption of DEERs in education. One of the significant challenges in implementing DEERs on a broader scale is the need for customizable and versatile design options that can be tailored to fit different educational objectives and learning styles. Digital tools like GDevelop address this challenge by offering flexibility in design, enabling educators to create DEERs that are not only pedagogically sound but also tailored to the unique needs of their student populations. This customization is essential in ensuring that DEERs are effective across various educational settings, from primary and secondary education to higher education, vocational training and life-long training (Cujdíková and Vankúš, 2024). By providing educators with the ability to design DEERs that align with specific curriculum goals and student learning outcomes, these platforms play a pivotal role in the successful integration of DEERs into mainstream education (Sidekerskiene and Damaševicius, 2023).

The use of digital platforms in the creation of DEERs also facilitates the incorporation of innovative educational technologies, such as VR and AR, into the learning experience. These technologies can be seamlessly integrated into DEERs to create more immersive and interactive environments that enhance student engagement and learning. For instance, by using VR and AR within DEERs, educators can create realistic simulations of real-world scenarios, allowing students to practice skills and apply knowledge in a controlled, yet highly realistic, setting. This integration not only enriches the learning experience but also helps to bridge the gap between theoretical knowledge and practical application, making DEERs a more effective educational tool.

The flexibility and versatility offered by digital tools like GDevelop also contribute to the scalability of DEERs, making it possible to adapt and implement these educational experiences across different cultural and institutional contexts. As educational institutions around the world continue to explore innovative teaching methods, the ability to customize DEERs to meet local educational standards, cultural expectations, and technological capacities is crucial. Platforms that support the easy adaptation of DEER content ensure that these learning environments can be effectively deployed in a wide range of educational systems, thereby promoting the global dissemination of DEERs as a cutting-edge educational methodology.

The role of digital tools and platforms in the creation and implementation of DEERs is indispensable. These technologies not only provide educators with the means to design and execute engaging and effective DEERs but also help to overcome the challenges associated with their widespread adoption in diverse educational settings. By offering customizable, versatile, and scalable design options, platforms like GDevelop ensure that DEERs can be tailored to meet the specific needs of different educational objectives and learning styles, making them a powerful tool in the future of education. As the adoption of DEERs continues to grow, the contribution of these digital platforms will be essential in driving their development and ensuring their success in enhancing educational outcomes worldwide.

Enhancing Soft Skills and Team Dynamics

DEERs have been recognized as an effective means of developing high-demand soft skills in learners. These interactive and immersive environments challenge participants to employ and enhance skills such as critical thinking, problem-solving, and decision-making. A study (Vekua et al., 2020) highlighted the effectiveness of escape rooms in fostering these competencies, which are necessary for graduates to solve future professional tasks and for successful socialization. The design of

DEERs, which often involves complex problem-solving under time constraints, naturally promotes the development of these skills. The collaborative nature of escape rooms requires participants to engage in teamwork, enhancing interpersonal skills and cooperation.

DEERs also promote effective communication and collaboration among participants, especially in professional education. They create scenarios where participants from diverse backgrounds must work together to achieve a common goal, thereby enhancing their communication skills and ability to collaborate effectively. This aspect of DEERs is particularly beneficial in interdisciplinary educational settings, where students from different fields come together, bringing varied perspectives and expertise. The collaborative challenges in DEERs provide a platform for these diverse participants to learn from each other and to appreciate the value of different viewpoints and skill sets.

Challenges in Cross-Industry Collaborations through Digital Educational Escape Rooms

The implementation of DEERs across various industries presents several significant challenges, particularly when it comes to aligning educational objectives with the diverse and often specialized needs of different sectors. Each industry, whether it be climate change education, healthcare, finance, or technology, operates under distinct goals, regulatory frameworks, and educational standards. This diversity requires a careful calibration of DEER content to ensure that it remains relevant, engaging, and effective for each specific audience. For example, in the realm of climate change education, DEERs are designed to foster experiential learning and enhance problem-solving skills (Ouariachi and Wim, 2020) with promising applications in environmental sustainability studies (Repetto et al., 2024). In marketing education, DEERs contribute with experiential learning and interactive pedagogical approaches (Shah et al., 2025). In programming education, DEERs support students' interest and facilitate active learning (Trigo and Antunes, 2024; Queiros et al., 2024). The challenge lies in making this content resonate with varied stakeholders, from students and educators to industry professionals and policymakers, each of whom may have different levels of expertise and perspectives on climate issues (Ouariachi and Wim, 2020). Tailoring the educational content to meet these diverse needs without diluting its core objectives requires a deep understanding of the specific sector and its unique demands.

Another critical challenge in cross-industry collaborations through DEERs is the complexity involved in measuring learning outcomes, particularly when it comes to soft skills development and long-term knowledge retention. Unlike traditional educational assessments, which often rely on quantifiable metrics such as test scores or grades, the success of DEERs is frequently gauged by more nuanced indicators, such as improvements in critical thinking, teamwork, creativity, and decision-making. These soft skills are inherently difficult to quantify, making it challenging to assess the true impact of DEERs on learners. The immersive and interactive nature of DEERs complicates the process of evaluating long-term retention of knowledge, as the learning experience is often personalized and dynamic, varying from one participant to another. This variability makes it difficult to establish standardized assessment methods that can be applied across different industries and educational contexts. Developing robust, scalable frameworks for measuring the efficacy of DEERs in enhancing both soft and hard skills remains a significant hurdle in their broader adoption and implementation (Hakshurian, 2023).

The lack of generic software tools for creating versatile digital escape room designs poses a substantial barrier to the widespread use of DEERs across industries such as energy industry (Arbesser-Rastburg et al., 2024). While some industries may have the resources to develop customized DEERs tailored to their specific needs, many do not, resulting in a reliance on generic or less flexible tools that may not fully capture the educational objectives or engagement potential of DEERs. The development of versatile, user-friendly software platforms that allow for the easy creation, customization, and deployment of DEERs across various sectors is essential for overcoming this challenge. Such tools would need to accommodate the diverse content requirements, interactive elements, and assessment mechanisms needed to address the specific goals of different industries. Creating such software requires significant investment in research and development, as well as collaboration between educational experts, industry leaders, and technology developers. Without these tools, the adoption of DEERs may be limited to industries with the capacity to invest in bespoke solutions, thereby restricting their potential impact on a wider scale (Huraj and Hudáková, 2022).

The challenge of fostering effective collaboration between educational institutions and industry partners cannot be overlooked. Cross-industry collaborations often involve stakeholders with varying expectations, priorities, and levels of commitment to the educational goals of DEERs. Aligning these differing perspectives to create a cohesive and impactful DEER experience requires strong communication, clear objectives, and a shared vision for the educational outcomes. There may be challenges related to intellectual property, data privacy, and the integration of proprietary content, all of which need to be carefully managed to ensure successful partnerships. Building trust and establishing effective channels of communication between educators and industry partners is critical for overcoming these challenges and realizing the full potential of DEERs in cross-industry educational initiatives (Makri et al., 2021).

Opportunities for DEERs in Educational Innovation and Professional Training

Despite the challenges associated with their implementation, DEERs present substantial opportunities for driving educational innovation and enhancing professional training. DEERs offer a dynamic and interactive platform for experiential learning, which is increasingly recognized as a critical component of effective education. Unlike traditional methods that often rely on passive learning, DEERs engage learners actively by placing them in scenarios that require problem-solving, critical thinking, and collaboration. This hands-on approach caters to diverse learning styles, making it possible to reach a wider range of students and professionals who may not thrive in conventional educational environments. The adaptability of DEERs allows them to be tailored to specific educational goals, whether in STEM education, where complex concepts can be explored in a controlled environment, or in corporate training, where soft skills such as leadership, communication, and teamwork can be honed through simulated real-world challenges (Mystakidis et al., 2021b).

The versatility of DEERs extends their applicability across various educational and professional contexts, providing a flexible tool that can be customized to meet the needs of different learning objectives. In the context of STEM education, DEERs can be designed to simulate scientific experiments, engineering challenges, or mathematical puzzles, allowing students to apply theoretical knowledge in practical, immersive settings. This not only reinforces learning but also enhances retention by making abstract concepts more tangible (Huraj and Hudáková, 2022). Similarly, in professional training, DEERs can replicate workplace scenarios, enabling participants to develop and refine essential soft skills. For instance, a DEER could simulate a crisis management situation, requiring participants to work together under pressure to find a solution. Such experiences are invaluable in preparing individuals for the complexities of modern professional environments, where the ability to think critically, collaborate effectively, and make informed decisions is paramount (Mystakidis et al., 2021b).

The integration of advanced technologies such as Virtual Reality (VR) and Augmented Reality (AR) significantly enhances the potential of DEERs to create more immersive and impactful learning experiences. VR and AR technologies can transform DEERs into fully immersive environments where learners can interact with the content in three-dimensional space, offering a level of engagement that traditional learning methods cannot match. For example, in a DEER focused on medical training, VR could be used to simulate a surgical procedure, allowing trainees to practice and perfect their skills in a risk-free environment. This kind of immersive learning experience not only improves the acquisition of practical skills but also builds confidence, as learners can repeatedly practice tasks until they achieve mastery (Chang and Hwang, 2021).

As educational institutions and industries continue to recognize the value of gamification in learning, DEERs are poised to play a more significant role in educational reform and professional development. Gamification elements such as challenges, rewards, and progression systems make learning more engaging and motivate learners to persist through difficulties. In the context of professional development, gamified DEERs can be particularly effective in motivating employees to participate in training programs that might otherwise be perceived as tedious or irrelevant. By making training sessions more interactive and enjoyable, DEERs can enhance the learning experience and lead to better outcomes, both in terms of knowledge retention and the application of skills in the workplace (Moseikina et al., 2022).

DEERs offer significant opportunities for innovation in education and professional training. Their ability to cater to different learning styles, their adaptability across various contexts, and their potential for integration with advanced technologies such as VR and AR make them a valuable tool for educators and trainers alike. As the demand for more engaging and effective learning experiences continues to grow, DEERs are likely to become increasingly important in educational reform and professional development, helping to prepare learners and professionals for the challenges of the 21st century. By harnessing the full potential of DEERs, educational institutions and industries can create learning environments that are not only more immersive and interactive but also more effective in achieving their educational and training objectives (Kim et al., 2023).

EVOLVING DEER DESIGN TRENDS AND PLAYER EXPECTATIONS

The evolution of DEER design trends focuses on enhancing cognitive, behavioral, and affective skills through engaging and versatile design. Research indicates a need for tools supporting the creation and adoption of digital educational escape rooms in teaching (Huraj and Hudáková, 2022).

Innovations in DEER Design

Recent innovations in digital technologies have greatly influenced the design and application of DEERs. The integration of advanced software tools and platforms allows for more dynamic, interactive, and engaging learning experiences. For instance, the use of platforms like GDevelop has enabled educators to create DEERs with a variety of tasks and learning activities, tailored to the specific needs of different educational settings (Huraj and Hudáková, 2022). These advancements facilitate the creation of DEERs that are not only pedagogically sound but also technologically sophisticated, enhancing the overall learning experience.

The integration of VR and AR technologies in DEERs marks a significant step forward in creating more immersive and interactive learning environments. VR technology allows learners to engage with content in a fully immersive three-dimensional virtual world, enhancing the depth and realism of educational experiences. This is particularly effective in fields such as STEM education, where complex concepts can be visualized and interacted with in a virtual space (Mystakidis et al., 2021b). AR, on the other hand, overlays digital information onto the physical world, providing a unique blend of real and virtual elements that enhance the learning process. These technologies contribute significantly to the metaverse design rooms (Damaševicius and Sidekerskiene, 2023), where virtual and augmented realities merge to create highly engaging and effective educational environments.

Adapting to Changing Player Expectations

The preferences and expectations of modern learners, especially those belonging to Generation Z, are significantly influenced by their immersion in a technology-rich environment. This demographic exhibits a preference for learning experiences that are not only educational but also engaging and interactive. Digital learning environments, including DEERs, are adapting to these changing expectations by integrating more dynamic and interactive elements. Studies show that learners now expect a blend of educational rigor and entertainment, often referred to as "edutainment," which DEERs can effectively provide. The integration of social and collaborative elements in these environments is also crucial, as it aligns with the social networking and communication platforms that are a significant part of these learners' lives (Sharma and Monteiro, 2012).

To meet the diverse and evolving needs of Generation Z learners, DEERs are increasingly incorporating advanced customization and personalization features (Dittman et al., 2021). This approach allows learners to engage with content that is tailored to their individual learning styles, preferences, and pace. The use of adaptive learning technologies and artificial intelligence in DEER design can facilitate this customization, offering a more personalized learning experience. These technologies can adjust the difficulty level of tasks, offer personalized feedback, and even alter the narrative of the escape room based on the learner's progress and responses. The aim is to create an inclusive learning environment that caters to a wide range of learners, including those with different abilities and learning preferences.

Increasing Role of Gamification in Digital Educational Escape Rooms

Importance of Advanced Gamification Elements

The incorporation of advanced gamification elements within DEERs is crucial for elevating student engagement, motivation, and overall learning outcomes. Gamification, which involves the application of game-design elements in non-game contexts, has become a powerful tool in education, transforming traditional learning environments into dynamic, interactive spaces that resonate more deeply with today's learners (Puritat, 2019). By integrating elements such as points, badges, leaderboards, and challenges, DEERs can effectively tap into the motivational structures that make games so compelling, thereby enhancing the educational experience (Alasmari, 2020; Dicheva et al., 2020). These gamified components introduce a competitive and rewarding dimension to learning, encouraging students to actively participate, persist through challenges, and strive for mastery (Huraj and Hudáková, 2022). This shift from passive reception to active engagement is key to fostering a deeper, more sustained interaction with the educational content, which is essential for effective learning (Denny et al., 2018).

The impact of gamification in DEERs extends beyond mere engagement; it also significantly improves learning outcomes by leveraging the psychological principles of motivation and reward (Kim and Castelli, 2021). For instance, the inclusion of points and badges provides immediate feedback and recognition for students' efforts, which can boost their intrinsic motivation (Ortiz-Rojas et al., 2019). Leaderboards introduce a social and competitive aspect that can spur students to excel, while challenges and quests offer structured goals that guide learning in a purposeful direction (Tasadduq et al., 2021). These elements not only make learning more enjoyable but also create a sense of accomplishment and progress, which is critical for maintaining motivation over time (Huraj and Hudáková, 2022). Research has shown that when students are engaged through gamification, they are more likely to retain information and apply it in meaningful ways (Dicheva et al., 2020). For example, in maritime higher education institutions, the use of gamified digital escape rooms—featuring quests, storytelling, and electronic testing—was found to significantly enhance both the learning process and student success rates (Huraj and Hudáková, 2022). Compared to traditional learning methods, students participating in these gamified activities demonstrated better knowledge retention and a higher level of academic achievement (Denny et al., 2018; Tasadduq et al., 2021; Voloshynov et al., 2019).

Advanced gamification elements in DEERs are particularly effective in promoting the development of critical thinking, problem-solving, and collaborative skills. The challenges and puzzles inherent in escape rooms require students to think creatively, analyze information critically, and work together to overcome obstacles (Denny et al., 2018). By embedding these cognitive tasks within a game-like framework, DEERs not only make learning more interactive but also mirror real-world scenarios where such skills are essential (Puritat, 2019). This experiential learning approach ensures that students are not just passive consumers of information but active participants in their own education, honing skills that are directly transferable to professional and everyday contexts (Huraj and Hudáková, 2022). The immediate and tangible rewards provided by gamification—whether through virtual badges, points, or the satisfaction of solving a complex puzzle—reinforce these skills and encourage students to apply them in increasingly sophisticated ways (Dicheva et al., 2020).

The flexibility of gamification allows DEERs to be customized to meet the specific learning objectives of diverse educational settings. Educators can adjust the difficulty of challenges, the structure of rewards, and the nature of the interactions to align with the goals of the curriculum and the needs of the students (Kim and Castelli, 2021). This adaptability makes gamification a versatile tool that can be used across various disciplines, from STEM education to humanities and vocational training (Tasadduq et al., 2021). In each case, the use of gamification transforms the learning process from a task-oriented activity into an immersive experience that fosters deeper engagement and more meaningful learning (Ortiz-Rojas et al., 2019). The positive outcomes associated with gamification in DEERs—improved motivation, better retention, and enhanced skill development—underscore its importance as a foundational element in modern educational design (Moseikina et al., 2022).

The importance of advanced gamification elements in DEERs cannot be overstated. These elements play a pivotal role in making learning more engaging, interactive, and effective by harnessing the motivational power of games (Huraj and Hudáková, 2022). Through the strategic use of points, badges, leaderboards, and challenges, DEERs create a learning environment that is not only enjoyable but also conducive to deep learning and skill acquisition (Denny et al., 2018). As educational institutions continue to explore innovative methods for enhancing student outcomes, the integration of advanced gamification techniques in DEERs stands out as a highly effective approach, capable of transforming the educational experience and better preparing students for the challenges of the future (Alasmari, 2020).

Complex Storytelling and Culturally-Situated Narrative Scenarios

Beyond the incorporation of traditional gamification elements, DEERs are increasingly leveraging complex storytelling and culturally-situated narrative scenarios to deepen learner engagement and educational impact. These narratives are crafted not only to entertain but to resonate profoundly with the values, concerns, and lived experiences of Generation Z learners, who are characterized by their heightened awareness of social issues and modern technology issues such as cybersecurity (Spatafora et al., 2024) and their desire for content that reflects their realities and aspirations. By embedding DEERs within rich, story-driven frameworks that tackle real-world challenges, educators can create learning experiences that are both relatable and transformative. This approach goes beyond mere knowledge transfer; it seeks to engage students on an emotional and intellectual level, fostering a deeper connection to the material and encouraging them to reflect on the broader implications of what they are learning (Paulauskas et al., 2023).

The use of complex storytelling in DEERs serves as a powerful pedagogical tool, as it not only captures the imagination of learners but also situates educational content within contexts that require critical thinking and ethical decision-making (Ouariachi and Wim, 2020). For instance, narrative scenarios might explore themes such as climate change, social justice, or global health, challenging students to solve problems that mirror the complexities of these issues in the real world. Through this immersive approach, learners are prompted to consider diverse perspectives, weigh the consequences of their actions, and develop empathy for those affected by the issues at hand (Huraj and Hudáková, 2022). Such experiences are particularly valuable in cultivating the critical thinking skills necessary for navigating an increasingly complex and interconnected world. By confronting students with scenarios that require them to engage with real-world dilemmas, DEERs not only enhance cognitive development but also foster a sense of social responsibility and ethical awareness (Mystakidis et al., 2021b).

The integration of culturally-situated narratives within DEERs ensures that these educational tools are inclusive and reflective of the diverse backgrounds and experiences of learners (Ouariachi and Wim, 2020). Generation Z, in particular, values content that is socially conscious and culturally relevant, making it imperative that DEERs address these preferences (Frania, 2021). Culturally-situated narratives can include stories that highlight issues of racial equality, gender rights, and environmental sustainability, all of which resonate strongly with contemporary youth (Huraj and Hudáková, 2022). By incorporating such themes into the fabric of DEERs, educators can create learning environments that are not only engaging but also empowering, providing students with the knowledge and tools they need to advocate for change in their communities and beyond (Frania, 2021). This approach also supports the development of global citizenship by encouraging learners to think critically about their role in addressing the challenges faced by societies around the world (Mystakidis et al., 2021b).

The effectiveness of DEERs in fostering critical thinking and empathy through storytelling is further enhanced by the interactive nature of these educational experiences (Ouariachi and Wim, 2020). Unlike traditional classroom settings where students might passively receive information, DEERs require active participation, with learners making decisions that directly influence the outcome of the narrative (Huraj and Hudáková, 2022). This interactivity transforms the learning process into a dynamic and participatory experience, where students are not merely observers but co-creators of the story (Frania, 2021). As they navigate through complex scenarios, learners must consider the ethical implications of their choices, collaborate with peers to solve problems, and adapt their strategies in response to new information (Mystakidis et al., 2021b). This process not only reinforces the educational content but also helps students develop a deeper understanding of the societal issues being explored, as well as the critical skills needed to address them in real life (Frania, 2021).

The incorporation of complex storytelling and culturally-situated narrative scenarios in DEERs represents a significant advancement in educational design, offering a more engaging, relevant, and impactful learning experience (Mystakidis et al., 2021b). By aligning the content of DEERs with the values and experiences of Generation Z learners, educators can create educational tools that are both inclusive and socially conscious, fostering critical thinking, empathy, and a sense of global responsibility (Ouariachi and Wim, 2020). As DEERs continue to evolve, the integration of rich narratives that address real-world challenges will remain a crucial element in their ability to engage and empower the next generation of learners (Huraj and Hudáková, 2022).

Challenges in Contemporary DEER Design

One of the significant challenges in DEER design is incorporating complex technologies like digital twins—virtual replicas of physical entities or systems (Damaševicius and Zailskaite-Jakšte, 2024). Integrating digital twins in DEERs can provide a highly realistic and interactive environment, but it also poses challenges in terms of technical expertise, resource allocation, and ensuring the virtual environment aligns accurately with real-world scenarios. Similarly, the incorporation of microlearning—delivering educational content in small, specific bursts—into DEERs presents challenges in terms of content design and ensuring engagement. DEERs need to balance the brief yet impactful nature of microlearning with the immersive and narrative-driven aspects of escape rooms. A study by Huang et al. (Huang et al., 2020) on applying digital escape rooms in elementary science teaching highlighted the effectiveness of such methods in improving learning motivation and problem-solving abilities, but also pointed out the need for careful design and implementation.

To address these challenges, DEER designers can adopt several strategies. For the integration of digital twins, a focus on user-centered design is essential. This involves understanding the end-user's needs and creating virtual environments that are not only technologically advanced but also intuitive and relevant to the learners.

Leveraging partnerships with technology experts and organizations can provide the necessary support and resources for implementing sophisticated technologies like digital twins. For microlearning (Damaševicius and Sidekerskiene, 2023), the key is to design content that is both engaging and educationally sound. Gamification techniques, story-driven narratives, and interactive challenges can be used to make microlearning components within DEERs more compelling and effective. Regular feedback and assessment mechanisms can also be integrated to track learning progress and ensure the microlearning objectives are being met.

GLOBAL PERSPECTIVES AND EMERGING CULTURAL INFLUENCES

DEERs are becoming a global phenomenon, with various cultural influences shaping their design and implementation. Studies indicate their potential in diverse educational settings, suggesting a culturally inclusive approach to their development.

Designing DEERs for Diverse Educational Contexts

Adoption and Adaptation of DEERs in Different Countries

The successful adoption and adaptation of DEERs in various countries and educational systems hinge on a deep understanding of the unique educational, cultural, and technological landscapes of each region (Makri et al., 2021). Implementing DEERs globally is not a one-size-fits-all endeavor; it necessitates a nuanced approach that takes into account the distinct educational norms, cultural values, and available technological resources that characterize different countries (Hakshurian, 2023). Each educational system has its own set of standards, goals, and pedagogical practices that must be considered when integrating DEERs into the curriculum (Repetto et al., 2023). This careful alignment with local educational objectives is crucial for ensuring that DEERs not only fit seamlessly into the existing educational framework but also enhance the learning experience in a meaningful and contextually relevant way (López-Pernas et al., 2021).

Studies have demonstrated that DEERs, when thoughtfully adapted, can significantly improve learning outcomes across a variety of educational settings (Moffett et al., 2023). For example, in maritime higher education institutions, DEERs have been employed to develop students' cognitive, behavioral, and affective skills in a manner that is both engaging and effective (Huang et al., 2020). These adaptations involve more than just superficial changes; they require a comprehensive rethinking of the content to ensure it aligns with the specific educational goals of the institution (Borrás-Gené et al., 2022). This might include modifying the language used in the DEERs to better suit the linguistic capabilities of the students, incorporating culturally relevant scenarios that resonate with the students' lived experiences, and ensuring that the educational objectives of the DEERs are consistent with the curriculum's focus on practical, hands-on learning (Voloshynov et al., 2019). Such localized adaptations are essential for making DEERs not only accessible but also impactful in diverse educational contexts.

Cultural adaptation is another critical component of successfully implementing DEERs in different countries (Veldkamp et al., 2020). Educational content that resonates with students in one cultural context may not be as effective in another due to differences in values, traditions, and social norms (Staneva et al., 2023). Therefore, it is essential to adapt DEERs to reflect the cultural realities of the students they are designed to serve (Vekua et al., 2020). This can involve incorporating culturally relevant examples, using culturally appropriate language, and designing scenarios that are meaningful within the local context (Kotzebue et al., 2022). By doing so, DEERs can foster a deeper connection between the learners and the material, making the educational experience more engaging and effective (Karageorgiou et al., 2019). Culturally adapted DEERs can help in promoting intercultural understanding and respect, as they provide students with the opportunity to explore and appreciate different cultural perspectives through immersive and interactive learning experiences.

Technological capabilities also play a significant role in the adaptation of DEERs (Dutra et al., 2023). The varying levels of technological infrastructure across different countries can present challenges to the implementation of DEERs (Moffett et al., 2023). In regions where access to advanced technology is limited, it may be necessary to adapt DEERs to work on less sophisticated platforms or to provide alternative methods of participation that do not rely heavily on high-tech solutions (Jacobs et al., 2023). This could involve simplifying the technological requirements of DEERs or developing low-tech versions that can be easily implemented in resource-constrained environments (Ouariachi and Wim, 2020). By considering these technological limitations and making appropriate adjustments, educators can ensure that DEERs are accessible to a broader range of students, regardless of their technological environment (Lathwesen and Belova, 2021).

The adoption and adaptation of DEERs in different countries require a multifaceted approach that takes into account local educational norms, cultural values, and technological capabilities (Howard et al., 2023). By carefully tailoring DEERs to fit the specific needs and contexts of different educational systems, educators can enhance the relevance and effectiveness of these innovative learning tools (Borrás-Gené et al., 2022). Such adaptations not only improve learning outcomes but also promote greater accessibility and inclusivity in global education (Mystakidis and Christopoulos, 2022). As DEERs continue to gain traction worldwide, the ability to adapt these tools to diverse educational contexts will be key to their long-term success and impact in the global educational landscape (Makri et al., 2021).

Application of DEERs in Various Cultural Settings

The application of DEERs in diverse cultural settings demands a nuanced approach that carefully considers local cultural norms, values, and educational expectations (Ouariachi and Wim, 2020). The effectiveness of DEERs hinges on their ability to engage students by presenting scenarios and narratives that are not only educational but also culturally resonant (Borrás-Gené et al., 2022). This requires designers to delve deeply into the cultural contexts of their target audience, ensuring that the DEERs they create are not only relevant but also respectful of the cultural specifics that characterize different societies (Makri et al., 2021). By doing so, DEERs can transcend mere educational tools to become powerful instruments for fostering deeper connections between learners and the subject matter, thereby enhancing the overall learning experience (Sidekerskiene and Damaševicius, 2023).

Incorporating culturally relevant scenarios into DEERs is particularly important in subjects like climate change education, where the impact of environmental issues is perceived differently across various regions (Ouariachi and Van Dam, 2022). For instance, a DEER designed for students in a coastal community might focus on the immediate threats posed by rising sea levels, whereas one designed for students in a desert region might emphasize water scarcity and its implications for agriculture (Liapi, 2020). By tailoring the content to reflect the specific environmental challenges faced by the learners' communities, DEERs can make the learning experience more immediate and relatable (Chineka and Yasukawa, 2020). This contextualization not only helps students grasp the relevance of the content to their own lives but also fosters a greater sense of responsibility and urgency regarding global environmental issues (Ouariachi and Wim, 2020). Such culturally sensitive approaches are crucial for ensuring that DEERs do more than just impart knowledge—they also inspire action and empower students to contribute to solutions in their own contexts.

Culturally tailored DEERs can enhance the depth and quality of student engagement by leveraging familiar cultural references and storytelling traditions (Purnama et al., 2022). For example, in cultures where oral storytelling is a significant aspect of tradition, DEERs might incorporate narrative-driven puzzles that align with local myths, legends, or historical events (Borrás-Gené et al., 2022). This not only makes the learning process more engaging by connecting with students on a cultural level but also preserves and celebrates cultural heritage within the educational framework (Sidekerskiene and Damaševicius, 2023). By embedding educational content within culturally relevant narratives, DEERs can facilitate a more profound and enduring understanding of the material, as students are more likely to remember and apply knowledge that is presented in a familiar and meaningful context (Ouariachi and Wim, 2020).

The application of DEERs in various cultural settings also requires careful consideration of local educational expectations and pedagogical practices (Dutra et al., 2023). Educational systems around the world differ in their approaches to teaching and learning, with some emphasizing rote memorization and others prioritizing critical thinking and problem-solving (Acosta and Carballo, 2022). To be effective, DEERs must align with these pedagogical expectations while also introducing innovative learning strategies (Makri et al., 2021). For example, in an educational context that traditionally emphasizes group learning and collaboration, DEERs might be designed to include team-based challenges that require students to work together to solve problems (Fotaris and Mastoras, 2019). In contrast, in a system that values individual achievement, DEERs might include more individualized tasks that allow students to demonstrate their knowledge independently (Jacobs et al., 2023). By aligning DEER design with local educational practices, educators can ensure that these tools are not only accepted within the educational system but also enhance the overall learning process (Christopoulos et al., 2022).

The application of DEERs in various cultural settings involves a deliberate and informed approach to design, one that prioritizes cultural relevance and educational effectiveness (Makri et al., 2021). By tailoring DEERs to the specific cultural and educational contexts of their target audiences, designers can create more engaging, meaningful, and impactful learning experiences (Veldkamp et al., 2020). These culturally sensitive DEERs not only improve student comprehension and engagement but also promote a deeper connection between learners and the content, ultimately leading to more effective and enduring educational outcomes (Huang et al., 2020). As DEERs continue to expand their reach globally, the ability to adapt these tools to diverse cultural settings will be key to maximizing their educational potential (Ouariachi and Wim, 2020).

Cultural Sensitivity in DEER Design

Importance of Culturally Sensitive Content in Digital Educational Escape Rooms

The integration of culturally sensitive content in the design of DEERs is not merely an adjunct consideration but a fundamental requirement for their success in a global educational landscape (Viken et al., 2021). The growing use of DEERs across various cultural contexts necessitates a design approach that goes beyond the superficial translation of content (Kartal et al., 2018). It demands a deep understanding of the diverse values, norms, and learning styles that characterize different cultures (van Boeijen et al., 2019). This approach ensures that DEERs resonate with learners from various backgrounds, making the educational experience not only accessible but also meaningful and effective (Garcia and Pantao, 2021). When DEERs are designed with cultural sensitivity, they can bridge the gap between global content and local relevance, thereby enhancing their overall educational impact (Baron, 2023).

Culturally sensitive content involves crafting scenarios, challenges, and narratives that reflect the lived experiences, historical contexts, and environmental realities of the target audience (Segura-Robles and Parra-González, 2019). This is particularly important in subjects like climate change education, where the impact of environmental issues can vary significantly across regions (Lal et al., 2021). A study on the use of educational escape rooms for teaching climate change demonstrated that when scenarios were tailored to the specific cultural and environmental contexts of the learners, the educational outcomes were markedly improved (Yli-Panula et al., 2022). By incorporating culturally relevant examples and local environmental challenges, the DEERs became more relatable to the participants, thereby increasing their engagement and facilitating a deeper understanding of the subject matter (Ouariachi and Wim, 2020). This not only underscores the importance of cultural sensitivity in content design but also highlights the role of DEERs in fostering global awareness while respecting local contexts.

The inclusion of culturally sensitive content in DEERs enhances their pedagogical effectiveness by aligning with the cognitive and affective dimensions of learning (Thapa, 2019). Cultural norms and values significantly influence how learners process information, interact with content, and engage with educational activities (Lee, 2021). For instance, in some cultures, learning is deeply communal, with a strong emphasis on collaboration and group consensus (Uibu and Tagamets, 2022). In others, individual achievement and critical thinking are prioritized (Melle, 2020). DEERs that reflect these cultural nuances can better support the learning preferences of their users, thereby improving both engagement and retention (de Vicente et al., 2021). By respecting and incorporating these cultural differences, DEERs can create a more inclusive learning environment that caters to the diverse needs of a global audience, thus maximizing their educational potential (Esser, 2019).

Culturally sensitive DEERs contribute to the development of intercultural competence among learners (Poole, 2019). As students engage with content that is informed by diverse cultural perspectives, they gain insights into different ways of thinking and problem-solving (Smith, 2018). This exposure helps to break down cultural barriers and fosters a greater appreciation for global diversity (Cojocariu, 2021). In a world that is increasingly interconnected, the ability to navigate and respect cultural differences is a crucial skill (Wrench et al., 2021). DEERs, when designed with cultural sensitivity, can play a pivotal role in developing this skill by providing learners with opportunities to engage in culturally diverse scenarios and challenges (Marrón-González, 2020). This not only enhances the educational value of DEERs but also prepares learners to function more effectively in multicultural environments (Caingcoy, 2023).

The importance of culturally sensitive content in DEERs cannot be overstated (Dyankova and Nikolova, 2023). It is essential for ensuring that these educational tools are both accessible and impactful for a diverse global audience (Demianenko and Moroz, 2023). By going beyond simple translation and integrating a deep respect for cultural differences in values, norms, and learning styles, DEERs can provide a more relevant and engaging learning experience (Smith, 2020). This approach not only improves educational outcomes but also promotes greater cross-cultural understanding, making DEERs a powerful tool for global education (Kieran and Anderson, 2018). As the use of DEERs continues to expand, the incorporation of culturally sensitive content will be crucial in ensuring their effectiveness and inclusivity across different cultural contexts (Gist et al., 2019).

Strategies for Creating Inclusive and Culturally Relevant DEER Experiences

Creating DEERs that are inclusive and culturally relevant requires a strategic and thoughtful approach, ensuring that these educational tools resonate with a diverse global audience (Elsayed-Ali et al., 2023). One of the foundational strategies for achieving this goal is conducting thorough research into the cultural backgrounds of the target audience (Alekhina et al., 2021). This research should encompass a broad spectrum of cultural elements, including values, beliefs, traditions, social norms, and educational practices (Steinböck et al., 2019). By gaining a deep understanding of these cultural dimensions, designers can create DEER experiences that are not only relevant but also respectful of the learners' cultural contexts (Coenraad et al., 2021). This approach helps to avoid the pitfalls of cultural insensitivity, such as the unintentional perpetuation of stereotypes or the overlooking of important cultural nuances that could alienate or disengage learners (Cumbo and Selwyn, 2021).

Consulting with cultural experts is a crucial step in the design process (Coenraad et al., 2019). Cultural experts bring valuable insights and perspectives that can inform the development of content that is both authentic and respectful (Ngwenya et al., 2021). Their expertise can help ensure that the scenarios, narratives, and challenges within DEERs accurately reflect the cultural realities of the target audience, thereby enhancing the educational impact (Coelho et al., 2022). Involving cultural experts early in the design process also allows for the identification and mitigation of potential cultural conflicts or misrepresentations, which could otherwise undermine the effectiveness of the DEER (Esteban-Guitart et al., 2020). Collaborating with members of the target audience in the design process can provide direct feedback on the relevance and appropriateness of the content, ensuring that it resonates with the intended users (Brinkley et al., 2021). This participatory design approach not only enhances the cultural relevance of the DEER but also empowers learners by giving them a voice in the creation of their educational experiences (Kee and Lai, 2022).

Avoiding stereotypes and generalizations is another critical strategy in designing culturally inclusive DEERs. Stereotypes can be harmful and reductive, leading to content that oversimplifies or misrepresents complex cultural identities (Caingcoy, 2023). Instead of relying on broad generalizations, designers should focus on creating content that reflects the diversity within cultures, recognizing the unique experiences and perspectives of different groups (Steen and Wilson, 2020). This can be achieved by incorporating a range of voices and viewpoints within the DEER, thereby presenting a more nuanced and accurate representation of the cultural context (Maduakolam et al., 2020). Authenticity in content creation is key to engaging learners meaningfully and ensuring that the DEER experience is both educational and empowering (Anker-Hansen and Andrée, 2019). By avoiding clichés and stereotypes, designers can foster a more inclusive and respectful learning environment (Caingcoy, 2023).

Another effective strategy for creating culturally relevant DEERs is to design them with adaptability and customization in mind. DEERs that are flexible and can be easily modified to suit different cultural contexts are more likely to be successfully implemented across diverse educational settings (Andini et al., 2020). This adaptability can be achieved by providing educators with tools and resources to customize the content according to the specific cultural needs of their learners (Hakshurian, 2023). For example, DEERs can include options for educators to adjust the language, symbols, and references used in the scenarios to better align with the cultural backgrounds of their students (Repetto et al., 2023). DEERs can be designed to allow for the inclusion of local or regional issues and challenges, making the learning experience more relevant and impactful for the learners (Makri et al., 2021). This flexibility not only enhances the cultural relevance of DEERs but also ensures their broader applicability and effectiveness in a global educational landscape (Lazarinis et al., 2022).

The creation of inclusive and culturally relevant DEER experiences requires a comprehensive and deliberate approach. By conducting thorough cultural research, consulting with experts, involving the target audience in the design process, and avoiding stereotypes, designers can develop DEERs that are both authentic and respectful of cultural differences. By designing DEERs that are adaptable and customizable, educators can ensure that these tools are effectively utilized in diverse educational settings. These strategies based on understanding and employing cultural factors (Table 1) contribute to the development of DEERs that engage learners and promote cross-cultural understanding and inclusivity, making them a valuable asset in the global educational landscape.

Table 1. Cultural factors in digital educational escape room design

Cultural Factor	Description
Language and Communication	The language used in the escape room, including instructions and puzzles, should be accessible to all participants, considering their language proficiency. This includes the use of clear, simple language or multilingual options.
Cultural References	The inclusion of culturally specific references, such as idioms, historical events, or local customs, should be carefully considered to avoid alienating participants unfamiliar with those references.
Visual and Symbolic Representation	The design should consider the cultural significance of colors, symbols, and imagery, ensuring they are appropriate and respectful for the target audience.
Social Interaction Norms	The escape room design should account for different cultural norms related to social interaction, such as communication styles, leadership roles, and group dynamics.
Educational Content	The educational material embedded in the escape room should be relevant and respectful of the cultural background of the participants, ensuring inclusivity and engagement.
Technology Accessibility	Consideration should be given to the technological familiarity and access levels of participants from different cultural backgrounds, ensuring the digital platform is user-friendly for all.
Time Perception	Different cultures perceive time differently, which can affect how they engage with time-limited tasks. Designers should consider flexible time management options.

Technology and Innovation in DEERs Across Diverse Cultures

Technology plays a crucial role in shaping DEER experiences across various cultures. The incorporation of advanced digital technologies in DEERs allows for the creation of culturally relevant and engaging learning experiences. For example, in a study focusing on digital escape rooms in maritime education, it was found that the integration of digital technologies enhanced the learning process by making it more interactive and engaging for students from diverse backgrounds (Voloshynov et al., 2019). This approach is particularly effective in bridging cultural gaps and providing a universal platform for learners from different cultural contexts. Technological advancements enable DEERs to be tailored to specific cultural narratives and educational standards, making them more relevant and accessible to a global audience.

Innovations in DEERs are increasingly emerging from diverse technological landscapes. The use of digital twins, virtual reality (VR), augmented reality (AR), and artificial intelligence (AI) in DEERs is transforming the traditional escape room concept into a more sophisticated educational tool. These technologies allow for the creation of highly immersive and interactive environments that cater to the learning styles and preferences of students from different cultural backgrounds. For instance, the use of VR in DEERs enables the creation of virtual environments that are culturally immersive and educational, enhancing the learning experience by providing realistic simulations of real-world scenarios (Mystakidis et al., 2021b). These innovations not only make DEERs more engaging and effective but also contribute to the global dissemination of knowledge and cultural understanding.

Challenges in Global DEER Implementation

Challenges in Implementing DEERs Globally

Implementing DEERs on a global scale presents a set of complex challenges that must be carefully navigated to ensure their effectiveness and inclusivity across diverse educational landscapes . One of the most prominent challenges is the issue of language diversity, which can significantly impact the universal applicability of DEERs. As these educational tools are designed to be immersive and interactive, the language used within the escape rooms is crucial for ensuring that learners can fully engage with the content. The linguistic diversity that characterizes many regions of the world necessitates careful consideration of how DEERs are translated and adapted to suit local contexts. This is not merely a matter of translating text from one language to another; it involves a deep understanding of cultural nuances, idiomatic expressions, and educational terminologies that may vary significantly across different languages. Effective translation and adaptation are essential to preserve the educational intent and immersive nature of DEERs while making them accessible to a broader audience. Without such adaptations, there is a risk that the educational impact of DEERs could be diminished, particularly in regions where learners may not be fluent in the original language of the content.

Another significant challenge in the global implementation of DEERs is the disparity in technological access across different regions (Makri et al., 2021). While DEERs rely on digital platforms that require specific hardware and software, not all educational institutions around the world have the resources to support these technological demands. In many developing regions, access to high-speed internet, up-to-date computer systems, and other necessary digital infrastructure is limited, which can hinder the effective deployment of DEERs. This technological divide exacerbates educational inequalities, as students in resource-constrained environments may be unable to benefit from the innovative learning experiences that DEERs offer. Addressing this challenge requires a multifaceted approach, including the development of DEERs that can function on lower-specification devices and the implementation of initiatives aimed at improving technological infrastructure in underserved regions. Educators and policymakers must work together to identify and mitigate the specific barriers to technological access within their local contexts, ensuring that all students have the opportunity to engage with these advanced educational tools.

In addition to language and technological challenges, the global implementation of DEERs must also account for the wide variation in educational standards and curricula across different countries. Educational systems around the world differ not only in the content they prioritize but also in the pedagogical approaches they employ. This diversity means that DEERs, which are designed to be immersive and context-specific, must be carefully adapted to align with local educational goals and standards. For instance, a DEER that is effective in one country may need significant modifications to be relevant and effective in another, particularly if there are differences in curricular content, assessment methods, or student learning styles. The adaptability of DEERs to these varied educational contexts is critical to their success. A study on the use of educational escape rooms in media literacy training during the COVID-19 pandemic highlighted this adaptability, demonstrating how DEERs can be tailored to fit the unique needs of different educational systems and cultural contexts. Such adaptability is essential for ensuring that DEERs can be integrated seamlessly into diverse educational frameworks, thereby maximizing their impact on student learning outcomes.

Overcoming Cultural Misconceptions and Stereotypes

Overcoming cultural misconceptions and stereotypes is a critical challenge in the globalization of DEERs, and addressing this issue is essential for ensuring their effectiveness and inclusivity in diverse educational contexts (Chamorro-Koc et al., 2021). As DEERs are increasingly adopted across the globe, the need for cultural sensitivity in their design becomes paramount (Linares, 2023). Cultural misconceptions and stereotypes, if not carefully managed, can not only diminish the educational impact of DEERs but also risk alienating or offending learners from different cultural backgrounds (van Boeijen et al., 2019). Therefore, it is essential that DEERs are developed with a deep respect for the values, traditions, and perspectives of the regions in which they are implemented (Heffernan and Pimpa, 2019). This cultural sensitivity ensures that the content resonates with the local audience, making the learning experience more relevant and engaging (Gregersen-Hermans, 2021). By incorporating diverse cultural viewpoints, DEERs can also serve as a platform for promoting cross-cultural understanding and respect, fostering a global educational environment that is inclusive and equitable (Swartz et al., 2019).

To achieve this level of cultural sensitivity, it is crucial for DEER designers to actively engage with cultural experts and conduct thorough research into the cultural contexts in which their content will be used (Garcia and Pantao, 2021). This involves not only understanding the explicit cultural markers, such as language and symbols, but also appreciating the deeper, more nuanced aspects of culture, including social norms, communication styles, and educational values (Kawaguchi-Suzuki et al., 2019). By collaborating with cultural consultants and educators from the target regions, designers can ensure that the DEER content is appropriate, respectful, and free from harmful stereotypes (Whinnery et al., 2020). This collaborative approach helps in creating educational experiences that are both meaningful and respectful, thus avoiding the pitfalls of cultural insensitivity that can undermine the educational objectives of DEERs (Jie and Pearlman, 2018).

International collaboration and exchanges play a vital role in the development of culturally sensitive DEERs (Williams et al., 2022). By working together across borders, educators and designers can share insights and best practices for integrating cultural nuances into DEERs (Chamorro-Koc et al., 2021). These collaborations can lead to the creation of DEERs that are not only tailored to specific cultural contexts but also have universal appeal, making them accessible to a broader audience (van Boeijen et al., 2019). For example, a DEER designed with input from multiple cultural perspectives is more likely to avoid the biases and assumptions that can inadvertently perpetuate stereotypes (Downey et al., 2018). Instead, such a DEER would celebrate cultural diversity and encourage learners to appreciate and understand perspectives different from their own (Höß et al., 2019). This approach not only enhances the educational value of DEERs but also contributes to a more harmonious and inclusive global learning environment (Gregersen-Hermans, 2021).

Addressing cultural misconceptions and stereotypes in DEERs has broader implications for global education (Sun and Kang, 2021). In an increasingly interconnected world, the ability to navigate and appreciate cultural diversity is an essential skill (Lees and Webb, 2018). DEERs, when designed with cultural sensitivity, can play a significant role in developing this skill by providing learners with experiences that challenge their preconceptions and broaden their understanding of different cultures (Moriarty et al., 2023). This can lead to a greater appreciation of global diversity and help learners develop the cultural competence necessary for success in a globalized world (Mackavey et al., 2023). By promoting cross-cultural understanding through DEERs, educators can contribute to the development of more tolerant and empathetic global citizens (Shrivastava, 2020).

Overcoming cultural misconceptions and stereotypes is a crucial step in the effective globalization of DEERs. By ensuring that DEERs are designed with cultural sensitivity and inclusivity, educators and designers can create educational experiences that are both impactful and respectful of the diverse cultural contexts in which they are used. Collaborative efforts and international exchanges are key to achieving this goal, as they provide the insights and expertise needed to navigate the complex landscape of global education. Ultimately, by addressing cultural challenges head-on, DEERs can become a powerful tool for fostering cross-cultural understanding and promoting a more inclusive and equitable global educational environment.

Future Directions for Global DEER Development

Future Directions in DEER Design

As we look towards the future of DEER design and development, several key trends emerge. One significant trend is the increasing integration of advanced digital technologies, such as Virtual Reality (VR) and Augmented Reality (AR), to create more immersive and interactive learning experiences. Studies indicate that these technologies can significantly enhance engagement and motivation in educational settings, making learning more effective and enjoyable for students (Mystakidis et al., 2021b). Also, there is a growing emphasis on designing DEERs that are adaptable and customizable, allowing them to meet the diverse learning needs of different student populations.

The future of DEERs is also shaped by the adoption of new pedagogical approaches that leverage the potential of emerging technologies. For instance, the concept of digital twins—virtual representations of real-world environments or systems—is being explored for its potential in creating highly realistic and engaging DEER scenarios. The incorporation of microlearning strategies, which involve delivering educational content in small, specific bursts, is becoming increasingly prevalent in DEER design. These approaches, combined with the advanced technological capabilities of DEERs, are set to transform the educational landscape, offering students more personalized, engaging, and effective learning experiences.

Predicting future trends in the development of DEERs reveals a trajectory marked by significant growth and innovation, driven by rapid technological advancements and the evolving needs of global education systems. As educational challenges become more complex and multifaceted, DEERs are increasingly being leveraged not only as tools for traditional academic learning but also as platforms for raising awareness and fostering understanding of critical global issues. One prominent trend is the integration of DEERs into educational strategies aimed at addressing pressing environmental concerns, such as climate change. The use of DEERs in this context, exemplified by initiatives like "Escape Global Warming," (Ouariachi and Van Dam, 2022) highlights the potential of these immersive environments to convey the urgency of climate change in a manner that is both engaging and educational. By embedding core environmental concepts within the narrative and challenges of the escape room, these DEERs encourage participants to think critically about global sustainability issues, making the learning experience both impactful and relevant. This shift underscores a broader trend in education towards experiential learning methods that not only impart knowledge but also inspire action and advocacy among learners.

In parallel with these thematic expansions, the field of DEERs is also witnessing a growing emphasis on international collaboration and knowledge exchange. As DEERs continue to gain traction across diverse educational contexts globally, the potential for cross-border partnerships among educators, researchers, and developers becomes increasingly evident. Such collaborations offer a unique opportunity to share insights, methodologies, and best practices, thereby enriching the design and implementation of DEERs with a diversity of perspectives and expertise. This global exchange is particularly crucial in ensuring that DEERs remain culturally sensitive and educationally effective across different regions. For example, the adaptation of DEERs for media literacy training during the COVID-19 pandemic demonstrated the flexibility of these tools in responding to the varying educational needs and cultural contexts of learners worldwide (Frania, 2021). Through collaborative projects, educators can tailor DEER content to better reflect local values and pedagogical approaches, thus enhancing their relevance and effectiveness in different cultural settings.

The benefits of international collaboration extend beyond the sharing of educational content; they also encompass the exchange of technological expertise and innovative pedagogical strategies. As DEERs evolve, the integration of advanced technologies such as VR, AR, and AI becomes increasingly common. These technologies not only enhance the immersive quality of DEERs but also allow for greater customization and adaptability in learning experiences. International partnerships can play a pivotal role in facilitating the development and dissemination of these technologies, ensuring that educators worldwide have access to the latest tools and techniques for creating effective DEERs. Collaborative efforts can lead to the development of standardized frameworks for DEER design and assessment, which can help to streamline the creation of these educational tools and ensure their quality and efficacy across different educational systems.

The future development of DEERs is poised to be shaped by both technological advancements and the growing recognition of their potential as tools for addressing complex global challenges. The trend towards using DEERs for environmental education, as well as other global issues, reflects a broader shift in educational priorities towards experiential and problem-based learning. At the same time, the increasing emphasis on international collaboration and knowledge exchange highlights the importance of cross-cultural dialogue in the evolution of DEERs. By fostering these collaborations, educators and researchers can ensure that DEERs continue to evolve in ways that are both innovative and inclusive, ultimately contributing to a more interconnected and informed global educational community. As these trends continue to unfold, DEERs are likely to become an integral part of educational strategies worldwide, offering new ways to engage learners and prepare them for the challenges of the future.

CONCLUDING THOUGHTS: THE NEXT FRONTIER IN DIGITAL ESCAPISM FOR EDUCATION

The field of DEERs has evolved significantly, influenced greatly by global perspectives and cultural diversity. This evolution not only signifies a shift in educational methodologies but also highlights the increasing importance of cultural inclusivity and adaptability in digital learning environments.

The landscape of DEERs has evolved rapidly, driven by technological advancements and changing player expectations. Modern DEERs incorporate advanced gamification elements and immersive technologies like VR and AR to create highly engaging and interactive learning experiences. There's been a notable shift towards designing DEERs that are not just educational tools but also sources of entertainment and engagement, catering to the demands of learners who seek experiences that are both informative and enjoyable. The integration of complex storytelling and contextually relevant scenarios has further enhanced the appeal and effectiveness of DEERs, making them more than just a learning platform but an immersive educational journey.

For educators, the evolution of DEERs presents an opportunity to engage students in novel ways, fostering learning through interactive, problem-solving activities. The challenge lies in effectively integrating these tools into the curriculum, ensuring they complement traditional teaching methods. Designers of DEERs must balance educational objectives with engaging gameplay, tailoring experiences to diverse learning styles and cultural backgrounds. Stakeholders in digital education, including policy makers and technology providers, need to support the development and implementation of DEERs through funding, research, and infrastructure development. Embracing these innovative educational tools can lead to more dynamic and effective learning environments.

The integration of global perspectives and cultural influences has been a driving force in the evolution of DEERs. As these educational tools have gained popularity across different countries, they have absorbed diverse cultural elements, leading to richer and more varied learning experiences. For instance, DEERs designed in one country often incorporate local historical, environmental, or social elements, making them uniquely relevant to learners in that context. As DEERs cross borders, they encounter new audiences with different cultural backgrounds and educational expectations. This has pushed DEER designers to think globally, creating content that is not only engaging but also respectful and reflective of diverse cultures. Such an approach ensures that DEERs do not just impart knowledge but also foster cross-cultural understanding and empathy.

The future of DEERs in education is promising, characterized by continuous innovation and adaptation. One important direction is the integration of advanced digital technologies like Virtual Reality (VR) and Augmented Reality (AR) to create more immersive learning experiences (Mystakidis et al., 2021b). The migration of DEERs to online environments, as necessitated by the COVID-19 pandemic, has opened new avenues for remote (Piernas et al., 2024) and blended learning (Neumann et al., 2020). DEERs are also evolving to encompass a broader range of subjects, including complex topics like climate change, offering a platform for experiential learning and awareness (Ouariachi and Wim, 2020).

REFERENCES

Abd-alrazaq, A. A., AlSaad, R., Alhuwail, D., Ahmed, A., Healy, P., Latifi, S., Aziz, S., Damseh, R., Alrazak, S. A., & Sheikh, J. (2023). Large language models in medical education: Opportunities, challenges, and future directions. *JMIR Medical Education*, 9, e48291. Advance online publication. DOI: 10.2196/48291 PMID: 37261894

Acosta, A. A. A., & Carballo, G. E. (2022). The use of immersive tools in higher education: Escape rooms. *InProceedings of the 2022 6th International Conference on Education and E-Learning*. DOI: DOI: 10.1145/3578837.3578848

Alasmari, T. (2020). Gamification effect on higher education students' motivation. *The Journal of Positive Psychology*, 57, 3009–3030. DOI: 10.17762/PAE.V57I9.891

Alekhina, S., Mel'nik, Y. V., Samsonova, E., and Shemanov, A. (2021). Assessment of inclusive process as a tool for designing inclusion in educational institution. DOI: .DOI: 10.17759/pse.2021260509

Algerafi, M. A. M., Zhou, Y., Oubibi, M., & Wijaya, T. T. (2023). Unlocking the potential: A comprehensive evaluation of augmented reality and virtual reality in education. *Electronics (Basel)*, 12(18), 3953. Advance online publication. DOI: 10.3390/electronics12183953

Almeida, F., & Simões, J. (2019). The role of serious games, gamification and industry 4.0 tools in the education 4.0 paradigm. *Contemporary Educational Technology*, 10(2), 120–136. Advance online publication. DOI: 10.30935/cet.554469

Amolloh, O. P., Lilian, G. K., & Wanjiru, K. G. (2018). Experiential learning, conditional knowledge and professional development at university of nairobi, kenya—Focusing on preparedness for teaching practice. *International Education Studies*, 11(7), 125. Advance online publication. DOI: 10.5539/ies.v11n7p125

Andini, D., Rahayu, A., Budiningsih, C. A., & Mumpuniarti, M. (2020). The curriculum adaptation model in fulfilling the learning need for diverse students at inclusive classroom. *Universal Journal of Educational Research*, 8(3D), 115–123. DOI: 10.13189/ujer.2020.081716

Anker-Hansen, J., & Andrée, M. (2019). *In pursuit of authenticity in science education*. Nordic Studies in Science Education., DOI: 10.5617/nordina.4723

Arbesser-Rastburg, G., Safikhani, S., Gustin, M., Hopfe, C., Schweiger, G., & Pirker, J. (2024). Project beyond: An escape room game in virtual reality to teach building energy simulations. *IEEE Conference on Computatonal Intelligence and Games*, DOI: DOI: 10.1109/CoG60054.2024.10645541

Arendale, D. (2018). Introduction to special issue on universal design for inclusive pedagogy and a future research agenda. *Education Sciences*, 8(4), 203. Advance online publication. DOI: 10.3390/educsci8040203

Azaiz, I., Deckarm, O., and Strickroth, S. (2023). Ai-enhanced auto-correction of programming exercises: How effective is gpt-3.5? *ArXiv*, abs/2311.10737, DOI: .DOI: 10.3991/ijep.v13i8.45621

Baron, J. V. (2023). Level of cultural sensitivity and its influence on teachers' performance. *Journal of Social, Humanity, and Education*, DOI: .DOI: 10.35912/jshe.v3i4.1392

Borrás-Gené, O., Diez, R., & Macías-Guillén, A. (2022). Digital educational escape room analysis using learning styles. *Information (Basel)*, 13(11), 522. DOI: 10.3390/info13110522

Brinkley, J., Huff, E. W., & Boateng, K. (2021). Tough but effective: Exploring the use of remote participatory design in an inclusive design course through student reflections. *Proceedings of the 52nd ACM Technical Symposium on Computer Science Education*, DOI: DOI: 10.1145/3408877.3432527

Burgstahler, S. (2022). Tips for designing an accessible and inclusive online course. *Proceedings of the International Conference on Society and Information Technologies*, DOI: DOI: 10.54808/ICSIT2022.01.79

Caingcoy, M. E. (2023). *Culturally responsive pedagogy*. Diversitas Journal., DOI: 10.48017/dj.v8i4.2780

Chamorro-Koc, M., Caldwell, G., Vasquez, A. V., Rodriguez, I. F., & Nates, C. (2021). Wearables of the past and the future: An immersive and cross-cultural learning experience of undergraduate students in an international design collaboration. *Creative Industries Journal*, 15(3), 293–316. DOI: 10.1080/17510694.2021.1939539

Chang, C.-C., & Hwang, G. (2021). An experiential learning-based virtual reality approach to fostering problem-resolving competence in professional training. *Interactive Learning Environments*, 31(8), 4713–4728. DOI: 10.1080/10494820.2021.1979049

Chen, Z., Jiang, Z., Yang, F., He, Z., Hou, Y., Cho, E., McAuley, J., Galstyan, A., Hu, X., & Yang, J. (2023). The first workshop on personalized generative ai @ cikm 2023: Personalization meets large language models. *Proceedings of the 32nd ACM International Conference on Information and Knowledge Management*, DOI: DOI: 10.1145/3583780.3615314

Cheng, S.-C., Hwang, G.-J., & Chen, C.-H. (2019). From reflective observation to active learning: A mobile experiential learning approach for environmental science education. *British Journal of Educational Technology*, 50(5), 2251–2270. DOI: 10.1111/bjet.12845

Chineka, R., & Yasukawa, K. (2020). Intergenerational learning in climate change adaptations; limitations and affordances. *Environmental Education Research*, 26(4), 577–593. DOI: 10.1080/13504622.2020.1733494

Christopoulos, A., Mystakidis, S., Cachafeiro, E., & Laakso, M. (2022). Escaping the cell: Virtual reality escape rooms in biology education. *Behaviour & Information Technology*, 42(9), 1434–1451. DOI: 10.1080/0144929X.2022.2079560

Coelho, C., Cordeiro, A., Alcoforado, L., & Moniz, G. (2022). Survey on student school spaces: An inclusive design tool for a better school. *Buildings (Basel, Switzerland)*, 12(4), 392. Advance online publication. DOI: 10.3390/buildings12040392

Coenraad, M., Palmer, J., Franklin, D., & Weintrop, D. (2019). Utilizing participatory design to develop a culturally relevant computer science curriculum. *Proceedings of the 50th ACM Technical Symposium on Computer Science Education*, DOI: DOI: 10.1145/3287324.3293803

Coenraad, M., Weintrop, D., Eatinger, D., Palmer, J., & Franklin, D. (2021). *Identifying youths' spheres of influence through participatory design*. Designs for Learning., DOI: 10.16993/dfl.163

Cojocariu, V.-M. (2021). *Pedagogical approaches to intercultural competence development*. Moldavian Journal for Education and Social Psychology., DOI: 10.18662/mjesp/4.2/24

Collins, C., Nanda, S., Palmer, B., Mohabbat, A., Schleck, C., Mandrekar, J., Mahapatra, S., Beckman, T., & Wittich, C. (2019). A cross-sectional study of learning styles among continuing medical education participants. *Medical Teacher*, 41(3), 318–324. DOI: 10.1080/0142159X.2018.1464134 PMID: 29703093

Craddock, G., Fitzpatrick, D., Hubbard, J., O'Reilly, R., & Wilkinson, M. (2022). Universal design in education (ude) across the life course: Applying a systems framework to create an inclusive experience for all. *InProceedings of the IASE 2021 Satellite Conference*. DOI: DOI: 10.52041/iase.kxvpc

Cressey, J. (2020). *Universal Design for Learning*. DOI: .DOI: 10.4018/978-1-7998-1770-3.ch008

Cristea, D., & Ivanov, M. catalina, R. C., renato, M. L., and Andrei, L. (2020). Immersive data analytics for enhancing organisational knowledge transfer processes through a custom developed virtual reality framework. *eLearning and Software for Education*, DOI: DOI: 10.12753/2066-026X-20-097

Cujdíková, M., & Vankúš, P. (2024). The use of educational escape rooms in the lifelong training of mathematics teachers. *Proceedings of the European Conference on Games-based Learning*, DOI: DOI: 10.34190/ecgbl.18.1.2653

Cumbo, B. J., & Selwyn, N. (2021). Using participatory design approaches in educational research. *International Journal of Research & Method in Education*, 45(1), 60–72. DOI: 10.1080/1743727X.2021.1902981

Dalton, E. M., Lyner-Cleophas, M., Ferguson, B., & Mckenzie, J. (2019). Inclusion, universal design and universal design for learning in higher education: South africa and the united states. *African Journal of Disability*, 8. Advance online publication. DOI: 10.4102/ajod.v8i0.519 PMID: 31392169

Damaševicius, R. (2023). *The rise of chatgpt and the demise of bloom's taxonomy of learning stages*. Creative AI Tools and Ethical Implications in Teaching and Learning., DOI: 10.4018/979-8-3693-0205-7.ch006

Damaševicius, R., & Sidekerskiene, T. (2023). *Designing metaverse escape rooms for microlearning in stem education. Fostering Pedagogy Through Micro and Adaptive Learning in Higher Education: Trends*. Tools, and Applications., DOI: 10.4018/978-1-6684-8656-6.ch009

Damaševicius, R., & Zailskaite-Jakšte, L. (2024). *Digital twin technology: Necessity of the future in education and beyond*. Automated Secure Computing for Next-Generation Systems.

de Vicente, D. P., del Olmo-Fernández, M. J. A., Santos-Villalba, M. J., & Leiva-Olivencia, J. J. (2021). A qualitative study on the intercultural educational sensitivity of the professors at the university of malaga (spain). *Education Sciences*, 11(2), 74. Advance online publication. DOI: 10.3390/educsci11020074

Debchaudhury, S. (2024). Impact of digital game-based learning experiences on reasoning skills and science engagement. DOI: .DOI: 10.34190/ecgbl.18.1.2690

Demianenko, O. and Moroz, T. (2023). Formation of intercultural competence in the process of implementing the multilingual education model. *PARADIGM OF KNOWLEDGE*, DOI: .DOI: 10.26886/2520-7474.4(58)2023.3

Denny, P., McDonald, F., Empson, R., Kelly, P., & Petersen, A. (2018). Empirical support for a causal relationship between gamification and learning outcomes. *In-Proceedings of the 2018 CHI Conference on Human Factors in Computing Systems*. DOI: DOI: 10.1145/3173574.3173885

Dicheva, D., Caldwell, R., & Guy, B. (2020). Do badges increase student engagement and motivation? *InProceedings of the 21st Annual Conference on Information Technology Education*. DOI: DOI: 10.1145/3368308.3415393

Dittman, J. M., Maiden, K. M., Matulewicz, A. T., Beaird, G., Lockeman, K. S., & Dow, A. (2021). A flexible customizable virtual escape room approach for interprofessional learners. *Journal of Interprofessional Education & Practice*, 24, 100455. DOI: 10.1016/j.xjep.2021.100455

Downey, G., Lloyd, K., Bilous, R. H., Hammersley, L., Rawlings-Sanaei, F., Amigó, M. F., Gilchrist, S., Baker, M., & Coffey, E. (2018). "classroom of many cultures": Educational design opportunities in intercultural co-creation. *Intercultural Studies in Higher Education*, 25, 45–59. DOI: 10.1007/978-3-319-74579-4_3

Dutra, S. V. O., Chee, V., & Clochesy, J. (2023). Adapting an educational software internationally: Cultural and linguistical adaptation. *Education Sciences*, 13(3), 237. Advance online publication. DOI: 10.3390/educsci13030237

Dyankova, G. and Nikolova, S. (2023). Multicultural competence as a teacher's metacognition to achieve a positive school climate. International Journal of Cognitive Research in Science, *Engineering and Education (IJCRSEE)*, DOI: .DOI: 10.23947/2334-8496-2023-11-2-257-265

Elsayed-Ali, S., Bonsignore, E. M., & Chan, J. (2023). Exploring challenges to inclusion in participatory design from the perspectives of global north practitioners. *Proceedings of the ACM on Human-Computer Interaction*, 7(CSCW1), 1–25. DOI: 10.1145/3579606

Esser, S. (2019). Reformulation of intercultural competence in the context of a diversity-conscious teacher education and pedagogical school development. *International Journal of Creativity and Innovation in Humanities and Education*, DOI: .DOI: 10.21608/ijcihe.2019.182878

Esteban-Guitart, M., Monreal-Bosch, P., Palma, M., & González-Ceballos, I. (2020). Sustaining students' identities within the context of participatory culture. designing, implementing and evaluating an interactive learning activity. *Sustainability (Basel)*, 12(12), 4870. DOI: 10.3390/su12124870

Eulenstein, M., Stiegel, S., Czech, S., Erkan, R., & Schoormann, T. (2025). Learning through designing: A visual inquiry tool for digital educational escape games. *Lecture Notes in Computer Science (including subseries Lecture Notes in Artificial Intelligence and Lecture Notes in Bioinformatics)*, DOI: DOI: 10.1007/978-3-031-78269-5_41

Foo, S. C., & Foo, K.-K. (2022). Purposeful field trip: Impact on experiential learning opportunities and critical thinking skills. *Pertanika Journal of Social Science & Humanities*, 30(1), Preprint. Advance online publication. DOI: 10.47836/pjssh.30.1.01

Fotaris, P., & Mastoras, T. (2019). Escape rooms for learning: A systematic review. *InProceedings of the 12th European Conference on Game Based Learning*. DOI: DOI: 10.34190/GBL.19.179

Frania, M. (2021). *Educational e-escape room as an educational method of media literacy training for future teachers during the covid-19 pandemic. Media Education*. Mediaobrazovanie., DOI: 10.13187/me.2021.3.452

Gan, W., Qi, Z., Wu, J., and Lin, C.-W. (2023). Large language models in education: Vision and opportunities. *ArXiv*, abs/2311.13160, DOI: .DOI: 10.48550/arXiv.2311.13160

Garcia, K. A. and Pantao, J. (2021). Cultural sensitivity and classroom management of teachers. *International Journal of Professional Development, Learners and Learning*, DOI: .DOI: 10.30935/ijpdll/11093

Gist, C. D., Jackson, I., Nightengale-Lee, B., & Allen, K. M. (2019). *Culturally responsive pedagogy in teacher education*. Oxford Research Encyclopedia of Education., DOI: 10.1093/acrefore/9780190264093.013.266

Gregersen-Hermans, J. (2021). Toward a curriculum for the future: Synthesizing education for sustainable development and internationalization of the curriculum. *Journal of Studies in International Education*, 25(4), 461–481. DOI: 10.1177/10283153211031033

Hakshurian, Y. (2023). Teachers' competencies in developing digital educational escape rooms. *Educatia*, 21(25), 66–71. Advance online publication. DOI: 10.24193/ed21.2023.25.07

Heffernan, M., & Pimpa, N. (2019). *Globally transformative student experience: Challenges and opportunities in learning and teaching in the transnational business education program*. Intercultural Studies in Higher Education., DOI: 10.1007/978-3-030-15758-6_9

Hevko, I. V., Siaonan, S., & Wu, H. (2023). Innovative interactive technologies in the training of professional education specialists by means of digital technologies. *Journal of Education, Health and Sport*, 37(1), 166–173. Advance online publication. DOI: 10.12775/JEHS.2023.37.01.013

Höß, B., Wasserman, M. E., and Fisher, S. L. (2019). Building a global education collaboration model using experiential learning: A fresh look at developing intercultural competence. *Journal of International Business Research And Marketing*, DOI: .DOI: 10.18775/jibrm.1849-8558.2015.51.3001

Howard, B., Ilyashenko, N., & Jacobs, L. (2023). Cross-cultural collaboration through virtual teaming in higher education. *Perspectives in Education*, 41(1), 74–87. Advance online publication. DOI: 10.38140/pie.v41i1.6319

Huang, S., Kuo, Y.-H., & Chen, H.-C. (2020). Applying digital escape rooms infused with science teaching in elementary school: Learning performance, learning motivation, and problem-solving ability. *Thinking Skills and Creativity*, 37, 100681. DOI: 10.1016/j.tsc.2020.100681

Huilan, S., Wei, W., Zhonghua, D., Xiaolong, Q., Dan, W., Shihai, Z., Qing, W., Liang, Y., & Youming, Q. (2020). Educational management in critical thinking training based on bloom's taxonomy and solo taxonomy. *2020 International Conference on Information Science and Education (ICISE-IE)*, pages 518–521, DOI: DOI: 10.1109/ICISE51755.2020.00116

Huraj, L., Hrmo, R., & Sejutová Sejutová Hudáková, M. (2022). The impact of a digital escape room focused on html and computer networks on vocational high school students. *Education Sciences*, 12(10), 682. Advance online publication. DOI: 10.3390/educsci12100682

Huraj, L., & Hudáková, M. S. (2022). Digital escape room for computer network mechanic students. In *2022 20th International Conference on Emerging eLearning Technologies and Applications (ICETA)*, pages 236–242. IEEE. DOI: 10.1109/ICETA57911.2022.9974737

Istiana, R., Herawati, D., Herniningtyas, F., Ichsan, I., & Ali, A. (2023). *Stem learning to improve problem solving ability on the topic of environmental education*. Jurnal Penelitian Pendidikan IPA., DOI: 10.29303/jppipa.v9i3.2979

Jacobs, S., Hardebusch, T., Gerhard, N., Franke, E., Peters, H., & Jaschke, S. (2023). Promoting competences for the digital world by an educational escape room. In *2023 IEEE 2nd German Education Conference (GECon)*. DOI: DOI: 10.1109/GECon58119.2023.10295150

Jantjies, M., Moodley, T., & Maart, R. (2018). Experiential learning through virtual and augmented reality in higher education. *Proceedings of the 23rd Annual International Conference on Engineering and Product Design Education (E&PDE 2019)*, pages 42–45, DOI: DOI: 10.1145/3300942.3300956

Jie, Z., & Pearlman, A. M. G. (2018). Expanding access to international education through technology enhanced collaborative online international learning (coil) courses. *International Journal of Technology in Teaching and Learning*, 14(1). Advance online publication. DOI: 10.37120/ijttl.2018.14.1.01

Joseph, S. (2023). *Large language model-based tools in language teaching to develop critical thinking and sustainable cognitive structures*. Rupkatha Journal on Interdisciplinary Studies in Humanities., DOI: 10.21659/rupkatha.v15n4.13

Kalvakolu, S., Söbke, H., Baalsrud Hauge, J., & Kraft, E. (2025). Combining 360° spaces and social vr. *Lecture Notes in Computer Science (including subseries Lecture Notes in Artificial Intelligence and Lecture Notes in Bioinformatics)*, DOI: DOI: 10.1007/978-3-031-78269-5_38

Karabacak, M., Ozkara, B. B., Margetis, K., Wintermark, M., & Bisdas, S. (2023). The advent of generative language models in medical education. *JMIR Medical Education*, 9, e48163. Advance online publication. DOI: 10.2196/48163 PMID: 37279048

Karageorgiou, Z., Mavrommati, I., & Fotaris, P. (2019). Escape room design as a game-based learning process for steam education. In *Proceedings of the 12th European Conference on Game Based Learning*. DOI: DOI: 10.34190/GBL.19.190

Kartal, G., Toprak, E., & Kumtepe, E. G. (2018). Culturally sensitive instructional design principles for online learning environments. *In Distance education and e-learning practices*. DOI: DOI: 10.4018/978-1-5225-3076-3.ch015

Kawaguchi-Suzuki, M., Law, M. G., Prisco, J., Head, K., Fu, L., Yumoto, T., Kamei, J., Yang, M., Cheng, K., & Hogue, M. (2019). Cultural sensitivity and global pharmacy engagement in asia: China, japan, south korea, and taiwan. *American Journal of Pharmaceutical Education*, 83(4), 7214. Advance online publication. DOI: 10.5688/ajpe7214 PMID: 31223161

Kazandis, I., & Pellas, N. (2019). Developing and assessing augmented reality applications for mathematics with trainee instructional media designers: An exploratory study on user experience. *Journal of Universal Computer Science*, 25, 489–514.

Kee, T. and Lai, A. (2022). Learning motivation and psychological empowerment of socioeconomically disadvantaged learners – an empirical study on inclusive project-based learning during covid-19. *International Journal of Inclusive Education*.

Kieran, L., & Anderson, C. (2018). Connecting universal design for learning with culturally responsive teaching. *Education and Urban Society*, 51(9), 1202–1216. DOI: 10.1177/0013124518785012

Kim, H., Hartleb, T., Bello, K., Aqlan, F., Zhao, R., & Yang, H. (2023). Behavioral modeling of collaborative problem solving in multiplayer virtual reality manufacturing simulation games. *Journal of Computing and Information Science in Engineering*, •••, 24.

Kim, J., & Castelli, D. (2021). Effects of gamification on behavioral change in education: A meta-analysis. *International Journal of Environmental Research and Public Health*, 18(7), 7350. DOI: 10.3390/ijerph18073550 PMID: 33805530

Kladchuen, R., & Srisomphan, J. (2021). The synthesis of a model of problem-based learning with the gamification concept to enhance the problem-solving skills for high vocational certificate. *International Journal of Emerging Technologies in Learning*, 16(14), 4. Advance online publication. DOI: 10.3991/ijet.v16i14.20439

Kotzebue, L., Zumbach, J., & Brandlmayr, A. (2022). Digital escape rooms as game-based learning environments: A study in sex education. *Multimodal Technologies and Interaction*, 6(2), 8. DOI: 10.3390/mti6020008

Laato, S., Morschheuser, B., Hamari, J., & Björne, J. (2023). Ai-assisted learning with chatgpt and large language models: Implications for higher education. *2023 IEEE International Conference on Advanced Learning Technologies (ICALT)*, pages 226–230, DOI: DOI: 10.1109/ICALT58122.2023.00072

Lal, A., Walsh, E., Wetherell, A., & Slimings, C. (2021). Climate change education in public health and medical curricula in australian and new zealand universities: a mixed methods study of barriers and areas for further action. DOI: DOI: 10.1101/2021.06.11.21258793

Latheesh, K. (2019). Leadership development through experiential learning. *International Education and Research Journal*, •••, 5.

Lathwesen, C. and Belova, N. (2021). Escape rooms in stem teaching and learning—prospective field or declining trend? a literature review. DOI: .DOI: 10.3390/educsci11060308

Lazarinis, F., Boididis, I., Kozanidis, L., & Kanellopoulos, D. (2022). *An adaptable multi-learner serious game for learning cultural heritage*. Advances in Mobile Learning Educational Research., DOI: 10.25082/AMLER.2022.01.004

Lee, A. (2021). *Teacher embodiment of culturally responsive pedagogies in a fifth grade classroom*. Profesorado, Revista de Currículum y Formación del Profesorado., DOI: 10.30827/profesorado.v25i3.21461

Lees, J., & Webb, G. (2018). A review of the literature to inform the development of a new model of global placement: The global learning partnership. *Physical Therapy Reviews : PTR*, 23(1), 40–49. DOI: 10.1080/10833196.2017.1341144

Lestari, I. (2021). Experiential learning using stem approach in improving students' problem solving ability. *Journal of Physics: Conference Series*, 1806, DOI: , V. (2020). Challenging escape from climate change: A learning experience. DOI: .DOI: 10.1088/1742-6596/1806/1/012005.Liapi

Linares, S. M. (2023). Intercultural sensitivity and cultural othering in english language pre-service teachers. *Teaching English as a Second or Foreign Language–TESL-EJ*, DOI: .DOI: 10.55593/ej.26104a5

Lindsay, E., Johri, A., and Bjerva, J. (2023). A framework for responsible development of automated student feedback with generative ai. *ArXiv*, abs/2308.15334, DOI: .DOI: 10.48550/arXiv.2308.15334

López-Pernas, S., Gordillo, A., Barra, E., & Quemada, J. (2021). Comparing face-to-face and remote educational escape rooms for learning programming. *IEEE Access : Practical Innovations, Open Solutions*, 9, 59270–59285. DOI: 10.1109/ACCESS.2021.3073601

Maas, M. and Hughes, J. (2020). Virtual, augmented and mixed reality in k–12 education: a review of the literature. *Technology, Pedagogy and Education*, 29:231 – 249, DOI : Mackavey, C. L., Henderson, C., and Stout, T. (2023). Stepping outside national borders: International active learning educational collaboration events. *International Nursing Review*, DOI: .DOI: 10.1080/1475939X.2020.1737210

Maduakolam, E., Madden, B., Kelley, T., & Cianciolo, A. (2020). Beyond diversity: Envisioning inclusion in medical education research and practice. *Teaching and Learning in Medicine*, 32(5), 459–465. DOI: 10.1080/10401334.2020.1836462 PMID: 33349086

Majid, R. (2021). *The significance of soft skills training in undergraduate education*. Higher Education, Skills and Work-Based Learning., DOI: 10.1108/heswbl-10-2020-0234

Makri, A., Vlachopoulos, D., & Martina, R. (2021). Digital escape rooms as innovative pedagogical tools in education: A systematic literature review. *Sustainability (Basel)*, 13(8), 4587. DOI: 10.3390/su13084587

Mancheno-Smoak, S. (2019). Experiential learning through the use of simulation and digital games in a classroom: An investigation of the effects of learning on student engagement, motivation, and learning outcomes. *The Journal of Scholarship of Teaching and Learning*. Advance online publication. DOI: 10.14434/josotl.v19i2.23842

Markowitz, D. M., Laha, R. R., Perone, B., Pea, R., & Bailenson, J. (2018). Immersive virtual reality field trips facilitate learning about climate change. *Frontiers in Psychology*, 9, 2364. Advance online publication. DOI: 10.3389/fpsyg.2018.02364 PMID: 30555387

Marrón-González, S. (2020). Building bridges to the integration of culture in the efl secondary classroom: Perspectives from teachers and learners. *Eighth International Conference on Technological Ecosystems for Enhancing Multiculturality*, DOI: DOI: 10.1145/3434780.3436670

Matveev, V., Nikolaevich, D. G., Lomakin, D. E., and Kharunzheva, E. (2021). The use of augmented reality for the formation of multicultural competence of undergraduates. *Perspectives of Science and Education*, DOI: .DOI: 10.32744/pse.2021.6.33

Mayombe, C. (2023a). Facilitating non-school-based technical and vocational training for disadvantaged youths in south africa. *Journal of Technical Education and Training*, DOI: .DOI: 10.30880/jtet.2023.15.01.016

Mayombe, C. (2023b). *Promoting youths' skills acquisition through experiential learning theory in vocational education and training in south africa*. Higher Education, Skills and Work-Based Learning., DOI: 10.1108/HESWBL-10-2022-0216

Mekacher, D. L. (2019). Augmented reality (ar) and virtual reality (vr): The future of interactive vocational education and training for people with handicap. *PUPIL: International Journal of Teaching, Education and Learning*, DOI: .DOI: 10.20319/pijtel.2019.31.118129

Melle, J. V. (2020). Developing students' intercultural sensitivity. a technical action research study at an international business school in the netherlands. (also known as 'the shoe sole project'). DOI: DOI: 10.17638/03090675

Moffett, J., Cassidy, D., Collins, N., Illing, J., de Carvalho Filho, M. A., & Bok, H. (2023). Exploring medical students' learning around uncertainty management using a digital educational escape room: A design-based research approach. *Perspectives on Medical Education*, 12(1), 86–98. DOI: 10.5334/pme.844 PMID: 36969324

Moriarty, S., Spruce, J., & Mathai, S. (2023). Critical pedagogies for collaborative innovation. *Proceedings of the International Conference on Engineering and Product Design Education, EPDE 2023*, DOI: DOI: 10.35199/EPDE.2023.51

Moseikina, M., Toktamysov, S., & Danshina, S. (2022). Modern technologies and gamification in historical education. *Simulation & Gaming*, 53(2), 135–156. DOI: 10.1177/10468781221075965

Mystakidis, S., & Christopoulos, A. (2022). Teacher perceptions on virtual reality escape rooms for stem education. *Information (Basel)*, 13(3), 136. DOI: 10.3390/info13030136

Mystakidis, S., Fragkaki, M., & Filippousis, G. (2021a). Ready teacher one: Virtual and augmented reality online professional development for k-12 school teachers. *Computers*, 10(10), 134. DOI: 10.3390/computers10100134

Mystakidis, S., Papantzikos, G., & Stylios, C. (2021b). Virtual reality escape rooms for stem education in industry 4.0: Greek teachers perspectives. In *2021 6th South-East Europe Design Automation, Computer Engineering, Computer Networks and Social Media Conference (SEEDA-CECNSM)*, pages 1–5. IEEE, DOI: DOI: 10.1109/SEEDA-CECNSM53056.2021.9566265

Neumann, K. L., Alvarado-Albertorio, F. M., & Ramírez-Salgado, A. (2020). Online approaches for implementing a digital escape room with preservice teachers. *Journal of Technology and Teacher Education*, 28, 415–424.

Ngwenya, N., Makoelle, T., & van der Merwe, M. (2021). Participatory action research as change strategy: A case of developing inclusive teaching and learning practices in an adult education centre in gauteng east district of south africa. *Interchange*, 52(3), 393–414. DOI: 10.1007/s10780-021-09435-3

Nieminen, J., & Pesonen, H. (2019). Taking universal design back to its roots: Perspectives on accessibility and identity in undergraduate mathematics. *Education Sciences*, 10(1), 12. Advance online publication. DOI: 10.3390/educsci10010012

O'shaughnessy, T. (2021). Universal Design for Learning and Accessibility: A Practitioner Approach. In Advances in Educational Technologies and Instructional Design (pp. 25–47). IGI Global. DOI: DOI: 10.4018/978-1-7998-7106-4.ch002

Ortiz-Rojas, M., Chiluiza, K., & Valcke, M. (2019). Gamification through leaderboards: An empirical study in engineering education. *Computer Applications in Engineering Education*, 27(4), 777–788. DOI: 10.1002/cae.12116

Oswald, G., Adams, R., & Hiles, J. (2018). Universal design for learning in rehabilitation education: Meeting the needs for equal access to electronic course resources and online learning. *Journal of Applied Rehabilitation Counseling*, 49(1), 19–22. DOI: 10.1891/0047-2220.49.1.19

Ouariachi, T., & Van Dam, M. (2022). Educational innovation to address climate change issues. *Advances in Game-Based Learning*, 263–278. Advance online publication. DOI: 10.4018/978-1-7998-8645-7.ch013

Ouariachi, T., & Wim, E. J. L. (2020). Escape rooms as tools for climate change education: An exploration of initiatives. *Environmental Education Research*, 26(8), 1193–1206. DOI: 10.1080/13504622.2020.1753659

Owan, V. J., Abang, K. B., Idika, D., Etta, E. O., & Bassey, B. (2023). Exploring the potential of artificial intelligence tools in educational measurement and assessment. *Eurasia Journal of Mathematics, Science and Technology Education*, 19(8), em2307. Advance online publication. DOI: 10.29333/ejmste/13428

Pamungkas, S. F., Widiastuti, I., & Suharno, S. (2020). 21st century learning: Experiential learning to enhance critical thinking in vocational education. *Universal Journal of Educational Research*, 8(4), 1345–1355. DOI: 10.13189/ujer.2020.080427

Pantelidis, P., Chorti, A., Papagiouvanni, G. P. I., & Drosos, C. (2018, March 28). ThrasyvoulosPanagiotakopoulos, Lales, G., and Sideris, M. (2018). Virtual and augmented reality in medical education. *International Journal of Technology in Teaching and Learning*. Advance online publication. DOI: 10.5772/intechopen.71963

Papanastasiou, G. P., Drigas, A., Skianis, C., Lytras, M. D., & Papanastasiou, E. (2018). Virtual and augmented reality effects on k-12, higher and tertiary education students' twenty-first century skills. *Virtual Reality (Waltham Cross)*, 23(4), 425–436. DOI: 10.1007/s10055-018-0363-2

Paralikar, S. J., Shah, C., Joshi, A., & Kathrotia, R. (2022). Acquisition of higher-order cognitive skills (hocs) using the flipped classroom model: A quasi-experimental study. *Cureus*. Advance online publication. DOI: 10.7759/cureus.24249 PMID: 35602838

Parody, L., Santos, J., Trujillo-Cayado, L. A., & Ceballos, M. (2022). Gamification in engineering education: The use of classcraft platform to improve motivation and academic performance. *Applied Sciences (Basel, Switzerland)*, 12(22), 11832. Advance online publication. DOI: 10.3390/app122211832

Patil, M., Balaji, R., and Shikhare, R. (2022). Smarter education methodologies for met: An approach with augmented reality. *OCEANS 2022 - Chennai*, pages 1–8, DOI: .DOI: 10.1109/OCEANSChennai45887.2022.9775457

Paulauskas, L., Paulauskas, A., Blažauskas, T., Damaševicius, R., & Maskeliunas, R. (2023). Reconstruction of industrial and historical heritage for cultural enrichment using virtual and augmented reality. *Technologies*, 11(2), 36. Advance online publication. DOI: 10.3390/technologies11020036

Phung, T., Padurean, ˘. V.-A., Cambronero, J., Gulwani, S., Kohn, T., Majumdar, R., Singla, A., & Soares, G. (2023). Generative ai for programming education: Benchmarking chatgpt, gpt-4, and human tutors. *Proceedings of the 2023 ACM Conference on International Computing Education Research-* Volume 2, DOI: DOI: 10.1145/3568812.3603476

Piernas, J., Meroño, M., and Asenjo, M. (2024). Virtual escape rooms: a gamification tool to enhance motivation in distance education [escape rooms virtuales: una herramienta de gamificación para potenciar la motivación en la educación a distancia]. *RIED-Revista Iberoamericana de Educacion a Distancia*, DOI: .DOI: 10.5944/ried.27.1.37685

Poole, A. (2019). *Negotiating intercultural spaces and teacher identity in an internationalised school in shanghai*. Intercultural Communication Education., DOI: 10.29140/ice.v2n2.128

Prasad, G. (2021). Evaluating student performance based on bloom's taxonomy levels. *Journal of Physics: Conference Series*, 1797(1), 012063. Advance online publication. DOI: 10.1088/1742-6596/1797/1/012063

Pujawan, I. G. N., Rediani, N. N., Antara, I. B. G. W., Putri, N. D. A., & Bayu, G. W. (2022). *Revised bloom taxonomy-oriented learning activities to develop scientific literacy and creative thinking skills*. Jurnal Pendidikan IPA Indonesia., DOI: 10.15294/jpii.v11i1.34628

Puritat, K. (2019). Enhanced knowledge and engagement of students through the gamification concept of game elements. *International Journal of Engineering Pedagogy*, 9(5), 41–54. DOI: 10.3991/ijep.v9i5.11028

Purnama, S., Ulfah, M., Ramadani, L., Rahmatullah, B., & Ahmad, I. (2022). *Digital storytelling trends in early childhood education in indonesia: A systematic literature review*. JPUD - Jurnal Pendidikan Usia Dini., DOI: 10.21009/JPUD.161.02

Purwaningtyas, D., Prabowo, H., Napitupulu, T. A., & Purwandari, B. (2022). *The integration of augmented reality and virtual laboratory based on the 5e model and vark assessment: A conceptual framework*. Jurnal Pendidikan IPA Indonesia., DOI: 10.15294/jpii.v11i3.36367

Putri, M. A. N., & Dwikoranto, D. (2022). *Implementation of stem integrated project based learning (pjbl) to improve problem solving skills*. Berkala Ilmiah Pendidikan Fisika., DOI: 10.20527/bipf.v10i1.12231

Qasrawi, R. and BeniAbdelrahman, A. (2020). The higher and lower-order thinking skills (hots and lots) in unlock english textbooks (1st and 2nd editions) based on bloom's taxonomy: An analysis study. *International Online Journal of Education and Teaching*, 7:744–758.

Queiros, R., Pinto, C., & Cruz, M. (2024). Work in progress: Leveraging virtual escape rooms for innovative computer programming learning environments. EDUNINE 2024 - 8th IEEE World Engineering Education Conference: Empowering Engineering Education: Breaking Barriers through Research and Innovation, Proceedings DOI: DOI: 10.1109/EDUNINE60625.2024.10500625

Queirós, R., Pinto, M., Simões, A., & Portela, C. F. (2022). A primer on gamification standardization. In *Advances in Human and Social Aspects of Technology*. DOI: DOI: 10.4018/978-1-7998-8089-9.ch001

Rajaram, S., & Nebeling, M. (2022). Paper trail: An immersive authoring system for augmented reality instructional experiences. *Proceedings of the 2022 CHI Conference on Human Factors in Computing Systems*, DOI: DOI: 10.1145/3491102.3517486

Repetto, M., Bianco Prevot, A., Brizio, A., Boldi, A., Talarico, M., Stanchi, S., Palma, D., & Acquadro Maran, D. (2024). Digital escape rooms: A resource for environmental education. [Switzerland]. *Sustainability (Basel)*, 16(19), 8525. Advance online publication. DOI: 10.3390/su16198525

Repetto, M., Bruschi, B., & Talarico, M. (2023). Key issues and pedagogical implications in the design of digital educational escape rooms. *Journal of E-Learning and Knowledge Society*, 19. Advance online publication. DOI: 10.20368/1971-8829/1135749

Saleem, A., Noori, N. M., & Ozdamli, F. (2021). Gamification applications in e-learning: A literature review. *Technology. Knowledge and Learning*, 27(1), 139–159. DOI: 10.1007/s10758-020-09487-x

Segura-Robles, A., & Parra-González, M. (2019). Analysis of teachers' intercultural sensitivity levels in multicultural contexts. *Sustainability (Basel)*, 11(11), 3137. Advance online publication. DOI: 10.3390/su11113137

Seo, J., & Richard, G. T. (2021). Scaffolding all abilities into makerspaces: A design framework for universal, accessible and intersectionally inclusive making and learning. *Information and Learning Science*, 122(11/12), 795–815. Advance online publication. DOI: 10.1108/ILS-10-2020-0230

Shah, P., Gadekar, N., & Taylor, S. (2025). Escape from passive learning: Harnessing gamified digital educational escape rooms for immersive marketing education. *Journal of Marketing Education*. Advance online publication. DOI: 10.1177/02734753241306093

Sharma, R., & Monteiro, S. (2012). Reflective blogging: Education in a social constructivist environment - a case study. *International Journal of e-Education, e-Business, e-. Management Learning*. Advance online publication. DOI: 10.7763/IJEEEE.2012.V2.87

Shrivastava, A. (2020). Weaving sustainability in indian higher education system. *Emerald Emerging Markets Case Studies*, 10(3), 1–14. DOI: 10.1108/EEMCS-01-2020-0015

Shutaleva, A. (2023). Experiential learning as principle of environmental education. *E3S Web of Conferences*, DOI: .DOI: 10.1051/e3sconf/202342010010

Sidekerskiene, T., & Damaševicius, R. (2023). Out-of-the-box learning: Digital escape rooms as a metaphor for breaking down barriers in stem education. *Sustainability (Basel)*, 15(9), 7393. Advance online publication. DOI: 10.3390/su15097393

Smith, E. C. (2018). *Fostering a global mind: Challenges and best practices in higher education*. Research Anthology on Business and Technical Education in the Information Era., DOI: 10.4018/978-1-5225-3776-2.ch004

Smith, E. C. (2020). *Fostering a global mind*. Multicultural Instructional Design., DOI: 10.4018/978-1-5225-9279-2.ch070

Smith-Carrier, T., Goulden, A., & Singh, R. C. B. (2021). *Implementing Universal Design for Learning in Social Work Education*., DOI: 10.4018/978-1-7998-7106-4.ch008

Spatafora, A., Wagemann, M., Sandoval, C., Leisenberg, M., & Vaz de Carvalho, C. (2024). An educational escape room game to develop cybersecurity skills. *Computers*, 13(8), 205. Advance online publication. DOI: 10.3390/computers13080205

Staneva, A., Ivanova, T., Rasheva-Yordanova, K., & Borissova, D. (2023). Gamification in education: Building an escape room using vr technologies. In *2023 46th MIPRO ICT and Electronics Convention (MIPRO)*, pages 678–683. DOI: DOI: 10.23919/MIPRO57284.2023.10159923

Steen, T., & Wilson, C. (2020). Individual and cultural factors in teachers' attitudes towards inclusion: A meta-analysis. *Teaching and Teacher Education*, 95, 103127. DOI: 10.1016/j.tate.2020.103127

Steinböck, M., Luckner, N., Kayali, F., Proyer, M., & Kremsner, G. (2019). Cologon: a game to foster communication skills in inclusive education. *Proceedings of the 14th International Conference on the Foundations of Digital Games*, DOI: DOI: 10.1145/3337722.3341834

Stolzenberger, C., Wolf, N., & Trefzger, T. (2021). Technology deployment in physics lessons: Understanding optics better with augmented reality. *Journal of Physics: Conference Series*, 1929(1), 012037. Advance online publication. DOI: 10.1088/1742-6596/1929/1/012037

Su, X., Wambsganss, T., Rietsche, R., Neshaei, S. P., & Käser, T. (2023). Reviewriter: Ai-generated instructions for peer review writing. *Proceedings of the 18th Workshop on Innovative Use of NLP for Building Educational Applications*, pages 57–71, DOI: DOI: 10.18653/v1/2023.bea-1.5

Sujathamalini, D. J., Gunasekaran, D., and R., V. T. (2022). Universal design for learning in inclusive class room to ensure sustainable development goals: Inclusive and equitable quality class room. *International Journal of Science and Healthcare Research*, DOI: .DOI: 10.52403/ijshr.20220434

Sun, J. (2023). *A project-based learning (pbl) event in a chinese rural school: How can pbl help teachers and students accomplish deep learning?* Research and Advances in Education., DOI: 10.56397/RAE.2023.04.06

Sun, Q., & Kang, H. (2021). Learning through academic collaborations in/with the east: North american adult education scholars' insights. *Adult Education Quarterly*, 72(1), 3–23. DOI: 10.1177/07417136211050119

Swartz, S., Barbosa, B., & Crawford, I. (2019). Building intercultural competence through virtual team collaboration across global classrooms. *Business and Professional Communication Quarterly*, 83(1), 57–79. DOI: 10.1177/2329490619878834

Tan, B. S., & Chong, K. S. (2023). Unlocking the potential of game-based learning for soft skills development: A comprehensive review. In Journal of ICT in Education (Vol. 10, Issue 2, pp. 29–54). Universiti Pendidikan Sultan Idris. DOI: DOI: 10.37134/jictie.vol10.2.3.2023

Tasadduq, M., Khan, M. S., Nawab, R. M. A., Jamal, M. H., & Chaudhry, M. (2021). Exploring the effects of gamification on students with rote learning background while learning computer programming. *Computer Applications in Engineering Education*, 29(6), 1871–1891. DOI: 10.1002/cae.22428

Thapa, S. (2019). Assessing Intercultural Competence in Teacher Education: A Missing Link. DOI: .DOI: 10.1007/978-3-030-21029-8_11

Trigo, A., & Antunes, M. (2024). Teaching programming courses with digital educational escape rooms (deer): A conceptual proposal conducive to learning by trial and error. DOI: DOI: 10.4230/OASIcs.ICPEC.2024.9

Tzirides, A., Saini, A., Zapata, G. C., Searsmith, D., Cope, B., Kalantzis, M., Castro, V., Kourkoulou, T., Jones, J., da Silva, R. A., Whiting, J. K., and Kastania, N. P. (2023). Generative ai: Implications and applications for education. *ArXiv*, abs/2305.07605, DOI: .DOI: 10.48550/arXiv.2305.07605

Uibu, K. and Tagamets, E. (2022). Promoting teachers' intercultural competences for teaching in the diverse classroom. *Education and New Developments 2022 – Volume 2*, DOI: .DOI: 10.36315/2022v2end038

van Boeijen, A., Sonneveld, M., Hao, C., & Khodadadeh, Y. (2019). Developing cultural sensitivity: A student's perspective. In DS 95: *Proceedings of the 21st International Conference on Engineering and Product Design Education (E&PDE 2019)*, University of Strathclyde, Glasgow. 12th -13th September 2019. DOI: DOI: 10.35199/epde2019.35

van Rensburg, T., & Matthee, M. (2024). A framework for creating virtual escape rooms to teach computational thinking. *Communications in Computer and Information Science*, 1862, 3–17. Advance online publication. DOI: 10.1007/978-3-031-48536-7_1

Vekua, N., Lubsky, A. A., Perevozchikova, M., & Folgerova, J. N. (2020). Peculiarities of forming high-demanded soft skills in the educational space of the escape room. *Pedagogical Skills and Education*, 48(6), 397–412. DOI: 10.32744/pse.2020.6.31

Veldkamp, A., Daemen, J., Teekens, S., Koelewijn, S., Knippels, M. P. J., & Joolingen, W. (2020). Escape boxes: bringing escape room experience into the classroom (extended abstract). pages 67–70. DOI: DOI: 10.1111/bjet.12935

Viken, A., Höckert, E., & Grimwood, B. (2021). Cultural sensitivity: Engaging difference in tourism. *Annals of Tourism Research*, 89, 103223. Advance online publication. DOI: 10.1016/j.annals.2021.103223

Voloshynov, S., Popova, H., Yurzhenko, A., and Shmeltser, E. (2019). The use of digital escape room in educational electronic environment of maritime higher education institutions. pages 347–359, DOI: .DOI: 10.31812/123456789/3869

Wart, A. V., O'Brien, T. C., Varvayanis, S., Alder, J., Greenier, J., Layton, R., Stayart, C., Wefes, I., & Brady, A. (2020). Applying experiential learning to career development training for biomedical graduate students and postdocs: Perspectives on program development and design. *CBE Life Sciences Education*, 19(3), es7. Advance online publication. DOI: 10.1187/cbe.19-12-0270 PMID: 32822277

Whinnery, S., Fogle, K. C., Stark, J., & Whinnery, K. (2020). Building collaborative teacher education: Integrating udl through a faculty learning community. *Journal of Practitioner Research*, 5(2). Advance online publication. DOI: 10.5038/2379-9951.5.2.1161

Wiesemüller, P., Mateen, S., Dengel, A., & Voß-Nakkour, S. (2024). Access to escape - an immersive game-based learning experience for accessibility education in virtual reality. *Communications in Computer and Information Science*, 1904, 273–288. Advance online publication. DOI: 10.1007/978-3-031-47328-9_21

Williams, C. C., Mostert, C., & van Vuuren, M. J. (2022). International collaboration in higher education: A reflection of student and lecturer experiences. *ECE Official Conference Proceedings*, DOI: DOI: 10.22492/issn.2188-1162.2022.34

Wirnata, D., Widiana, I. W., & Bayu, G. W. (2021). *The effectiveness of learning activities based on revised bloom taxonomy on problem-solving ability*. Indonesian Journal Of Educational Research and Review., DOI: 10.23887/ijerr.v4i2.37370

Wogu, I. A., Misra, S., Assibong, P. A., Olu-Owolabi, E. F., Maskeliūnas, R., & Damasevicius, R. (2019). Artificial intelligence, smart classrooms and online education in the 21st century: Implications for human development. *Journal of Cases on Information Technology*, 21(3), 66–79. DOI: 10.4018/JCIT.2019070105

Wrench, A., Neill, B., & Diamond, A. (2021). International service-learning: Possibilities for developing intercultural competence and culturally responsive pedagogies. *Asia-Pacific Journal of Teacher Education*, 50(2), 215–228. DOI: 10.1080/1359866X.2021.2010275

Yan, L., Sha, L., Zhao, L., Li, Y., Martínez-Maldonado, R., Chen, G., Li, X., Jin, Y., & Gaševic, ́. D. (2023). Practical and ethical challenges of large language models in education: A systematic scoping review. *British Journal of Educational Technology*. Advance online publication. DOI: 10.1111/bjet.13370

Yli-Panula, E., Jeronen, E. K., & Mäki, S. (2022). School culture promoting sustainability in student teachers' views. *Sustainability (Basel)*, 14(12), 7440. Advance online publication. DOI: 10.3390/su14127440

Zatarain-Cabada, R., Estrada, M. L. B., Cárdenas-Sainz, B. A., & Chavez-Echeagaray, M. E. (2022). Experiences of web-based extended reality technologies for physics education. *Computer Applications in Engineering Education*, 31(1), 63–82. DOI: 10.1002/cae.22571

KEY TERMS AND DEFINITIONS

21st-Century Skills: Competencies such as creativity, collaboration, and digital literacy, emphasized in DEER-based education for workforce readiness.

Accessibility in DEERs: The design of DEERs to be inclusive, incorporating features like subtitles, adjustable difficulty levels, and multimodal content for learners with diverse needs.

Adaptive Learning: Educational methods that dynamically adjust to the learner's progress, performance, and preferences, often powered by AI in DEERs.

Augmented Reality (AR): A technology that overlays digital information onto the physical world, creating an enriched interactive learning environment.

Cultural Sensitivity: Designing DEER content to respect and integrate the diverse cultural contexts of learners, promoting inclusivity and global understanding.

Culturally-Situated Narrative Scenarios: DEER content that reflects the cultural backgrounds, values, and experiences of learners to enhance relevance and engagement.

Digital Educational Escape Rooms (DEERs): Interactive, immersive learning environments designed to teach through problem-solving, storytelling, and collaboration, often utilizing gamification and advanced technologies like VR and AR.

Experiential Learning: Learning through active participation in tasks, scenarios, or simulations within DEERs, encouraging the application of knowledge and skills.

Gamification: The integration of game-like elements such as points, rewards, and challenges into DEERs to increase engagement and motivation among learners.

Generative AI: Artificial intelligence used in DEERs to create dynamic scenarios and challenges, enhancing personalized and engaging learning pathways.

Immersive Learning: A hands-on approach facilitated by VR/AR in DEERs, allowing learners to interact deeply with educational material in simulated environments.

Interdisciplinary Applications: The use of DEERs across multiple educational fields such as STEM, environmental education, and vocational training, demonstrating their versatility.

Large Language Models (LLMs): AI-driven models capable of generating personalized, contextually relevant educational content, enabling adaptive learning experiences in DEERs.

Microlearning: Delivering educational content in short, focused bursts within DEERs, aligning with modern learner preferences for quick and targeted learning.

Problem-Based Learning: An educational strategy where learners acquire knowledge by solving complex, real-world problems, integral to the DEER framework.

Soft Skills: Non-technical skills like problem-solving, communication, and collaboration, which are crucial for 21st-century workforce readiness and actively developed in DEERs.

Virtual Reality (VR): A fully immersive digital environment that allows users to interact with 3D content in a simulated space, enhancing learning experiences in fields like STEM education.

Compilation of References

Abbasi, S., Burkhardt, J.-M., & Denis, M. (2009). Spatial problem solving: Assembling three-dimensional puzzles in real and virtual environments. *Cognitive Processing*, 10(S2), 179–181. DOI: 10.1007/s10339-009-0298-3 PMID: 19693560

Abd-alrazaq, A. A., AlSaad, R., Alhuwail, D., Ahmed, A., Healy, P., Latifi, S., Aziz, S., Damseh, R., Alrazak, S. A., & Sheikh, J. (2023). Large language models in medical education: Opportunities, challenges, and future directions. *JMIR Medical Education*, 9, e48291. Advance online publication. DOI: 10.2196/48291 PMID: 37261894

Abdul Jabbar, A. I., & Felicia, P. (2015). Gameplay engagement and learning in game-based learning: A systematic review. *Review of Educational Research*, 85(4), 740–779. DOI: 10.3102/0034654315577210

Abdul Rahim, A. S. (2023). Escape the desert island: Blended escape rooms in the first-semester problem-based learning. *Journal of Chemical Education*, 100(6), 2459–2465. DOI: 10.1021/acs.jchemed.3c00119

Abensur Vuillaume, L., Laudren, G., Bosio, A., Thévenot, P., Pelaccia, T., & Chauvin, A. (2021). A didactic escape game for emergency medicine aimed at learning to work as a team and making diagnoses: Methodology for game development. *JMIR Serious Games*, 9(3), 9. DOI: 10.2196/27291 PMID: 34463628

Abrams, S. S., and Gerber, H. R. (2021). Videogames, libraries, and the feedback loop: Learning beyond the stacks. DOI: .DOI: 10.1108/9781800715059

Abrams, S. S., & Gerber, H. R. (2021). Videogames, feedback loops, and classroom practice. In *Teaching in the Game-Based Classroom* (1st ed., p. 9). Routledge.

Acosta, A. A. A., & Carballo, G. E. (2022). The use of immersive tools in higher education: Escape rooms. *InProceedings of the 2022 6th International Conference on Education and E-Learning*. DOI: DOI: 10.1145/3578837.3578848

Adams, E. (2010). *Fundamentals of Game Design*. New Riders.

Adwiah, A. R., Tantia, A. F., and Rantikassari, I. A. (2023). Implementation of storytelling method with folktales in instilling character values in children: A study at aba warungboto kindergarten. *Golden Age: Jurnal Ilmiah Tumbuh Kembang Anak Usia Dini*, DOI: .DOI: 10.14421/jga.2023.81-05

Akcaoglu, M., Jensen, L. J., and Gonzalez, D. (2021). Understanding children's problem-solving strategies in solving game-based logic problems. *International Journal of Technology in Education and Science*, DOI: .DOI: 10.46328/ijtes.98

Alasmari, T. (2020). Gamification effect on higher education students' motivation. *The Journal of Positive Psychology*, 57, 3009–3030. DOI: 10.17762/PAE.V57I9.891

Aldalur, I. (2025). Enhancing software development education through gamification and experiential learning with genially. *Software Quality Journal*, 33(1), 1. Advance online publication. DOI: 10.1007/s11219-024-09699-9

Alekhina, S., Mel'nik, Y. V., Samsonova, E., and Shemanov, A. (2021). Assessment of inclusive process as a tool for designing inclusion in educational institution. DOI: .DOI: 10.17759/pse.2021260509

Alexandropoulou, A., & Tsezou, A. 2023. Teaching science, sustainability and technology through gamification: The approach of escape rooms. DOI: DOI: 10.5194/egusphere-egu23-7397

Algerafi, M. A. M., Zhou, Y., Oubibi, M., & Wijaya, T. T. (2023). Unlocking the potential: A comprehensive evaluation of augmented reality and virtual reality in education. *Electronics (Basel)*, 12(18), 3953. Advance online publication. DOI: 10.3390/electronics12183953

Alha, K., Loerke, D., Koskinen, E., & Paavilainen, J. (2023). Augmented play: An analysis of augmented reality features in location-based games. *Convergence (London)*, 29(2), 342–361. Advance online publication. DOI: 10.1177/13548565231156495

Al-Jibouri, S., & Mawdesley, M. (2001). Design and experience with a computer game for teaching construction project planning and control. *Engineering, Construction, and Architectural Management*, 8(5/6), 418–427. DOI: 10.1108/eb021201

Alkhathlan, M., Tlachac, M. L., Harrison, L., & Rundensteiner, E. (2022). Improving image accessibility by combining haptic and auditory feedback. In *Proceedings of the 24th International ACM SIGACCESS Conference on Computers and* Accessibility. DOI: DOI: 10.1145/3517428.3550362

Almeida, F., & Simões, J. (2019). The role of serious games, gamification and industry 4.0 tools in the education 4.0 paradigm. *Contemporary Educational Technology*, 10(2), 120–136. Advance online publication. DOI: 10.30935/cet.554469

Almeida, R., Marques, D. F., Durán, R. L., Aguado, Y. B., Cid-Bartolomé, T., González-González, S., & Hatzikou, M. (2023). A game changer joint design journey of prems and proms in the hosmartai virtual coach study. *International Journal of Integrated Care*, 23(S1), 659. DOI: 10.5334/ijic.ICIC23586

Amolloh, O. P., Lilian, G. K., & Wanjiru, K. G. (2018). Experiential learning, conditional knowledge and professional development at university of nairobi, kenya—Focusing on preparedness for teaching practice. *International Education Studies*, 11(7), 125. Advance online publication. DOI: 10.5539/ies.v11n7p125

Anderson, L. W., & Krathwohl, D. R. (2001). *A Taxonomy for Learning, Teaching, and Assessing: A Revision of Bloom's Taxonomy of Educational Objectives*. Longman.

Andini, D., Rahayu, A., Budiningsih, C. A., & Mumpuniarti, M. (2020). The curriculum adaptation model in fulfilling the learning need for diverse students at inclusive classroom. *Universal Journal of Educational Research*, 8(3D), 115–123. DOI: 10.13189/ujer.2020.081716

Angelo, T., & Cross, P. (1993). *Classroom Assessment Techniques: A Handbook for College Teachers*. Jossey-Bass.

Ang, J. W. J., Ng, Y. N. A., & Liew, R. S. (2020). Physical and digital educational escape room for teaching chemical bonding. *Journal of Chemical Education*, 97(9), 2849–2856. Advance online publication. DOI: 10.1021/acs.jchemed.0c00612

Anker-Hansen, J., & Andrée, M. (2019). *In pursuit of authenticity in science education*. Nordic Studies in Science Education., DOI: 10.5617/nordina.4723

Antoniou, P. E., Papamalis, F., Dafli, E., Poultourtzidis, I., Schwarz, D., Woodham, L., Dimitriadis, S., Tagaras, K., Kyriakidis, N., David, P., Nikolaidou, M., Skríšovská, T., Poulton, T., & Bamidis, P. D. (2024). Medical education escape room aligned with flipped classroom and powered by mobile augmented reality. *Electronics (Basel)*, 13(12), 2367. Advance online publication. DOI: 10.3390/electronics13122367

Arbesser-Rastburg, G., Safikhani, S., Gustin, M., Hopfe, C., Schweiger, G., & Pirker, J. (2024). Project beyond: An escape room game in virtual reality to teach building energy simulations. *IEEE Conference on Computatonal Intelligence and Games*, DOI: DOI: 10.1109/CoG60054.2024.10645541

Arendale, D. (2018). Introduction to special issue on universal design for inclusive pedagogy and a future research agenda. *Education Sciences*, 8(4), 203. Advance online publication. DOI: 10.3390/educsci8040203

Arnold, S., Fujima, J., & Jantke, P. K. (2013). Storyboarding Serious Games for Large-scale Training Applications. In Proceedings of the 5th International Conference on Computer Supported Education - CSEDU; SciTePress, pages 651-655. DOI: DOI: 10.5220/0004415606510655

Asok, D., Abirami, A., Angeline, N., & Lavanya, R. (2016). Active learning environment for achieving higher-order thinking skills in engineering education. In *2016 IEEE 4th International Conference on MOOCs, Innovation and Technology in Education (MITE)*, pages 47–53. DOI: DOI: 10.1109/MITE.2016.020

Asteroid escape. (2024). https://www.smartgames.eu/uk/one-player-games/asteroid-escape. Accessed: 2024-01-01.

Ayas, A. Y., Aydın, H., Çetinkaya, A., & Güney, Z. (2023). *Artificial intelligence (ai)-based self-deciding character development application in two-dimensional video games.* Bilgi ve Ileti¸sim Teknolojileri Dergisi., DOI: 10.53694/bited.1247338

Azaiz, I., Deckarm, O., and Strickroth, S. (2023). Ai-enhanced auto-correction of programming exercises: How effective is gpt-3.5? *ArXiv*, abs/2311.10737, DOI: .DOI: 10.3991/ijep.v13i8.45621

Back, J., Back, S., Bexell, E., Stanisic, S., & Rosqvist, D. (2019). The quest: An escape room inspired interactive museum exhibition. In *Extended Abstracts of the Annual Symposium on Computer-Human Interaction in Play Companion Extended Abstracts.* ACM, DOI: DOI: 10.1145/3341215.3356987

Bado, N. (2022). Game-based learning pedagogy: A review of the literature. *Interactive Learning Environments*, 30(5), 936–948. DOI: 10.1080/10494820.2019.1683587

Baeva, L. (2020). Existential risks and problem of escapism under conditions of e-culture. *Przeglad Wschodnioeuropejski*, 11(1), 299–313. DOI: 10.31648/pw.5991

Baker, R., & Siemens, G. (2014). *Educational Data Mining and Learning Analytics.* Cambridge University Press., DOI: 10.1017/CBO9781139519526.016

Bakhsheshi, F. F. (2019). Serious games and serious gaming in escape rooms. In *2019 International Serious Games Symposium (ISGS)*, pages 42–47. IEEE, DOI: DOI: 10.1109/ISGS49501.2019.9047019

Balaskas, A., Schueller, S. M., Cox, A. L., Rashleigh, C., & Doherty, G. (2023). Examining young adults daily perspectives on usage of anxiety apps: A user study. *PLOS Digital Health*, 2(1), e0000185. DOI: 10.1371/journal.pdig.0000185 PMID: 36812622

Bandura, A. (1977). *Social Learning Theory*. Prentice Hall.

Bangun, A. V., & Pragholapati, A. (2021). *Enhancing critical thinking skills in nursing higher education in preparation for the industrial revolution 4.0*. KnE Life Sciences., DOI: 10.18502/kls.v6i1.8756

Banks, J. A. (2008). *An Introduction to Multicultural Education*. Pearson.

Barab, S. A., Thomas, M. K., Dodge, T., Carteaux, R., & Tuzun, H. (2005). Making learning fun: Quest atlantis, a game without guns. *Educational Technology Research and Development*, 53(1), 86–107. DOI: 10.1007/BF02504859

Baron, J. V. (2023). Level of cultural sensitivity and its influence on teachers' performance. *Journal of Social, Humanity, and Education*, DOI: .DOI: 10.35912/jshe.v3i4.1392

Baudrillard, J. (1981). *Simulacra and Simulations* (1st ed.). Routledge., DOI: 10.4324/9780367809195

Beatty, I. D. (2014). Gaming the system: Video games as a theoretical framework for instructional design. *ArXiv*. DOI: /arXiv.1401.6716DOI: 10.48550

Beatty, I. D. (2014). *Gaming the system: Video games as a theoretical framework for instructional design*. ArXiv.

Bégoin-Augereau, S., & Caron-Parque, J. (2010). Modified decision processes marked by linguistic forms in a problem solving task. *Cognitive Systems Research*, 11(3), 260–286. DOI: 10.1016/j.cogsys.2009.12.004

Benassi, A. (2019). Escape room a scuola: Ambienti fisici e virtuali per l'apprendimento. *Journal on Educational Technology*, 27, 174–185.

Benavente, B. (2022). *Physical, Hybrid, and Digital Escape Rooms*. In Advances in Game-Based Learning (pp. 340–355). IGI Global. DOI: DOI: 10.4018/978-1-7998-8645-7.ch017

Bender, E. M., Gebru, T., McMillan-Major, A., & Shmitchell, S. (2021). On the dangers of stochastic parrots: Can language models be too big? *Proceedings of the 2021 ACM Conference on Fairness, Accountability, and Transparency*. DOI: 10.1145/3442188.3445922

Bharti, M. K. (2023). *Exploring the impact of gamification on students' motivation, and learning outcomes in secondary education.* International Journal For Multidisciplinary Research., DOI: 10.36948/ijfmr.2023.v05i05.7877

Bhugaonkar, K., Bhugaonkar, R., & Masne, N. (2022). The trend of metaverse and augmented & virtual reality extending to the healthcare system. *Cureus*, 14. Advance online publication. DOI: 10.7759/cureus.29071 PMID: 36258985

Black, P., & Wiliam, D. (1998). Assessment and classroom learning. *Assessment in Education: Principles, Policy & Practice*, 5(1), 7–74. DOI: 10.1080/0969595980050102

Blahova, L., & Kostolny, J. (2023). Development of a module for collecting students' feedback in an interactive manner. In *2023 21st International Conference on Emerging eLearning Technologies and Applications (ICETA)*, pages 62–67. DOI: DOI: 10.1109/ICETA61311.2023.10343671

Bloom, B. S. (1956). *Taxonomy of Educational Objectives, Handbook I: The Cognitive Domain.* David McKay Co Inc.

Bolivar, S., Ortega, F., Zock-Obregon, M., & Rishe, N. (2018). *3d spatial gaming interaction to broad cs participation. In 3D Spatial Gaming Interaction to Broad CS Participation.* Springer.

Bonwell, C. C., & Eison, J. A. (1991). *Active Learning: Creating Excitement in the Classroom.* ASHE-ERIC Higher Education Report. School of Education and Human Development, George Washington University, Washington DC.

Borrás-Gené, O., Diez, R., & Macías-Guillén, A. (2022). Digital educational escape room analysis using learning styles. *Information (Basel)*, 13(11), 522. DOI: 10.3390/info13110522

Borrego, C., Fernández, C., Blanes, I., & Robles, S. (2017). Room escape at class: Escape games activities to facilitate the motivation and learning in computer science. *Journal of Technology and Science Education*, 7(2), 162–171. DOI: 10.3926/jotse.247

Botelho, F. H. F. (2021). Accessibility to digital technology: Virtual barriers, real opportunities. *Assistive Technology*, 33(sup1), 27–34. DOI: 10.1080/10400435.2021.1945705 PMID: 34951832

Botev, J., Mayer, J., & Rothkugel, S. (2019). Immersive mixed reality object interaction for collaborative context-aware mobile training and exploration. In *Proceedings of the 11th ACM Workshop on Immersive Mixed and Virtual Environment Systems.* DOI: DOI: 10.1145/3304113.3326117

Brady, S. C., & Andersen, E. (2019). An escape-room inspired game for genetics review. *Journal of Biological Education*, 55(4), 406–417. DOI: 10.1080/00219266.2019.1703784

Breien, F. S., & Wasson, B. (2020). Narrative categorization in digital game-based learning: Engagement, motivation amp; learning. *British Journal of Educational Technology*, 52(1), 91–111. DOI: 10.1111/bjet.13004

Brinkley, J., Huff, E. W., & Boateng, K. (2021). Tough but effective: Exploring the use of remote participatory design in an inclusive design course through student reflections. *Proceedings of the 52nd ACM Technical Symposium on Computer Science Education*, DOI: DOI: 10.1145/3408877.3432527

Bruckner, F., Baumann, C., Husinsky, M., Jakl, A., Püringer, J., Suess, R. V., Vogt, G., & Wintersberger, M. (2022). Interactive storytelling in extended reality: Concepts for the design. In *Interactive Film & Media Journal*. DOI: DOI: 10.32920/ifmj.v2i4.1693

Bruner, J. (1991). The narrative construction of reality. *Critical Inquiry*, 18(1), 1–21. DOI: 10.1086/448619

Bruner, J. (1996). *The Culture of Education*. Harvard University Press. DOI: 10.4159/9780674251083

Brusilovsky, P., & Millán, E. (2007). *User Models for Adaptive Hypermedia and Adaptive Educational Systems*. Springer Berlin Heidelberg., DOI: 10.1007/978-3-540-72079-9_1

Burgstahler, S. (2022). Tips for designing an accessible and inclusive online course. *Proceedings of the International Conference on Society and Information Technologies*, DOI: DOI: 10.54808/ICSIT2022.01.79

Bybee, R. W. (2002). *Learning Science and the Science of Learning*. NSTA press. DOI: 10.2505/9780873552080

Byrne, R. M., & Johnson-Laird, P. (2009). "if" and the problems of conditional reasoning. *Trends in Cognitive Sciences*, 13(7), 282–287. DOI: 10.1016/j.tics.2009.04.003 PMID: 19540792

Cabioch, T., Champagnat, R., Bosser, A.-G., Chiganne, J.-N., & Diéguez, M. (2019). Timing interactive narratives. *In2019 IEEE Conference on Games (CoG)*, pages 1–8. IEEE.

Caeiro, M., Vázquez, M. M., Nistal, M., Mikic-Fonte, F. A., Iglesias, M. J. F., Tsalapata, H., Heidmann, O., Carvalho, C. V., Jesmin, T., Terasmaa, J., & Tolstrup, L. (2021). A collaborative city-based game to support soft skills development in engineering and economics. In *2021 International Symposium on Computers in Education (SIIE)*, pages 1–4. DOI: DOI: 10.1109/SIIE53363.2021.9583639

Caingcoy, M. E. (2023). *Culturally responsive pedagogy*. Diversitas Journal., DOI: 10.48017/dj.v8i4.2780

Cairns, P., Power, C., Barlet, M., Haynes, G., Kaufman, C., & Beeston, J. (2019). Enabled players: The value of accessible digital games. *Games and Culture*, 16(2), 262–282. DOI: 10.1177/1555412019893877

Cai, W., Chi, Y., Zhou, C., Zhu, C., & Leung, V. C. M. (2018). Ubgaming: Ubiquitous cloud gaming system. *IEEE Systems Journal*, 12(3), 2483–2494. DOI: 10.1109/JSYST.2018.2797080

Calvert, J., & Hume, M. (2022). Immersing learners in stories: A systematic literature review of educational narratives in virtual reality. *Australasian Journal of Educational Technology*, 55–71. Advance online publication. DOI: 10.14742/ajet.7032

Cardoso, P., Morgado, L., & Coelho, A. (2020). Authoring game-based learning activities that are manageable by teachers. *ERCIM News*, 2020, •••.

Carpenter, K. M., Cunningham, A. D., Smith, L., & Krigbaum, K. (2023). The great escape: A novel approach to collaborative learning (pilot). *Journal of Occupational Therapy Education*, 7(2). Advance online publication. DOI: 10.26681/jote.2023.070209

CAST. (2018). Universal design for learning guidelines version 2.2. http://udlguidelines.cast.org

Cesário, V., Ribeiro, M., & Coelho, A. (2023). Design recommendations for improving immersion in role-playing video games. A focus on storytelling and localisation. [s]. *Interaction Design and Architecture*, (58), 207–225. DOI: 10.55612/s-5002-058-009)

Cezarotto, M., & Armstrong, A. L. (2023). *Designing inclusive educational games: accessibility rubric*. InfoDesign - Revista Brasileira de Design da Informação., DOI: 10.51358/id.v20i3.1064

Chamorro-Koc, M., Caldwell, G., Vasquez, A. V., Rodriguez, I. F., & Nates, C. (2021). Wearables of the past and the future: An immersive and cross-cultural learning experience of undergraduate students in an international design collaboration. *Creative Industries Journal*, 15(3), 293–316. DOI: 10.1080/17510694.2021.1939539

Chang, C.-C., & Hwang, G. (2021). An experiential learning-based virtual reality approach to fostering problem-resolving competence in professional training. *Interactive Learning Environments*, 31(8), 4713–4728. DOI: 10.1080/10494820.2021.1979049

Chao, T. (2023). The influence of cross-cultural adaptation on two language acquisition strategy. *Journal of Education and Educational Research*, 3(2), 2. DOI: 10.54097/jeer.v3i2.9038

Chen, Z., Jiang, Z., Yang, F., He, Z., Hou, Y., Cho, E., McAuley, J., Galstyan, A., Hu, X., & Yang, J. (2023). The first workshop on personalized generative ai @ cikm 2023: Personalization meets large language models. *Proceedings of the 32nd ACM International Conference on Information and Knowledge Management*, DOI: DOI: 10.1145/3583780.3615314

Cheng, R., Wu, N., Varvello, M., Chen, S., & Han, B. (2022). Are we ready for metaverse?: a measurement study of social virtual reality platforms. *Proceedings of the 22nd ACM Internet Measurement Conference, DOI:* DOI: 10.1145/3517745.3561417

Cheng, S.-C., Hwang, G.-J., & Chen, C.-H. (2019). From reflective observation to active learning: A mobile experiential learning approach for environmental science education. *British Journal of Educational Technology*, 50(5), 2251–2270. DOI: 10.1111/bjet.12845

Chen, H., Zhang, X., Xu, Y., Ren, J., Fan, J., Ma, Z., & Zhang, W. (2019). T-gaming: A cost-efficient cloud gaming system at scale. *IEEE Transactions on Parallel and Distributed Systems*, 30(12), 2849–2865. DOI: 10.1109/TPDS.2019.2922205

Chen, S.-Y., Tsai, J.-C., Liu, S.-Y., & Chang, C.-Y. (2021). The effect of a scientific board game on improving creative problem solving skills. *Thinking Skills and Creativity*, 41, 100921. DOI: 10.1016/j.tsc.2021.100921

Chen, Y., & Cheng, H. (2022). *The economics of the metaverse: A comparison with the real economy*. Metaverse., DOI: 10.54517/met.v3i1.1802

Chi, M. T. H. (2009). Active-constructive-interactive: A conceptual framework for differentiating learning activities. *Topics in Cognitive Science*, 1(1), 73–105. DOI: 10.1111/j.1756-8765.2008.01005.x PMID: 25164801

Chineka, R., & Yasukawa, K. (2020). Intergenerational learning in climate change adaptations; limitations and affordances. *Environmental Education Research*, 26(4), 577–593. DOI: 10.1080/13504622.2020.1733494

Cho, E. (2012). Meta-representation of video game through the cross-media storytelling: Focusing on the animated motion picture game over. *12*:25–36, DOI: .DOI: 10.7583/JKGS.2012.12.3.25

Choi, Y., Lee, S. M., & Li, H. (2013). Audio and visual distractions and implicit brand memory: A study of video game players. *Journal of Advertising*, 42(2-3), 219–227. DOI: 10.1080/00913367.2013.775798

Christopoulos, A., Mystakidis, S., Cachafeiro, E., & Laakso, M. (2022). Escaping the cell: Virtual reality escape rooms in biology education. *Behaviour & Information Technology*, 42(9), 1434–1451. DOI: 10.1080/0144929X.2022.2079560

Clapson, M. L., Gilbert, B. C. T., Mozol, V., Schechtel, S., Tran, J., & White, S. (2020). Chemescape: Educational battle box puzzle activities for engaging outreach and active learning in general chemistry. *Journal of Chemical Education*, 97(1), 125–131. Advance online publication. DOI: 10.1021/acs.jchemed.9b00612

Clark, R. C., & Mayer, R. E. (2011). *E-Learning and the Science of Instruction: Proven Guidelines for Consumers and Designers of Multimedia Learning*. Pfeiffer. DOI: 10.1002/9781118255971

Clune, M. (2019). Digital escape game. *Pacific Journal of Technology Enhanced Learning, 2.*

Coelho, C., Cordeiro, A., Alcoforado, L., & Moniz, G. (2022). Survey on student school spaces: An inclusive design tool for a better school. *Buildings (Basel, Switzerland)*, 12(4), 392. Advance online publication. DOI: 10.3390/buildings12040392

Coenraad, M., Palmer, J., Franklin, D., & Weintrop, D. (2019). Utilizing participatory design to develop a culturally relevant computer science curriculum. *Proceedings of the 50th ACM Technical Symposium on Computer Science Education*, DOI: DOI: 10.1145/3287324.3293803

Coenraad, M., Weintrop, D., Eatinger, D., Palmer, J., & Franklin, D. (2021). *Identifying youths' spheres of influence through participatory design*. Designs for Learning., DOI: 10.16993/dfl.163

Coffman-Wolph, S., Gray, K., and Pool, M. (2018). Designing an escape room game to develop problem solving and spatial reasoning skills. DOI: DOI: 10.5703/1288284316855

Cojocariu, V.-M. (2021). *Pedagogical approaches to intercultural competence development*. Moldavian Journal for Education and Social Psychology., DOI: 10.18662/mjesp/4.2/24

Cole, J. (2009). *Using Moodle: Teaching with the Popular Open-source Course Management System*. O'Reilly Media, Inc.

Collins, C., Nanda, S., Palmer, B., Mohabbat, A., Schleck, C., Mandrekar, J., Mahapatra, S., Beckman, T., & Wittich, C. (2019). A cross-sectional study of learning styles among continuing medical education participants. *Medical Teacher*, 41(3), 318–324. DOI: 10.1080/0142159X.2018.1464134 PMID: 29703093

Colzato, L., van Leeuwen, P. J. A., van den Wildenberg, W. P., & Hommel, B. (2010). Doom'd to switch: Superior cognitive flexibility in players of first person shooter games. *Frontiers in Psychology*, 1. Advance online publication. DOI: 10.3389/fpsyg.2010.00008 PMID: 21833191

Conati, C. (2009). Intelligent tutoring systems: New challenges and directions. *Annual Review of Computer Science*, 3, 555–580.

Cordova-Rangel, J., & Caro, K. (2021). Designing a serious game to promote visitors' engagement in a science museum exhibition. In *Proceedings of the 8th Mexican Conference on Human-Computer Interaction*. DOI: DOI: 10.1145/3492724.3492732

Costa, D. M., Francisco, A. P., and Russo, L. M. S. (2020). Hardness of modern games. *ArXiv, abs/2005.10506.*

Coughlan, T., & Lister, K. (2022). *Creating stories of learning, for learning: Exploring the potential of self-narrative in education with 'our journey.* Creativity and Cognition., DOI: 10.1145/3527927.3535208

Craddock, G., Fitzpatrick, D., Hubbard, J., O'Reilly, R., & Wilkinson, M. (2022). Universal design in education (ude) across the life course: Applying a systems framework to create an inclusive experience for all. *InProceedings of the IASE 2021 Satellite Conference*. DOI: DOI: 10.52041/iase.kxvpc

Cressey, J. (2020). *Universal Design for Learning*. DOI: .DOI: 10.4018/978-1-7998-1770-3.ch008

Cristea, D., & Ivanov, M. catalina, R. C., renato, M. L., and Andrei, L. (2020). Immersive data analytics for enhancing organisational knowledge transfer processes through a custom developed virtual reality framework. *eLearning and Software for Education*, DOI: DOI: 10.12753/2066-026X-20-097

Csikszentmihalyi, M. (1990). *Flow: The Psychology of Optimal Experience*. Harper & Row.

Cujdíková, M., & Vankúš, P. (2024). The use of educational escape rooms in the lifelong training of mathematics teachers. *Proceedings of the European Conference on Games-based Learning*, DOI: DOI: 10.34190/ecgbl.18.1.2653

Cumbo, B. J., & Selwyn, N. (2021). Using participatory design approaches in educational research. *International Journal of Research & Method in Education*, 45(1), 60–72. DOI: 10.1080/1743727X.2021.1902981

Dalton, E. M., Lyner-Cleophas, M., Ferguson, B., & Mckenzie, J. (2019). Inclusion, universal design and universal design for learning in higher education: South africa and the united states. *African Journal of Disability*, 8. Advance online publication. DOI: 10.4102/ajod.v8i0.519 PMID: 31392169

Damaševičius, R., & Sidekerskienė, T. (2023). Designing Metaverse Escape Rooms for Microlearning in STEM Education. In Advances in Higher Education and Professional Development (pp. 192–211). IGI Global. DOI: DOI: 10.4018/978-1-6684-8656-6.ch009

Damaševicius, R. (2023). *The rise of chatgpt and the demise of bloom's taxonomy of learning stages.* Creative AI Tools and Ethical Implications in Teaching and Learning., DOI: 10.4018/979-8-3693-0205-7.ch006

Damaševicius, R., & Zailskaite-Jakšte, L. (2024). *Digital twin technology: Necessity of the future in education and beyond.* Automated Secure Computing for Next-Generation Systems.

Darejeh, A. (2023). Empowering education through eerp: A customizable educational vr escape room platform. In *2023 IEEE International Symposium on Mixed and Augmented Reality Adjunct (ISMAR-Adjunct), pages 764–766.* DOI: DOI: 10.1109/ISMAR-Adjunct60411.2023.00166

Darling-Hammond, L., Hyler, M. E., & Gardner, M. (2017). *Effective Teacher Professional Development.* Learning Policy Institute. DOI: 10.54300/122.311

David, D., Edwin, , Arman, E., Hikari, , Chandra, N., & Nadia, N. (2019). Development of escape room game using vr technology. *Procedia Computer Science*, 157, 646–652. DOI: 10.1016/j.procs.2019.08.223

De Castell, S., Perry, N., Bailey, L., & Jenson, J. (2023). Microgames and language learning: Performance before competence? *European Conference on Games Based Learning*, 17(1):143–153, DOI: DOI: 10.34190/ecgbl.17.1.1766

de Vicente, D. P., del Olmo-Fernández, M. J. A., Santos-Villalba, M. J., & Leiva-Olivencia, J. J. (2021). A qualitative study on the intercultural educational sensitivity of the professors at the university of malaga (spain). *Education Sciences*, 11(2), 74. Advance online publication. DOI: 10.3390/educsci11020074

Debchaudhury, S. (2024). Impact of digital game-based learning experiences on reasoning skills and science engagement. DOI: .DOI: 10.34190/ecgbl.18.1.2690

Deci, E. L., & Ryan, R. M. (2000). The "what" and "why" of goal pursuits: Human needs and the self-determination of behavior. *Psychological Inquiry*, 11(4), 227–268. DOI: 10.1207/S15327965PLI1104_01

Deci, E. L., Vallerand, R. J., Pelletier, L. G., & Ryan, R. M. (1991). Motivation and education: The self-determination perspective. *Educational Psychologist*, 26(3–4), 325–346. DOI: 10.1080/00461520.1991.9653137

Délécraz, C. (2023). Scoring the original soundtrack of an escape room. *Journal of Sound and Music in Games*, 4(1), 26–71. DOI: 10.1525/jsmg.2023.4.1.26

Demianenko, O. and Moroz, T. (2023). Formation of intercultural competence in the process of implementing the multilingual education model. *PARADIGM OF KNOWLEDGE*, DOI: .DOI: 10.26886/2520-7474.4(58)2023.3

Denny, P., McDonald, F., Empson, R., Kelly, P., & Petersen, A. (2018). Empirical support for a causal relationship between gamification and learning outcomes. *In-Proceedings of the 2018 CHI Conference on Human Factors in Computing Systems*. DOI: DOI: 10.1145/3173574.3173885

Deterding, S., O'Hara, K., Sicart, M., Dixon, D., and Nacke, L. (2011). Gamification: Using game design elements in non-gaming contexts. page 2425 – 2428. DOI: .DOI: 10.1145/1979742.1979575

Dicheva, D., Caldwell, R., & Guy, B. (2020). Do badges increase student engagement and motivation? *InProceedings of the 21st Annual Conference on Information Technology Education*. DOI: DOI: 10.1145/3368308.3415393

Dicheva, D., Dichev, C., Agre, G., & Angelova, G. (2015). Gamification in education: A systematic mapping study. *Journal of Educational Technology & Society*, 18(3), 75–88.

Dickey, M. D. (2005). Engaging by design: How engagement strategies in popular computer and video games can inform instructional design. *Educational Technology Research and Development*, 53(2), 67–83. DOI: 10.1007/BF02504866

Dimitriadou, A., Djafarova, N., Turetken, O., Verkuyl, M., & Ferworn, A. (2020). Challenges in serious game design and development: Educators' experiences. *Simulation & Gaming*, 52(2), 132–152. DOI: 10.1177/1046878120944197

Dittman, J. M., Maiden, K. M., Matulewicz, A. T., Beaird, G., Lockeman, K. S., & Dow, A. (2021). A flexible customizable virtual escape room approach for interprofessional learners. *Journal of Interprofessional Education & Practice*, 24, 100455. DOI: 10.1016/j.xjep.2021.100455

Donohue, S., Woldorff, M., & Mitroff, S. (2010). Video game players show more precise multisensory temporal processing abilities. *Attention, Perception & Psychophysics*, 72(4), 1120–1129. DOI: 10.3758/APP.72.4.1120 PMID: 20436205

dos Santos Nunes, E. P., da Conceição Júnior, V. A., & Borges, L. C. L. F. (2018). Interaction techniques to promote accessibility in games for touchscreen mobile devices: A systematic review. In *Proceedings of the 17th International Conference on Human-Computer Interaction, pages* 178–191. DOI: DOI: 10.1007/978-3-319-92049-8_13

Downey, G., Lloyd, K., Bilous, R. H., Hammersley, L., Rawlings-Sanaei, F., Amigó, M. F., Gilchrist, S., Baker, M., & Coffey, E. (2018). "classroom of many cultures": Educational design opportunities in intercultural co-creation. *Intercultural Studies in Higher Education*, 25, 45–59. DOI: 10.1007/978-3-319-74579-4_3

Drosos, V., Alexakos, C., Alexandri, A., & Tsolis, D. (2018). Evaluating 3d serious games on cultural education. *In 2018 9th International Conference on Information, Intelligence, Systems and Applications (IISA)*, pages 1–5. IEEE, DOI: DOI: 10.1109/IISA.2018.8633689

Dunleavy, M., Dede, C., & Mitchell, R. (2008). Affordances and limitations of immersive participatory augmented reality simulations for teaching and learning. *Journal of Science Education and Technology*, 18(1), 7–22. DOI: 10.1007/s10956-008-9119-1

Durmaz, Y., & Gokalp, D. (2023). *Introduction of metaverse to our lives and unlimited services in the world of metaverse*. International Business & Economics Studies., DOI: 10.22158/ibes.v6n1p1

Dutra, S. V. O., Chee, V., & Clochesy, J. (2023). Adapting an educational software internationally: Cultural and linguisticual adaptation. *Education Sciences*, 13(3), 237. DOI: 10.3390/educsci13030237

Dweck, C. (2006). *Mindset: The New Psychology of Success*. Random House.

Dyankova, G. and Nikolova, S. (2023). Multicultural competence as a teacher's metacognition to achieve a positive school climate. International Journal of Cognitive Research in Science, *Engineering and Education (IJCRSEE)*, DOI: .DOI: 10.23947/2334-8496-2023-11-2-257-265

Edyburn, D. L. (2010). Would you recognize universal design for learning if you saw it? ten propositions for new directions for the second decade of udl. *Learning Disability Quarterly*, 33(1), 33–41. DOI: 10.1177/073194871003300103

Efe, A. (2022). *Taking virtual reality and augmented reality to the next level: Artificial intelligence with mixed reality*. Uluslararası Yönetim Bilişim Sistemleri ve Bilgisayar Bilimleri Dergisi., DOI: 10.33461/uybisbbd.1190125

Egan, K. (1986). *Teaching as Story Telling*. Althouse Press.

Egenfeldt-Nielsen, S. (2007). Educational Potential of Computer Games. *Continuum*.

Eggen, P., & Kauchak, D. (2010). *Educational Psychology: Windows on Classrooms*. Pearson.

Egidi, M. (2006). Chapter 1 decomposition patterns in problem solving. *Contributions to Economic Analysis*, 280:15–46, DOI: DOI: 10.1016/S0573-8555(06)80003-1

Ekin, C. C., Polat, E., & Hopcan, S. (2023). Drawing the big picture of games in education: A topic modeling-based review of past 55 years. *Computers & Education*, 194, 104700. DOI: 10.1016/j.compedu.2022.104700

Elsayed-Ali, S., Bonsignore, E. M., & Chan, J. (2023). Exploring challenges to inclusion in participatory design from the perspectives of global north practitioners. *Proceedings of the ACM on Human-Computer Interaction*, 7(CSCW1), 1–25. DOI: 10.1145/3579606

Esser, S. (2019). Reformulation of intercultural competence in the context of a diversity-conscious teacher education and pedagogical school development. *International Journal of Creativity and Innovation in Humanities and Education*, DOI: .DOI: 10.21608/ijcihe.2019.182878

Esteban-Guitart, M., Monreal-Bosch, P., Palma, M., & González-Ceballos, I. (2020). Sustaining students' identities within the context of participatory culture. designing, implementing and evaluating an interactive learning activity. *Sustainability (Basel)*, 12(12), 4870. DOI: 10.3390/su12124870

Eukel, H. N., Frenzel, J. E., Frazier, K., & Miller, M. (2020). Unlocking student engagement: Creation, adaptation, and application of an educational escape room across three pharmacy campuses. *Simulation & Gaming*, 51(2), 167–179. DOI: 10.1177/1046878119898509

Eukel, H. N., & Morrell, B. (2020). Ensuring educational escape-room success: The process of designing, piloting, evaluating, redesigning, and re-evaluating educational escape rooms. *Simulation & Gaming*, 52(1), 18–23. DOI: 10.1177/1046878120953453

Eulenstein, M., Stiegel, S., Czech, S., Erkan, R., & Schoormann, T. (2025). Learning through designing: A visual inquiry tool for digital educational escape games. *Lecture Notes in Computer Science (including subseries Lecture Notes in Artificial Intelligence and Lecture Notes in Bioinformatics)*, DOI: DOI: 10.1007/978-3-031-78269-5_41

Falk, J. H., & Dierking, L. D. (2016). *The Museum Experience Revisited*. Routledge., DOI: 10.4324/9781315417851

Fenwick, T. J. (2000). Expanding conceptions of experiential learning: A review of the five contemporary perspectives on cognition. *Adult Education Quarterly*, 50(4), 243–272. DOI: 10.1177/07417130022087035

Fernandes, F., & Werner, C. (2022). Accessibility in the metaverse: Are we prepared? *Anais do XIII Workshop sobre Aspectos da Interação Humano-Computador na Web Social (WAIHCWS 2022), DOI:* DOI: 10.5753/waihcws.2022.226618

Ferrari, A., Huichapa, T., Spoletini, P., Novielli, N., Fucci, D., & Girardi, D. (2023). Using voice and biofeedback to predict user engagement during product feedback interviews. *ACM Transactions on Software Engineering and Methodology*. Advance online publication. DOI: 10.1145/3635712

Foo, S. C., & Foo, K.-K. (2022). Purposeful field trip: Impact on experiential learning opportunities and critical thinking skills. *Pertanika Journal of Social Science & Humanities*, 30(1), Preprint. Advance online publication. DOI: 10.47836/pjssh.30.1.01

Foster, C., Urdl, E., Steller, C., Sharaf, S., Rößner, R., Mystakidis, S., Ingram, J., Ramamoorthy, H. K., & Mentzelopoulos, M. (2024). Work-in-progress—telecommunications history escape room puzzle game: Become a phyctoria archmaster. In *Academic Proceedings of the 10th International Conference of the Immersive Learning Research Network (iLRN2024)*, iLRN2024, page 90–95. The Immersive Learning Research Network, DOI: DOI: 10.56198/U6C0WURU6

Fotaris, P., & Mastoras, T. (2019). Escape rooms for learning: A systematic review. *InProceedings of the 12th European Conference on Game Based Learning*. DOI: DOI: 10.34190/GBL.19.179

Fotaris, P., & Mastoras, T. (2022). Room2educ8: A framework for creating educational escape rooms based on design thinking principles. *Education Sciences*, 12(11), 768. Advance online publication. DOI: 10.3390/educsci12110768

Frania, M. (2021). *Educational e-escape room as an educational method of media literacy training for future teachers during the covid-19 pandemic. Media Education*. Mediaobrazovanie., DOI: 10.13187/me.2021.3.452

Friedmann, J. (2021). Storytelling for media: Introduction to the theory and practice of narrative design. ISBN: 9783838557649, DOI: DOI: 10.36198/9783838557649

Friedrich, C., Teaford, H., Taubenheim, A., Boland, P., & Sick, B. (2018). Escaping the professional silo: An escape room implemented in an interprofessional education curriculum. *Journal of Interprofessional Care*, 33(5), 573–575. DOI: 10.1080/13561820.2018.1538941 PMID: 30362849

Friedrich, C., Teaford, H., Taubenheim, A., & Sick, B. (2020). Interprofessional health care escape room for advanced learners. *The Journal of Nursing Education*, 59(1), 46–50. DOI: 10.3928/01484834-20191223-11 PMID: 31945176

Fullan, M. (2007). *The New Meaning of Educational Change*. Teachers College Press.

Gagné, R. M. (1985). *The Conditions of Learning and Theory of Instruction*. Holt, Rinehart & Winston.

Gan, W., Qi, Z., Wu, J., and Lin, C.-W. (2023). Large language models in education: Vision and opportunities. *ArXiv*, abs/2311.13160, DOI: .DOI: 10.48550/arXiv.2311.13160

Garcez, L., Thiry, M., & Fernandes, A. M. R. (2020). Accessible features to support visually impaired people in game development: A systematic literature review of the last 15 years. In *2020 15th Iberian Conference on Information Systems and Technologies (CISTI), pages 1–6*. DOI: DOI: 10.23919/CISTI49556.2020.9140904

Garcia, K. A. and Pantao, J. (2021). Cultural sensitivity and classroom management of teachers. *International Journal of Professional Development, Learners and Learning*, DOI: .DOI: 10.30935/ijpdll/11093

Gardner, H. (1983). *Frames of Mind: The Theory of Multiple Intelligences*. Basic Books.

Gardner, M. (1960). Mathematical games. *Scientific American*, 202(3), 172–182. DOI: 10.1038/scientificamerican0360-172

Garrand, T. (2020). *Interactive multimedia narrative and linear narrative. Writing for Multimedia and the Web*. Elsevier., DOI: 10.4324/9780080474861

Garris, R., Ahlers, R., & Driskell, J. E. (2002). Games, motivation, and learning: A research and practice model. *Simulation & Gaming*, 33(4), 441–467. DOI: 10.1177/1046878102238607

Garwood, J. (2020). Escape to learn! an innovative approach to engage students in learning. *The Journal of Nursing Education*, 59(5), 278–282. DOI: 10.3928/01484834-20200422-08 PMID: 32352543

Gay, G., & Banks, J. A. (2010). *Culturally Responsive Teaching: Theory, Research, and Practice*. Teachers College Press.

Gee, J. P. (2003). *What Video Games Have to Teach Us About Learning and Literacy*. Palgrave Macmillan. DOI: 10.1145/950566.950595

Ge, M. (2023). *Review on the application of deep learning algorithms in video game ai agent*. Applied and Computational Engineering., DOI: 10.54254/2755-2721/4/2023322

Geremias, M. S., Dutra, T. C., Gasparini, I., & Maschio, E. (2023). An educational game to promote the development of computational thinking in children both neurotypical and with intellectual disabilities. *2023 IEEE International Conference on Advanced Learning Technologies (ICALT)*, DOI: DOI: 10.1109/ICALT58122.2023.00036

Gil Ruiz, P. (2024). Gamification as a methodology to enhance analytical and sustainable engagement on social media. *Discover Education*, 3(1), 4. Advance online publication. DOI: 10.1007/s44217-023-00074-7

Gist, C. D., Jackson, I., Nightengale-Lee, B., & Allen, K. M. (2019). *Culturally responsive pedagogy in teacher education*. Oxford Research Encyclopedia of Education., DOI: 10.1093/acrefore/9780190264093.013.266

Goh, E. S., Sunar, M. S., & Ismail, A. (2019). 3d object manipulation techniques in handheld mobile augmented reality interface: A review. *IEEE Access : Practical Innovations, Open Solutions*, 7, 40581–40601. DOI: 10.1109/ACCESS.2019.2906394

Gomes, G., Vidal, C., Neto, J. B. C., & Nogueira, Y. L. (2021). Two level control of non-player characters for navigation in 3d games scenes: A deep reinforcement learning approach. In *2021 20th Brazilian Symposium on Computer Games and Digital Entertainment (SBGames), pages 182–190*. DOI: DOI: 10.1109/SBGames54170.2021.00030

Gómez-Cambronero, A., González-Pérez, A., Miralles, I., & Casteleyn, S. (2019). Mixed reality escape room video game to promote learning in indoor environments. *In EDULEARN19 Proceedings*, pages 6098–6104.

Gómez-Cambronero, Á., González-Pérez, A., Miralles, I., & Casteleyn, S. (2019). Mixed reality escape room video game to promote learning in indoor environments. In *EDULEARN19 Proceedings*, pages 6098–6104. DOI: 10.21125/edulearn.2019.1412

Gong, J., Shi, Y., Wang, J., Shi, D., and Xu, Y.-Q. (2018). Escape from the dark jungle: A 3d audio game for emotion regulation.

Grabska-Gradzinska, I., Grabska, E., Nowak, L., & Palacz, W. (2020). Towards automatic generation of storyline aided by collaborative creative design. In *2020 International Conference on Interactive Collaborative Learning (ICL)*, pages 47–56. DOI: DOI: 10.1007/978-3-030-60816-3_6

Grande-de Prado, M., García-Martín, S., Baelo, R., & Abella-García, V. (2020). Edu-escape rooms. In *Encyclopedia*. DOI: DOI: 10.3390/encyclopedia1010004

Gregersen-Hermans, J. (2021). Toward a curriculum for the future: Synthesizing education for sustainable development and internationalization of the curriculum. *Journal of Studies in International Education*, 25(4), 461–481. DOI: 10.1177/10283153211031033

Grönberg, N., Knutas, A., Hynninen, T., & Hujala, M. (2020). An online tool for analyzing written student feedback. *Proceedings of the 20th Koli Calling International Conference on Computing Education Research*, DOI: DOI: 10.1145/3428029.3428565

Grönberg, N., Knutas, A., Hynninen, T., & Hujala, M. (2021). Palaute: An online text mining tool for analyzing written student course feedback. *IEEE Access : Practical Innovations, Open Solutions*, 9, 134518–134529. DOI: 10.1109/ACCESS.2021.3116425

Guo, X., Rees, D., & Richards, M. (2020). A scalable interactive mixed reality escape room simulation for anatomy learning. *International Journal of Modeling and Optimization*, 10, 41–46. DOI: 10.7763/IJMO.2020.V10.744

Guo, Y. (2021). Why are cultural adaptations such a prevalent strategy in the media business? *Scientific and Social Research*, 3(5), 1253. DOI: 10.36922/ssr.v3i5.1253

Gupta, R., Nawani, R., & Talreja, V. (2016). Virtual reality content creation using unity 3d and blender. *International Journal of Computer Applications*, 156(3), 8–12. DOI: 10.5120/ijca2016912395

Guskey, T. R. (2000). *Evaluating Professional Development*. Corwin Press.

Gyaurov, D., Fabricatore, C., & Bottino, A. (2022). Features of entertainment digital games for learning and developing complex problem-solving skills: A protocol for a systemic review. *International Journal of Qualitative Methods*, 21, 16094069221128491. Advance online publication. DOI: 10.1177/16094069221128491

Hajj, M. H. E. (2023). The future of digital arts. *International Journal of Education and Learning Research, DOI:*.DOI: 10.21608/ijelr.2023.211965.1002

Hakshurian, Y. (2023). Teachers' competencies in developing digital educational escape rooms. *Educatia*, 21(25), 66–71. Advance online publication. DOI: 10.24193/ed21.2023.25.07

Halimah, L., Arifin, R. R. M., Yuliariatiningsih, M. S., Abdillah, F., & Sutini, A. (2020). Storytelling through "wayang golek" puppet show: Practical ways in incorporating character education in early childhood. *Cogent Education*, 7(1), 1794495. Advance online publication. DOI: 10.1080/2331186X.2020.1794495

Halloran, J., & Minerva, A. (2019). Touch and play? investigating the value of touchscreens for gamer experience. *Entertainment Computing*, 32, 100312. Advance online publication. DOI: 10.1016/j.entcom.2019.100312

Hannafin, M., Hill, J., & Land, S. (1997). Student-centered learning and interactive multimedia: Status, issues, and implications. *Contemporary Education*, 68, 94–99.

Hanus, A., Hoover, M., Lim, A., & Miller, J. (2019). A collaborative virtual reality escape room with passive haptics. In *IEEE Conference on Virtual Reality and 3D User Interfaces (VR)*, pages 1413–1414. DOI: DOI: 10.1109/VR.2019.8798241

Harley, D., Grambart, S., Bergel, R. S., & Mazalek, A. (2022). Together alone: A tangible online narrative. *InSixteenth International Conference on Tangible, Embedded, and Embodied Interaction*. ACM. DOI: 10.1145/3490149.3501311

Hassan, L. (2023). Accessibility of games and game-based applications: A systematic literature review and mapping of future directions. *New Media & Society*. Advance online publication. DOI: 10.1177/14614448231204020

Hattie, J., & Timperley, H. (2007). The power of feedback. *Review of Educational Research*, 77(1), 81–112. DOI: 10.3102/003465430298487

Heffernan, M., & Pimpa, N. (2019). *Globally transformative student experience: Challenges and opportunities in learning and teaching in the transnational business education program*. Intercultural Studies in Higher Education., DOI: 10.1007/978-3-030-15758-6_9

Hermanns, M., Deal, B., Campbell, A., Hillhouse, S., Opella, J., Faigle, C., & Campbell, R.IV. (2017). Using an "escape room" toolbox approach to enhance pharmacology education. *Journal of Nursing Education and Practice*, 8(4), 89. DOI: 10.5430/jnep.v8n4p89

Hernández-Peña, H., Lagomarsino-Montoya, M., Aguirre-Martínez, G., Mansilla-Sepúlveda, J., Estay-Sepúlveda, J. G., & Ganga-Contreras, F. (2022). *On the digital age from humanistic psychology*. Environment and Social Psychology., DOI: 10.18063/esp.v7.i1.1401

Hevko, I. V., Siaonan, S., & Wu, H. (2023). Innovative interactive technologies in the training of professional education specialists by means of digital technologies. *Journal of Education, Health and Sport*, 37(1), 166–173. Advance online publication. DOI: 10.12775/JEHS.2023.37.01.013

Hmelo-Silver, C., Duncan, R. G., & Chinn, C. A.HMELO-SILVER. (2007). Scaffolding and achievement in problem-based and inquiry learning: A response to kirschner, sweller, and clark (2006). *Educational Psychologist*, 42(2), 99–107. DOI: 10.1080/00461520701263368

Hong, V., & Reiser, E. (2020). Make your escape: An active learning approach. In *Proceedings of the 51st ACM Technical Symposium on Computer Science Education*. DOI: DOI: 10.1145/3328778.3372557

Horváth, I. (2021). An analysis of personalized learning opportunities in 3d vr. *Frontiers of Computer Science*, 3, 673826. Advance online publication. DOI: 10.3389/fcomp.2021.673826

Höß, B., Wasserman, M. E., and Fisher, S. L. (2019). Building a global education collaboration model using experiential learning: A fresh look at developing intercultural competence. *Journal of International Business Research And Marketing*, DOI: .DOI: 10.18775/jibrm.1849-8558.2015.51.3001

Howard, B., Ilyashenko, N., & Jacobs, L. (2023). Cross-cultural collaboration through virtual teaming in higher education. *Perspectives in Education*, 41(1), 74–87. Advance online publication. DOI: 10.38140/pie.v41i1.6319

Huang, O. W. S., Cheng, H. N. H., & Chan, T. (2007). Number jigsaw puzzle: A mathematical puzzle game for facilitating players' problem-solving strategies. *In2007 First IEEE International Workshop on Digital Game and Intelligent Toy Enhanced Learning (DIGITEL'07)*, pages 130–134. DOI: DOI: 10.1109/DIGITEL.2007.37

Huang, S., Kuo, Y.-H., & Chen, H.-C. (2020). Applying digital escape rooms infused with science teaching in elementary school: Learning performance, learning motivation, and problem-solving ability. *Thinking Skills and Creativity*, 35, 100635. DOI: 10.1016/j.tsc.2020.100681

Huang, Y., Zhu, M., Wang, J., Pathak, N., Shen, C., Keegan, B., Williams, D., & Contractor, N. (2009). The formation of task-oriented groups: Exploring combat activities in online games. *In2009 International Conference on Computational Science and Engineering,* volume 4, *pages* 122–127. *DOI:* DOI: 10.1109/CSE.2009.465

Huerta-Cardoso, O., Unver, E., Kuş, A., & Arslan, R. (2023). Enhancing the pedagogic impact of augmented reality experiences: Storyboarding as a human-centred design approach. In *CAD'23*, pages 350–356. DOI: DOI: 10.14733/cadconfP.2023.350-356

Huilan, S., Wei, W., Zhonghua, D., Xiaolong, Q., Dan, W., Shihai, Z., Qing, W., Liang, Y., & Youming, Q. (2020). Educational management in critical thinking training based on bloom's taxonomy and solo taxonomy. *2020 International Conference on Information Science and Education (ICISE-IE)*, pages 518–521, DOI: DOI: 10.1109/ICISE51755.2020.00116

Huizinga, J. (2014). *Homo Ludens Ils 86*. Routledge., DOI: 10.4324/9781315824161

Huraj, L., & Hudáková, M. S. (2022). Digital escape room for computer network mechanic students. *In 2022 20th International Conference on Emerging eLearning Technologies and Applications (ICETA)*, DOI: DOI: 10.1109/ICETA54985.2022.983967

Huraj, L., & Hudáková, M. S. (2022). Digital escape room for computer network mechanic students. In *2022 20th International Conference on Emerging eLearning Technologies and Applications (ICETA)*, pages 236–242. IEEE. DOI: 10.1109/ICETA57911.2022.9974737

Huraj, L., Hrmo, R., & Sejutová Hudáková, M. (2022). The impact of a digital escape room focused on HTML and computer networks on vocational high school students. *Education Sciences*, 12(10), 682. DOI: 10.3390/educsci12100682

Huynh, E., Nyhout, A., Ganea, P. A., & Chevalier, F. (2020). Designing narrative-focused role-playing games for visualization literacy in young children. *IEEE Transactions on Visualization and Computer Graphics*, 27(2), 924–934. DOI: 10.1109/TVCG.2020.3030464 PMID: 33048745

Ifenthaler, D., Eseryel, D., & Ge, X. (Eds.). (2012). *Assessment in Game-Based Learning: Foundations, Innovations, and Perspectives*. Springer. DOI: 10.1007/978-1-4614-3546-4

Iqbal, M. Z., & Campbell, A. G. (2023). *Metaverse as tech for good: Current progress and emerging opportunities*. Virtual Worlds., DOI: 10.3390/virtualworlds2040019

Istiana, R., Herawati, D., Herniningtyas, F., Ichsan, I., & Ali, A. (2023). *Stem learning to improve problem solving ability on the topic of environmental education*. Jurnal Penelitian Pendidikan IPA., DOI: 10.29303/jppipa.v9i3.2979

Itenge, H., Sedano, C. I., & Winschiers-Theophilus, H. (2021). Virtually escaping lockdown - co-designing a mixed reality escape room narrative with Namibian learners. *In3rd African Human-Computer Interaction Conference: Inclusiveness and Empowerment*, DOI: DOI: 10.1145/3448696.3448700

Ito, M. (2009). *Engineering Play: A Cultural History of Children's Software*. The MIT Press., DOI: 10.7551/mitpress/7939.001.0001

Ivory, J. D., & Kalyanaraman, S. (2007). The effects of technological advancement and violent content in video games on players' feelings of presence, involvement, physiological arousal, and aggression. *Journal of Communication*, 57(3), 532–555. DOI: 10.1111/j.1460-2466.2007.00356.x

Jacobs, S., Hardebusch, T., Gerhard, N., Franke, E., Peters, H., & Jaschke, S. (2023). Promoting competences for the digital world by an educational escape room. In *2023 IEEE 2nd German Education Conference (GECon)*. DOI: DOI: 10.1109/GECon58119.2023.10295150

Jacobs, H. H. (1989). *Interdisciplinary Curriculum: Design and Implementation*. Association for Supervision and Curriculum Development.

Jantjies, M., Moodley, T., & Maart, R. (2018). Experiential learning through virtual and augmented reality in higher education. *Proceedings of the 23rd Annual International Conference on Engineering and Product Design Education (E&PDE 2019)*, pages 42–45, DOI: DOI: 10.1145/3300942.3300956

Jantke, K. P., & Spundflasch, S. (2013). Storyboarding pervasive learning games. In *Proceedings of the 2013 International Conference on Advanced ICT, ICAICTE-13*. Atlantis Press, DOI: DOI: 10.2991/icaicte.2013.9

Jassawalla, A., Sashtital, H., & Witt, A. (2022). Increasing students' felt need for teamwork: An experiential learning exercise that sets the stage. *In8th International Conference on Higher Education Advances (HEAd'22)*, DOI: DOI: 10.4995/HEAd22.2022.14590

Jenkins, H. (2004). Game design as narrative architecture. *Computer*, 44, 118–130.

Jie, Z., & Pearlman, A. M. G. (2018). Expanding access to international education through technology enhanced collaborative online international learning (coil) courses. *International Journal of Technology in Teaching and Learning*, 14(1). Advance online publication. DOI: 10.37120/ijttl.2018.14.1.01

Jiménez, C., Arís, N., Magreñán Ruiz, Á., & Orcos, L. (2020). Digital escape room, using genial.ly and a breakout to learn algebra at secondary education level in spain. *Education Sciences*, 10(10), 271. DOI: 10.3390/educsci10100271

Johnson, C., Urazov, M., and Zanoli, E. (2022). Escapeling: A gamified, ai-supported chatbot for collaborative language practice. *Lecture Notes in Networks and Systems*, 349 LNNS:141 – 148, DOI: .DOI: 10.1007/978-3-030-90677-1_14

Johnson, D. W., & Johnson, R. T. (1994). *Learning together and alone: Cooperative, competitive, and individualistic learning*. Allyn and Bacon.

Johnson, D. W., & Johnson, R. T. (1994). *Learning Together and Alone: Cooperative, Competitive, and Individualistic Learning*. Allyn and Bacon.

Jonassen, D. H. (1999). Designing constructivist learning environments. In Reigeluth, C. M. (Ed.), *Instructional-design Theories and Models: A New Paradigm of Instructional Theory* (Vol. II, pp. 215–239). Lawrence Erlbaum Associates.

Jones, V., & Jones, L. (2016). *Comprehensive Classroom Management: Creating Communities of Support and Solving Problems*. Pearson.

Joo, H.-T., & Kim, K.-J. (2019). Visualization of deep reinforcement learning using grad-cam: How ai plays atari games? In *2019 IEEE Conference on Games (CoG), pages* 1–2. DOI: DOI: 10.1109/CIG.2019.8847950

Joseph, S. (2023). *Large language model-based tools in language teaching to develop critical thinking and sustainable cognitive structures*. Rupkatha Journal on Interdisciplinary Studies in Humanities., DOI: 10.21659/rupkatha.v15n4.13

Joyce, B., & Showers, B. (2002). *Student Achievement through Staff Development*. ASCD.

Kaczmarczyk, K., & Salvoni, M. (2016). Hedge mazes and landscape gardens as cultural boundary objects. *Sign Systems Studies, 44*(1/2):53–68, ISSN: 1406-4243, DOI: DOI: 10.12697/SSS.2016.44.1-2.04

Kaczmarek, L. D., & Drążkowski, D. (2014). MMORPG escapism predicts decreased well-being: Examination of gaming time, game realism beliefs, and online social support for offline problems. *Cyberpsychology, Behavior, and Social Networking*, 17(5), 298–302. DOI: 10.1089/cyber.2013.0595 PMID: 24605951

Kagan, S., Kyle, P., & Scott, S. (2009). *Managing Diverse Classrooms: How to Build on Students' Cultural Strengths*. ASCD.

Kahila, J., Parkki, T., Gröhn, A., Karvinen, A., Telimaa, E., Riikonen, P., Tiitta, R., Haantio, P., Keinänen, A., Kerkkänen, T., Jormanainen, I., Penttinen, S., & Tedre, M. (2020). Escape room game for CT learning activities in the primary school. *InProceedings of the 20th Koli Calling International Conference on Computing Education Research*. DOI: 10.1145/3428029.3428063

Kalantzis, M., & Cope, B. (2004). Designs for learning. *E-Learning and Digital Media*, 1(1), 38–93. DOI: 10.2304/elea.2004.1.1.7

Kalvakolu, S., Söbke, H., Baalsrud Hauge, J., & Kraft, E. (2025). Combining 360° spaces and social vr. *Lecture Notes in Computer Science (including subseries Lecture Notes in Artificial Intelligence and Lecture Notes in Bioinformatics)*, DOI: DOI: 10.1007/978-3-031-78269-5_38

Kanareiko, D. (2023). *Issue 3* (Vol. 2023). Narrative design in pedagogical scheming. *Ergodesign*. Bryansk State Technical University BSTU., DOI: 10.30987/2658-4026-2023-3-262-267

Kapp, K. (2012). *The Gamification of Learning and Instruction: Game-Based Methods and Strategies for Training and Education*. Pfeiffer.

Karabacak, M., Ozkara, B. B., Margetis, K., Wintermark, M., & Bisdas, S. (2023). The advent of generative language models in medical education. *JMIR Medical Education*, 9, e48163. Advance online publication. DOI: 10.2196/48163 PMID: 37279048

Karageorgiou, Z., Mavrommati, I., & Fotaris, P. (2019). Escape room design as a game-based learning process for steam education. In *Proceedings of the 13th European Conference on Game Based Learning* (ECGBL 2019), Odense, Denmark.

Karageorgiou, Z., Mavrommati, I., & Fotaris, P. (2019). Escape room design as a game-based learning process for STEAM education. *InProceedings of the 12th European Conference on Game Based Learning*.

Karageorgiou, Z., Mavrommati, I., & Fotaris, P. (2019). Escape room design as a game-based learning process for steam education. In *Proceedings of the 12th European Conference on Game Based Learning*. DOI: DOI: 10.34190/GBL.19.190

Kartal, G., Toprak, E., & Kumtepe, E. G. (2018). Culturally sensitive instructional design principles for online learning environments. *In Distance education and e-learning practices*. DOI: DOI: 10.4018/978-1-5225-3076-3.ch015

Kaul, V., Morris, A., Chae, J. M., Town, J., & Kelly, W. F. (2021). Delivering a novel medical education "escape room" at a national scientific conference: First live, then pivoting to remote learning because of COVID-19. *Chest*, 160(4), 1424–1432. Advance online publication. DOI: 10.1016/j.chest.2021.04.069 PMID: 34029564

Kawaguchi-Suzuki, M., Law, M. G., Prisco, J., Head, K., Fu, L., Yumoto, T., Kamei, J., Yang, M., Cheng, K., & Hogue, M. (2019). Cultural sensitivity and global pharmacy engagement in asia: China, japan, south korea, and taiwan. *American Journal of Pharmaceutical Education*, 83(4), 7214. Advance online publication. DOI: 10.5688/ajpe7214 PMID: 31223161

Kazandis, I., & Pellas, N. (2019). Developing and assessing augmented reality applications for mathematics with trainee instructional media designers: An exploratory study on user experience. *Journal of Universal Computer Science*, 25, 489–514.

Kazem, M. M. A. (2022). The effects of educational games on the development of some cognitive skills in children aged (4-5) years. *Journal of STEPS for Humanities and Social Sciences*, DOI: .DOI: 10.55384/2790-4237.1147

Kazimoglu, C., Kiernan, M., & Bacon, L. (2010). Enchanting e-learning through the use of interactive-feedback loop in digital games. *In3rd International Conference on Human System Interaction*, DOI: DOI: 10.1109/HSI.2010.5514523

Kee, T. and Lai, A. (2022). Learning motivation and psychological empowerment of socioeconomically disadvantaged learners – an empirical study on inclusive project-based learning during covid-19. *International Journal of Inclusive Education.*

Keller, J. M. (1987). Development and use of the arcs model of instructional design. *Journal of Instructional Development : JID*, 10(3), 2–10. DOI: 10.1007/BF02905780

Kerr, A. (2003). Live life to the power of PS2: Locating the digital games industry in the new media environment. *Irish Communication Review, 9*(2).

Khundam, C. (2020). Storytelling platform for interactive digital content in virtual museum. *ECTI-CIT*, 15(1), 34–49. DOI: 10.37936/ecti-cit.2021151.239948

Kieran, L., & Anderson, C. (2018). Connecting universal design for learning with culturally responsive teaching. *Education and Urban Society*, 51(9), 1202–1216. DOI: 10.1177/0013124518785012

Kilanioti, I. (2022). Teaching a serious game for the sustainable development goals in the scratch programming tool. *European Journal of Engineering and Technology Research*, DOI: .DOI: 10.24018/ejeng.2022.1.CIE.2957

Kim, H., Hartleb, T., Bello, K., Aqlan, F., Zhao, R., & Yang, H. (2023). Behavioral modeling of collaborative problem solving in multiplayer virtual reality manufacturing simulation games. *Journal of Computing and Information Science in Engineering*, •••, 24.

Kim, J. (2021). Advertising in the metaverse: Research agenda. *Journal of Interactive Advertising*, 21(3), 141–144. DOI: 10.1080/15252019.2021.2001273

Kim, J., & Castelli, D. (2021). Effects of gamification on behavioral change in education: A meta-analysis. *International Journal of Environmental Research and Public Health*, 18(7), 7350. DOI: 10.3390/ijerph18073550 PMID: 33805530

Kirschner, P. A., Sweller, J., & Clark, R. E. (2006). Why minimal guidance during instruction does not work: An analysis of the failure of constructivist, discovery, problem-based, experiential, and inquiry-based teaching. *Educational Psychologist*, 41(2), 75–86. DOI: 10.1207/s15326985ep4102_1

Kiruthika, V., Jagadeeswari, M., Sneha Prabha, C., Vaishnavi, V., & Sreejaa, H. (2022). Virtual reality based escape room. *In2022 International Conference on Advanced Computing Technologies and Applications (ICACTA)*, pages 1–5. IEEE, DOI: DOI: 10.1109/ICACTA54488.2022.9753531

Kladchuen, R., & Srisomphan, J. (2021). The synthesis of a model of problem-based learning with the gamification concept to enhance the problem-solving skills for high vocational certificate. *International Journal of Emerging Technologies in Learning*, 16(14), 4. Advance online publication. DOI: 10.3991/ijet.v16i14.20439

Klamma, R., Sous, D., Hensen, B., and Koren, I. (2020). Educational escape games for mixed reality. DOI: .DOI: 10.1007/978-3-030-57717-9_41

Knox, A. (2023). *Game-Based Learning Design Optimized for Cognitive Load.* Springer International Publishing., DOI: 10.1007/978-3-031-25102-3_20

Koenig, N., Denk, N., Wimmer, S., & Prandstaetter, H. (2022). Creating an escape room for cultural mediation: Insights from "the archivist's dream". *European Conference on Games Based Learning, 16*(1):297–306, DOI: DOI: 10.34190/ecgbl.16.1.682

Kohn, J., & Rank, S. (2016). Evaluating physical movement as trigger for transitioning between environments in virtual reality. *InProceedings of the 2016 CHI Conference Extended Abstracts on Human Factors in Computing Systems.* DOI: 10.1145/2851581.2892506

Kolb, D. A. (1984). *Experiential Learning: Experience as the Source of Learning and Development.* Prentice Hall.

Kolb, D. A., Boyatzis, R. E., & Mainemelis, C. (2014). *Experiential learning theory: Previous research and new directions.* Perspectives on Thinking, Learning, and Cognitive Styles., DOI: 10.4324/9781410605986-9

Kole, K. (2023). E-learning material for computational thinking in secondary education. Bachelor Thesis, University of Twente, The Netherlands.

Koo, D.-M. (2009). The moderating role of locus of control on the links between experiential motives and intention to play online games. *Computers in Human Behavior*, 25(2), 466–474. DOI: 10.1016/j.chb.2008.10.010

Kopel, M., & Hajas, T. (2018). Implementing ai for non-player characters in 3d video games. In *Springer, pages 610–619*. DOI: DOI: 10.1007/978-3-319-75417-8_57

Kosmadoudi, Z., Lim, T., Ritchie, J., Louchart, S., Liu, Y., & Sung, R. (2013). Engineering design using game-enhanced CAD: The potential to augment the user experience with game elements. *Computer Aided Design*, 45(3), 777–795. DOI: 10.1016/j.cad.2012.08.001

Kotsonas, A. (2018). A cultural history of the Cretan labyrinth: Monument and memory from prehistory to the present. *American Journal of Archaeology*, 122(3), 367–396. DOI: 10.3764/aja.122.3.0367

Kotzebue, L., Zumbach, J., & Brandlmayr, A. (2022). Digital escape rooms as game-based learning environments: A study in sex education. *Multimodal Technologies and Interaction*, 6(2), 8. DOI: 10.3390/mti6020008

Krekhov, A., Emmerich, K., Rotthaler, R., & Krueger, J. (2021). Puzzles unpuzzled. *Proceedings of the ACM on Human-Computer Interaction*, 5(CHI PLAY), 1–24. DOI: 10.1145/3474696

Kuo, H., Pan, A.-J., Lin, C.-S., & Chang, C.-Y. (2022). Let's escape! The impact of a digital-physical combined escape room on students' creative thinking, learning motivation, and science academic achievement. *Education Sciences*, 12(9), 615. Advance online publication. DOI: 10.3390/educsci12090615

Kutzin, J., Sanders, J., & Strother, C. (2021). Transitioning escape rooms to a virtual environment. *Simulation & Gaming*, 52(6), 796–806. DOI: 10.1177/10468781211035171

Laal, M., & Ghodsi, S. M. (2012). Benefits of collaborative learning. *Procedia: Social and Behavioral Sciences*, 31, 486–490. DOI: 10.1016/j.sbspro.2011.12.091

Laato, S., Morschheuser, B., Hamari, J., & Björne, J. (2023). Ai-assisted learning with chatgpt and large language models: Implications for higher education. *2023 IEEE International Conference on Advanced Learning Technologies (ICALT)*, pages 226–230, DOI: DOI: 10.1109/ICALT58122.2023.00072

Lal, A., Walsh, E., Wetherell, A., & Slimings, C. (2021). Climate change education in public health and medical curricula in australian and new zealand universities: a mixed methods study of barriers and areas for further action. DOI: DOI: 10.1101/2021.06.11.21258793

Lama, P., Payal, S., Bisht, S., & Gurung, A. (2021). *Escapism in the works of existentialist writers*. Webology., DOI: 10.29121/WEB/V18I2/3

Langley, P., & Cooper, R. (2017). Symposium on problem solving and goal-directed sequential activity. In: Gunzelmann, G. and Howes, A. and Tenbrink, T. and Davelaar, Eddy J. (eds.) Proceedings of the 39th Annual Conference of the Cognitive Science Society. Austin, U.S., pp. 29-30.

LaPolla, M., & Baars, B. (1992). A psychologically implausible architecture that is always conscious, always active. *Behavioral and Brain Sciences*, 15(3), 448–449. DOI: 10.1017/S0140525X00069582 PMID: 24924012

Latheesh, K. (2019). Leadership development through experiential learning. *International Education and Research Journal*, •••, 5.

Lathwesen, C., & Belova, N. (2021). Escape rooms in stem teaching and learning—Prospective field or declining trend? a literature review. *Education Sciences*, 11(6), 308. DOI: 10.3390/educsci11060308

Lathwesen, C., & Eilks, I. (2024). Can you make it back to earth? a digital educational escape room for secondary chemistry education to explore selected principles of green chemistry. *Journal of Chemical Education*, 101(8), 3193–3201. DOI: 10.1021/acs.jchemed.4c00149

Lauriola, I., Lavelli, A., & Aiolli, F. (2021). An introduction to deep learning in natural language processing: Models, techniques, and tools. *Neurocomputing*. Advance online publication. DOI: 10.1016/j.neucom.2021.05.103

Lave, J., & Wenger, E. (1991). *Situated Learning: Legitimate Peripheral Participation*. Cambridge University Press., DOI: 10.1017/CBO9780511815355

Lazarinis, F., Boididis, I., Kozanidis, L., & Kanellopoulos, D. (2022). *An adaptable multi-learner serious game for learning cultural heritage*. Advances in Mobile Learning Educational Research., DOI: 10.25082/AMLER.2022.01.004

Lee, A. (2021). *Teacher embodiment of culturally responsive pedagogies in a fifth grade classroom*. Profesorado, Revista de Currículum y Formación del Profesorado., DOI: 10.30827/profesorado.v25i3.21461

Lees, J., & Webb, G. (2018). A review of the literature to inform the development of a new model of global placement: The global learning partnership. *Physical Therapy Reviews : PTR*, 23(1), 40–49. DOI: 10.1080/10833196.2017.1341144

Lekea, D. I. K., Stamatelos, D., & Raptis, P. (2021). Learning how to escape the unthinkable with virtual reality: The case of pilots' training on emergency procedures. *InIOP Conference Series: Materials Science and Engineering*, volume 1024. DOI: DOI: 10.1088/1757-899X/1024/1/012098

Lestari, I. (2021). Experiential learning using stem approach in improving students' problem solving ability. *Journal of Physics: Conference Series*, 1806, DOI: , V. (2020). Challenging escape from climate change: A learning experience. DOI: .DOI: 10.1088/1742-6596/1806/1/012005.Liapi

Li, C.-T., Wang, C.-P., & Hou, H. (2022). Applying an escape room game in synchronous online learning activity: A preliminary analysis of learners' learning achievement, flow and technology acceptance. *In 2022 12th International Congress on Advanced Applied Informatics (IIAI-AAI)*, pages 659–660. IEEE.

Lim Qian Pink, M., Darun, M. R., & Nawani, G. (2019). *Decrypting the real life escape room experience*. KnE Social Sciences., DOI: 10.18502/kss.v3i22.5063

Lima, E. S. D., Feijó, B., & Furtado, A. (2021). Adaptive branching quests based on automated planning and story arcs. In *2021 20th Brazilian Symposium on Computer Games and Digital Entertainment (SBGames)*, pages 9–18. DOI: DOI: 10.1109/SBGames54170.2021.00012

Lima, E. S. D., Feijó, B., & Furtado, A. (2022). A character-based model for interactive storytelling in games. In *2022 21st Brazilian Symposium on Computer Games and Digital Entertainment (SBGames)*, pages 1–6. DOI: DOI: 10.1109/SBGAMES56371.2022.9961071

Linares, S. M. (2023). Intercultural sensitivity and cultural othering in english language pre-service teachers. *Teaching English as a Second or Foreign Language–TESL-EJ*, DOI: .DOI: 10.55593/ej.26104a5

Lindsay, E., Johri, A., and Bjerva, J. (2023). A framework for responsible development of automated student feedback with generative ai. *ArXiv*, abs/2308.15334, DOI: .DOI: 10.48550/arXiv.2308.15334

Liu, J., Snodgrass, S., Khalifa, A., Risi, S., Yannakakis, G. N., & Togelius, J. (2020). Deep learning for procedural content generation. *Neural Computing & Applications*, 33(1), 19–37. DOI: 10.1007/s00521-020-05383-8

Liu, Z., Gong, S., Tan, Z., & Demian, P. (2023). Immersive technologies-driven building information modeling (bim) in the context of metaverse. *Buildings (Basel, Switzerland)*, 13(6), 1559. Advance online publication. DOI: 10.3390/buildings13061559

Liu, Z.-Y., Shaikh, Z. A., & Gazizova, F. (2020). Using the concept of game-based learning in education. *International Journal of Emerging Technologies in Learning*, 15(14), 53–64. DOI: 10.3991/ijet.v15i14.14675

Löffler, E., Schneider, B., Zanwar, T., & Asprion, P. M. (2021). Cysececscape 2.0 - a virtual escape room to raise cybersecurity awareness. *International Journal of Serious Games*, 8(3), 51–69. DOI: 10.17083/ijsg.v8i1.413

López López, C. M. and Lázaro, M. (2019). Escape folroom: Propuesta de juego integrada en clase de formación y orientación laboral. *2:91–105, DOI*: .DOI: 10.24310/IJNE2.2.2019.7453

Lopez-Bermudez, V., & Martinez, A. (2023). Maruxaína: The sea nymph who will be judged. In *Proceedings of the 17th European Conference on Game-Based Learning*, pages 467–476. DOI: DOI: 10.4018/978-1-6684-6081-8.ch008

López-Faican, L., & Martínez, J. (2020). Emofindar: Evaluation of a mobile multi-player augmented reality game for primary school children. *Computers & Education*, 149, 103814. DOI: 10.1016/j.compedu.2020.103814

López-Pernas, S., Gordillo, A., Barra, E., & Quemada, J. (2021). Comparing face-to-face and remote educational escape rooms for learning programming. *IEEE Access : Practical Innovations, Open Solutions*, 9, 59270–59285. DOI: 10.1109/ACCESS.2021.3073601

Loureiro, S., Guerreiro, J., & Japutra, A. (2021). How escapism leads to behavioral intention in a virtual reality store with background music. *Journal of Business Research*, 134, 288–300. DOI: 10.1016/j.jbusres.2021.05.035

Lowood, H. (2009). Videogames in computer space: The complex history of pong. *IEEE Annals of the History of Computing*, 31(3), 31. DOI: 10.1109/MAHC.2009.53

Lu, A. (2015). Narrative in exergames: Thoughts on procedure, mechanism, and others. *Games for Health Journal*, 4(1), 19–24. DOI: 10.1089/g4h.2014.0090 PMID: 26181676

Luhova, T. (2020). Narrative and storytelling in the knowledge structure of the educational business video games as factors of the synergy of information technologies and spiritually-oriented pedagogy, pages 42–59, DOI: .DOI: 10.28925/2414-0325.2020.8.6

Lyons, R., Sweeney, E. M., and Lynch, C. (2023). Applied gamification using virtual escape rooms: Case studies in entrepreneurial and anatomy classes. *Irish Journal of Technology Enhanced Learning*, DOI: .DOI: 10.22554/ijtel.v7i1.103

Maas, M. and Hughes, J. (2020). Virtual, augmented and mixed reality in k–12 education: a review of the literature. *Technology, Pedagogy and Education*, 29:231 – 249, DOI : Mackavey, C. L., Henderson, C., and Stout, T. (2023). Stepping outside national borders: International active learning educational collaboration events. *International Nursing Review*, DOI: .DOI: 10.1080/1475939X.2020.1737210

Macan, T. H. (1994). Time management: Test of a process model. *The Journal of Applied Psychology*, 79(3), 381–391. DOI: 10.1037/0021-9010.79.3.381

Macapugay, K., & Nakamura, B. (2024). The student empowerment through narrative, storytelling, engagement, and identity framework for student and community empowerment: A culturally affirming pedagogy. *Genealogy*, 8(3), 94. DOI: 10.3390/genealogy8030094

Maduakolam, E., Madden, B., Kelley, T., & Cianciolo, A. (2020). Beyond diversity: Envisioning inclusion in medical education research and practice. *Teaching and Learning in Medicine*, 32(5), 459–465. DOI: 10.1080/10401334.2020.1836462 PMID: 33349086

Magdy, R., Rashad, S., Hany, S., Tarek, M., Hassan, M. A., & Mohammed, A. (2021). Deep reinforcement learning approach for augmented reality games. In *2021 International Mobile, Intelligent, and Ubiquitous Computing Conference (MIUCC), pages 330–336*. DOI: DOI: 10.1109/MIUCC52538.2021.9447671

Ma, J.-P., Chuang, M.-H., & Lin, R. (2018). An innovated design of escape room game box through integrating steam education and PBL principle. In Rau, P. L. (Ed.), Lecture Notes in Computer Science: Vol. 10912. *Cross-Cultural Design. Applications in Cultural Heritage, Creativity and Social Development. CCD 2018*. Springer., DOI: 10.1007/978-3-319-92252-2_6

Majid, R. (2021). *The significance of soft skills training in undergraduate education*. Higher Education, Skills and Work-Based Learning., DOI: 10.1108/heswbl-10-2020-0234

Makri, A., Vlachopoulos, D., & Martina, R. (2021). Digital escape rooms as innovative pedagogical tools in education: A systematic literature review. *Sustainability (Basel)*, 13(8), 4587. DOI: 10.3390/su13084587

Maksimenko, A. (2023). Experience of the narrative approach in foreign educational practice. *Vestnik Orenburgskogo gosudarstvennogo universiteta*, DOI: .DOI: 10.25198/1814-6457-238-33

Makuch, G. (2023). A linguistic analysis of translating intertextuality in "the witcher" game. BA thesis, University of Silesia in Katowice. DOI: DOI: 10.13140/RG.2.2.13032.03842

Malegiannaki, I., Daradoumis, T., & Retalis, S. (2020). Teaching cultural heritage through a narrative-based game. [JOCCH]. *Journal on Computing and Cultural Heritage*, 13(4), 1–28. DOI: 10.1145/3414833

Mancheno-Smoak, S. (2019). Experiential learning through the use of simulation and digital games in a classroom: An investigation of the effects of learning on student engagement, motivation, and learning outcomes. *The Journal of Scholarship of Teaching and Learning*. Advance online publication. DOI: 10.14434/josotl.v19i2.23842

Manojlovic, H. (2022). *Escape room as a teaching method.* Opus et Educatio., DOI: 10.3311/ope.504

Marchand, A., & Hennig-Thurau, T. (2013). Value creation in the video game industry: Industry economics, consumer benefits, and research opportunities. *Journal of Interactive Marketing*, 27(3), 141–157. DOI: 10.1016/j.intmar.2013.05.001

Markowitz, D. M., Laha, R. R., Perone, B., Pea, R., & Bailenson, J. (2018). Immersive virtual reality field trips facilitate learning about climate change. *Frontiers in Psychology*, 9, 2364. Advance online publication. DOI: 10.3389/fpsyg.2018.02364 PMID: 30555387

Marrón-González, S. (2020). Building bridges to the integration of culture in the efl secondary classroom: Perspectives from teachers and learners. *Eighth International Conference on Technological Ecosystems for Enhancing Multiculturality*, DOI: DOI: 10.1145/3434780.3436670

Martinez, E., Alvarez, J., Carbonell, V., Florez, M., & Montoya, E. (2023). Escaping the lab: Increasing motivation in the physics classroom through educational escape rooms. *Romanian Reports in Physics*, 75(1).

Martín-Gutiérrez, A., Leal-Vega, L., Alcoceba-Herrero, I., Herrera-Medina, J., Arenillas-Lara, J. F., & Coco-Martín, M. B. (2022). *Educational Opportunities of Virtual Game-Based Initiatives for Students With Disabilities*. IGI Global., DOI: 10.4018/978-1-6684-4287-6.ch021

Marzano, R. J. (2007). *The Art and Science of Teaching: A Comprehensive Framework for Effective Instruction*. ASCD.

Mat Sin, N., Talib, O., Norishah, T., Ishak, A., & Baki, R. (2014). Male students and digital game: Reason, motivation and feeling. *International Journal of Information and Education Technology (IJIET).*

Mateen, S., Wiesemüller, P., & Voß-Nakkour, S. (2023). Design of an accessible vr-escape room for accessibility education. In *SMART ACCESSIBILITY 2023,The Eighth International Conference on Universal Accessibility in the Internet of Things and Smart Environments*, Venice, Italy.

Mateos-Aparicio, I., Sánchez-Paniagua, M., Garcia-Alonso, A., López-Ruíz, B., Molina, M., Moreno-Guzmán, M., Pérez-Rodríguez, M., Raggio, J., Rueda, C., Sevilla, P., Bodas-Folguera, C., and PEÑA-ARMADA, R. D. L. (2020). Students perceive the gamification through activities based on escape rooms very positively for learning at university. *IJAEDU- International E-Journal of Advances in Education*, 6:281–286, DOI: .DOI: 10.18768/ijaedu.819338

Mateos-Aparicio, I., Sánchez-Paniagua, M., García-Alonso, A., López-Ruiz, B., Molina, M., Moreno-Guzmán, M., Pérez-Rodríguez, M., Raggio, J., Rueda, C., Sevilla, P., Bodas-Folguera, C., & de la Peña-Armada, R. (2020). Student perception of an activity based on escape rooms for learning at university. In *Proceedings of ADVED 2020- 6th International Conference on Advances in Education*. DOI: 10.47696/adved.202018

Matthee, M., & Turpin, M. (2019). Invited paper: Teaching critical thinking, problem solving, and design thinking: Preparing its students for the future. *Journal of Information Systems Education*, 30, 242–252.

Matveev, V., Nikolaevich, D. G., Lomakin, D. E., and Kharunzheva, E. (2021). The use of augmented reality for the formation of multicultural competence of undergraduates. *Perspectives of Science and Education*, DOI: .DOI: 10.32744/pse.2021.6.33

Mauri-Medrano, M., González-Yubero, S., Falcón-Linares, C., & Cardoso-Moreno, M. J. (2024). Gamifying the university classroom: A comparative analysis of game dimensions through educational escape room and a digital board game. *Frontiers in Education*, 9, 1354674. Advance online publication. DOI: 10.3389/feduc.2024.1354674

Mayer, R. E. (2001). *Multimedia Learning*. Cambridge University Press. DOI: 10.1017/CBO9781139164603

Mayombe, C. (2023a). Facilitating non-school-based technical and vocational training for disadvantaged youths in south africa. *Journal of Technical Education and Training*, DOI: .DOI: 10.30880/jtet.2023.15.01.016

Mayombe, C. (2023b). *Promoting youths' skills acquisition through experiential learning theory in vocational education and training in south africa*. Higher Education, Skills and Work-Based Learning., DOI: 10.1108/HESWBL-10-2022-0216

Mcelreath, R. (2010). The coevolution of genes, innovation, and culture in human evolution. In Kappeler, P., & Silk, J. (Eds.), *Mind the Gap*. Springer., DOI: 10.1007/978-3-642-02725-3_21

McGuire, M., & Jenkins, O. C. (2008). *Creating games - mechanics, content, and technology*. CRC Press. DOI: 10.1201/b10678

McIntyre, P., Kerrigan, S., Fulton, J., King, E., & Williams, C. (2023). *Electronic Games and Interactive Content.In Palgrave Studies in Creativity and Culture, page 349–371*. Springer International Publishing, DOI: DOI: 10.1007/978-3-031-45972-6_14

Mekacher, D. L. (2019). Augmented reality (ar) and virtual reality (vr): The future of interactive vocational education and training for people with handicap. *PUPIL: International Journal of Teaching, Education and Learning*, DOI: .DOI: 10.20319/pijtel.2019.31.118129

Meletiadou, E. (2023). Applying Digital Escape Rooms Infused With Cross-Cultural Management Theories in Higher Education: Academic Performance, Student Engagement, and Intercultural Intelligence Development. In *Handbook of Research on Redesigning Teaching*. Learning, and Assessment in the Digital Era., DOI: 10.4018/978-1-6684-8292-6.ch001

Melle, J. V. (2020). Developing students' intercultural sensitivity. a technical action research study at an international business school in the netherlands. (also known as 'the shoe sole project'). DOI: DOI: 10.17638/03090675

Melo, L. B. (2021). Playing with escape room games for children: An online experience at home in times of pandemic. In *2021 16th Iberian Conference on Information Systems and Technologies (CISTI)*. DOI: DOI: 10.23919/CISTI52073.2021.9476536

Menon, D., & Romero, M. (2020). Game mechanics supporting a learning and playful experience in educational escape games. In Global Perspectives on Gameful and Playful Teaching and Learning. DOI: DOI: 10.4018/978-1-7998-2015-4.ch007

Mercado, M. A. T. (2023). Css quest: Gamifying computer system servicing module using gamified progress tracking and interactive storytelling algorithm for grade 12 students in calamba city. *International Journal of Advanced Research in Computer Science*, DOI: .DOI: 10.26483/ijarcs.v14i3.6996

Meyer, A., Rose, D. H., & Gordon, D. (2014). *Universal Design for Learning: Theory and Practice*. CAST Professional Publishing.

Mijal, M., Ciesla, M., and Gromadzka, M. (2020). Escape room as an adult education tool. pages 163–178, DOI: .DOI: 10.14746/HL.2020.13.10

Minor, E., Oliver, K., & Fleming, K. (2024). From glyphs to bytes: Ancient egypt and the future of digital humanities in museum-based learning. In Minor, E. (Ed.), *Teaching Ancient Egypt in Museums* (1st ed.). Routledge. DOI: 10.4324/9781003348221-18

Min, W., Mott, B. W., Park, K., Taylor, S., Akram, B., Wiebe, E., Boyer, K., & Lester, J. C. (2020). Promoting computer science learning with block-based programming and narrative-centered gameplay. In *2020 IEEE Conference on Games (CoG)*, pages 654–657. DOI: DOI: 10.1109/CoG47356.2020.9231881

Moffett, J. and Cassidy, D. (2023). Building a digital educational escape room using an online design-thinking process. *Online Learning* (Vol. 27, Issue 2). The Online Learning Consortium. DOI: .DOI: 10.24059/olj.v27i2.3279

Moffett, J., Cassidy, D., Collins, N., Illing, J., de Carvalho Filho, M. A., & Bok, H. (2023a). Exploring medical students' learning around uncertainty management using a digital educational escape room: A design-based research approach. *Perspectives on Medical Education*, 12(1), 86–98. DOI: 10.5334/pme.844 PMID: 36969324

Moffett, S., Torres, C., & Brooks, C. (2023b). Building effective educational escape rooms: Insights from iterative design and user feedback. *British Journal of Educational Technology*, 54(1), 114–128.

Momenzad, M., Majidi, B., & Eshghi, M. (2018). Deep summarization of academic textbooks for adaptive gamified virtual learning environments. In *2018 2nd National and 1st International Digital Games Research Conference: Trends, Technologies, and Applications (DGRC), pages 88–94*. DOI: DOI: 10.1109/DGRC.2018.8712065

Monaghan, S., & Nicholson, S. (2017). Bringing escape room concepts to pathophysiology case studies. *HAPS Educator*, 21(2), 49–65. DOI: 10.21692/haps.2017.015

Monnot, M., Laborie, S., Hébrard, G., & Dietrich, N. (2020). New approaches to adapt escape game activities to large audience in chemical engineering: Numeric supports and students' participation. *Education for Chemical Engineers*, 32, 50–58. DOI: 10.1016/j.ece.2020.05.007

Montebello, M. (2021). Personalized learning environments. In *2021 International Symposium on Educational Technology (ISET), pages* 134–138. DOI: DOI: 10.1109/ISET52350.2021.00036

Moreno, R., & Mayer, R. E. (2007). Interactive multimodal learning environments. *Educational Psychology Review*, 19(3), 309–326. DOI: 10.1007/s10648-007-9047-2

Moriarty, S., Spruce, J., & Mathai, S. (2023). Critical pedagogies for collaborative innovation. *Proceedings of the International Conference on Engineering and Product Design Education, EPDE 2023*, DOI: DOI: 10.35199/EPDE.2023.51

Morrell, B., & Ball, H. M. (2020). Can you escape nursing school?: Educational escape room in nursing education. *Nursing Education Perspectives*, 41(3), 197–198. DOI: 10.1097/01.NEP.0000000000000441 PMID: 30614962

Morrison, T. (2021). *Labyrinth*. Springer International Publishing., DOI: 10.1007/978-3-319-57072-3_4

Morris, T. H. (2020). Experiential learning–a systematic review and revision of kolb's model. *Interactive Learning Environments*, 28(8), 1064–1077. DOI: 10.1080/10494820.2019.1570279

Moseikina, M., Toktamysov, S., & Danshina, S. (2022). Modern technologies and gamification in historical education. *Simulation & Gaming*, 53(2), 135–156. DOI: 10.1177/10468781221075965

Mozelius, P., Sällvin, L., and Humble, N. (2023). An educational escape room for computational thinking - defining the requirements. *Education and New Developments 2023 – Volume 2*, DOI: .DOI: 10.36315/2023v2end040

Mulyana, A., & Nurcahyani, N. (2022). *The effect of the puzzle playing method on improving the cognitive development of children aged 4-6 years*. KnE Life Sciences., DOI: 10.18502/kls.v7i2.10354

Murray, J. H. (1997). *Hamlet on the Holodeck: The Future of Narrative in Cyberspace*. The MIT Press.

Mustafa, B. (2022). *Analyzing education based on metaverse technology*. Technium Social Sciences Journal., DOI: 10.47577/tssj.v32i1.6742

Mystakidis, S., Cachafeiro, E., & Hatzilygeroudis, I. (2019). Enter the serious e-scape room: A cost-effective serious game model for deep and meaningful e-learning. *In 2019 10th International Conference on Information, Intelligence, Systems and Applications (IISA).*

Mystakidis, S., Cachaferio, E., & Hatzilygeroudis, I. (2019). Enter the serious e-scape room: A cost-effective serious game model for deep and meaningful e-learning. In *2019 10th International Conference on Information, Intelligence, Systems and Applications (IISA), pages 1–6*. DOI: DOI: 10.1109/IISA.2019.8900673

Mystakidis, S., & Christopoulos, A. (2022). Teacher perceptions on virtual reality escape rooms for stem education. *Information (Basel)*, 13(3), 136. DOI: 10.3390/info13030136

Mystakidis, S., Fragkaki, M., & Filippousis, G. (2021a). Ready teacher one: Virtual and augmented reality online professional development for k-12 school teachers. *Computers*, 10(10), 134. DOI: 10.3390/computers10100134

Mystakidis, S., Papantzikos, G., & Stylios, C. (2021). Virtual reality escape rooms for stem education in industry 4.0: Greek teachers perspectives. *In 2021 6th South-East Europe Design Automation, Computer Engineering,Computer Networks and Social Media Conference (SEEDA-CECNSM)*, pages 1–5. IEEE, DOI: DOI: 10.1109/SEEDA-CECNSM53056.2021.9566265

Nah, F., Telaprolu, V. R., Rallapalli, S., & Rallapalli Venkata, P. (2013). Gamification of education using computer games. In Lecture Notes in Computer Science, pages 99–107. DOI: DOI: 10.1007/978-3-642-39226-9_12

Nair, V., Garrido, G. M., and Song, D. (2022). Exploring the privacy risks of adversarial vr game design. *ArXiv, abs/2207.13176*, DOI: .DOI: 10.48550/arXiv.2207.13176

Najm, A., Christofi, M., Hadjipanayi, C., Kyrlitsias, C., & Michael-Grigoriou, D. (2022). *The iterative development of an online multiplayer escape room game for improving social interaction through edutainment.* CSRN. DOI: 10.24132/CSRN.3201.8

Nakano, Y. (2017). Cognitive and attentional process in insight problem solving of the puzzle game "tangram.". *Cognitive Science.*

Narciss, S. (2007). Feedback strategies for interactive learning tasks. In Spector, J. M., Merrill, M. D., van Merrienboer, J. J. G., & Driscoll, M. P. (Eds.), *Handbook of Research on Educational Communications and Technology* (3rd ed.). Lawrence Erlbaum Assoc.

Nasir, M., Lyons, K. A., Leung, R., & Moradian, A. (2013). *Cooperative games and their effect on group collaboration.*, DOI: 10.1007/978-3-642-38827-9_43

Naul, E., & Liu, M. (2019). Why story matters: A review of narrative in serious games. *Journal of Educational Computing Research*, 58(3), 687–707. DOI: 10.1177/0735633119859904

Neto, L. V., Junior, P. H. F., Bordini, R. A., & Otsuka, J. (2019). Design and implementation of an educational game considering issues for visually impaired people inclusion. *Smart Learning Environments*, 7(1), 4. Advance online publication. DOI: 10.1186/s40561-019-0103-4

Neumann, K. L., Alvarado-Albertorio, F. M., & Ramírez-Salgado, A. (2020). Online approaches for implementing a digital escape room with preservice teachers. *Journal of Technology and Teacher Education*, 28, 415–424.

Newcombe, N. S., & Stieff, M. (2012). Six myths about spatial thinking. *International Journal of Science Education*, 34(6), 955–971. DOI: 10.1080/09500693.2011.588728

Ngwenya, N., Makoelle, T., & van der Merwe, M. (2021). Participatory action research as change strategy: A case of developing inclusive teaching and learning practices in an adult education centre in gauteng east district of south africa. *Interchange*, 52(3), 393–414. DOI: 10.1007/s10780-021-09435-3

Nicholson, S. (2014). *A RECIPE for Meaningful Gamification*. Springer International Publishing., DOI: 10.1007/978-3-319-10208-5_1

Nicol, D. J., & Macfarlane-Dick, D. (2006). Formative assessment and self-regulated learning: A model and seven principles of good feedback practice. *Studies in Higher Education*, 31(2), 199–218. DOI: 10.1080/03075070600572090

NIELSEN, J. (1993). *The Usability Engineering Lifecycle*, page 71–114. Elsevier, DOI: .DOI: 10.1016/B978-0-08-052029-2.50007-3

Nieminen, J., & Pesonen, H. (2019). Taking universal design back to its roots: Perspectives on accessibility and identity in undergraduate mathematics. *Education Sciences*, 10(1), 12. Advance online publication. DOI: 10.3390/educsci10010012

Nikkarikoski, I., Pehkonen, J., Kivelä, I., Alavesa, P., & Korkiakoski, M. (2024). Galaxy escape: A game as a tabletop holographic digital twin interface. In Companion Proceedings of the 2024 Annual Symposium on Computer-Human Interaction in Play (pp. 215–220). CHI PLAY '24: The Annual Symposium on Computer-Human Interaction in Play. ACM. DOI: DOI: 10.1145/3665463.3678810

Nimpattanavong, C., Choensawat, W., & Sookhanaphibarn, K. (2022). Action prediction of ai bot in fightingice by using deep learning model. In *2022 IEEE 4th Global Conference on Life Sciences and Technologies (LifeTech), pages 259–262*. DOI: DOI: 10.1109/LifeTech53646.2022.9754922

Norman, D. (2004). *Emotional Design: Why We Love (or Hate) Everyday Things*. Basic Books.

Nouchi, R., Taki, Y., Takeuchi, H., Hashizume, H., Nozawa, T., Kambara, T., Sekiguchi, A., Miyauchi, C., Kotozaki, Y., Nouchi, H., & Kawashima, R. (2013). Brain training game boosts executive functions, working memory and processing speed in young adults: A randomized controlled trial. *PLoS One*, 8(2), e55518. Advance online publication. DOI: 10.1371/journal.pone.0055518 PMID: 23405164

Novak, D., Nagle, A., & Riener, R. (2014). Linking recognition accuracy and user experience in an affective feedback loop. *IEEE Transactions on Affective Computing*, 5(2), 168–172. Advance online publication. DOI: 10.1109/TAFFC.2014.2326870

O'shaughnessy, T. (2021). Universal Design for Learning and Accessibility: A Practitioner Approach. In Advances in Educational Technologies and Instructional Design (pp. 25–47). IGI Global. DOI: DOI: 10.4018/978-1-7998-7106-4.ch002

Offord, M., Honeychurch, S., Quinn, N., Barr, M., and Mullen, H. (2022). Escape room: using a simple text-based game to promote business undergraduates' digital self-reliance. *Journal of Learning Development in Higher Education,* DOI: .DOI: 10.47408/jldhe.v123.855

Oh, I.-S., Rho, S., Moon, S., Son, S., Lee, H., & Chung, J. (2019). Creating pro-level ai for a real-time fighting game using deep reinforcement learning. *IEEE Transactions on Games*, 14(2), 212–220. DOI: 10.1109/TG.2021.3049539

Oktorino, K., & Chong, H. (2015). Developing a 3D action puzzle game application "the mechanic" using Unity. *In 2015 IEEE Conference on e-Learning, e-Management and e-Services (IC3e),* pages 120–124. IEEE.

Olin, P. A., Issa, A. M., Feuchtner, T. M., & Grønbæk, K. (2020). Designing for heterogeneous cross-device collaboration and social interaction in virtual reality. In *Proceedings of the 32nd Australian Conference on Human-Computer Interaction*. DOI: DOI: 10.1145/3441000.3441070

Oliveira, I., Carvalho, V., Soares, F., Novais, P., Oliveira, E., & Gomes, L. (2023). Development of a virtual reality escape room game for emotion elicitation. *Information (Basel)*, 14(9), 514. Advance online publication. DOI: 10.3390/info14090514

Omarali, P. S. P. (2022). Considerations of game dynamics and mechanics in college students' use of mobile games for learning. *International Journal of Mobile and Blended Learning*, 14(2), 1–12. DOI: 10.4018/IJMBL.309739

Ong, Y., Ismail, A., Iahad, N. A., Rameli, M., & Dollah, R. (2020). A mobile game sdk for remote collaborative between two users in augmented and virtual reality. In *IOP Conference Series: Materials Science and Engineering,* volume 979. DOI: DOI: 10.1088/1757-899X/979/1/012003

Ortiz-Rojas, M., Chiluiza, K., & Valcke, M. (2019). Gamification through leaderboards: An empirical study in engineering education. *Computer Applications in Engineering Education*, 27(4), 777–788. DOI: 10.1002/cae.12116

Osuna-Acedo, S. (2021). Gamification and MOOCs. In *MOOCs and the Participatory Challenge* (pp. 89–101). Springer International Publishing., DOI: 10.1007/978-3-030-67314-7_6

Oswald, G., Adams, R., & Hiles, J. (2018). Universal design for learning in rehabilitation education: Meeting the needs for equal access to electronic course resources and online learning. *Journal of Applied Rehabilitation Counseling*, 49(1), 19–22. DOI: 10.1891/0047-2220.49.1.19

Ouankhamchan, P., and Fujinami, T. (2018). Effects of casual computer game on cognitive performance through hemodynamic signals. DOI: .DOI: 10.1007/978-3-030-31605-1_35

Ouariachi, T., & Van Dam, M. (2022). Educational innovation to address climate change issues. *Advances in Game-Based Learning*, 263–278. Advance online publication. DOI: 10.4018/978-1-7998-8645-7.ch013

Ouariachi, T., & Wim, E. J. L. (2020). Escape rooms as tools for climate change education: An exploration of initiatives. *Environmental Education Research*, 26(8), 1193–1206. DOI: 10.1080/13504622.2020.1753659

Owan, V. J., Abang, K. B., Idika, D., Etta, E. O., & Bassey, B. (2023). Exploring the potential of artificial intelligence tools in educational measurement and assessment. *Eurasia Journal of Mathematics, Science and Technology Education*, 19(8), em2307. Advance online publication. DOI: 10.29333/ejmste/13428

Palioura, M.-M., & Dimoulas, C. A. (2022). Digital storytelling in education: A transmedia integration approach for the non-developers. *Education Sciences*, 12(8), 559. Advance online publication. DOI: 10.3390/educsci12080559

Pamungkas, S. F., Widiastuti, I., & Suharno, S. (2020). 21st century learning: Experiential learning to enhance critical thinking in vocational education. *Universal Journal of Educational Research*, 8(4), 1345–1355. DOI: 10.13189/ujer.2020.080427

Panchal, K., & Sharma, R. (2022). The effect of gamification on students' engagement and motivation in higher education: A meta-analysis. *Journal of Educational Technology & Society*, 25, 103–118. DOI: 10.1504/IJTEL.2021.10037644

Pan, R., Lo, H., & Neustaedter, C. (2017). Collaboration, awareness, and communication in real-life escape rooms. *In Proceedings of the 2017 Conference on Designing Interactive Systems,* DOI: DOI: 10.1145/3064663.3064767

Pantelidis, P., Chorti, A., Papagiouvanni, G. P. I., & Drosos, C. (2018, March 28). ThrasyvoulosPanagiotakopoulos, Lales, G., and Sideris, M. (2018). Virtual and augmented reality in medical education. *International Journal of Technology in Teaching and Learning*. Advance online publication. DOI: 10.5772/intechopen.71963

Papanastasiou, G. P., Drigas, A., Skianis, C., Lytras, M. D., & Papanastasiou, E. (2018). Virtual and augmented reality effects on k-12, higher and tertiary education students' twenty-first century skills. *Virtual Reality (Waltham Cross)*, 23(4), 425–436. DOI: 10.1007/s10055-018-0363-2

Paralikar, S. J., Shah, C., Joshi, A., & Kathrotia, R. (2022). Acquisition of higher-order cognitive skills (hocs) using the flipped classroom model: A quasi-experimental study. *Cureus*. Advance online publication. DOI: 10.7759/cureus.24249 PMID: 35602838

Paris, D. (2012). Culturally sustaining pedagogy: A needed change in stance, terminology, and practice. *Educational Researcher*, 41(3), 93–97. DOI: 10.3102/0013189X12441244

Park, E., Rishika, R., Janakiraman, R., Houston, M., & Yoo, B. (2018). Social dollars in online communities: The effect of product, user, and network characteristics. *Journal of Marketing*, 82(1), 93–114. DOI: 10.1509/jm.16.0271

Parker, K., & Welch, T. D. (2021). Innovative teaching strategies to escape the anxiety in simulation. *Teaching and Learning in Nursing*, 16(4), 414–417. Advance online publication. DOI: 10.1016/j.teln.2021.06.005

Parody, L., Santos, J., Trujillo-Cayado, L. A., & Ceballos, M. (2022). Gamification in engineering education: The use of classcraft platform to improve motivation and academic performance. *Applied Sciences (Basel, Switzerland)*, 12(22), 11832. Advance online publication. DOI: 10.3390/app122211832

Parra-Martínez. (2023). The limits of immersion: Case Study of Desolatium a VR Graphic Adventure. *Proceedings of DiGRA 2023*.

Patil, M., Balaji, R., and Shikhare, R. (2022). Smarter education methodologies for met: An approach with augmented reality. *OCEANS 2022 - Chennai*, pages 1–8, DOI: .DOI: 10.1109/OCEANSChennai45887.2022.9775457

Paton, J. (2022). Place, space and object oriented learning methods with the library using ghosts and monsters to create safe spaces for learning. *The Journal of Play in Adulthood.*

Paulauskas, L., Paulauskas, A., Blažauskas, T., Damaševičius, R., & Maskeliūnas, R. (2023). Reconstruction of industrial and historical heritage for cultural enrichment using virtual and augmented reality. *Technologies*, 11(2), 36. DOI: 10.3390/technologies11020036

Pecheranskiy, I. (2023). Technical peculiarities of the narrative action games of the 1990s–2010s as a factor of their development in the context of digital art. *Culture of Ukraine*, DOI: .DOI: 10.31516/2410-5325.081.02

Perkins, D. N., & Salomon, G. (1992). Transfer of learning. In Husén, T. and Postlethwaite, T. N., editors, *The International Encyclopedia of Education*, pages 425–441. Pergamon, Oxford, 2 edition.

Pfau, J., Smeddinck, J. D., & Malaka, R. (2018). Towards deep player behavior models in mmorpgs. In *Proceedings of the 2018 Annual Symposium on Computer-Human Interaction in Play*. DOI: DOI: 10.1145/3242671.3242706

Phung, T., Padurean, ˘. V.-A., Cambronero, J., Gulwani, S., Kohn, T., Majumdar, R., Singla, A., & Soares, G. (2023). Generative ai for programming education: Benchmarking chatgpt, gpt-4, and human tutors. *Proceedings of the 2023 ACM Conference on International Computing Education Research*- Volume 2, DOI: DOI: 10.1145/3568812.3603476

Piaget, J. (1954). *The construction of reality in the child.* Routledge., DOI: 10.1037/11168-000

Piaget, J. (1970). Piaget's theory. In Mussen, P. H. (Ed.), *Carmichael's Manual of Child Psychology* (3rd ed., Vol. 1). Wiley.

Piernas, J., Meroño, M., and Asenjo, M. (2024). Virtual escape rooms: a gamification tool to enhance motivation in distance education [escape rooms virtuales: una herramienta de gamificación para potenciar la motivación en la educación a distancia]. *RIED-Revista Iberoamericana de Educacion a Distancia*, DOI: .DOI: 10.5944/ried.27.1.37685

Pirrone, M., & D'Ulizia, A. (2024). The localization of software and video games: Current state and future perspectives. *Information (Basel)*, 15(10), 648. DOI: 10.3390/info15100648

Polydorou, D. (2024). Immersive storytelling experiences: A design methodology. *Digital Creativity (Exeter)*, 35(4), 301–320. DOI: 10.1080/14626268.2024.2389886

Poole, A. (2019). *Negotiating intercultural spaces and teacher identity in an internationalised school in shanghai.* Intercultural Communication Education., DOI: 10.29140/ice.v2n2.128

Poon Teng Fatt, J. (2000). Understanding the learning styles of students: Implications for educators. *The International Journal of Sociology and Social Policy*, 20(11/12), 31–45. DOI: 10.1108/01443330010789269

Popham, W. J. (2008). *Transformative Assessment*. ASCD.

Pornsakulpaisal, R., Ahmed, Z., Bok, H., de Carvalho Filho, M. A., Goka, S., Li, L., Patki, A., Salari, S., Sooknarine, V., Yap, S. W., & Moffett, J. (2023). Building digital escape rooms for learning: From theory to practice. *The Clinical Teacher*, 20(2), e13559. Advance online publication. DOI: 10.1111/tct.13559 PMID: 36691337

Prasad, B., Reddy, M., Reddy, V., & Kishore, R. (2017). Game development - bounty rescue step. *International Journal of Engineering*, 02, 25–31.

Prasad, G. (2021). Evaluating student performance based on bloom's taxonomy levels. *Journal of Physics: Conference Series*, 1797(1), 012063. Advance online publication. DOI: 10.1088/1742-6596/1797/1/012063

Pratiwi, R., Andriati, R., & Indah, F. P. S. (2020). The positive effect of educative game tools (puzzle) on cognitive levels of pre-school children (4-5 years). *The Malaysian Journal of Nursing*, 11(3), 35–41. Advance online publication. DOI: 10.31674/mjn.2020.v11i03.006

Prattico, F. G., Strada, F., Lamberti, F., & Bottino, A. (2019). Asteroid escape: A serious game to foster teamwork abilities. DOI: DOI: 10.2312/egs.2019013

Prensky, M. (2001). *Digital Game-Based Learning*. McGraw-Hill.

Puga, R. U. (2022). Game-based learning: A tool that enhances the collaborative work. case study: undergraduate students. *European Conference on Games Based Learning*, DOI: DOI: 10.34190/ecgbl.16.1.419

Pujawan, I. G. N., Rediani, N. N., Antara, I. B. G. W., Putri, N. D. A., & Bayu, G. W. (2022). *Revised bloom taxonomy-oriented learning activities to develop scientific literacy and creative thinking skills*. Jurnal Pendidikan IPA Indonesia., DOI: 10.15294/jpii.v11i1.34628

Puritat, K. (2019). Enhanced knowledge and engagement of students through the gamification concept of game elements. *International Journal of Engineering Pedagogy*, 9(5), 41–54. DOI: 10.3991/ijep.v9i5.11028

Purnama, S., Ulfah, M., Ramadani, L., Rahmatullah, B., & Ahmad, I. (2022). *Digital storytelling trends in early childhood education in indonesia: A systematic literature review*. JPUD - Jurnal Pendidikan Usia Dini., DOI: 10.21009/JPUD.161.02

Purwaningtyas, D., Prabowo, H., Napitupulu, T. A., & Purwandari, B. (2022). *The integration of augmented reality and virtual laboratory based on the 5e model and vark assessment: A conceptual framework*. Jurnal Pendidikan IPA Indonesia., DOI: 10.15294/jpii.v11i3.36367

Pusey, M., Wong, K. W., and Rappa, N. (2021). Using case studies to explore need satisfaction and frustration in puzzle video games. *Games and Culture, 17:*752–772, DOI: .DOI: 10.1177/15554120211056126

Putri, M. A. N., & Dwikoranto, D. (2022). *Implementation of stem integrated project based learning (pjbl) to improve problem solving skills*. Berkala Ilmiah Pendidikan Fisika., DOI: 10.20527/bipf.v10i1.12231

Qasrawi, R. and BeniAbdelrahman, A. (2020). The higher and lower-order thinking skills (hots and lots) in unlock english textbooks (1st and 2nd editions) based on bloom's taxonomy: An analysis study. *International Online Journal of Education and Teaching*, 7:744–758.

Queiros, R., Pinto, C., & Cruz, M. (2024). Work in progress: Leveraging virtual escape rooms for innovative computer programming learning environments. EDUNINE 2024 - 8th IEEE World Engineering Education Conference: Empowering Engineering Education: Breaking Barriers through Research and Innovation, Proceedings DOI: DOI: 10.1109/EDUNINE60625.2024.10500625

Queirós, R., Pinto, M., Simões, A., & Portela, C. F. (2022). A primer on gamification standardization. In *Advances in Human and Social Aspects of Technology*. DOI: DOI: 10.4018/978-1-7998-8089-9.ch001

Rahim, A. S. A., Wahab, M. S. A., Ali, A. A., & Hanafiah, N. H. (2022). Educational escape rooms in pharmacy education: A narrative review. *Pharmacy Education*, 22(1), 540–557. Advance online publication. DOI: 10.46542/pe.2022.221.540557

Rahman, N. H. A., Mansor, S. N. A., Saad, S. M., Yahaya, S. H., & Omar, M. (2023). Stem-modules using gamification as induction teaching aids for non-academic tahfiz students. In *International Teaching Aid Competition 2023* (pp. 153–160). Universiti Teknologi MARA. [et al.]

Rajaram, S., & Nebeling, M. (2022). Paper trail: An immersive authoring system for augmented reality instructional experiences. *Proceedings of the 2022 CHI Conference on Human Factors in Computing Systems*, DOI: DOI: 10.1145/3491102.3517486

Ranković, M., Dedić, V., Jovanović, M., Petrovic, D., & Ilić, M. (2020). Escape rooms in virtual reality, the concept and application in electronic business doing. *Independent Journal of Management & Production*, 11(4), 1247–1261. DOI: 10.14807/ijmp.v11i4.1117

Rawlinson, R. E., & Whitton, N. (2024). Escape rooms as tools for learning through failure. *Electronic Journal of e-Learning*, 00. Advance online publication. DOI: 10.34190/ejel.21.7.3182

Recke, M., & Perna, S. (2021). Emergent narratives in remote learning experiences for project based education. *Electronic Journal of e-Learning*, 19(2), 59–70. DOI: 10.34190/ejel.19.2.2142

Reed, J. and Ferdig, R. (2021). Gaming and anxiety in the nursing simulation lab: A pilot study of an escape room. *Journal of professional nursing: official journal of the American Association of Colleges of Nursing*, 37 2:298–305, DOI: .DOI: 10.1016/j.profnurs.2021.01.006

Reeves, L. E., Bolton, E., Bulpitt, M., Scott, A., Tomey, I., Gates, M., & Baldock, R. A. (2021). Use of augmented reality (ar) to aid bioscience education and enrich student experience. *Research in Learning Technology*, 29, 2572. DOI: 10.25304/rlt.v29.2572

Reis, S., Reis, L., & Lau, N. (2021). Game adaptation by using reinforcement learning over meta games. *Group Decision and Negotiation*, 30(2), 321–340. DOI: 10.1007/s10726-020-09652-8

Repetto, M., Bianco Prevot, A., Brizio, A., Boldi, A., Talarico, M., Stanchi, S., Palma, D., & Acquadro Maran, D. (2024). Digital escape rooms: A resource for environmental education. [Switzerland]. *Sustainability (Basel)*, 16(19), 8525. Advance online publication. DOI: 10.3390/su16198525

Repetto, M., Bruschi, B., & Talarico, M. (2023). Key issues and pedagogical implications in the design of digital educational escape rooms. *Journal of E-Learning and Knowledge Society*, 19. Advance online publication. DOI: 10.20368/1971-8829/1135749

Reuter, J., Dias, M. F., Amorim, M., Figueiredo, C., & Veloso, C. (2020). How to create educational escape rooms? strategies for creation and design. *InEighth International Conference on Technological Ecosystems for Enhancing Multiculturality.* DOI: 10.1145/3434780.3436664

Revilla, A., Zamarvide, S., Lacosta, I., Pérez, F., Lajara, J., Kevelham, B., Juillard, V., Rochat, B., Drocco, M., Devaud, N., Barbeau, O., Charbonnier, C., Lange, P., Li, J., Mei, Y., Lawicka, K., Jansen, J., Reimat, N., Subramanyam, S., & César, P. (2021). A collaborative vr murder mystery using photorealistic user representations. In *2021 IEEE Conference on Virtual Reality and 3D User Interfaces Abstracts and Workshops (VRW), pages 766–766*. DOI: DOI: 10.1109/VRW52623.2021.00266

Risi, S. and Togelius, J. (2019). Procedural content generation: From automatically generating game levels to increasing generality in machine learning. *ArXiv, abs/1911.13071*.

Rocha, S., & Lopes, A. (2020). Navigation based application with augmented reality and accessibility. In *Extended Abstracts of the 2020 CHI Conference on Human Factors in Computing Systems*. DOI: DOI: 10.1145/3334480.3383004

Roohi, S., Guckelsberger, C., Relas, A., Heiskanen, H., Takatalo, J., & Hämäläinen, P. (2021). Predicting game difficulty and engagement using ai players. *Proceedings of the ACM on Human-Computer Interaction*, 5(CHI PLAY), 1–17. DOI: 10.1145/3474658

Ross, R., & Bennett, A. (2022). Increasing engagement with engineering escape rooms. *IEEE Transactions on Games*, 14(2), 161–169. DOI: 10.1109/TG.2020.3025003

Ross, S., Sabol, T., & Walker, A. (2022). Increasing engagement in stem education through narrative and game-based learning. *Journal of Science Education and Technology*, 31(5), 675–688.

Rostami, S., & Maier, M. (2022). The metaverse and beyond: Implementing advanced multiverse realms with smart wearables. *IEEE Access : Practical Innovations, Open Solutions*, 10, 110796–110806. DOI: 10.1109/ACCESS.2022.3215736

Rouse, R. (2005). *Game Design: Theory and Practice*. Wordware Publishing, Inc.

Ruiz, L. M. S., López-Alfonso, S., Moll-López, S., Moraño-Fernández, J. M., & Vega-Fleitas, E. (2022). Educational digital escape rooms footprint on students' feelings: A case study within aerospace engineering. *Inf.*, 13(2), 92. DOI: 10.3390/info13020092

Russell, J. (2020). Telling better stories: Toward critical, place-based, and multispecies narrative pedagogies in hunting and fishing cultures. *The Journal of Environmental Education*, 51(3), 232–245. DOI: 10.1080/00958964.2019.1641064

Ryan, M.-L. (2001). *Narrative as Virtual Reality: Immersion and Interactivity in Literature and Electronic Media*. Johns Hopkins University Press.

Rychert, A., Ganuza, M. L., & Selzer, M. N. (2024). Integrating gpt as an assistant for low-cost virtual reality escape-room games. *IEEE Computer Graphics and Applications*, 44(4), 14–25. DOI: 10.1109/MCG.2024.3426314 PMID: 38990754

Sadler, D. R. (1989). Formative assessment and the design of instructional systems. *Instructional Science*, 18(2), 119–144. DOI: 10.1007/BF00117714

Sahin, G. (2023). Designing a digital escape room game: An experience of a digital learning tool in basic education. *Journal of Educational Technology and Online Learning*, DOI: .DOI: 10.31681/jetol.1334912

Saleem, A., Noori, N. M., & Ozdamli, F. (2021). Gamification applications in e-learning: A literature review. *Technology. Knowledge and Learning*, 27(1), 139–159. DOI: 10.1007/s10758-020-09487-x

Salen, K., & Zimmerman, E. (2003). *Rules of Play: Game Design Fundamentals*. The MIT Press.

Salvador-Ullauri, L., Acosta-Vargas, P., González, M., & Luján-Mora, S. (2020). Combined method for evaluating accessibility in serious games. *Applied Sciences (Basel, Switzerland)*, 10(18), 6324. DOI: 10.3390/app10186324

Sánchez, R. M. (2023). *Transforming online education: the impact of gamification on teacher training in a university environment*. Metaverse Basic and Applied Research., DOI: 10.56294/mr202347

Sanders, J. E., Kutzin, J. M., & Strother, C. (2020). Escape the simulation room. *Simulation & Gaming*, 52(1), 62–71. DOI: 10.1177/1046878120963591

Santiago López, N. R. (2024). Gamification for analytical chemistry: A digital escape room. *Educación en la Química*, 35(2), 187–195. DOI: 10.22201/fq.18708404e.2024.2.86355

Sarage, D., O'Neill, B., & Eaton, C. (2020). There is no i in escape: Using an escape room simulation to enhance teamwork and medication safety behaviors in nursing students. *Simulation & Gaming*, 52(1), 40–53. DOI: 10.1177/1046878120976706

Saripudin, D., Komalasari, K., & Anggraini, D. (2021). Value-based digital storytelling learning media to foster student character. *International Journal of Instruction*, 14(2), 369–384. Advance online publication. DOI: 10.29333/iji.2021.14221a

Schäfer, S. (2018). *Agency and Authorship in Ludic Narrative Environments*. Springer International Publishing., DOI: 10.1007/978-3-030-04028-4_19

Schaller, M., Simpson, J., & Kenrick, D. (2006). *Evolution and social psychology*. Psychology Press.

Schell, J. (2008). *The Art of Game Design: A Book of Lenses*. Morgan Kaufmann. DOI: 10.1201/9780080919171

Schimpf, C., Lachmann, J., Wetzel, M. H., Fischer, P. D. B., Leineweber, A., & Rafaja, D. (2023). Escape our lab: Creating an escape room game in the field of materials science and crystallography. *Journal of Applied Crystallography*, 56(5), 1544–1556. DOI: 10.1107/S1600576723006714 PMID: 37791356

Schön, D. (1983). *The Reflective Practitioner: How Professionals Think in Action*. Temple Smith.

Seaborn, K., & Fels, D. I. (2015). Gamification in theory and action: A survey. *International Journal of Human-Computer Studies*, 74, 14–31. DOI: 10.1016/j.ijhcs.2014.09.006

Segura-Robles, A., & Parra-González, M. (2019). Analysis of teachers' intercultural sensitivity levels in multicultural contexts. *Sustainability (Basel)*, 11(11), 3137. Advance online publication. DOI: 10.3390/su11113137

Selwyn, N. (2019). *Should robots replace teachers?: Ai and the future of education*. Polity Press.

Seo, J., & Richard, G. T. (2021). Scaffolding all abilities into makerspaces: A design framework for universal, accessible and intersectionally inclusive making and learning. *Information and Learning Science*, 122(11/12), 795–815. Advance online publication. DOI: 10.1108/ILS-10-2020-0230

Seo, K., & Tindall, D. (2011). *Inspiring personal and social transformation through avatar role play in an online immersive virtual environment.*, DOI: 10.4018/978-1-60960-046-4.ch021

Sereno, M., Wang, X., Besançon, L., McGuffin, M. J., & Isenberg, T. (2020). Collaborative work in augmented reality: A survey. *IEEE Transactions on Visualization and Computer Graphics*, 28, 2530–2549. DOI: 10.1109/TVCG.2020.3032761 PMID: 33085619

Shah, P., Gadekar, N., & Taylor, S. (2025). Escape from passive learning: Harnessing gamified digital educational escape rooms for immersive marketing education. *Journal of Marketing Education*. Advance online publication. DOI: 10.1177/02734753241306093

Shakeri, H., Singhal, S., Pan, R., Neustaedter, C., & Tang, A. (2017). Escaping together: The design and evaluation of a distributed real-life escape room. In *Proceedings of the Annual Symposium on Computer-Human Interaction in Play*. DOI: DOI: 10.1145/3116595.3116601

Shang, J., Chen, S., Wu, J., & Yin, S. (2020). Arspy: Breaking location-based multi-player augmented reality application for user location tracking. *IEEE Transactions on Mobile Computing*, 21(2), 433–447. DOI: 10.1109/TMC.2020.3007740

Shanta, S., & Wells, J. (2020). Tle design based learning: Assessing student critical thinking and problem solving abilities. *International Journal of Technology and Design Education*, 32(1), 267–285. DOI: 10.1007/s10798-020-09608-8

Sharma, R., & Monteiro, S. (2012). Reflective blogging: Education in a social constructivist environment - a case study. *International Journal of e-Education, e-Business, e-. Management Learning*. Advance online publication. DOI: 10.7763/IJEEEE.2012.V2.87

Sheldon, L. (2020). *The Multiplayer Classroom: Designing Coursework as a Game*. CRC Press., DOI: 10.1201/9780429285035

Sherstyuk, A., Vincent, D., & Treskunov, A. (2009). Towards virtual reality games. *InProceedings of the 8th International Conference on Virtual Reality Continuum and its Applications in Industry, VRCAI '09,* page 315–316. ACM, DOI: DOI: 10.1145/1670252.1670322

Shin, Y., Kim, J., Jin, K., & Kim, Y. B. (2020). Playtesting in match 3 game using strategic plays via reinforcement learning. *IEEE Access : Practical Innovations, Open Solutions*, 8, 51593–51600. DOI: 10.1109/ACCESS.2020.2980380

Shi, Y.-R., & Shih, J.-L. (2015). Game factors and game-based learning design model. *International Journal of Computer Games Technology*, 2015, 1–11. Advance online publication. DOI: 10.1155/2015/549684

Shmelev, B. A. (2021). Linguodidactic potential of educational computer games. *Tambov University Review. Series. Humanities (Washington)*, (192), 58–69. Advance online publication. DOI: 10.20310/1810-0201-2021-26-192-58-69

Shneiderman, B. (2020). Human-centered artificial intelligence: Reliable, safe and trustworthy. *International Journal of Human-Computer Interaction*, 36(6), 495–504. DOI: 10.1080/10447318.2020.1741118

Shoulson, A. (2021). Producing immersive interactive narratives in complex 3d open worlds. In *Proceedings of the AAAI Conference on Artificial Intelligence and Interactive Digital Entertainment*. DOI: DOI: 10.1609/aiide.v9i6.12604

Shrivastava, A. (2020). Weaving sustainability in indian higher education system. *Emerald Emerging Markets Case Studies*, 10(3), 1–14. DOI: 10.1108/EEMCS-01-2020-0015

Shutaleva, A. (2023). Experiential learning as principle of environmental education. *E3S Web of Conferences*, DOI: .DOI: 10.1051/e3sconf/202342010010

Shute, V. J. (2008). Focus on formative feedback. *Review of Educational Research*, 78(1), 153–189. DOI: 10.3102/0034654307313795

Shute, V. J., & Zapata-Rivera, D. (2012). Adaptive educational systems. In Durlach, P. J., & Lesgold, A. M. (Eds.), *Adaptive Technologies for Training and Education* (pp. 7–27). Cambridge University Press. DOI: 10.1017/CBO9781139049580.004

Sidekerskienė, T., & Damaševičius, R. (2023). Out-of-the-box learning: Digital escape rooms as a metaphor for breaking down barriers in stem education. *Sustainability (Basel)*, 15(9), 7393. Advance online publication. DOI: 10.3390/su15097393

Siemens, G., & Long, P. (2011). Penetrating the fog: Analytics in learning and education. *TD Tecnologie Didattiche*, 22, 132–137.

Simon, H. A., & Simon, P. A. (2007). Trial and error search in solving difficult problems: Evidence from the game of chess. *Systems Research and Behavioral Science*, 7, 425–429. DOI: 10.1002/sres.3850070507

Singanamalla, S., Thies, W., & Scott, C. (2018). Avatar: Enabling immersive collaboration via live mobile video. In *Proceedings of the 3rd International Workshop on Multimedia Alternate Realities*. DOI: DOI: 10.1145/3268998.3269003

Skinner, B. F. (1957). *Verbal Behavior*. Appleton-Century-Crofts. DOI: 10.1037/11256-000

Skult, N. and Smed, J. (2020). Interactive storytelling in extended reality: Concepts for the design. pages 449–467, DOI: .DOI: 10.1007/978-3-030-37643-7_21

Slavin, R. E. (1995). *Cooperative Learning: Theory, Research, and Practice*. Allyn and Bacon.

Smith-Carrier, T., Goulden, A., & Singh, R. C. B. (2021). *Implementing Universal Design for Learning in Social Work Education*., DOI: 10.4018/978-1-7998-7106-4.ch008

Smith, E. C. (2018). *Fostering a global mind: Challenges and best practices in higher education*. Research Anthology on Business and Technical Education in the Information Era., DOI: 10.4018/978-1-5225-3776-2.ch004

Smith, E. C. (2020). *Fostering a global mind*. Multicultural Instructional Design., DOI: 10.4018/978-1-5225-9279-2.ch070

Soares, E. S., & Bulitko, V. (2019). Deep variational autoencoders for npc behaviour classification. In *2019 IEEE Conference on Games (CoG), pages* 1–4. DOI: DOI: 10.1109/CIG.2019.8848095

Spatafora, A., Wagemann, M., Sandoval, C., Leisenberg, M., & Vaz de Carvalho, C. (2024). An educational escape room game to develop cybersecurity skills. *Computers*, 13(8), 205. Advance online publication. DOI: 10.3390/computers13080205

Squire, K. (2006). From content to context: Videogames as designed experience. *Educational Researcher*, 35(8), 19–29. DOI: 10.3102/0013189X035008019

Squire, K. (2011). *Video Games and Learning: Teaching and Participatory Culture in the Digital Age*. Teachers College Press.

Staneva, A., Ivanova, T., Rasheva-Yordanova, K., & Borissova, D. (2023). Gamification in education: Building an escape room using vr technologies. In *2023 46th MIPRO ICT and Electronics Convention (MIPRO)*, pages 678–683. DOI: DOI: 10.23919/MIPRO57284.2023.10159923

Steen, T., & Wilson, C. (2020). Individual and cultural factors in teachers' attitudes towards inclusion: A meta-analysis. *Teaching and Teacher Education*, 95, 103127. DOI: 10.1016/j.tate.2020.103127

Steinböck, M., Luckner, N., Kayali, F., Proyer, M., & Kremsner, G. (2019). Cologon: a game to foster communication skills in inclusive education. *Proceedings of the 14th International Conference on the Foundations of Digital Games*, DOI: DOI: 10.1145/3337722.3341834

Stellmacher, C., Ternieten, J., Soroko, D., & Schöning, J. (2022). Escaping the privacy paradox: Evaluating the learning effects of privacy policies with serious games. *Proceedings of the ACM on Human-Computer Interaction,* 6. DOI: 10.1145/3549495

Stenseng, F., Rise, J., & Kraft, P. (2012). Activity engagement as escape from self: The role of self-suppression and self-expansion. *Leisure Sciences*, 34(1), 19–38. DOI: 10.1080/01490400.2012.633849

Stokes, J. and Pike, J. (2022). Future ready? engaging learners and building transferable skills through authentic assessment and digital literacy. *Pacific Journal of Technology Enhanced Learning*, DOI: .DOI: 10.24135/pjtel.v4i1.139

Stolzenberger, C., Wolf, N., & Trefzger, T. (2021). Technology deployment in physics lessons: Understanding optics better with augmented reality. *Journal of Physics: Conference Series*, 1929(1), 012037. Advance online publication. DOI: 10.1088/1742-6596/1929/1/012037

Stork, M. (2020). Supporting twenty-first century competencies using robots and digital storytelling. *Journal of Formative Design in Learning*, 4(1), 43–50. DOI: 10.1007/s41686-019-00039-w

Sujathamalini, D. J., Gunasekaran, D., and R., V. T. (2022). Universal design for learning in inclusive class room to ensure sustainable development goals: Inclusive and equitable quality class room. *International Journal of Science and Healthcare Research*, DOI: .DOI: 10.52403/ijshr.20220434

Sun, J. (2023). *A project-based learning (pbl) event in a chinese rural school: How can pbl help teachers and students accomplish deep learning?* Research and Advances in Education., DOI: 10.56397/RAE.2023.04.06

Sun, K.-T., Lin, Y.-C., Wu, C.-Y., & Huang, Y.-M. (2009). An application of the genetic programming technique to strategy development. *Expert Systems with Applications*, 36(3), 5157–5161. DOI: 10.1016/j.eswa.2008.06.066

Sun, Q., & Kang, H. (2021). Learning through academic collaborations in/with the east: North american adult education scholars' insights. *Adult Education Quarterly*, 72(1), 3–23. DOI: 10.1177/07417136211050119

Sun, Y., Yuan, B., Xiang, Q., Zhou, J., Yu, J., Dai, D., & Zhou, X. (2023). Intelligent decision-making and human language communication based on deep reinforcement learning in a wargame environment. *IEEE Transactions on Human-Machine Systems*, 53(1), 201–214. DOI: 10.1109/THMS.2022.3225867

Surapaneni, K. M. (2023). Creating an immersive learning experience through digital escape rooms in physiology education. DOI: .DOI: 10.1152/advan.00102.2023

Suter, B., Bauer, R., & Kocher, M. (2021). *Narrative mechanics*. Edition Medienwissenschaft., DOI: 10.14361/9783839453452

Su, X., Wambsganss, T., Rietsche, R., Neshaei, S. P., & Käser, T. (2023). Reviewriter: Ai-generated instructions for peer review writing. *Proceedings of the 18th Workshop on Innovative Use of NLP for Building Educational Applications*, pages 57–71, DOI: DOI: 10.18653/v1/2023.bea-1.5

Swartz, S., Barbosa, B., & Crawford, I. (2019). Building intercultural competence through virtual team collaboration across global classrooms. *Business and Professional Communication Quarterly*, 83(1), 57–79. DOI: 10.1177/2329490619878834

Sweller, J. (1988). Cognitive load during problem-solving: Effects on learning. *Cognitive Science*, 12(2), 257–285. DOI: 10.1207/s15516709cog1202_4

Taber, K. S. (2011). Constructivism as educational theory: Contingency in learning, and optimally guided instruction. *Educational Theory*, •••, 39–61.

Talbot, T., Thiry, K. E., and Jenkins, M. P. (2020). Storyboarding the virtuality: Methods and best practices to depict scenes and interactive stories in virtual and mixed reality. pages 129–135, DOI: .DOI: 10.1007/978-3-030-51828-8_17

Tan, B. S., & Chong, K. S. (2023). Unlocking the potential of game-based learning for soft skills development: A comprehensive review. In Journal of ICT in Education (Vol. 10, Issue 2, pp. 29–54). Universiti Pendidikan Sultan Idris. DOI: DOI: 10.37134/jictie.vol10.2.3.2023

Tang, C., Wang, Z., Sima, X., & Zhang, L. (2020). Research on artificial intelligence algorithm and its application in games. In *2020 2nd International Conference on Artificial Intelligence and Advanced Manufacture (AIAM), pages 386–389*. DOI: DOI: 10.1109/AIAM50918.2020.00085

Tang, T., Vezzani, V., & Eriksson, V. (2020). Developing critical thinking, collective creativity skills and problem solving through playful design jams. *Thinking Skills and Creativity*, 37, 100696. DOI: 10.1016/j.tsc.2020.100696

Tan, T. F., Li, Y., Lim, J. S., Gunasekeran, D., Teo, Z. L., Ng, W. Y., & Ting, D. (2022). Metaverse and virtual health care in ophthalmology: Opportunities and challenges. *Asia-Pacific Journal of Ophthalmology*, 11(3), 237–246. DOI: 10.1097/APO.0000000000000537 PMID: 35772084

Tasadduq, M., Khan, M. S., Nawab, R. M. A., Jamal, M. H., & Chaudhry, M. (2021). Exploring the effects of gamification on students with rote learning background while learning computer programming. *Computer Applications in Engineering Education*, 29(6), 1871–1891. DOI: 10.1002/cae.22428

Teixeira, I., Lima, D., & Vasconcelos, C. (2023). Storytelling as an interdisciplinary strategy in geoscience education. *International Conference on Education and New Developments 2023 – Volume 2*, DOI: DOI: 10.36315/2023v2end117

Teófilo, M., Lucena, V., Nascimento, J., Miyagawa, T., & Maciel, F. (2018). Evaluating accessibility features designed for virtual reality context. In *2018 IEEE International Conference on Consumer Electronics (ICCE), pages* 1–6. DOI: DOI: 10.1109/ICCE.2018.8326167

Terlouw, G., Kuipers, D., van 't Veer, J. T., Prins, J., & Pierie, J. (2021). The development of an escape room–based serious game to trigger social interaction and communication between high-functioning children with autism and their peers: Iterative design approach. *JMIR Serious Games*, 9(1), e19765. Advance online publication. DOI: 10.2196/19765 PMID: 33755023

Thapa, S. (2019). Assessing Intercultural Competence in Teacher Education: A Missing Link. DOI: .DOI: 10.1007/978-3-030-21029-8_11

Tian, R., Saito, D., Washizaki, H., Fukazawa, Y., Kobayashi, H., & Tsuji, A. (2024). Enhancing programming education through game-based learning: Design and implementation of a puyo puyo-inspired teaching tool. In *Proceedings of the 55th ACM Technical Symposium on Computer Science Education* V. 2, *SIGCSE 2024*, page 1838–1839. ACM, DOI: DOI: 10.1145/3626253.3635477

Ticala, R., Ciupe, A., Meza, S., & Orza, B. (2020). Augmenting learning through vr storytelling. In *2020 International Symposium on Electronics and Telecommunications (ISETC)*, pages 1–4. DOI: DOI: 10.1109/ISETC50328.2020.9301040

Tomlinson, C. A. (1999). *The Differentiated Classroom: Responding to the Needs of All Learners*. ASCD.

Tomlinson, C. A., & McTighe, J. (2006). *Integrating Differentiated Instruction & Understanding by Design*. ASCD.

Torres, C., Liberato, H., Fernandes, A., Brito, L., & Covas, M. (2023). A microlearning approach by the# estudoemcasa apoia project. In INTED2023 Proceedings (pp. 5470-5474). IATED.

To, W. M., Yu, B. T. W., Chung, A., & Chung, D. W. K. (2023). Metaverse: Trend, emerging themes, and future directions. *Transactions on Emerging Telecommunications Technologies*. Advance online publication. DOI: 10.1002/ett.4912

Tregel, T. (2021). Online content generation in mobile applications - adaptation and personalization for location-based game systems. *Technische Universität Darmstadt, DOI:* DOI: 10.26083/TUPRINTS-00017499

Treglia, E., Magnanini, A., & Caione, G. (2022). *Video games and accessibility*. Research Anthology on Developments in Gamification and Game-Based Learning., DOI: 10.4018/978-1-6684-3710-0.ch059

Tricariaco, M. (2021). The educational value of the escape room in virtual environments. In *Encyclopedia of Information Science and Technology, Fifth Edition, pages 341–372*. DOI: DOI: 10.4018/978-1-7998-7638-0.ch015

Trigo, A., & Antunes, M. (2024). Teaching programming courses with digital educational escape rooms (deer): A conceptual proposal conducive to learning by trial and error. DOI: DOI: 10.4230/OASIcs.ICPEC.2024.9

Trinh, M. P. (2021). Can you escape? using mobile app escape games to create a holistic experiential learning process online. *Management Teaching Review*, 7(3), 236–244. DOI: 10.1177/23792981211056921

Tsai, C.-Y., & Hsu, K.-H. (2016). An application of using bluetooth indoor positioning, image recognition and augmented reality. *In 2016 IEEE 13th International Conference on e-Business Engineering (ICEBE),* pages 276–281. IEEE.

Turner, A., Tichter, A., & Pillow, T. (2021). Let's escape didactics: Virtual escape room as a didactic modality in residency. *Journal of Education & Teaching in Emergency Medicine*, 6(2). Advance online publication. DOI: 10.5070/M562052905 PMID: 37465713

Tzima, S., Styliaras, G., and Bassounas, A. (2020). Revealing hidden local cultural heritage through a serious escape game in outdoor settings. *Inf., 12*(10).

Tzima, S., Styliaras, G., & Bassounas, A. (2020). Revealing hidden local cultural heritage through a serious escape game in outdoor settings. *Information (Basel)*, 12(1), 10. DOI: 10.3390/info12010010

Tzirides, A., Saini, A., Zapata, G. C., Searsmith, D., Cope, B., Kalantzis, M., Castro, V., Kourkoulou, T., Jones, J., da Silva, R. A., Whiting, J. K., and Kastania, N. P. (2023). Generative ai: Implications and applications for education. *ArXiv*, abs/2305.07605, DOI: .DOI: 10.48550/arXiv.2305.07605

Uddin, M., Manickam, S., Ullah, H., Obaidat, M. A., & Dandoush, A. (2023). Unveiling the metaverse: Exploring emerging trends, multifaceted perspectives, and future challenges. *IEEE Access : Practical Innovations, Open Solutions*, 11, 87087–87103. DOI: 10.1109/ACCESS.2023.3281303

Uibu, K. and Tagamets, E. (2022). Promoting teachers' intercultural competences for teaching in the diverse classroom. *Education and New Developments 2022 – Volume 2*, DOI: .DOI: 10.36315/2022v2end038

Usmani, S., Sharath, M., & Mehendale, M. (2022). Future of mental health in the metaverse. *General Psychiatry*, 35(4), e100825. Advance online publication. DOI: 10.1136/gpsych-2022-100825 PMID: 36189180 van Boeijen, A., Sonneveld, M., Hao, C., & Khodadadeh, Y. (2019). Developing cultural sensitivity: A student's perspective. In DS 95: *Proceedings of the 21st International Conference on Engineering and Product Design Education (E&PDE 2019)*, University of Strathclyde, Glasgow. 12th -13th September 2019. DOI: DOI: 10.35199/epde2019.35

van der Boon, M., Fermoselle, L., Haar, F. T., Dijkstra-Soudarissanane, S., & Niamut, O. (2022). Deep learning augmented realistic avatars for social vr human representation. In *ACM International Conference on Interactive Media Experiences*. DOI: DOI: 10.1145/3505284.3532976

Van Eck, R. (2006). Digital game-based learning: It's not just the digital natives who are restless. *EDUCAUSE Review*, 41, 1–16.

van Merrienboer, J. J. G., Kirschner, P. A., & Kester, L. (2003). Taking the load off a learner's mind: Instructional design for complex learning. *Educational Psychologist*, 38(1), 5–13. DOI: 10.1207/S15326985EP3801_2

Van Merriënboer, J. J. G., & Sweller, J. (2005). Cognitive load theory and complex learning: Recent developments and future directions. *Educational Psychology Review*, 17(2), 147–177. DOI: 10.1007/s10648-005-3951-0

van Rensburg, T., & Matthee, M. (2024). A framework for creating virtual escape rooms to teach computational thinking. *Communications in Computer and Information Science*, 1862, 3–17. Advance online publication. DOI: 10.1007/978-3-031-48536-7_1

VanLehn, K. (2006). The behavior of tutoring systems. *International Journal of Artificial Intelligence in Education*, •••, 227–265.

Vaswani, A., Shazeer, N., Parmar, N., Uszkoreit, J., Jones, L., Gomez, A. N., Kaiser, L., & Polosukhin, I. (2017). Attention is all you need. *Advances in Neural Information Processing Systems*, •••, 30.

Vekua, N., Lubsky, A. A., Perevozchikova, M., & Folgerova, J. N. (2020). *Peculiarities of forming high-demanded soft skills in the educational space of the escape room*. Pedagogical Skills and Education. DOI: 10.32744/pse.2020.6.31

Veldkamp, A., Daemen, J., Teekens, S., Koelewijn, S., Knippels, M. P. J., and Joolingen, W. (2020a). Escape boxes: bringing escape room experience into the classroom (extended abstract). *British Journal of Educational Technology*, pages 67–70, DOI: .DOI: 10.1111/bjet.12935

Veldkamp, A., Knippels, M.-C. P. J., & van Joolingen, W. R. (2021). Beyond the early adopters: Escape rooms in science education. *Frontiers in Education*, 6, 622860. Advance online publication. DOI: 10.3389/feduc.2021.622860

Veldkamp, A., Merx, S., & van Winden, J. (2020b). Educational escape rooms: Challenges in aligning game and education. DOI: DOI: 10.20944/preprints202010.0344.v1

Veldkamp, A., van de Grint, L., Knippels, M. C. P. J., & van Joolingen, W. V. (2020). Escape education: A systematic review on escape rooms in education. *Educational Research Review*, 31, 100364. Advance online publication. DOI: 10.1016/j.edurev.2020.100364

Veldkamp, A., van de Grint, L., Knippels, M. P. J., & van Joolingen, W. V. (2020c). Escape education: A systematic review on escape rooms in education. *Educational Research Review*, DOI: DOI: 10.20944/preprints202003.0182.v1

Vemula, S. (2021). Leveraging vr/ar/mr and ai as innovative educational practices for "igeneration" students. pages 265–277. DOI: DOI: 10.4018/978-1-7998-4739-7.ch015

Vera-Morales, M., García, G. C., Lucas, V. S., & León-Castro, L. (2023). Gamification in the teaching of ecology for university students. In 2023 IEEE 3rd International Conference on Advanced Learning Technologies on Education & Research (ICALTER), DOI: DOI: 10.1109/ICALTER61411.2023.10372921

Vergne, M., Simmons, J. D., & Bowen, R. S. (2019). Escape the lab: An interactive escape-room game as a laboratory experiment. *Journal of Chemical Education*, 96(5), 985–991. DOI: 10.1021/acs.jchemed.8b01023

Verswijvelen, M., Sosa, R., & Martini, N. (2020). Designing game-inspired narratives for learning. In *DRS2020*. Synergy., DOI: 10.21606/drs.2020.294

Veugen, C. (2016). Assassin's creed and transmedia storytelling. *International Journal of Gaming and Computer-Mediated Simulations*, 8(2), 1–19. DOI: 10.4018/IJGCMS.2016040101

Vicente, K. J., & Kirlik, A. (1992). On putting the cart before the horse: Taking perception seriously in unified theories of cognition. *Behavioral and Brain Sciences*, 15(3), 461–462. DOI: 10.1017/S0140525X00069715 PMID: 24924025

Videnovik, M., Vold, T., Dimova, G., Kiønig, L., & Trajkovik, V. (2022). Migration of an escape room–style educational game to an online environment: Design thinking methodology. *JMIR Serious Games*, 10(3), 10. DOI: 10.2196/32095 PMID: 36155979

Viken, A., Höckert, E., & Grimwood, B. (2021). Cultural sensitivity: Engaging difference in tourism. *Annals of Tourism Research*, 89, 103223. Advance online publication. DOI: 10.1016/j.annals.2021.103223

Viudes-Carbonell, S. J., Gallego-Durán, F. J., Llorens-Largo, F., & Molina-Carmona, R. (2021). Towards an iterative design for serious games. *Sustainability (Basel)*, 13(6), 3290. Advance online publication. DOI: 10.3390/su13063290

Vlachopoulos, D. (2021). Sustainable educational escape rooms for entrepreneurship students. In *INTED2021 Proceedings*, volume 1 of *INTED2021*, page 10240–10246. IATED, DOI: DOI: 10.21125/inted.2021.2136

Vodă, A. I., Căuti anu, C., Grădinaru, C., Tanasescu, C., & de Moraes, G. H. S. M. (2022). Exploring digital literacy skills in economics and social sciences and humanities students. *Sustainability (Basel)*, 14(5), 2483. DOI: 10.3390/su14052483

Voloshynov, S., Popova, H., Yurzhenko, A., and Shmeltser, E. (2019). The use of digital escape room in educational electronic environment of maritime higher education institutions. pages 347–359, DOI: .DOI: 10.31812/123456789/3869

Voreopoulou, A., Mystakidis, S., & Tsinakos, A. (2024). Augmented reality escape classroom game for deep and meaningful english language learning. *Computers*, 13(1), 24. DOI: 10.3390/computers13010024

Vörös, A. I. V., & Sárközi, Z. (2017). Physics escape room as an educational tool. AIP Conference Proceedings, Volume 1916, Issue 1, id.050002. DOI: DOI: 10.1063/1.5017455

Vygotsky, L. S. (1978). *Mind in Society: The Development of Higher Psychological Processes*. Harvard University Press.

Vygotsky, L. S. (1978). *Mind in society: The development of higher psychological processes*. Harvard University Press., DOI: 10.2307/j.ctvjf9vz4

Wang, C., Lee, B., & Lo, T. (2023). The design of a novel digital puzzle gaming system for young children's learning by interactive multi-sensing and tangible user interfacing techniques. *Sustainability (Basel)*, 15(4), 3036. DOI: 10.3390/su15043036

Wang, Y., & Chiew, V. (2010). On the cognitive process of human problem solving. *Cognitive Systems Research*, 11(1), 81–92. DOI: 10.1016/j.cogsys.2008.08.003

Wart, A. V., O'Brien, T. C., Varvayanis, S., Alder, J., Greenier, J., Layton, R., Stayart, C., Wefes, I., & Brady, A. (2020). Applying experiential learning to career development training for biomedical graduate students and postdocs: Perspectives on program development and design. *CBE Life Sciences Education*, 19(3), es7. Advance online publication. DOI: 10.1187/cbe.19-12-0270 PMID: 32822277

Weinstein, C. S., & Mignano, A. J. (2003). *Elementary Classroom Management: Lessons for Research and Practice* (3rd ed.). McGraw-Hill.

Wells, A. K., Mayer, R., Plass, J., & Homer, B. (2021). Playing a video game and learning to think: What's the connection? *Journal of Cognitive Enhancement: Towards the Integration of Theory and Practice*, 5(4), 459–467. DOI: 10.1007/s41465-021-00214-7

Wenger, E. (1998). *Communities of Practice: Learning, Meaning, and Identity*. Cambridge University Press. DOI: 10.1017/CBO9780511803932

Whinnery, S., Fogle, K. C., Stark, J., & Whinnery, K. (2020). Building collaborative teacher education: Integrating udl through a faculty learning community. *Journal of Practitioner Research*, 5(2). Advance online publication. DOI: 10.5038/2379-9951.5.2.1161

Wiesemüller, P., Mateen, S., Dengel, A., & Voß-Nakkour, S. (2024). Access to escape - an immersive game-based learning experience for accessibility education in virtual reality. *Communications in Computer and Information Science*, 1904, 273–288. Advance online publication. DOI: 10.1007/978-3-031-47328-9_21

Wiggins, G., & McTighe, J. (2005). *Understanding by Design*. ASCD.

Williams, C. C., Mostert, C., & van Vuuren, M. J. (2022). International collaboration in higher education: A reflection of student and lecturer experiences. *ECE Official Conference Proceedings*, DOI: DOI: 10.22492/issn.2188-1162.2022.34

Williams, J., & Kirschner, D. (2012). Coordinated action in the massively multi-player online game world of warcraft. *Symbolic Interaction*, 35(3), 340–367. DOI: 10.1002/symb.22

Wilms, I., Petersen, A., & Vangkilde, S. (2013). Intensive video gaming improves encoding speed to visual short-term memory in young male adults. *Acta Psychologica*, 142(1), 108–118. DOI: 10.1016/j.actpsy.2012.11.003 PMID: 23261420

Wirnata, D., Widiana, I. W., & Bayu, G. W. (2021). *The effectiveness of learning activities based on revised bloom taxonomy on problem-solving ability*. Indonesian Journal Of Educational Research and Review., DOI: 10.23887/ijerr.v4i2.37370

Wogu, I. A., Misra, S., Assibong, P. A., Olu-Owolabi, E. F., Maskeliūnas, R., & Damasevicius, R. (2019). Artificial intelligence, smart classrooms and online education in the 21st century: Implications for human development. *Journal of Cases on Information Technology*, 21(3), 66–79. DOI: 10.4018/JCIT.2019070105

Wood, D., Bruner, J. S., & Ross, G. (1976). The role of tutoring in problem solving*. *Journal of Child Psychology and Psychiatry, and Allied Disciplines*, 17(2), 89–100. DOI: 10.1111/j.1469-7610.1976.tb00381.x PMID: 932126

Wrench, A., Neill, B., & Diamond, A. (2021). International service-learning: Possibilities for developing intercultural competence and culturally responsive pedagogies. *Asia-Pacific Journal of Teacher Education*, 50(2), 215–228. DOI: 10.1080/1359866X.2021.2010275

Xanthidis, D., Manolas, C., Paul, S., & Xanthidou, O. K. (2020). Virtual and augmented reality: Enhancing the learning experience in higher education in the u.a.e. current standing & research directions. In *2020 Seventh International Conference on Information Technology Trends (ITT)*, pages 206–211. DOI: DOI: 10.1109/ITT51279.2020.9320882

Xie, H., Zhu, W., Vong, L., Wong, T.-L., & Liao, S. S. (2019). Ai in education: Current insights and future perspectives. *Journal of Computer Science and Technology*, 34(4), 869–889.

Xu, J. (2022). From augmented reality location-based games to the real-world metaverse. *Extended Abstracts of the 2022 Annual Symposium on Computer-Human Interaction in Play, DOI:* DOI: 10.1145/3505270.3558363

Yan, L., Sha, L., Zhao, L., Li, Y., Martínez-Maldonado, R., Chen, G., Li, X., Jin, Y., & Gaševic, ´. D. (2023). Practical and ethical challenges of large language models in education: A systematic scoping review. *British Journal of Educational Technology*. Advance online publication. DOI: 10.1111/bjet.13370

Yeasmin, S., & Albabtain, L. A. (2020). Escape the countries: A vr escape room game. In *2020 3rd International Conference on Computer Applications & Information Security (ICCAIS)*, pages 1–6. IEEE. DOI: 10.1109/ICCAIS48893.2020.9096727

Yeasmin, S., & Albattain, L. A. (2020a). Escape the countries: A vr escape room game. *In 2020 3rd International Conference on Computer Applications & Information Security (ICCAIS)*, pages 1–6. IEEE.

Yeasmin, S., & Albattain, L. A. (2020b). *Implementation of a virtual reality escape room game. 2020 IEEE Graphics and Multimedia*. GAME., DOI: 10.1109/GAME50158.2020.9315039

Yin, W. (2022). An artificial intelligent virtual reality interactive model for distance education. *Journal of Mathematics*, 2022(1), 7099963. Advance online publication. DOI: 10.1155/2022/7099963

Yli-Panula, E., Jeronen, E. K., & Mäki, S. (2022). School culture promoting sustainability in student teachers' views. *Sustainability (Basel)*, 14(12), 7440. Advance online publication. DOI: 10.3390/su14127440

Yousof, S. M. (2023). *Crossword Puzzle Games, Short Stories, and Mind Maps Assignments as Innovative Online Teaching Methods: Three Promising Applied Experiences During the COVID-19 Pandemic*. Emerald Publishing Limited., DOI: 10.1108/S2055-364120230000052005

Zahedi, L., Batten, J., Ross, M. S., Potvin, G., Damas, S., Clarke, P. J., & Davis, D. (2021). Gamification in education: A mixed-methods study of gender on computer science students' academic performance and identity development. *Journal of Computing in Higher Education*, 33(2), 441–474. DOI: 10.1007/s12528-021-09271-5

Zanto, T. P., Giannakopoulou, A., Gallen, C. L., Ostrand, A. E., Younger, J. W., Anguera-Singla, R., Anguera, J. A., & Gazzaley, A. (2024). Digital rhythm training improves reading fluency in children. *Developmental Science*, 27(3), e13473. Advance online publication. DOI: 10.1111/desc.13473 PMID: 38193394

Zataraín-Cabada, R., Estrada, M. L. B., Cárdenas-Sainz, B. A., & Chavez-Echeagaray, M. E. (2022). Experiences of web-based extended reality technologies for physics education. *Computer Applications in Engineering Education*, 31(1), 63–82. DOI: 10.1002/cae.22571

Zeng, H., He, X., & Pan, H. (2020). Implementation of escape room system based on augmented reality involving deep convolutional neural network. *Virtual Reality (Waltham Cross)*, 25(3), 585–596. DOI: 10.1007/s10055-020-00476-0

Zhang, J., Zou, G., & Zhang, G. (2021). "meet the deer king": "splash-ink" interaction in the innovative vr game based on dunhuang art and culture. In *SIGGRAPH Asia 2021 XR*. DOI: DOI: 10.1145/3478514.3487618

Zhang, X., Chen, H., Zhao, Y., Ma, Z., Xu, Y., Huang, H., Yin, H., & Wu, D. O. (2019). Improving cloud gaming experience through mobile edge computing. *IEEE Wireless Communications*, 26(4), 178–183. DOI: 10.1109/MWC.2019.1800440

Zhong, H., Wang, L., & Zhang, H. (2021). The application of virtual reality technology in the digital preservation of cultural heritage. *Computer Science and Information Systems*, 18(2), 535–551. DOI: 10.2298/CSIS200208009Z

Žilak, M., Car, Z., & Čuljak, I. (2022). A systematic literature review of handheld augmented reality solutions for people with disabilities. *Sensors (Basel)*, 22(20), 7719. Advance online publication. DOI: 10.3390/s22207719 PMID: 36298070

Zook, A., & Riedl, M. O. (2014). *Automatic game design via mechanic generation.* AAAI. DOI: 10.1609/aaai.v28i1.8788

Zou, R., Yin, S., Song, T., Qin, P., and Lee, Y.-C. (2024). Mitigating ageism through virtual reality: Intergenerational collaborative escape room design. DOI: .DOI: 10.1145/3613905.3650914

About the Authors

Robertas Damaševičius received his Ph.D. in Informatics Engineering from Kaunas University of Technology (KTU) in 2005. Currently, he is a Professor at Software Engineering Department, KTU. He is also Professor at Department of Applied Informatics, Vytautas Magnus University, Kaunas; and Adjunct Professor at Faculty of Applied Mathematics, Silesian University of Technology, Poland. He is the author of over 400 papers as well as a monograph published by Springer. He is also the Editor-in-Chief of the Information Technology and Control journal and has been Guest Editor of several invited issues of international journals (IEEE Access, Biomed Research International, Computational Intelligence and Neuroscience, Journal of Universal Computer Science).

Tatjana Sidekerskienė obtained a Bachelor of Science degree in Mathematics from the Faculty of Fundamental Sciences at Kaunas University of Technology (KTU) in Kaunas, Lithuania, in 2003. She went on to complete her Master of Science studies in 2006. She is presently employed as a Lecturer at the Department of Applied Mathematics at KTU, where she teaches mathematics courses. Her primary research interest was in the field of time series analysis. Currently, her research is focused on STEAM education. Tatjana Sidekerskienė has achieved notable success, having won the internal competition twice for her contributions to enhancing the quality of education and promoting innovative teaching practices at the Kaunas University of Technology. She is an author of 20 research articles. Her recent research interests are related to smart education and educational games.

Index

Symbols

A

B

C

D

E

F

R

S

T

V

www.ingramcontent.com/pod-product-compliance
Lightning Source LLC
LaVergne TN
LVHW080613300926
849162LV00005B/1146